Plant-Microbe Interactions

A constant research effort to understand plant-microbe interactions makes it indispensable to keep abreast of the latest research developments. Researchers from a range of disciplines have used multiple approaches to infer this field. With the advent of next-generation techniques, both molecular and computational, the field has entered a new phase. These approaches often result in massive information, which is sometimes tangled and in need of further analysis. These types of analyses also require cutting-edge data analytics as well as efficient statistical models.

Plant-Microbe Interactions: Harnessing Next-Generation Molecular Technologies for Sustainable Agriculture provides a comprehensive picture of the modern-day analytics and approaches being used to provide insights into the interactions between plant and microbe. A wide range of technologies are explored along with practical guides toward these techniques. A detailed understanding of omics data in various areas could be obtained from this compilation.

Key Features:

- Crosstalk between plant and microbe
- Overview of advanced molecular techniques used to study plant-microbe interaction
- Practical guide to technologies such as NGS
- Omics data analysis used to study plant-microbe interaction
- Role of soil metagenomics
- Advanced technologies such as nanotechnology and CRISPR serving to study plant-microbe interaction

This book will serve as a great reference to various next-generation techniques in the field of plant-microbe interaction, thereby helping to better understand the mechanism. This will also help budding researchers shape their research in similar areas.

Advances and Applications in Biotechnology

Series Editor:

Dr. Gustavo Molina,
Dr. Vijai Kumar Gupta

Plant-Microbe Interactions: Harnessing Next-Generation Molecular Technologies for Sustainable Agriculture

Edited by Jagajjit Sahu, Anukool Vaishnav, and Harikesh Bahadur Singh

For more information about this series, please visit:
https://www.routledge.com/Advances-and-Applications-in-Biotechnology/book-series/AABT

Plant-Microbe Interactions
Harnessing Next-Generation Molecular Technologies for Sustainable Agriculture

Edited by
Jagajjit Sahu, Anukool Vaishnav, and
Harikesh Bahadur Singh

CRC Press
Taylor & Francis Group
Boca Raton London New York

CRC Press is an imprint of the
Taylor & Francis Group, an **informa** business

First edition published 2023
by CRC Press
6000 Broken Sound Parkway NW, Suite 300, Boca Raton, FL 33487-2742

and by CRC Press
2 Park Square, Milton Park, Abingdon, Oxon, OX14 4RN

Library of Congress Cataloging-in-Publication Data
Names: Sahu, Jagajjit, editor. | Vaishnav, Anukool, editor. | Singh, H. B., Dr., editor.
Title: Plant-microbe interactions : harnessing next-generation molecular technologies for sustainable agriculture / edited by Jagajjit Sahu, Anukool Vaishnav, and Harikesh Bahadur Singh.
Other titles: Harnessing next-generation molecular technologies for sustainable agriculture
Description: First edition | Boca Raton, FL : CRC Press, 2022. |
Includes bibliographical references and index.
Identifiers: LCCN 2022002806 (print) | LCCN 2022002807 (ebook) |
ISBN 9780367774424 (hardback) | ISBN 9780367774448 (paperback) | ISBN 9781003171416 (ebook)
Subjects: LCSH: Plant-microbe relationships. | Sustainable agriculture. | Alternative agriculture.
Classification: LCC QR351 .P574 2022 (print) | LCC QR351 (ebook) |
DDC 579/.178–dc23/eng/20220228
LC record available at https://lccn.loc.gov/2022002806
LC ebook record available at https://lccn.loc.gov/2022002807

ISBN: 9780367774424 (hbk)
ISBN: 9780367774448 (pbk)
ISBN: 9781003171416 (ebk)

DOI: 10.1201/9781003171416

Typeset in Times
by codeMantra

Contents

Preface

The interplay between microorganisms and plants has been fluctuating between boon and bane. The illustrious behavior of microorganisms to maintain plant defense mechanism and furthermore its potential to sustain plant health has been evident in innumerable publications. However, the real-world implementation of such hypothetic information is still scanty. The use of traditional methods, which are time-consuming, labor-intensive, and costly in the long run, is an impediment to the realization of these hypotheses. The conventional approaches fail to perceive empirical scenario inside the laboratories. Scientists had been conducting such practices for quite a long period until the next-generation approaches came into picture. These emerging perspectives are not only efficient and robust in understanding the existent circumstances, but also less time-consuming. There has been an upsurge in the interest toward these pragmatic approaches among the scientific communities in the recent past. Interpretation of plant–microbe relationship is not a new realm; however, the amount of unfolded information has unlimited potential. To bring into play the next-generation technologies and approaches, the study of plant–microbe interaction has reinforced some important information.

This book focuses on bringing together the latest findings on various advanced molecular techniques and other cutting-edge approaches along with their uses across various cases as well as future prospects. The forethought is to cover all the available mainstream next-generation technologies for plant–microbe interaction studies in this book. It talks about the advanced molecular techniques in crop improvement such as RNA interference, CRISPR/Cas, and nanotechnology along with cutting-edge technologies such as next-generation sequencing (NGS) and omics approaches. Most of the omics approaches find many useful applications in bioinformatics support. Along with the advanced techniques, some of the chapters also revisit stable techniques, which have potential use in studying plant–microbe interaction in effective ways. This book also emphasizes the molecular basis of plant resistance mechanisms, the chemical molecules and signals during interaction between plant and microbe, and the multi-trophic interactions inducing secondary metabolites. Chapter 2 briefly covers about the beneficial microbes and how they interact with plants through soil and root. This book also provides a detailed practical aspect of the NGS technology and how it can be used to unravel the hidden information regarding plant–microbe interactions. In another effort to understand the genetic makeup in rice thereby improving its productivity, one chapter is completely dedicated to providing valuable functional genomics information. Almost all the omics approaches have been discussed across several chapters to have a rigorous insight into the available information and research gaps about the mechanism of plant–microbe interaction. A few chapters provide very particular information such as metaomics approach to understand specific interactions between ethnomedicinal plants and endophyte microbiome, which can be of interest to many plant biologists. It will mostly focus on the pitfalls of traditional technologies, advantages of the available cutting-edge technologies, and future prospects of the emerging technologies in studying plant–microbe interrelation. We hope that this volume will serve as a reference book for students, scientists, teachers, and researchers to understand the real scenario and prospects that can be achievable by organizing their research on various aspects.

Editors

Dr Jagajjit Sahu is currently the director of a start-up called GyanArras Academy, which he co-founded. His primary area of expertise is bioinformatics, especially in the field of genomics and data science. His research interests focus on the improvement and development of efficient approaches for data analysis. He received his PhD from the Department of Life Science and Bioinformatics, Assam University, India, and has postdoctoral research experiences from University of Aberdeen, Banaras Hindu University, Assam Agricultural University, and National Centre for Cell Science. Dr Sahu has a background in mathematics, biotechnology, and bioinformatics. Having more than 12 years of experience, he has been able to publish 37 journal articles and a few book chapters. Dr Sahu has edited 16 manuscripts for the journal *Archives of Phytopathology and Plant Protection* and two special issues for *Current Genomics.* He has presented his research at several international conferences and acted as a resource person in many workshops. He has also reviewed several manuscripts for such reputed journals as *Briefings in Bioinformatics, International Journal of Molecular Sciences, Genes,* and *Sensors.*

Dr. Anukool Vaishnav is working as a Post-Doctoral Scientist (Swiss excellence fellow) at University of Zurich, Switzerland. He is also associated with Department of Biotechnology, GLA University, India as an Assistant Professor. He received his Ph.D. in Biosciences from Mody University of Science & Technology, Sikar, India. He has operated SERB-National Postdoctoral Fellowship (NPDF) at the Department of Mycology and Plant Pathology, Institute of Agricultural Sciences, Banaras Hindu University, Varanasi, India. He has nine years of research experience in the field of plant–microbe interaction. His research is mainly focused on microbial-mediated plant protection against biotic and abiotic stresses and characterization of signaling molecules and secondary metabolites (soluble and VOCs) in plant holobionts. As an active researcher, he has about 50 publications including research articles, review articles and book chapters for reputed journals and edited books. He is also editor of five books published with Springer and Elsevier. Recently, he has awarded one Indian patent and nine patents are still under examination. He is an editorial member of the journal Frontiers in Microbiology and Current Genomics He has been awarded with Swiss Excellence Fellowship and Young Scientist Awards from different agencies.

Professor Harikesh Bahadur Singh is a Distinguished Professor in the Department of Biotechnology, GLA University, Mathura, India. He has served as Professor & Head, Department of Mycology and Plant Pathology, at Institute of Agricultural Sciences, Banaras Hindu University, India. He received his Ph.D. in Botany from Department of Mycology and Plant Pathology, Banaras Hindu University, Varanasi, India. His major research focus is on bioinoculants, biological control of plant pathogens, and nanobiotechnology. Prof. Singh has been decorated with several prestigious national awards and honors for his scientific contributions in the field of organic farming and translational agriculture, with the notable being CSIR Technology Prize for Biological Sciences by Hon'ble Prime Minister of India, M.S. Swaminathan Award by Society for Plant Research, Vigyan Bharti Award, and Prof. V.P. Bhide Memorial Award by Society for Plant Research. He has 20 US patents, which have successfully been transferred for commercial production of biopesticides to several industrial houses in India. He has contributed to more than 300 publications and 27 books.

Contributors

Hafiza Iqra Almas
Department of Botany
University of Agriculture
Faisalabad, Pakistan

K. Anandhi
Department of Genetics and Plant Breeding
Tamil Nadu Agricultural University
Coimbatore, Tamil Nadu, India

Eduardo J. Azevedo Correa
Empresa de Pesquisa Agropecuária de Minas Gerais
Campo Experimental de Pitangui
Pitangui, Brazil

Madhumita Barooah
Department of Agricultural Biotechnology
Assam Agricultural University
Jorhat, Assam, India

Sudipta Sankar Bora
DBT-North East Centre for Agricultural Biotechnology (DBT-NECAB)
Assam Agricultural University
Jorhat, Assam, India

Cristine Marie B. Brown
National Institute of Molecular Biology and Biotechnology
University of the Philippines
Los Baños, Philippines

Marilyn B. Brown
National Institute of Molecular Biology and Biotechnology
University of the Philippines
Los Baños, Philippines

Ankita Chatterjee
Principal Project Associate
National Chemical Laboratory
Pune, Maharashtra, India

Channakeshavaiah Chikkaputtaiah
Biological Sciences and Technology Division
CSIR-North East Institute of Science and Technology
Jorhat, Assam, India

Prassan Choudhary
Amity Institute of Biotechnology
Amity University
Lucknow, Uttar Pradesh, India

Debajit Das
Biological Sciences and Technology Division
CSIR-North East Institute of Science and Technology
Jorhat, Assam, India

Jugabrata Das
College of Horticulture
Assam Agricultural University
Jorhat, Assam, India

Sampa Das
Division of Plant Biology
Bose Institute
Kolkata, West Bengal, India

Khumbarti Debbarma
Department of Agricultural Microbiology
University of Agricultural Sciences, GKVK
Bangalore, Karnataka, India

Ashim Debnath
Department of Genetics and Plant Breeding, College of Agriculture
Acharya Narendra Deva University of Agriculture & Technology, Kumarganj
Ayodhya, Uttar Pradesh, India

Yengkhom Linthoingambi Devi
Department of Genetics & Plant Breeding
College of Agriculture, IGKV
Raipur, Chhattisgarh, India

Kuntal Kumar Dey
Department of Agricultural Biotechnology
Assam Agricultural University
Jorhat, Assam, India

Xiongming Du
Zhengzhou Research Base, State Key Laboratory of Cotton Biology
Zhengzhou University
Zhengzhou, China
and
State Key Laboratory of Cotton Biology
Institute of Cotton Research Chinese Academy of Agricultural Science
Anyang, China

Neimar F. Duarte
Instituto Federal de Minas Gerais
Belo Horizonte, Brazil

Bahman Fazeli-Nasab
Research Department of Agronomy and Plant Breeding, Agricultural Research Institute
University of Zabol
Zabol, Iran

Sebantee Ganguly
Department of Botany
University of Calcutta
Kolkata, West Bengal, India

Gunajit Goswami
DBT-North East Centre for Agricultural Biotechnology (DBT-NECAB)
Assam Agricultural University
Jorhat, Assam, India

S. Harish
Department of Plant Pathology
Tamil Nadu Agricultural University
Coimbatore, Tamil Nadu, India

Shoupu He
Zhengzhou Research Base, State Key Laboratory of Cotton Biology
Zhengzhou University
Zhengzhou, China
and
State Key Laboratory of Cotton Biology
Institute of Cotton Research Chinese Academy of Agricultural Science
Anyang, China

Akansha Jain
Division of Plant Biology
Bose Institute
Kolkata, West Bengal, India

Raheba Jami
Department of Horticulture, Faculty of Agriculture
Herat University
Herat, Afghanistan

Yinhua Jia
State Key Laboratory of Cotton Biology
Institute of Cotton Research Chinese Academy of Agricultural Science
Anyang, China

S.R. Joshi
Microbiology Laboratory, Department of Biotechnology and Bioinformatics
North-Eastern Hill University
Shillong, Meghalaya, India

Prakash Jyoti Kalita
Division of Bioscience and Bioinformatics, College of Natural Science
Myongji University
Yongin, South Korea

Manmeet Kaur
Department of Microbiology
Punjab Agricultural University
Ludhiana, Punjab, India

Sanjeev Kumar
Department of Biosciences & Bioengineering
Indian Institute of Technology Guwahati
Guwahati, Assam, India

Asha Kumari
Department of Plant Physiology,
ICAR-IARI, Hazaribagh
Hazaribagh, Jharkhand, India

Dharmendra Singh Lagoriya
Department of Agricultural Biotechnology
DBT-AAU Centre, A.A.U
Jorhat, Assam, India

Hongge Li
State Key Laboratory of Cotton Biology
Institute of Cotton Research Chinese Academy of Agricultural Science
Anyang, China

Arabinda Mahanty
Crop Protection Division
ICAR – National Rice Research Institute
Cuttack, Odisha, India

Robert Nepomuceno
National Institute of Molecular Biology and Biotechnology
University of the Philippines
Los Baños, Philippines

Fenella M.W. Nongkhlaw
Microbiology Laboratory, Department of Biotechnology and Bioinformatics
North-Eastern Hill University
Shillong, Meghalaya, India

Marcela Claudia Pagano
Institute of Sciences
Federal University of Minas Gerais
Belo Horizonte, Brazil

Siddhartha Pal
National Centre for Cell Science
Pune, Maharashtra, India

Zhaoe Pan
State Key Laboratory of Cotton Biology
Institute of Cotton Research Chinese Academy of Agricultural Science
Anyang, China

S. Parthasarathy
Department of Plant Pathology
Amrita School of Agricultural Sciences, Amrita Vishwa Vidyapeetham
Coimbatore, Tamil Nadu, India

Ricky Raj Paswan
Department of Agricultural Biotechnology
Assam Agricultural University
Jorhat, Assam, India

Jayakumar Pathma
Department of Entomology, School of Agriculture
Lovely Professional University
Phagwara, Punjab, India

Sathi Paul
Division of Plant Biology
Bose Institute
Kolkata, West Bengal, India

Zhen Peng
State Key Laboratory of Cotton Biology
Institute of Cotton Research Chinese Academy of Agricultural Science
Anyang, China

Ramin Piri
Department of Agronomy and Plant Breeding, Faculty of Agriculture
University of Tehran
Tehran, Iran

S.R. Prabhukarthikeyan
Crop Protection Division
ICAR – National Rice Research Institute
Cuttack, Odisha, India

Bishal Pun
Microbiology Laboratory, Department of Biotechnology and Bioinformatics
North-Eastern Hill University
Shillong, Meghalaya, India

Abdul Qayyum
Department of Plant Breeding and Genetics
Bahauddin Zakariya University
Multan, Pakistan

Guangyong Qin
Zhengzhou Research Base, State Key Laboratory of Cotton Biology
Zhengzhou University
Zhengzhou, China

Ratul Moni Ram
KVK Saran
Bihar, India

P.C. Rath
Crop Protection Division
ICAR – National Rice Research Institute
Cuttack, Odisha, India

Sunayana Rathi
Department of Biochemistry & Agricultural Chemistry
Assam Agricultural University
Jorhat, Assam, India

Abdul Rehman
Zhengzhou Research Base, State Key Laboratory of Cotton Biology
Zhengzhou University
Zhengzhou, China
and
State Key Laboratory of Cotton Biology
Institute of Cotton Research Chinese Academy of Agricultural Science
Anyang, China
and
Department of Plant Breeding and Genetics
Bahauddin Zakariya University
Multan, Pakistan

Abdolhosein Rezaei
Department of Agronomy, Faculty of Plant Production
Gorgan University of Agriculture Science and Natural Resources, Golestan University
Golestan, Iran

Paromita Saikia
Department of Agricultural Biotechnology
DBT-AAU Centre, A.A.U
Jorhat, Assam, India

Trishna Sarmah
Department of Soil Science
Assam Agricultural University
Jorhat, Assam, India

Surojit Sen
Department of Zoology
Mariani College
Jorhat, Assam, India

Kriti Sengupta
Bioenergy Group
Agharkar Research Institute
Pune, Maharashtra, India

Surbhi Shriti
Division of Plant Biology
Bose Institute
Kolkata, West Bengal, India

H.B. Singh
Department of Plant Pathology
Somvanshi Research Foundation
Lucknow, Uttar Pradesh, India

Dhanawantari L. Singha
Biological Sciences and Technology Division
CSIR-North East Institute of Science and Technology
Jorhat, Assam, India

H.S. Sodhi
Department of Microbiology
Punjab Agricultural University
Ludhiana, Punjab, India

Basavaraj Teli
Crop Protection Division
ICAR – National Rice Research Institute
Cuttack, Odisha, India

Naser Vahabi
Department of Gardening and Greenery, Faculty of Agriculture
University of Zabol
Zabol, Iran

Moharam Valizadeh
Research Center of Medicinal Plants
University of Sistan and Baluchestan
Zahedan, Iran

Bakhytzhan Yelikbayev
Chemical Processes and Industrial Ecology
Kazakh National Agrarian University
Almaty, Kazakhstan

1

Novel Approaches and Advanced Molecular Techniques for Crop Improvement

Dhanawantari L. Singha
CSIR-North East Institute of Science and Technology

Debajit Das, Ricky Raj Paswan
CSIR-North East Institute of Science and Technology

Channakeshavaiah Chikkaputtaiah*
CSIR-North East Institute of Science and Technology

Sanjeev Kumar*
Indian Institute of Technology Guwahati

CONTENTS

* Corresponding author's email: channakeshav@neist.res.in & sanjeev.bt@iitg.ac.in

DOI: 10.1201/9781003171416-1

1.1 Introduction

Food is utmost necessity for sustainability of human life on earth. Varieties of crops have been harvested since greater than thousands of years. With the continuously increasing global population, which is expected to increase from 7.4 billion in 2016 to more than 9–9.3 billion by 2050, the demand of food will rise up to 70% and henceforth the need to produce more food. Our agriculture will face enormous challenges to feed the global population, which will be fulfilled by developing climate-resilient crops with higher yields and improved quality (Tilman, 2012). However, currently, conventional breeding approaches are most widely used for crop improvement, which is more labor intensive and takes several years to develop commercial varieties. Unfortunately, the conventional methods are no more serviceable toward the current needs. To fulfill the global population's food demand in the present scenario, new methods need to be introduced for better production, improved nutrient content, and enhanced disease resistance. Although, since last five decades, global food grain production is continuously growing proportionate to increasing population, still more than 2 billion people of the world suffer hidden hunger or malnutrition caused by the deficiency of micronutrients and proteins (Ruel-Bergeron et al., 2015). Recent studies on global food security focus on probable solutions to provide a future balance between consumption and supply of food, which is a reductionist perspective of food security (Calzadilla et al., 2011). It has been assumed that the production of food grain worldwide must be increased by ~60%–70% by the year 2050 to fulfill the demand of expanding population and growing consumption of food (Godfray et al., 2010; Bruinsma, 2009; FAO, 2015).

In this age of technology, modern biotechnology has opened up new horizons in the field of science, which can provide improved genotypes in several of domesticated crops that can survive under climate change. Recent advancements in the fields of genetic engineering, genomics, and bioinformatics can help in the development of stress- and climate-resilient crops, which can sustain in adverse conditions. In this chapter, we mainly focus on advanced molecular biology applications for crop improvement, such as plant tissue culture, mutagenesis, RNA interference, metabolic engineering, genome editing, various transformation methods, next-generation sequencing (NGS), and omics approaches. We also highlight advanced bioinformatics tools, role of nanotechnology in crop improvement, allele mining, gene pyramiding, linkage and association mapping, molecular breeding (MB), marker-assisted backcrossing (MABC), marker-assisted recurrent selection (MARS), and genome-wide selection (GWS) for crop improvement.

1.2 Plant Tissue Culture in Crop Improvement

Plant tissue culture (PTC) is an *in vitro* technology, which has been well recognized and extensively used to regenerate various plant parts and seeds in a nutrient medium (Reddy et al., 2013) and sterile conditions. PTC has become advanced in the recent past to regenerate any kind of plant materials in an artificial nutrient medium with plant hormones and growth regulators with favorable conditions through an *in vitro* technique such as micropropagation. These techniques are widely adapted to improve a variety of crops with desired characters. The major things required for the plant tissue culture are various plant tissues (explants), suitable medium containing both organic and inorganic compounds as nutrient supplements on which the plantlets could grow and develop further, and various kinds of plant growth hormones, particularly auxin, cytokinin, and gibberellin. PTC method has widely been adapted to create genetic variability from which crop plants can be improved. By associating advanced molecular biology techniques with plant tissue culture, the transfer of desirable trait into crop plants becomes easier. The various ornamental and clonally propagated plant industries are massively working to improve crop cultivars. The genetic variation induced at chromosomal level and transposable variations have widely been seen by tissue culture, which is beneficial for crop improvement. In the recent past, various attempts have been made to produce crops by introducing somaclonal variations. Therefore, a large number of cytoplasmic and nuclear genetic alterations have been made to introduce phenotypic variations. PTC has been considered the safest technique to produce plants with desired traits. The major advantages of PTC on crop improvement are the following: (i) The improved crops from this technology can facilitate the interspecific and intergeneric crosses to overcome physiological barriers-based self-incompatibility (Brown and Thorpe, 1995); (ii) a large number of crop varieties have been recovered through pollination of pistils and ovules through either self-pollination or cross-pollination; (iii) various agricultural crops such as corn, canola, and cotton tobacco have been developed by implying haploidy; (iv) various economically important plants such as orchids, roses, and bananas have been developed through embryo culture; (v) by using plant tissue culture techniques, micropropagation of ornamental, horticultural plants, secondary metabolites production, and conservation of some endangered medicinal crops can be done; (vi) salinity tolerance has been developed through *in vitro* selection in tobacco cell lines; (vii) various other varieties with resistance to drought and heat have been developed; (viii) *in vitro* propagation via cell, tissue, meristem, and organ culture, organogenesis, and somatic embryogenesis have been done in the recent past; (ix) by adapting bio-farming of some economically important plants, a vast variety of recombinant proteins and number of crucial drugs could be produced; (x) yield and quality of the crops are getting massively increased by using this technology; (xi) various methods of PTC played a dominant role in the second green revolution, in which plant biotechnology was considered to make desirable crops. Various important PTC techniques widely used for crop improvement are wide hybridization, haploidy, somaclonal variation, micropropagation, synthetic seed, pathogen eradication, and germplasm preservation.

1.3 Crop Improvement by Genetic Engineering

1.3.1 Mutagenesis

Mutagenesis is defined as the phenomenon in which sudden heritable changes occur within the genome of an organism. Its occurrence can be spontaneous or can be on exposure to different chemical, physical, or biological agents (Oladosu et al., 2016). In plant science, it has been considered a powerful strategy for bordering genetic variability in various species (Kumawat et al., 2019a). It has great significance, especially in crops where natural sources for the genetic variations are limited. Mutation breeding and plant mutagenesis assume a huge part in expanding the genetic variability for desired traits in various food crops (Chaudhary et al., 2019). In plant breeding programs, physical and chemical mutagens are effectively applied for the advancement of new varieties with improved characteristics (Kodym and Afza, 2003). Now, it's a mainstay of modern plant breeding, alongside recombinant breeding and transgenic breeding (Shu et al., 2012). In plant science research, different mutagenesis approaches have been utilized to distinguish novel genes and their functional regulations.

In mutation breeding study, three known kinds of mutagenesis are used. The first is radiation-induced mutagenesis, in which mutations occur as a result of exposure to radiation (gamma rays, X-rays, or ion beams); the second is chemically induced mutagenesis; and the third is insertional mutagenesis (site-directed mutagenesis, a result of DNA insertions either through the genetic transformation and through the addition of T-DNA or the activation of transposable elements) (Forster et al., 2012). Induced mutagenesis is considered as one of the most effective tools for the detection and elucidation of key regulatory genes and molecular mechanisms. It is a promising methodology for delivering new varieties with improved agronomic traits, such as biofortification and higher stress tolerance (biotic and abiotic stresses) (Chaudhary et al., 2019).

Mutation breeding is a three-step process for direct release of improved crops, which is comprised of (i) inducing mutations which may take up to a year, (ii) screening for putative mutant candidates, and (iii) mutant varietal release. The foremost complicated and time-devouring step is mutant selection.

Generally, several years are required to identify useful traits that are stable throughout the propagation cycles, and the third step, mutant varietal release, follows the standardized procedures of the country where the material is developed. This regularly requires multi-locational trials with farmer contribution (Jankowicz-Cieslak et al., 2017). Several years are ordinarily required to recognize valuable characteristics that are uniform through propagation cycles. Whereas the timing of this may shift, it more frequently requires a shorter duration than the selection and testing stage. The procedure gets to be longer and more complicated in case the selected mutants are utilized as pre-breeding material in hybridizations.

In molecular biology, scientific advancements have re-enhanced mutation breeding by making it more effective and productive than ever before. With new innovative technological developments, mutation screening by genotype became feasible. The common strategies, Targeting Induced Local Lesions IN Genomes (TILLING) and EcoTILLING, where mutagens are used to induce mutation randomly in the genome to cause a high density of triggered mutations, can specifically distinguish allelic changes in the genome (Wang et al., 2012; Kurowska et al., 2011). TILLING utilizes large offspring populations through chemical or irradiation mutagenesis, but only the gene of interest is studied instead of phenotypic screening (Jung et al., 2018). This involves substantial knowledge of the underlying genetic processes, which, for many agronomic characteristics, are notable today. Genotype-dependent mutation screens have been applied in all major crop species, and multiple mutants have been identified and recognized. Physical mutagens, such as fast neutron, UV, X-ray, and gamma radiations, and chemical mutagens, including N-methyl-N-nitrosourea (MNU), sodium azide, hydrogen fluoride (HF), methyl methanesulfonate (MMS), and ethyl methanesulfonate (EMS), have broadly been investigated over the last century. In addition, biological mutagens comprise *Agrobacterium* and transposon-based chromosomal integration. The mutation induced by EMS is a profoundly powerful technique and is therefore commonly used to develop improved crop varieties in crop breeding. Recently, in plant science, the use of fast neutron (FN) bombardment to create a mutagenized population has been gaining prominence. FN results in significant deletions from a few bases to a million bases (sometimes > 1 Mb) and a greater proportion of double lesions that are not repairable, as well as chromosome alterations in the genome. FN has been shown to be a very powerful mutagen in plants and the FN-treated lines are easily generated and deletion library is quickly assembled, which helps us to locate deletion mutant (Li et al., 2001). A random deletion library generated by FN mutagenesis lines may provide valuable and significant information for the reverse genetic approaches. Recently, Kumawat and colleagues (2019a) have highlighted the use of FN mutagenesis to build a resource of gene deletion lines. For functional genomics and even to perform reverse genomics in plants, such tools would be useful. Till date, FN mutagenesis has effectively been used in rice, *Arabidopsis*, tomato, wheat, cotton, barley, soybean, and peanut (Chaudhary et al., 2019; Kumawat et al., 2019b). A photoperiod-insensitive rice has recently been developed through FN technology.

The decision on the type of mutagen to be used for mutation breeding is also based on previous achievements announced for the species and different contemplations, for example the accessibility of mutagens, expenses, and foundation (Bado et al., 2015; Mba, 2013; MVD, 2020). In the database of registered mutant varieties, mutant varieties produced with ionizing radiation, specifically gamma rays, dominate (MVD, 2020). More than 232 unique yields and plant species have been subjected to mutation breeding, including diverse fundamental harvests, for instance rice, wheat, rapeseed, sunflower, cotton, and banana (MVD, 2020). The most favored varieties are the induced mutant varieties with improved agronomic and nutritional quality traits.

Several varieties of rice, maize, wheat, sugarcane, lentil, cotton, chickpea, rapeseed, mung bean, and durum wheat have been produced with resistance to biotic and abiotic stress factors through mutation breeding (Kharkwal and Shu, 2009, Ahloowalia et al., 2004, Nakagawa, 2009, Moustafa, 2009, Suprasanna et al., 2009). Recently, a high-yielding, short-duration dwarf rice variety has been developed through induced mutagenesis (Sharma et al., 2019). Improvements of crop quality and different nutritional characteristics, such as oil and protein quality and protein content, are also other targets of mutation breeding programs in various plants (Oladosu et al., 2016). Induced mutagenesis is also being used for the enhancement of medicinal plants. For instance, through induced mutagenesis, *Nigella sativa* mutant lines were produced to improve plant height, seed yield, and thymoquinone content (Chaudhury et al., 2019).

Induced mutagenesis and its breeding methods are potential tools for enhancing the quantitative and qualitative characteristics of crops within a much shorter time frame than conventional breeding.

Mutagenic treatment of seeds and various plant parts remains a beneficial instrument for isolating the optimal variations and improving tolerance to biotic and abiotic stresses in different crops due to its relative straightforwardness, low cost, and simplicity.

The crop varieties produced worldwide through mutation breeding demonstrate its ability as a versatile and feasible solution that is applicable to any crop if suitable targets and selection methods are pursued as needed.

1.3.2 Genome Editing

In India, Europe, and various parts of the world, the use of hereditary-adjusted plants Genetically Modified Organism (GMO) is typically hindered by prohibitive enactment. Therefore, novel genome altering strategies open up new viewpoints for the pragmatic application of mutants in crop breeding. As a preferred method for engineering desired mutations, the availability of whole-genome sequence information for large numbers of crops has allowed target-specific genome editing techniques. Currently, the available genome editing methods such as zinc finger nucleases (ZFNs), transcription activator-like effector nucleases (TALENs), and clustered regularly interspaced short palindromic repeat (CRISPR)/CRISPR-associated protein 9 (Cas9) endonuclease are used to execute site-specific mutations in many plant species (Chaudhary et al., 2019). Presently, due to its simplicity and robustness, the CRISPR/Cas9 technology has transformed genome editing and has therefore been used to improve biotic and abiotic stress tolerance. Another primitive technology, i.e., ZFNs-based genome editing, paved the way for targeting every gene for the very first time. ZFNs are chimeric molecules composed of three to four binding domains of zinc finger DNA, each of which recognizes a triplet of nucleotides by binding and a FokI nuclease (Smith et al., 2000). Two chimeric molecules with neighboring target sites on both strands of the DNA are essential for targeted mutagenesis, both of which possess a FokI nuclease domain to create a double-strand break (Jung et al., 2018). This process has been used in several plant species and plant cultivars. In 2009, it was first reported that the tobacco and maize genes could be modified with ZFNs (Shukla et al., 2009; Townsend et al., 2009). However, it is a time-consuming method to produce zinc finger domains for target sites, and studies of the inefficiency of recognition domains have been reported. The use of ZFN, however, has become less common (Chuadhary et al., 2019).

Sequence-specific nuclease-based technique like TALENs gradually became an effective genome editing method and has been used by a vast variety of methods in a wide range of organisms (Sprink et al., 2015). The combination of a customized TALE DNA-binding domain (DBD) produced by plant pathogens of the genus *Xanthomonas* and the domain FokI restriction endonuclease (FokI-R) generates TALENs (Christian et al., 2010). Since 26–56 bp can be identified by a TALEN pair, specific genomic targets can be easily selected. The TALE protein DNA-binding domain consists of highly conserved repeats of usually 34 amino acids that conform to the target DNA sequence nucleotide (Boch et al., 2009). At positions 13 and 14, the amino acid residues are referred to as repeat variable diresidues (RVDs) and determine the attachment to each of the four DNA bases such that the DNA sequence correlates one by one (Christian et al., 2010; Boch et al., 2009; Moscou et al., 2009). TALENs can also be engineered and assembled with readily available kits, such as the Golden Gate Kit, which uses ligation-based cloning processes in a single stage with several repeats (Cermak et al., 2011). TALENs are cost-effective and much simpler to handle than ZFNs (Beumer et al., 2013). Successful application of TALENs for site-directed mutagenesis has been reported in a number of model and crop plants such as *Arabidopsis*, tobacco, *Nicotiana benthamiana*, *Brachypodium distachyon*, rice, *Brassica oleracea*, soybean, wheat, maize, and barley (Wendt et al., 2013; Gurushidze et al., 2017; Jankowicz-Cieslak et al., 2017). However, TALENs comprise of large and repetitive constructs and designing TALENs for multiple targets demands more time and precision to alter the target sequence. Prior to the advent of game-changing genome editing tools like CRISPR/Cas9, plant research was confined to picking ZFNs and TALENs. CRISPR/Cas9 functions in bacteria and archaea as an adaptive immune system, where it targets foreign viral or plasmid DNA degradation. CRISPR RNA (crRNA)—a small RNA molecule—binds to a target sequence of DNA and recruits the Cas9 endonuclease along with the trans-activating RNA (tracrRNA), resulting in a double-strand break. Both short RNAs are fused into a chimeric single-guide RNA (sgRNA) for effective application in eukaryotes without hampering the function of both individual RNAs (Jinek et al., 2012). The presence of a protospacer adjacent motif (PAM) of NGG flanking the 3′ end of the target sequence, the so-called protospacer, is necessary to design sgRNA. The PAM interacts with the Cas9 PAM-interacting domain (PI domain) (Jinek et al., 2014; Nishimasu et al., 2014). The sgRNA gene and the Cas9 endonuclease gene are necessary for genome editing via CRISPR/Cas9. Stably transformed plants expressing Cas9 and sgRNA can be identified with a selectable marker, and editing can take place at any time of plant growth (Jung et al., 2018). Direct RNA–DNA identification permits fast and easy synthesis of new sgRNAs for practically every target of interest, which is one of the most important advantages of the CRISPR/Cas9 system. For genome editing of phytoene desaturase (PDS) gene in *Arabidopsis*, tobacco, and rice, direct DNA transfer to protoplasts for transient expression of Cas9 and sgRNA was also used (Nekrasov et al., 2013; Li et al., 2013; Shan et al., 2013). Recently, CRISPR-associated endonuclease Cpf1 derived from Prevotella and Francisella has emerged as a modern method for accurate genome editing, which entails DNA-free dissection of plant material, with higher specificity, and has broad applications (Zaidi et al., 2017; Li et al., 2018).

1.3.3 RNA Interference (RNAi)

Post-transcriptional gene silencing (PTGS) or RNA interference (RNAi) is a natural process associated with developmental regulation, preservation of genome integrity, and protection against foreign nucleic acids. Compared to genome editing technology such as CRISPR/Cas9 or TALENs, it provides new possibilities for plant scientists for improvement of plant cultivars. RNAi may also be used to contribute to integrated pest control and sustainable solutions for agricultural production to ensure global food stability. Plant varieties developed through RNAi technology come within the ambit of existing regulatory frameworks for GMO (Mezzetti et al., 2020). The pathway functions at the molecular level by the conversion of long dsRNA into 21- to 26-bp small interfering RNA (siRNA) molecules processed from larger dsRNA. These siRNAs explicitly identify target messenger RNA (mRNA) complementary sequences, contributing to post-transcriptional silencing by both targeted degradation and translational inhibition. Therefore, in order to eliminate unwanted metabolites or increase beneficial nutrients in crops, plant genes may also be targeted. Also, RNAi can be exploited to inhibit targeted genes in pests and pathogens, contributing to effective plant defense. Since siRNAs identify their own complementary mRNAs, systems with high precision can be programmed to target genes with homologous sequences in a narrow range of organisms. Available evidence from genomic and transcriptomic sequences can boost the design of extremely precise targeting dsRNAs, thus reducing the chance of off-target effects or silencing effects in non-target species (Christiaens et al., 2018).

RNAi is currently being exploited for endogenous gene expression in plants and furthermore to target pest and pathogen genes both within plants (i.e., host-induced gene silencing, HIGS) and as topical applications (e.g., spray-induced gene silencing, SIGS) (Mezzetti et al., 2020). Raju et al. (2018) reported the silencing of the MutS HOMOLOG1 (MSH1) gene in different plant species via RNAi, which acts in a variety of developmental modifications combined with plant defense modifications, phytohormones, and abiotic stress response pathways combined with methylome repatterning. Globally, several virus-resistant plants such as squash and papaya have been approved for cultivation (Mezzetti et al., 2020). Also, many more solutions for virus control are now being developed (Khalid et al., 2017; Limera et al., 2017). In addition, plant resistance is being studied against a wide range of pests and fungal pathogens, particularly pathogen insect vectors and a variety of diseases such as cereal rusts or fruit gray mold (Andrade and Hunter 2016; McLoughlin et al., 2018; Wang et al., 2016). RNAi technology has additionally been utilized to improve crops' nutritional value. For example, RNAi-mediated repression of the caffeine synthase gene has contributed to a substantial decrease in caffeine production in coffee plants (Ogita et al., 2003). An increased seed lysine content was reported in maize plants by inhibiting the expression of low lysine content maize zein storage proteins (Segal et al., 2003; Huang et al., 2006; Houmard et al., 2007). A more recent breakthrough is the use of gene silencing as a modern plant defense or growth regulator technique, i.e., the "biopesticide" use of dsRNA (Worrall et al., 2019). It can very well be applied as foliar sprays, root drenching, seed treatments, or trunk infusions, and because of the processing costs, specificity, and improved biosafety compared to conventional pesticides and some alternative bio-control methodologies, there is a remarkable commercial interest in this method (Rodrigues and Petric 2020; Bramlett et al., 2019; Cagliari et al., 2019; Zotti et al., 2018). In addition, spray-induced gene silencing

(SIGS) is usually used by concentrating particular genes in a weed that do not exist in crops or other weeds for weed control. For the control of grass weeds in a variety of graminaceous crops, such as wheat and rice, such a system would be necessary, but formulations and techniques that permit the entry into weed cells are currently very difficult (Jiang et al., 2014; Dalakouras et al., 2016). Other specific aspects of RNAi are high mobility of the siRNA molecules through plant's vascular system and its movement from source to sink to inside the plant (Molnar et al., 2011). Thus, dsRNA produced in parts of the plant, such as rootstock, can then spread into the grafted parts of the plant, including fruit, thereby conferring disease resistance to the whole plant. This results in non-genetically modified (non-GM) fruits that are instead guarded by target-specific degradable small RNA molecules (Limera et al., 2017; Zhao and Song, 2014). Recently, De Francesco et al. (2020) have reported resistance to citrus psorosis virus (CPsV) through the above-mentioned technique in sweet orange. Sustainable agriculture, advanced pest control, and agricultural sustainability are currently in the spotlight globally. Thus, with numerous developments in pest and pathogen resistance management technologies, crop improvement is crucial and modern crop safety applications need to be followed by effective strategies for testing and resistance control.

1.3.4 Metabolic Engineering

As an effective method to increase crop production and productivity, genetic modification of food crops has thoroughly been explored. During abiotic stress, the unraveling of whole genomes and the identification of regulation of gene families has rendered a vital contribution to extending agricultural yields and giving plants resistance to abiotic stress. Conventional breeding strategies, however, are less fruitful, subject to complex processes and their interplay in response to abiotic stress. Modern biotechnological methods (metabolic engineering and synthetic biology) to produce better varieties with stress tolerance, secondary metabolites production, and biofortification of crops are gaining momentum (Kour et al., 2019c; Yadav et al., 2019, Garg et al., 2018). It also offers an alternative, reliable, and precise means of synthesizing specific micronutrients, phytonutrients, and/or bioactive components in crops and enhancing health-promoting properties in food crops (Zhu et al., 2020; Garg et al., 2018). The method of enhancing or implementing the production of target compounds in vivo by modulating one or more genes or gene networks is metabolic engineering (Farre et al., 2014; Fu et al., 2018). Currently, the synthesis of phytohormones in plants and their functional involvement in key biological processes illustrate a possible method of growing crops tolerant to abiotic stress (Tiwari et al., 2020). Phytohormones thus represent a stated objective for crop metabolic engineering. Phytohormones are synthesized in low concentrations during abiotic stress responses in plants, regulate cell signalling pathways, are transported to different plant organs, and play an important role in plant responses to varied environmental circumstances (Fahad et al., 2015; Kazan, 2015). The role of abscisic acid in abiotic stress is well documented, affecting different processes such as plant development, regulation of growth, seed dormancy, and stomatal closure (Li et al., 2010). Furthermore, a few phytohormones have been demonstrated to exhibit defensive role in plants, specifically in response to abiotic stress through regulation of plant growth and nutrition and activation of signalling pathways. For instance, plant genetic manipulations for abscisic acid biosynthetic enzymes and their role in resistance to abiotic stress have widely been considered (Jewell et al., 2010). A study found, for example, that over-expression of ABA-related stress in *Arabidopsis thaliana* increased osmotic stress resistance in the plant (Park et al., 2008). A comparative study of genetic engineering in the *Lycopersicum esculentum* gene of dehydration responsive element-binding (DREB) showed increased resistance to water-deficit stress (Lee et al., 2003). Transgenic maize with drought tolerance was developed by Habben and colleagues (2014) by engineering the biosynthesis of ethylene. Peterhansel et al. (2008) summarized some crucial metabolic engineering strategies that resolve all of the photosynthesis limitations listed in the C3 plant.

The work of the International Rice Research Institute and a group of researchers with the goal of incorporating a true C4 cycle into the essential C3 crop rice is a development of metabolic engineering toward global food security (Mitchell and Sheehy 2006; Normile, 2006). In the biofortification of crops, metabolic engineering plays a key role in enhancing nutritional efficiency in order to combat nutrient deficits by alleviating the levels of mineral nutrients components. A variety of biofortified crops have been produced over the last 20 years since the β-carotene-enriched Golden Rice generation. Few examples are anthocyanin-enriched "Purple Tomatoes," "Purple Endosperm Rice" (Butelli et al., 2008, Zhu et al., 2017), and astaxanthin-enriched "aSTARice" (Zhu et al., 2018). Recently, Yazdani et al. (2019) have reported higher and more stable carotenoid levels in biofortified tomatoes. Notable progress has been made in the use of metabolic engineering to biofortify crops till date, and there are still a few challenges (Garcı´a-Granados et al., 2019). The main challenge is the absence of knowledge of metabolic pathways and key regulators in an organism. The combined research of genomics, transcriptomics, proteomics, and metabolomics would strengthen the understanding of metabolic pathways and their critical functions and components (Zhu et al., 2020). Another limitation of molecular techniques is that major metabolic pathways involve multiple regulatory factors and enzymes. Using high-efficiency multigene expression vector systems (such as the TGS II system) in conjunction with the genome editing tool CRISPR allows whole metabolic pathways to express and modify upstream and downstream genes in more flexible and specific ways. A deep understanding of biosynthetic pathways is the need of hour, and more specifically, an improved development in metabolic engineering technology will achieve the restoration and regulation of complex multi-stage metabolic networks. This would contribute to the production of new biofortified crop varieties with various nutrients (such as phytonutrients, vitamins, minerals, and beneficial nutraceuticals) as well as stress-resistant crops, eventually leading to high crop yield and agricultural productivity and thus meeting the criteria for improved human diet and well-being. Expanding agricultural yield using genetic engineering techniques in the current scenario provides an upcoming solution to rising global nutritional demands.

1.4 Novel Genomics Technologies

Upgrading on a regular basis NGS and genomics technologies has significantly helped the scientific community resolve the barriers to crop improvement. Deep characterization at genomics, transcriptomics, proteomics, and metabolomics levels has helped researchers to decipher the complex cellular response of crop species to a variety of environmental pressures, including both biotic and abiotic stresses. The study and manipulation of economically important traits to enhance the genetic potential and nutritional qualities of crop plants has now become feasible, thanks to the commendable advances in the field of "omics technologies." This section of the chapter highlights the applications of "NGS and omics technologies" to achieve the objectives of food security.

1.4.1 Application of Next-Generation Sequencing (NGS) Technologies to Crop Improvement

For any crop improvement program, the availability of genome sequence information is of greatest significance. Over the last few decades, major advances in the field of NGS technologies have mesmerized the plant science researchers with the remarkable pace at which whole genome/transcriptome sequencing projects of many crop species have been completed within a short time frame. Several NGS platforms have been launched, including 454 FLX (Roche) (Margulies et al., 2005), HiSeq/genome analyzer (Illumina Solexa) (Bennett, 2004; Bennett et al., 2005), the SoLiD (Life Technologies), and some other newer platforms such as HeliScope (Helicos) (Milos, 2008) and Ion Torrent (Life Technologies) (Rothberg et al., 2011). The NGS technologies for whole genome sequencing (WGS), whole genome re-sequencing (WGRS) (essential for the identification of whole genome nucleotide variation), *de novo* sequencing, genotyping by sequencing (GBS), and transcriptomic and epigenetic analysis (Varshney et al., 2009) are now widely applicable. Nevertheless, NGS technologies have several technological challenges, which include the requirement of a great amount of time for processing and analyzing the data, adequate bioinformatics expertise to derive definitive conclusions from the sequence data, and short read length (<300 bp). Moreover, second-generation sequencing (SGS) technologies have few other drawbacks such as intrinsic biases and unclear alignment of repetitive elements, leading to extremely fragmented draft genome assembly, which renders it more challenging to study hidden insertions and deletions (indels) and structural variants (Sedlazeck et al., 2018). The dynamic existence of the polyploidy and the presence of repetitive and transposable elements increase the difficulties in genome assembly at plant genome level as opposed to animal genome (Salzberg and Yorke, 2005; Schnable et al., 2009). As a result, third-generation sequencing (TGS) technologies such as Pacific Biosciences (PacBio) single-molecule real-time sequencing (Eid et al., 2009) and Oxford Nanopore Technologies (ONT) sequencing have been introduced over the last few years to address the limitations of NGS technologies. In short periods and at a reduced cost, the TGS technologies produce relatively longer read sequences (length >10 kb), improving accuracy and allowing contiguous genome assemblies (Chen et al., 2017; Vlk and Repkova, 2017). These developments in the field of genome sequencing technologies opened the door to high-quality crop reference genome and therefore aided downstream analyses in breeding schemes, including association mapping and variant calling, which help in the identification of agronomically important traits in the genome. TGS is anticipated to replace SGS by 47% in the next few years (Peterson et al., 2010).

To date, a number of plant genomes (Huang et al., 2009, Sato et al., 2011, Shulaev et al., 2011, Wang et al., 2011, Xu et al., 2011) have been sequenced. *De novo* sequencing of complex plant genome is still challenging, so a combination of different approaches such as Sanger sequencing and/or Roche Pyrosequencing along with other NGS platforms makes it easier to assemble accurately and efficiently than a single NGS platform. WGRS provides better resolution in combination with data generated from reference-based sequencing and is regarded as an important application that meets the basic attributes of NGS technologies. An overview of application of NGS technologies in different crop improvement programs is shown in Figure 1.1. Rapid acquisition of vast quantities of genome-scale variable data sets not only enables the detection of unique mutations in target traits, but also allows identification of the allelic variants responsible for phenotypic diversity (DePristo et al., 2011). Via NGS technologies, the

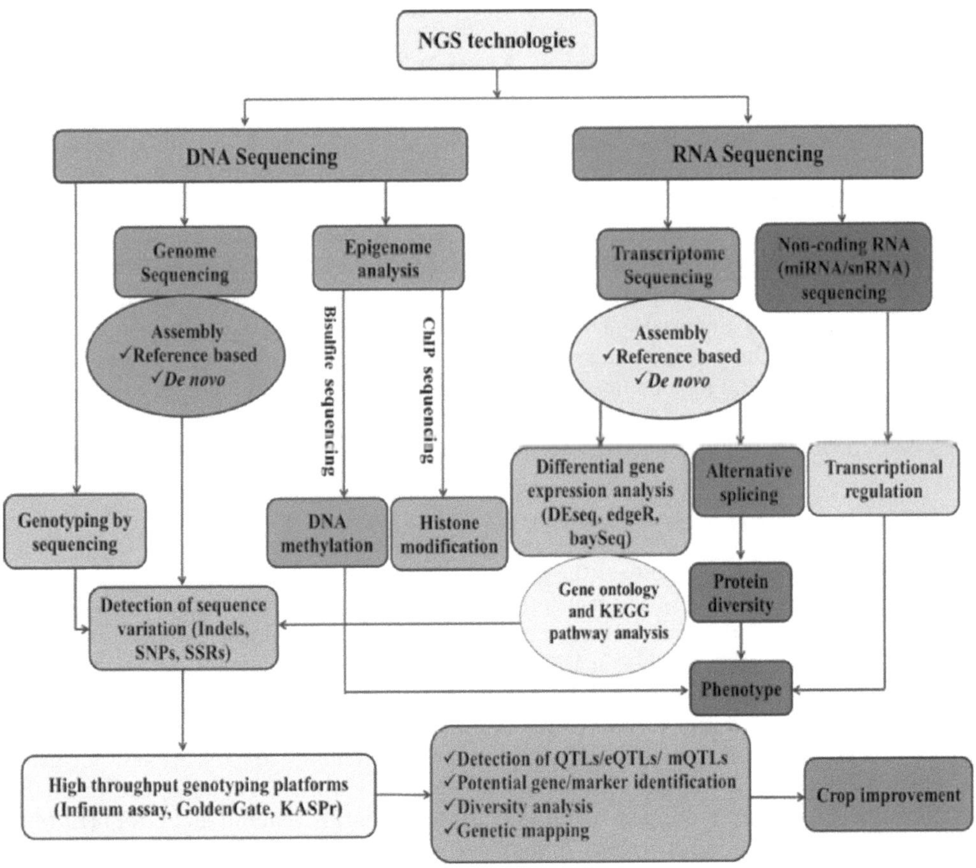

FIGURE 1.1 Application of various NGS technologies in crop improvement programs.

enormous sequence data thus obtained enabled the breeders to improve the genetic map by increasing the marker density. In addition, NGS technologies have also permitted the discovery of DNA polymorphisms such as single nucleotide polymorphisms (SNPs) and insertion/deletion polymorphisms (indels) even in closely related cultivars and ecospecies (Yamamoto et al., 2010; Arai-Kichise et al., 2011). Gene-based functional nucleotide polymorphism, if found within the target trait, has been shown to be more effective and authentic. Among several other applications of NGS in breeding programs, the genome-wide association study (GWAS) is quickly picking up prevalence as it enables the dissection of genetic architecture of complex attributes in plants and offers greater resolution to identify multiple recombination events and explore natural variations linked with phenotypical differences (Atwell et al., 2010; Huang and Han, 2014). A variety of selective marker discovery strategies have been developed using various NGS platforms, such as complexity reduction of polymorphic sequences (CRoPS), restriction site associated DNA sequencing (RADseq), double-digest RADseq (ddRADseq), diversity arrays technology sequencing (DArTseq), genome reduction on restriction site conservation (GR-RSC), reduced representation libraries (RRLs), sequence-based polymorphic marker technology, multiplexed shotgun genotyping (MSG), genotyping by sequencing (GBS), molecular inversion probe, and solution hybrid selection and microarray-based GS, which require partial genome information that can also be used without prior WGS knowledge (Toonen et al., 2013; Ray and Satya, 2014). RADseq (or its variants) and GBS were found to be suitable for genomics-assisted breeding (GAB) among these various genotyping platforms and are most often used in GWAS and genomic selection (GS) studies (Yang et al., 2012; Glaubitz et al., 2014). Rapid innovations of genome sequencing technologies indicate that the cost of WGS or NGS would fall by several times and WGS would be the preferred method over targeted genome sequencing in the coming years (Marroni et al., 2012).

1.4.2 Implications of Different "Omics" Approaches in Crop Improvement

New plant breeding approaches such as "omics approach" have become the chosen method for the plant breeders to improve crop productivity because of the overwhelming advances in sequencing technologies. Omics technological tools are the latest progress in the field of molecular biology, which involves the study of genomics, transcriptomics, proteomics, and metabolomics to define plant adaptability in various stressful environments. The study of omics would help to understand how the genetic information encoded in three-letter "codons" governs the phenotype of the organism. Furthermore, omics approaches in the context of agricultural biotechnology have enabled the identification of a number of novel genes responsible for important plant physiological features, such as grain yield and biotic and abiotic stress resistance. Omics enables a system biology approach to understand the complex interplay between genes, proteins, and metabolites of the resulting phenotype. This integrated approach relies primarily on various chemical analytical methods, profound knowledge of computational and bioinformatics analysis, and different biological disciplines, which lead to crop improvements and protection.

1.4.2.1 Genomics in Crop Improvement

In a broad sense, genomics is the study of the genome of any organism, including identification of gene sequences, intragenic sequences, gene architecture, and functional analysis, which has always been a fascinating field of biological research (Duque et al., 2013). The science of genomics comes under the purview of genetics that use recombinant DNA, DNA sequencing techniques, and bioinformatics for the sequencing, processing, assembly, and interpretation of function (functional genomics) and structure of genomes (structural genomics). Genome sequencing, which started with the first generation of techniques (Sanger sequencing) back in 1970, is the most commonly adopted method for genomics, followed by eventual upgradation to NGS techniques (Roche 454 FLX, Illumina, and SoLiD) in the mid-1990s and more recently third-generation sequencing platforms (PacBio and ONT) (El-Metwally et al., 2014b, c). The information generated through genomics study would offer plant breeders with the following advantages: (i) sequencing and *de novo* assembly in case of non-model crop species, (ii) a detailed inventory of genes with their functional annotation and ontology, (iii) identification of multiple

SNP/indel markers to enable fine mapping and selection of superior genotypes, and (iv) identification of specific genes/quantitative trait loci (QTLs)/alleles/mutations pertaining to the target traits along with the markers linked to such traits (through GWAS) (Edward and Batley, 2010; Rafalski, 2002; Bundock et al., 2009). The discovery of novel genes underpinning various important traits would, in fact, facilitate their introgression in other cultivars/species by either genetic modification or marker-assisted selection (Edward and Batley, 2010). (v) "Marker-Chip Panel" design for effective genotyping and genomic selection, which may also be used for other closely related varieties or cultivars, and (vi) to understand the evolutionary history of genome within and among population (population genomics). DNA in association with the histone protein forms the chromatins and the structure of chromatins whether in relaxed form (euchromatin) or in rigid form (heterochromatin), has an important impact on gene expression. The histone proteins are vulnerable to undergo epigenetic modifications such as cytosine DNA methylation, histone modification, and small RNA-induced methylation. The status of methylation in the genome at the whole plant level could be analyzed by bisulfite sequencing using NGS technology. Analogously, chromatin immunoprecipitation (ChIP) along with NGS techniques, the so-called ChIP-seq, is being used to generate genome-scale maps of histone modifications. All of these genome-level studies are called epigenomics. It would be painstaking to enlist all the contributions that genomics brings to study the mechanism of plant stress response and tolerance research. WGR-based GWAS is an aspect of genomics that may potentially be used to identify different important mutations in crop genome. GBS has successfully been used in wheat crop (*Triticum aestivum*) for the identification of 19,992 SNPs associated with fusarium head blight resistance (Arruda et al., 2016). Similarly, 23,154 DArTseq markers linked with ear rot disease was identified in maize (*Zea mays*) (dos Santos et al., 2016).

1.4.2.2 Transcriptomics in Crop Improvement

The word transcriptome is used to denote the study of transcripts, i.e., to capture, under certain environmental conditions, the global RNA expression profile in various plant organs/tissues/cells. The transcriptome is highly dynamic in nature, as opposed to the genome, which remains persistent despite growth conditions, age, or organs (El-Metwally et al., 2014a). While the genomics study lists many valuable details at the genome level, it is mandatory that the transcriptome study be performed because (i) transcription is always under the influence of magnitude of stress since not all the genes of the genome are expressed throughout a plant's growth and development; (ii) alternate splicing plays a very important role in the regulation of protein diversity and post-translational modifications; and (iii) by studying genomics, other RNA classes, such as miRNA, snoRNA, lincRNA, and piRNA, are practically impossible to study. In a particular stressful environment, the expression profile of different genes of a plant is generally achieved with techniques such as microarray, serial analysis of gene expression (SAGE), expressed sequence tag sequencing (EST sequencing), and RNA-seq through NGS. The transcriptome study is helpful in identifying different genes that govern the resistance and/or susceptibility of a plant to certain environmental cues. In transcriptomic study, for instance, candidate genes associated with plant's stress tolerance/resistance mechanism are identified by comparing the transcriptome of the same plant under optimal and stress conditions (Le et al., 2012; Zhang et al., 2014). Furthermore, transcriptomics study is extremely helpful in crop improvement programs, as this would allow researchers to identify different biotic stress- and abiotic stress-responsive candidate genes, to find the signaling pathways involved in stress response, to predict the likely gene functions, and to help understand the mechanisms underlying plant–biotic stress and plant–abiotic stress interactions (Kawahara et al., 2012; Jogaiah et al., 2013; De Cremer et al., 2013; Agarwal et al., 2014). The ever increasing availability of online resources and transcriptome databases alongside simultaneous advances in the field of bioinformatics would make it much easier to understand the plants' stress tolerance/resistance mechanism through novel, in-depth genome-wide analysis (Mochida and Shinozaki, 2011; Duque et al., 2013; Jogaiah et al., 2013). RNA-seq approach has been used in rice (*Oryza sativa*) crop to identify the differentially expressed candidate genes against wild and mutant strains of *Xanthomonas oryzae* (Wang et al., 2019), herbicide (metribuzin)-resistant genes in wheat (*Triticum aestivum*) (Pilcher et al., 2018), and cadmium stress-responsive genes in rice (Sun et al., 2015).

1.4.2.3 Proteomics in Crop Improvement

Proteins, among all the biomolecules, play the most important role in a living system as they catalyze all the biochemical, physiological, and molecular reactions necessary for performing basic cell functions. Proteomics is concerned with the analysis of all types of proteins expressed in certain conditions in a given organ/tissue/cell of an organism. Analogous to transcriptome, proteome is often influenced by multiple temporal and environmental factors, which are therefore variable in nature and subject to various translational and post-translational modifications (Nat et al., 2007). The proteomics research also helps to explain the fundamental processes of diverse biological and cellular responses to different biotic and abiotic stresses (Renaut et al., 2006). Hundreds of thousands of distinct protein subsets in plants detected by a range of techniques, including two-dimensional polyacrylamide gel electrophoresis (2D-PAGE), mass spectrometry (MS), liquid chromatography–mass spectrometry (LC–MS), and matrix-assisted laser desorption/ionization–mass spectrometry (MALDI–MS), have shown that they play a significant functional role in the determination of flavor, texture, yield, and nutritional content of virtually all the food products (Roberts, 2002). There are various types of proteomes that can be studied; detailed information about the proteins expressed is given by each of them. The study of whole proteome and phosphoproteome, however, has revealed critical details about the stress tolerance mechanisms of plants (Helmy et al., 2011). Proteomics has proven to be an important crop improvement strategy because it provides better explanation of how protein helps to sustain cellular homeostasis, involves itself with pathways for signal transduction, and hence activates the expression of various stress-responsive genes that are essential to preserve cellular integrity when plants face different stress conditions. 2DE proteomic approach was used to study the differentially expressed proteins during oxidative stresses using soybean (*Glycine max*) leaves samples (Galant et al., 2012). Likewise, in wheat (*T. aestivum*), the expression pattern of different proteins in leaves in response to drought stress was studied using 2D DIGE and iTRAQ (Ford et al., 2011).

1.4.2.4 Metabolomics in Crop Improvement

More than 200,000 different metabolites are synthesized by plants. They are crucial for maintenance of cell/tissue/organ structural integrity and are necessary for the diverse physiological processes associated with growth, development, reproduction, and plant defense response that make up the metabolome. Metabolomics applies to all the strategies involved in detecting and quantifying the entire set of metabolites generated in a biological sample, which, of course, are extremely variable in nature and affected by both internal and external factors. There are two distinct forms of metabolites: primary metabolites (required for normal plant physiological functioning such as growth, development, and reproduction) and secondary metabolites (mainly used as an arsenal of protection against biotic and abiotic stresses). Continuous metabolic network monitoring will allow the scientists to track biotic and abiotic stress response-related changes, while helping to develop improved crop species and a basic understanding of system biology (Aliferis and Chrysayi, 2011). Metabolomics, in many instances, gives us quick information about what actually occurs inside the cells during plant growth, differentiation, fruit ripening, pests and diseases defense mechanisms, stress resistance, etc. The study of metabolomics has become much simpler with the advent of modern biochemical techniques such as MS, nuclear magnetic resonance (NMR), Fourier transform infrared spectroscopy (FTIR), and gas chromatography (GC) combined with MS (GC–MS) (Duque et al., 2013; Jogaiah et al., 2013). Metabolomics study is typically conducted in combination with other omics experiments (transcriptomics/proteomics) to approximate the link between the metabolite level and the expression level of genes/proteins (Srivastava et al., 2013). Generally speaking, in response to a particular environmental stress, the specific pattern of gene expression elucidates the precise composition of metabolites in a plant species. Since the combined transcriptomics, proteomics, and metabolomics data sets have become widely available, the researchers extensively use these approaches in epigenomic QTL (eQTL), proteomic QTL (pQTL), and metabolic QTL (mQTL), to survey the modifications at the levels of mRNA, protein, and metabolite, respectively. Kazmi et al. (2017) studied the metabolic profiles of tomato (*Solanum lycopersicum*) germinating seeds of RIL population through GC–TOF MS. Alseekh and coworkers (2017) studied the accumulation of secondary metabolites in tomato fruits of introgression lines through UPLC–MS.

1.5 Role of Bioinformatics in Crop Improvement

Bioinformatics plays a crucial role with variety of applications in today's plant and crop science. The advanced tools and methods are much needed to organize the huge data and extend our ability to analyze complex biological systems. Bioinformatics resources and various web databases are providing vast information about the genomic data that are largely required for the research purpose. Bioinformatics develops rapid and advanced software, databases, algorithms, and other various useful tools of data analysis to make novel discoveries. The advanced bioinformatics tools and databases enable the researchers to store the information and analyze the big data, annotation of various target genes, and retrieval of outcomes, which also improve the understanding for living system research.

Using bioinformatics tools for the crop improvement has become more promising for the translational agriculture in the recent era. It provides crucial information from the big genomic data of plants, and it also provides the sequence of many genes that can be explored by using various bioinformatics tools. Bioinformatics tools are also quite useful to sequence the economically important crops to know the detailed information about various traits that are beneficial for the study.

Agricultural bioinformatics, also known as agri-informatics, plays a crucial role in crop improvement program; for example, the rice and corn (Poaceae family) have generated a large amount of genomic data that could be used to develop biotic resistance. The discovery of novel genes by using computer-based software has also aimed for improved seed quality, nutritional content of plants for better human health (nutritional genomics), and crop engineering to deal with heavy metal stress (Aslam et al., 2017). The genomic data are also being utilized in gene mining to find genes associated with desirable phenotypes, which are called economic trait loci (ETLs), hereditary disorders, and infectious disease resistance.

Various omics researches are also based on the prediction of candidate genes and consequently on the predicted functions (Lockhart and Winzeler, 2000). Transcriptomic and metabolomic data also elucidated regulatory networks that are crucial against plant stressors. Therefore, various crops have been protected from both biotic and abiotic stressors and their potential yield has been restored. In the recent era, the use of bioinformatics tools boosts the innovations and also improves different crops varieties of economic importance.

1.6 Nanotechnology in Crop Improvement

Agriculture has been considered as the major backbone of most of the developing countries, on which more than 60% of the global population is dependent for their livelihood. The recent scenario implies that this sector is facing acute challenges such as climate change, scarcities of useful resources, and usage of too much chemical fertilizer and pesticides (Raliya et al., 2017). Nanotechnology has now emerged as the greatest imperative tool in recent agriculture system, and it is widely predicted to become a driving economic force in the near future. Nanotechnology employs various chemical agents and novel delivery systems to boost crop productivity, and it also has high potential to decrease the bulk use of agrochemicals. Nanotechnology can provide better possible solutions to the current existing problems in the field of agriculture. Nanotechnology provides various applications for the benefit of agriculture, such as delivery of agrochemicals, improvement of pesticides, development of nanoscale carriers, smart packing of the products, development of nanosensors for various applications in the agricultural field, and detection of nutrient deficiencies in the soil. In the recent scenario, the demand for nanofertilizers has become very high, which are being used as alternates to bulk fertilizers and reduce pollution of soil and water by different agrochemicals. The nanofertilizer also facilitates the slow and steady release of nutrients, which reduces the loss of nutrients and enhances the nutrient use efficiency in the agricultural field. It has reduced the costs of environmental protection by slow-release fertilizers (replacement of soluble fertilizers). It has been assumed that our future food would be in such a way that we will have the ability to detect the presence of contaminants and spoiling agents. There are many such reports that have shown the involvement of nanoparticles or nanotechnology in crop improvement in various aspects. The beneficial effects observed by using these nanoparticles include enhanced seed germination, enhancement of root and shoot length, and increased vegetative biomass of seedlings in variety of crops. The use of nanoparticles in many crop plantations such as soybean, spinach, and peanut enhanced many physiological parameters such as photosynthetic activity and nitrogen metabolism. In a recent report, the seed germination of tomato plant was enhanced by penetrance of carbon nanotubes (CNTs). The mechanism behind the increase in the seed germination was due to water uptake ability of CNTs. Titanium dioxide (TiO_2) nanoparticles have been known to enhance the growth of spinach, and the reason behind this was that these nanoparticles have enhanced the Rubisco activase activity and have improved the light absorbance. Similarly, in 2010, there was a report that ZnO nanoparticles had retarded corn and rye grass seed germination. It was also reported that the use of silicon nanoparticles increased disease and stress resistance. Recently, it has been discovered that by using SWCNTs containing cerium nanoparticles, photosynthetic activity of plants can be increased threefold. In another report, plant metabolism was regenerated, which increased the nucleic acid content when the germinated seeds were exposed to the magnetic field. An increase in root length was observed when iron oxide was used in the growth medium (Bombin et al., 2015), may be due to dissolution of iron nanoparticles-induced positive changes have been observed in the form of decreased oxidative stress to spinach chloroplast under ultraviolet-B radiation by titanium dioxide nanoparticles. Similar phenomena have been also observed in rice by transmission of fullerol through seeds for generations. By using nanoparticle-mediated carbon nanotubes, changes in genetic expression in potato and tomato have been observed. Although the implementation of nanotechnology for agriculture sustainability enhanced the yield, the increase in biomass and secondary metabolites is at preliminary stage. Further research is needed to meet the global requirement of improved agriculture. It is evident from the past research that more investigations are urgently required to know the various types of nanoparticles, appropriate size, their proximate concentrations, and mode of application to make them more applicable on a large scale for the benefit of agriculture. Nanotechnology has many applications in agriculture, such as crop improvement, post-harvest technology, water management, detection of pests and diseases, soil management, development of efficient fertilizers and pesticides, and gene expression and their regulation analysis.

1.7 Modern Breeding Techniques for Crop Improvement

The development of conventional plant breeding into modern breeding techniques increases the scope of sustainable productivity of cultivated crops. Modern breeding techniques help in the development of superior and high-yielding crop varieties in field conditions. Through those techniques, targeted traits for selective physiological and genetic characters are able to be identified. Furthermore, those selected desired traits are introgressed to develop elite hybrid crops.

1.7.1 Allele Mining for Crop Improvement

The identification of alleles responsible for improved quality in both cultivated and wild relatives remains a challenge. Allele mining (AM) has proved to be an important approach in dissecting those naturally occurring superior traits for crop improvement by identifying the origin of those responsible alleles and its associated haplotypes and developing allele-specific markers for marker-assisted breeding (Kumar et al., 2010). The introduction of such superior alleles from wild and cultivated relatives has proved to improve the traits of economically important cultivated crops such as rice (Xiao et al., 1996, 1998; Pidon et al., 2020; Kaur et al., 2020), wheat (Kroupin et al., 2020), and tomato (deVicente and Tanksley, 1993) through AM. The two major approaches of AM are shown below.

1.7.1.1 EcoTILLING-Based Allele Mining

EcoTILLING is the use of TILLING technique for identifying natural variation within genes. It is rapid, inexpensive, and can be utilized for large-scale mining of novel alleles corresponding to an agronomically important trait (Wang et al., 2010; Xia et al., 2012). Most of the AM strategies require prior information of the nature/type or flanking regions of the allele. EcoTILLING (Ecotype Targeting Induced Local Lesions IN Genomes), on the other hand, does not necessarily require prior information about SNPs. After PCR amplification, the heteroduplexes region involving SNPs is cleaved with a CEL-I nuclease that is specific to mismatches (Figure 1.2). For resolution of fluorescent dye-labeled CEL-I-cleaved heteroduplex fragments, EcoTILLING utilizes advanced tools such as LICOR NEN Model 4300 DNA Analyzer, ABI 377 sequencer, Transgenomic WAVE-HS denaturing high-performance liquid chromatography, and eGene capillary electrophoresis for genotyping. EcoTILLING-based allele mining is often used for crops with narrow genetic base and low genetic polymorphism, such as chickpea. In chickpea, agarose gel-based EcoTILLING mining was used for identifying novel allelic variants of candidate genes corresponding to desired agronomic traits (Bajaj et al., 2016).

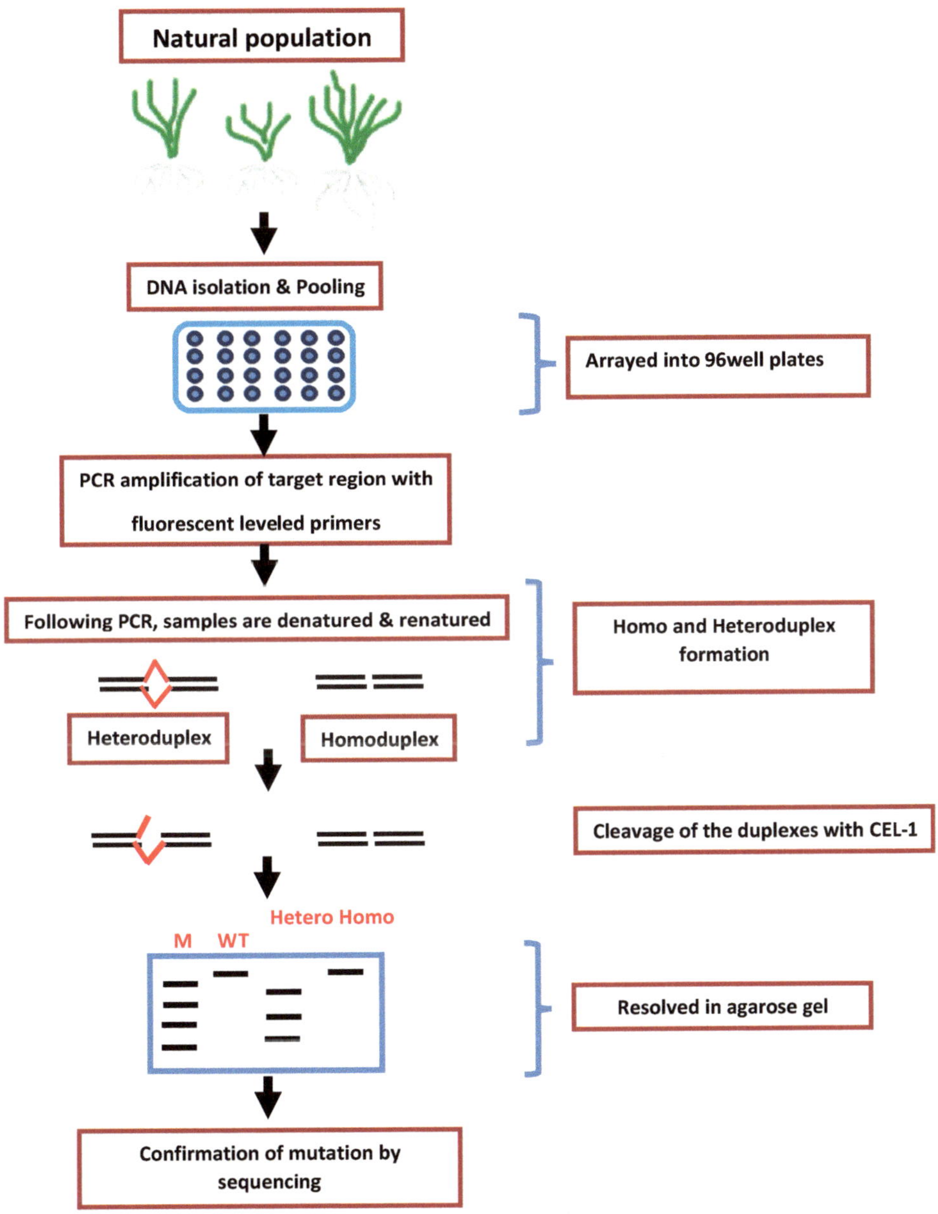

FIGURE 1.2 Schematic representation of the steps followed in agarose gel-based EcoTILLING assay for mining novel SNPs.

1.7.1.2 Sequencing-Based Allele Mining

Sequencing-based AM helps in identifying the nucleotide sequence changes associated with targeted alleles through PCR followed by the identification of nucleotide variation by DNA sequencing techniques. Sequencing-based allele mining has advantages over EcoTILLING as it does not require laborious steps. However, it involves huge sequencing costs. The steps followed in the sequencing-based AM are shown in Figure 1.3.

Sequencing-based AM was used for improving the rice bran quality (Kaur et al., 2020). Rice bran is a by-product obtained after milling and consists of valuable substances such as phytonutrients, 12%–23% oil, 14%–16% protein, and 8%–10% crude fiber. Therefore, the rice bran oil is preferable than other vegetable oils. However, rice bran contains high levels of free fatty acids, which makes it not preferable for human consumption or other applications such as production of high-quality edible oil. OsPLDα1-encoding lipolytic enzyme phospholipase D alpha 1 is responsible for such rancidity. To identify the novel sources of alleles with lower or null activity of the enzyme, a detailed analysis of DNA sequence variation was performed in wild and cultivated rice cultivars. Therefore, the identified novel allele could further be deployed in the breeding programs to overcome rice bran rancidity in elite cultivars.

1.7.1.3 Haplotype-Based AM

Haplotype-based AM is an improved method of AM. In genetic studies, the use of multiparental population gives more diverse alleles than the biparental population. However, the inheritance pattern of the biparental population is simpler than that of the multiparental population. In plants, multiparental intermated populations, known as multi-parent advanced generation inter-cross (MAGIC) lines, have been developed (Kover et al., 2009; Pascual et al., 2015; Sallam et al., 2015; Huang et al., 2012; Mackay et al., 2014; Sannemann et al., 2015; Meng et al., 2016). In those lines, the effects of haplotypes from

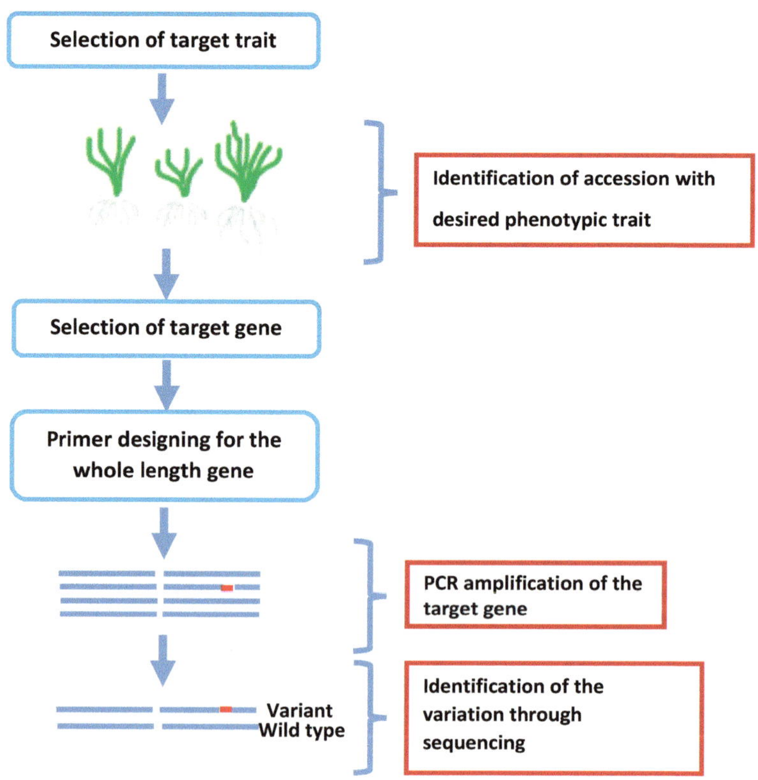

FIGURE 1.3 Steps followed in sequencing-based AM.

multiple founders on phenotypes are not considered. In MAGIC population, GWAS is used for AM. The haplotypes in a MAGIC population might be informative for GWAS because they represent recombined chromosomal segments derived from founders. Haplotype-based AM was used for identifying allelic variation in genes controlling agronomic traits in Japan-MAGIC (JAM) populations (Ogawa et al., 2018).

1.7.2 Gene Pyramiding for Crop Improvement

Gene pyramiding (GP) is a method of stacking multiple desirable genes from different parents simultaneously into a single genotype developing elite lines or varieties. GP improves the efficiency in plant breeding by identifying large genetic stocks and delivering those stocks for the development of improved desired multi-trait capabilities. However, it depends on critical factors such as targeted gene number, the gap between the genes and its markers, the number of genotypes selected, and the nature of germplasm. The usage of advanced tools such as DNA chips, microarrays, and SNPs makes the GP easier. However, the process of GP requires long time with high cost in addition to the epistatic effect. On the other hand, GP based on MAS effectively increases the pyramiding of genes into a single genetic background.

1.7.2.1 Marker-Assisted Gene Pyramiding

Pyramiding of genes is difficult through conventional breeding due to the dominance and epistatic effects and linkage drag of genes governing the desired trait. MAS, on the other hand, enables to attain the desired trait by overcoming the limitations of conventional breeding, which involves indirect selection of traits by selecting the marker linked to the gene of interest. MAS effectively facilitates GP into a single genetic background. It accelerates the process of GP by identifying plants with desired trait combination from large population at a very early stage. Furthermore, MAS-based GP has become a reality with the development and availability of an array of DNA-based markers, also known as molecular markers, and dense molecular genetic maps in crop plants. Molecular markers are tools for the study of polymorphism and identify the gene of interest in the chromosome segments. Examples of traits improved through MAS are given in Table 1.1.

TABLE 1.1

Crop Improvement through MAS-Based Gene Pyramiding in Major Crops

Important Crop	Area	Target Trait	Target Gene	References
Rice	Biotic stress	Bacterial blight resistance	*Xa4, xa5, xa13, Xa21*	Huang et al. (1997), Singh et al. (2001), Narayanan et al. (2002), Kottapalli et al. (2010), Singh et al. (2011), Dokku et al. (2013), Suh et al. (2013), Jamaloddin et al. (2020)
		Blast resistance	*Pi54, Pi1, Pi(2)t, Piz5, Pi(t)a*	Hittalmani et al. (2000), Singh et al. (2011,2013), Jamaloddin et al. (2020)
		Gall midge resistance	*Gm1, Gm4*	Kumaravadivel et al. (2006)
	Abiotic stress	Submergence	*Sub1*	Nandi et al. (1997), Reddy et al. (2009), Iftekharuddaula et al. (2015)
	Others	Low amylose content	*Wx-mq*	Tao et al. (2016)
Wheat	Biotic	Leaf rust resistance	*Lr41, Lr42, Lr43*	Cox et al. (1993)
		Powdery mildew resistance	*Pm-1, Pm-2*	Liu et al. (2000)
Cotton	Biotic	Insect pest resistance	*Cry 1Ac, Cry 2Ac*	Jackson et al. (2003), Gahan et al. (2005)
Pea	Biotic	Nodulation ability	*Sym9, Sym10*	Schneider et al. (2002)
Barley	Biotic	Yellow mosaic virus resistance	*rym4, rym5, rym9, rym11*	Werner et al. (2005)
Soybean	Biotic	Soybean mosaic virus resistance	*Rsv1, Rsv3, Rsv4*	Shi et al. (2009)

1.7.2.2 Marker-Assisted Backcrossing

An alternative to MAS-based GP could be the marker-assisted backcrossing (MABC), where the desired genotype is obtained by crossing the root genotype with a wild parent without the desired allele. Furthermore, crossing the root genotype with any one of the founding parents will be more ideal instead of crossing with any parent without desired trait. This method is the most efficient method of GP where the desired genotype can be obtained after two generations from the root genotype. This method reduces the donor genome content of the progenies by repeated backcrosses. Moreover, the linkage phase of the offspring is known. Further, due to the availability of linked markers for disease resistance, MABC is easier to pyramid various genes for the said trait. In high-yielding rice cultivars, consistent blast resistance varieties can be obtained through MABC. Popular Indian cultivars such as Pusa Basmati, Samba Mahsuri, PR106, MTU1010, and Tellahamsa were successfully made resistant through MABC with one or two resistance genes of either bacterial blight and fungal blast disease (Joseph et al., 2004; Sundaram et al., 2008; Singh et al., 2001; Arunakumari et al., 2016; Jamaloddin et al., 2020).

1.7.3 Implication of Marker-Assisted Recurrent Selection (MARS) in Crop Improvement

MARS is a development over marker-assisted backcrossing (MABC), which is a frequent and pursuit practice among the molecular breeders. MARS is a recurrent method of selection involving the use of molecular markers to identify and evaluate the complex traits controlled by multiple genomic regions (QTLs) with the goal of selecting the best performing genotype within or across a similar population (Ribaut et al., 2010). MARS is an effective method that allows the breeders to conduct genotypic selection and inter-crossing between selected plant populations usually for one selection period in the same crop cycle (Jiang et al., 2007a). By following MARS, the restrictions faced by MABC in introducing several genes/QTLs (of minor effects) could be solved. Essentially, MARS is based on an *ad hoc* significance analysis in which the detection of a target trait connected to particular marker is the first step and calculation of its influence is the second. Several cycles of marker-based selection are used in MARS to identify F_2 populations that possess favorable alleles for most, if not all, QTLs and recombination of the selected progenies with the selfed one. The use of MARS can therefore contribute to an increase in the efficacy of recurrent selection and may improve the integration of multiple beneficial genes/QTLs from different origins by recurrent selection based on multi-parent population (Asima Gazalet al., 2015). For each breeding population, *de novo* QTL mapping is performed separately in MARS, in order to identify the complex traits regulated by multiple QTLs with low genetic effect. The selected individuals are subjected to multiple pollination following the detection of major QTLs of interest in the MARS breeding population, with the priority being on producing lines that optimally recover their target QTLs from both the parents. MARS has been suggested to "forward breeding" native genes and multi-QTL pyramiding of attributes such as grain yield and biotic and abiotic stress resistance (Crosbie et al., 2006, Ribaut et al., 2010).

Due to the benefits it provides compared to other MAS methods, MARS have widely been used in crop improvement projects. Latest studies indicate that the grain production of tropical maize populations with MARS during dry season has improved dramatically (Semagn et al., 2015; Beyene et al., 2016a). In biparental mapping populations, Semagn et al. (2015) recorded an average grain yield of 184 kg/ha cycle under well-watered conditions and 45 kg/ha under dry-season testing. Beyene et al. (2016a) reported an average maize grain yield of 105 kg/ha year under well-watered conditions and 51 kg/ha year under dry-season trial.

1.7.4 Implication of Genome-Wide Selection or Genomic Selection (GWS or GS) in Crop Improvement

A strategy that has been shown to resolve the bottlenecks of MAS for quantitative traits (Heffner et al., 2009) is genome-wide selection (GWS) or genomic selection (GS). It is a variation of marker-based selection approach that utilizes all the available molecular markers to quantify the impact of all loci and that helps predict the genetic values of an unexplored population, in order to ensure correct and effective selection and increase plant genetic potential in breeding projects (Lorenz, 2013). Instead of defining particular QTLs, the main purpose of GS approach is to assess the genetic ability of an individual. Each marker is viewed as a putative QTL in this particular method, and most, if not all, of the genes or QTLs present in the genome are in linkage disequilibrium with at least one marker, thus minimizing the risk of escaping small-effect QTLs (Guo et al., 2012). In implementation, multiple variables such as the size of the training population, data used in GS model, marker density, relationships between individuals, and the use of pedigree knowledge determine the accuracy of GS method. GS is essentially used to estimate additive genetic values, while the non-additive genetics are mostly ignored. In GS, using individuals with both genotypic and phenotypic information, the marker effects are estimated. To measure the genomic estimated breeding value (GEBV), the marker effects thus estimated are combined with marker information from an individual. GS in plant breeding programs may be employed in three separate ways: (i) within-breeding-cycle GS, (ii) through-the-breeding-cycle GS, and (iii) in serious situations, using phenotypically untested parents, depending solely on their GEBV. Across the breeding period, GS allows for the direct selection of traits that are not easy to quantify in early generations. Most plant GS experiments have tested the precision of the LOOCV or k-fold system within-breeding-cycle GS. The outcomes of untested parents were not as comparable across the breeding cycle GS as they were within-breeding-cycle GS. In short, the GS makes it easy to choose an individual without providing any sort of phenotypic data by using the best suited model to predict the GEBV of the individual. However, if our goal is to optimize the precision of GEBV in GS, it is very important to choose the required training population that could be used to build a model for the estimation of GEBV that is indicative of the selection candidates in the breeding program to which GWS will be applied. In several cases, GS has been used in crop breeding systems; for example, Heffner et al. (2010) reported that GS genetic gain in maize was better than that of MAS, while Albrecht et al. (2011) claimed that GS genetic improvement in maize was higher than that of traditional pedigree breeding. Correspondingly, Song et al. (2017) reported a substantial decrease in predictive precision when estimating yields of double haploid (DH) winter wheat across cycles relative to within cycles.

1.8 Summary and Future Prospects

The global world population is rapidly increasing, and in the next two decades, it is expected to cross 9 billion. Therefore, in the upcoming days, it is going to be the greatest challenge to feed over 9 billion people and to deal with hunger of such a huge population. To fulfill the food demand, new methods need to be introduced for better production, disease resistance, and improved nutrient content of the crop plants. The biggest hurdle to fulfill this challenge is the rapidly changing climate with time. It is essential to introduce better crop varieties that can survive under this rapidly changing climate and provide maximum yield. Adapting the modern techniques and novel approaches is the key element of doing advanced agriculture, and still, there are many areas that need to be worked on, in the field of crop improvement. In future, it is expected that gene transfer or transfer of desirable traits to the target plant might be done through transfer of complete chromosome via microinjection, which can confer multigenic traits. In the recent past, NGS technology has made access to genomic resources of various plants, including some lesser studied orphan crops. This technology will also facilitate the identification and confirmation of introgression lines for generating desirable traits.

Recent advances in sequencing and genotyping have brought major breakthroughs in developing molecular markers and large-scale genotyping in both major and minor crops. It can be used to generate high-quality transcriptome apart from developing high-density genetic and physical maps. Omics approaches such as metabolomics, proteomics, transcriptomics, and genomics can be effectively used for the identification of various genomic regions or genes that are involved in expression of different traits and that are useful for breeding community. At the same time, high-throughput sequencing and genotyping approaches can also be utilized for the detection of genetic variation that exists in germplasm collection in the cultivated gene pool and other various landraces. Furthermore, the genes or QTLs for various traits identified through linkage mapping or omics approaches can also be introgressed in elite varieties or genotype of interest by using GWS, MARS, and MABC approaches.

In summary, the advanced molecular biology tools and novel approaches described in this chapter have a great potential to impact crop improvement through either advanced breeding or genetic engineering and genomics-based approaches. However, at this stage, it is really important that different technologies/novel approaches should be brought into practice from theory to the laboratory practices and finally to the field; only then the potential of genomics, NGS, metabolic engineering, RNAi, genome editing, GAB techniques, and other nanotechnology-based approaches can be realized for the improvement of crop plants. The integration of modern approaches with conventional breeding methods should also prove quite useful for enhancing the genetic gain leading to crop improvement. At present, owing to the reduced costs of high-throughput sequencing and genotyping technologies integrated with advanced bioinformatics, we can assume a bright future on application of these novel tools/approaches in sustainable crop improvement programs.

ABBREVIATIONS

CRISPR:	clustered regularly interspaced short palindromic repeats
FTIR:	Fourier transform infrared spectroscopy
GBS:	genotyping by sequencing
GE:	genetic engineering
GWAS:	genome-wide association study
GWS:	genome-wide selection
LC–MS:	liquid chromatography–mass spectrometry
MABC:	marker-assisted backcrossing
MALDI–MS:	matrix-assisted laser desorption/ionization–mass spectrometry
MARS:	marker-assisted recurrent selection
MAS:	marker-assisted selection
NGS:	next-generation sequencing
NMR:	nuclear magnetic resonance
PTC:	plant tissue culture
PTGS:	post-transcriptional gene silencing
QTLs:	quantitative trait loci
RNAi:	RNA interference
TALEN:	transcription activator-like effector nuclease
TILLING:	Targeting Induced Local Lesions IN Genomes
ZFN:	zinc finger nuclease

REFERENCES

Agarwal et al. 2014. Expanding frontiers in plant transcriptomics in aid of functional genomics and molecular breeding. *Biotechnology Journal*, 9(12): 1480–1492.

Ahloowalia et al. 2004. Global impact of mutation-derived varieties. *Euphytica*, 135: 187–204.

Albrecht et al. 2011. Genome-based prediction of testcross values in maize. *Theoritical and Applied Genetics* 123: 339–350.

Aliferis, K. A. and M. Chrysayi-Tokousbalides. 2011. Metabolomics in pesticide research and development: review and future perspectives. *Metabolomics*, 7(1): 35–53.

Alseekh et al. 2017. Canalization of tomato fruit metabolism. *The Plant Cell*, 29(11): 2753–2765.

Andrade, E.C. and Hunter, W.B. 2016. RNA interference – natural gene-based technology for highly specific pest control (HiSPeC). In: Abdurakhmonov IY (ed) *RNA Interference*. InTech, Rijeka, pp. 391–409.

Arai-Kichise et al. 2011. Discovery of genome-wide DNA polymorphisms in a landrace cultivar of japonica rice by whole-genome sequencing. *Plant and Cell Physiology*, 52(2):274–282.

Arruda et al. 2016. Comparing genomic selection and marker-assisted selection for Fusarium head blight resistance in wheat (Triticum aestivum L.). *Molecular Breeding*, 36(7): 84.

Arunakumari et al. 2016. Marker-assisted pyramiding of genes conferring resistance against bacterial blight and blast diseases into Indian rice variety MTU1010. *Rice Science*, 23: 306–316.

Asima Gazal et al. 2015. Molecular breeding for resilience in maize. *Journal of Applied and Natural Science*, 7(2):1057–1063.

Aslam et al. 2017. A role of bioinformatics in agriculture. In: Ahmed M, Stockle C. editors *Quantification of Climate Variability, Adaptation and Mitigation for Agricultural Sustainability*. Cham: Springer, pp. 413–434.

Atwell et al. Vilhjálmsson et al. 2010. Genome-wide association study of 107 phenotypes in Arabidopsis thaliana inbred lines. *Nature*, 465(7298): 627–631.

Bado et al. 2015. Plant mutation breeding current progress and future assessment. *Plant Breeding Reviews* 39: 23–87

Bajaj et al. 2016. EcoTILLING-based association mapping efficiently delineates functionally relevant natural allelic variants of candidate genes governing agronomic traits in chickpea. *Frontiers in Plant science*, 7:450.

Bennett, S. 2004 Solexa ltd. *Pharmacogenomics*, 5(4):433–438.

Bennett et al. 2005. Toward the $1000 human genome, *Pharmacogenomics*, 6(4): 373–382.

Beumer et al. 2013. Comparing zinc finger nucleases and transcription activator-like effector nucleases for gene targeting in Drosophila. *G3 (Bethesda)*, 3(10): 1717–1725.

Beyene et al. 2016a. Improving maize grain yield under drought stress and non-stress environments in sub-Saharan Africa using marker-assisted recurrent selection. *Crop Science*. 56:344–353.

Boch et al. 2009. Breaking the code of DNA binding specificity of TAL-type III effectors. *Science*, 326(5959): 1509–1512.

Bombin et al. 2015. Developmental and reproductive effects of iron oxide nanoparticles in Arabidopsis thaliana. *International Journal of Molecular Sciences*, 16(10): 24174–24193.

Bramlett et al. 2019. RNA-based biocontrols: a new paradigm in crop protection. *Engineering*. https://doi.org/10.1016/j.eng.2019.09.008.

Brown, D. C. W. and Thorpe, T. A. 1995. Crop improvement through tissue culture. *World Journal of Microbiology and Biotechnology*, 11(4): 409–415.

Bruinsma, J. 2009. The resource outlook to 2050: by how much do land, water and crop yields need to increase by 2050. *Expert Meeting on How to Feed the World*, 2050: 24–26.

Bundock et al. 2009. Targeted single nucleotide polymorphism (SNP) discovery in a highly polyploid plant species using 454 sequencing. *Plant Biotechnology Journal*, 7(4): 347–354.

Butelli et al. 2008. Enrichment of tomato fruit with health-promoting anthocyanins by expression of select transcription factors. *Nature of Biotechnology*, 26:1301–1308.

Cagliari et al. 2019. Management of pest insects and plant diseases by non-transformative RNAi. *Frontiers Plant Science*, 10:1319.

Calzadilla et al. 2011. Water scarcity and the impact of improved irrigation management: a computable general equilibrium analysis. *Agricultural Economics*, 42(3): 305–323.

Cermak et al. 2011. Efficient design and assembly of custom TALEN and other TAL effector-based constructs for DNA targeting. *Nucleic Acids Research*, 39(17): 7879.

Chaudhary et al. 2019. Mutation breeding in tomato: advances, applicability and challenges. *Plants*, 8: 128.

Chen et al. 2017. The performance of activated sludge exposed to arsanilic acid and amprolium hydrochloride in sequencing batch reactors. *International Biodeterioration & Biodegradation*, 116:260–265.

Christiaens et al. 2018. Literature review of baseline information on RNAi to support the environmental risk assessment of RNAi based GM plants. *EFSA Supporting Publications*, 15(5): 1424.

Christian et al. 2010. Targeting DNA double-strand breaks with TAL effector nucleases. *Genetics*, 186(2): 756–761.

Cox et al. 1993. Leaf rust-resistance genes, Lr41, Lr42 and Lr43 transferred from Triticum tauschii to common wheat. *Crop Science*, 34: 339–343.

Crosbie et al. 2006. Plant breeding: past, present and future. In: K.R. Lamkey and M. Lee editors. *Plant Breeding: The Arnel R. Hallauer International Symposium*. Oxford, UK: Blackwell Publishing, pp. 3–50.

Dalakouras et al. 2016. Induction of silencing in plants by high-pressure spraying of in vitro synthesized small RNAs. *Front Plant Science*, 7: 1327.

De Cremer et al. 2013. RNA seq-based transcriptome analysis of L actuca sativa infected by the fungal necrotroph *Botrytis cinerea. Plant, Cell & Environment*, 36(11): 1992–2007.

De Francesco et al. (2020). Transgenic Sweet Orange expressing hairpin CP-mRNA in the interstock confers tolerance to citrus psorosis virus in the nontransgenic scion. *Transgenic Research*, 29: 215.

DePristo et al. 2011. A framework for variation discovery and genotyping using next-generation DNA sequencing data. *Nature Genetics*, 43(5):491.

deVicente et al. 1993. QTL analysis of transgressive segregation in an interspecific tomato cross. Genetics, 134:585–596.

Dokku et al. 2013. Genetic enhancement of host plant-resistance of the Lalat cultivar of rice against bacterial blight employing marker-assisted selection. *Biotechnology Letters* 35:1339–1348.

dos Santos et al. 2016. Genomic selection to resistance to *Stenocarpella maydis* in maize lines using DArTseq markers. *BMC Genetics*, 17(1): 86.

Duque et al. 2013. Chapter 3 Abiotic stress responses in plants: unraveling the complexity of genes and networks to survive. In K. Vahdati and C. Leslie editors. *Abiotic Stress: Plant Responses and Applications in Agriculture.* Croatia: INTECH Open, pp. 49–102.

Edward, D. and Batley, J. 2010. Plant genome sequencing: applications for crop improvement. *Plant Biotechnology Journal*, 8(1): 2–9.

Eid et al. 2009. Real-time DNA sequencing from single polymerase molecules. *Science*, 323(5910):133–138.

El-Metwally et al. 2014a. Basics of molecular biology for next-generation sequencing. In *Next Generation Sequencing Technologies and Challenges in Sequence Assembly.* doi:10.1007/978-1-4939-0715-1_1.

El-Metwally et al. 2014b. First- and next-generations sequencing methods. In *Next Generation Sequencing Technologies and Challenges in Sequence Assembly.* doi:10.1007/978-1-4939-0715-1_3.

El-Metwally et al. 2014c. New horizons in next-generation sequencing. In *Next Generation Sequencing Technologies and Challenges in Sequence Assembly.* doi:10.1007/978-1-4939-0715-1_6.

Fahad et al. 2015. Potential role of phytohormones and plant growth-promoting rhizobacteria in abiotic stresses: consequences for changing environment. *Environmental Science and Pollution Research*, 22: 4907–4921.

FAO, I. 2015. The state of food insecurity in the world 2015. In *Meeting* (pp. 107–119).

Farre et al. 2014. Engineering complex metabolic pathways in plants. *Annual Review of Plant Biology*, 65: 187–223.

Ford et al. Bacic. 2011. Quantitative proteomic analysis of wheat cultivars with differing drought stress tolerance. *Frontiers in Plant Science*, 2, p.44.

Forster et al. 2012. Plant mutagenesis in crop improvement: basic terms and applications. In: Shu QY, Forster BP, Nakagawa H, editors. *Plant mutation breeding and biotechnology.* Wallingford: CABI, pp. 9–20.

Fu et al. 2018. Next-generation plant metabolic engineering, inspired by an Ancient Chinese irrigation system. *Molecular Plant*, 11: 47–57.

Gahan et al. 2005. Genetic basis of resistance to Cry 1Ac and Cry 2Aa in Heliothis virescens (Lepidoptera: Noctuidae). *Journal of Economic Entomology*, 98:1357–1368.

Galant et al. 2012. From climate change to molecular response: redox proteomics of ozone-induced responses in soybean. *New Phytologist*, 194(1): 220–229.

García-Granados et al. 2019. Metabolic engineering and synthetic biology: synergies, future, and challenges. *Frontiers in Bioengineering and Biotechnology*, 7: 36.

Garg et al. 2018. Biofortified crops generated by breeding, agronomy, and transgenic approaches are improving lives of millions of people around the world. *Frontiers in Nutrition*, 5: 12.

Glaubitz et al. 2014. TASSEL-GBS: a high capacity genotyping by sequencing analysis pipeline. *PLoS ONE*, 9:e90346.

Godfray et al. 2010. The future of the global food system.

Guo et al. 2012. Evaluation of genome-wide selection efficiency in maize nested association mapping populations. *Theoretical and Applied Genetics*, 124:261–275.

Gurushidze et al. 2017. Site-Directed Mutagenesis in Barley by Expression of TALE Nuclease in Embryogenic Pollen. In: Jankowicz-Cieslak J, Tai T, Kumlehn J, Till B. editors. *Biotechnologies for Plant Mutation Breeding.* Cham: Springer, pp. 113–128.

Habben et al. 2014. Transgenic alteration of ethylene biosynthesis increases grain yield in maize under field drought-stress conditions. *Plant Biotechnol Journal*, 12: 685–693

Heffner et al. 2009. Genomic selection for crop improvement. *Crop Science*, 49:1–12.

Heffner et al. 2010. Plant breeding with genomic selection: Gain per unit time and cost. *Crop Science*, 50:1681–1690.

Helmy et al. 2011. OryzaPG-DB: rice proteome database based on shotgun proteogenomics. *BMC Plant Biology*, 11(1): 1–9.

Hittalmani et al. 2000. Fine mapping and DNA marker-assisted pyramiding of the three major genes for blast resistance in rice. *Theoretical and Applied Genetics*, 100: 1121–1128.

Houmard et al. 2007. High-lysine corn generated by endosperm specific suppression of lysine catabolism using RNAi. *Plant Biotechnology Journal*, 5: 605–614.

Huang, X., and B. Han. 2014. Natural variations and genome-wide association studies in crop plants. *Annual Review of Plant Biology*, 65:531–551.

Huang et al. 1997. Pyramiding of bacterial blight resistance genes in rice: marker-assisted selection using RFLP and PCR. *Theoretical and Applied Genetics*, 95:313–320.

Huang et al. 2006. High lysine and high tryptophan transgenic maize resulting from the reduction of both 19- and 22-kD a-zeins. *Plant Molecular Biology*, 61: 525–535.

Huang et al. 2009. The genome of the cucumber, *Cucumis sativus L. Nature Genetics*, 41(12):1275–1281.

Huang et al. 2012. A multiparent advanced generation inter-cross population for genetic analysis in wheat. *Plant Biotechnology Journal*, 10:826–839.

Iftekharuddaula et al. 2015. Development of new submergence tolerant rice variety for Bangladesh using marker-assisted backcrossing. *Rice Science*, 22(1): 16–26.

Jackson et al. 2003. Field performance of transgenic cottons expressing one or two Bacillus thuringiensis endotoxins against bollworm, Helicoverpa zea (Boddie). *Journal of Cotton Science*, 7: 57–64.

Jamaloddin et al. 2020. Marker Assisted Gene Pyramiding (MAGP) for bacterial blight and blast resistance into mega rice variety "Tellahamsa". *PLoS ONE*, 15(6): e0234088.

Jankowicz-Cieslak et al. 2017. Biotechnologies for plant mutation breeding: protocols. p. 340.

Jewell et al. 2010. Transgenic plants for abiotic stress resistance . In *Transgenic Crop Plants*. In: Kole C, Michler, CH, Abbott, AG, Hall TC. editors. Berlin, Heidelberg: Springer-Verlag, pp. 67–132.

Jiang et al. 2007a. QTL analysis of resistance to Fusarium head blight in the novel wheat germplasm CJ 9306. I. Resistance to fungal spread. *Theoretical and Applied Genetics*, 116:3–13.

Jiang et al. 2014. Systemic gene silencing in plants triggered by fluorescent nanoparticle-delivered double-stranded RNA. *Nanoscale*, 6: 9965–9969.

Jinek et al. 2012. A programmable dual-RNA – guided DNA endonuclease in adaptive bacterial immunity. *Science*, 337: 816–822.

Jinek et al. 2014. Structures of Cas9 endonucleases reveal RNA-mediated conformational activation. *Science*, 343: 1247997.

Jogaiah et al. 2013. Systems biology-based approaches toward understanding drought tolerance in food crops. *Critical Reviews in Biotechnology*, 33(1):23–39.

Joseph et al. 2004. Combining bacterial blight resistance and Basmati quality characteristics by phenotypic and molecular marker-assisted selection in rice. *Molecular Breeding*, 13: 377–387.

Jung et al. 2018. Recent developments in genome editing and applications in plant breeding. *Plant Breeding*, 137(1): 1–9.

Kaur et al. 2020. Novel allelic variation in the Phospholipase D alpha1 gene (OsPLDα1) of wild Oryza species implies to its low expression in rice bran. *Scientific Reports*, 10: 6571.

Kawahara et al. 2012. Simultaneous RNA-Seq analysis of a mixed transcriptome of rice and blast fungus interaction. *PLoS ONE*, 7(11): e49423.

Kazan, K. 2015. Diverse roles of jasmonates and ethylene in abiotic stress tolerance. *Trends Plant Science*, 20: 219–229.

Kazmi et al. 2017. Metabolomic analysis of tomato seed germination. *Metabolomics*, 13(12): 145.

Khalid et al. 2017. Small RNA based genetic engineering for plant viral resistance: application in crop protection. *Frontiers in Microbiology*, 8: 43.

Kharkwal, M.C. and Shu, Q.Y. 2009. The role of induced mutations in world food security. In: Shu QY, editor. *Induced Plant Mutations in the Genomics Era*. Rome: Food and Agriculture Organization of the United Nations, pp. 33–38.

Kodym, A. and Afza, R. 2003. Physical and chemical mutagenesis. In *Methods in Molecular Biology*, Ed. Clifton, NJ, 236, 189–204.

Kottapalli et al. 2010. Effective strategy for pyramiding three bacterial blight resistance genes into fine grain rice cultivar, Samba mahsuri, using sequence tagged site markers. *Biotechnology Letters*, 32:989–996.

Kour et al. 2019c. Agriculturally and industrially important fungi: current developments and potential biotechnological applications.

Kover et al. 2009. A multiparent advanced generation inter-cross to fine-map quantitative traits in Arabidopsis thaliana. *PLoS Genetics*, 5: e1000551.

Kroupin et al. 2020. Allele mining of TaGRF-2D gene 5'-UTR in Triticum aestivum and Aegilops tauschii genotypes. *PLoS ONE*, 15(4): e0231704.

Kumar et al. 2010. Allele mining in crops: Prospects and potentials. *Biotechnology Advances* 28(4): 451–461.
Kumaravadivel et al. 2006. Molecular marker-assisted selection and pyramiding genes for gall midge resistance in rice suitable for Tamil Nadu Region. In: ABSTRACTS-2nd International Rice Congress 2006. p. 257.
Kumawat et al. 2019a. Fast neutron mutagenesis in plants: advances, applicability and challenges. *Plants*, 8: 164.
Kumawat et al. 2019b. Expanding avenue of fast neutron mediated mutagenesis for crop improvement. *Plants*, 8(6): 164.
Kurowska et al. 2011. TILLING a shortcut in functional genomics. *Journal of Applied Genetics*, 52(4): 371–390.
Le et al. 2012. Differential gene expression in soybean leaf tissues at late developmental stages under drought stress revealed by genome-wide transcriptome analysis. *PLoS ONE*, 7:e49522.
Lee et al. 2003. Expression of *Arabidopsis* CBF1 regulated by an ABA/stress inducible promoter in transgenic tomato confers stress tolerance without affecting yield. *Plant, Cell & Environment*, 26: 1181–1190.
Li et al. 2001. Fast neutron deletion mutagenesis-based reverse genetics system for plants. *Plant Journal*, 27: 235–242.
Li et al. 2010. Abscisic acid pretreatment enhances salt tolerance of rice seedlings: proteomic evidence. *Biochim Biophys Acta Proteins Proteomics*, 1804: 929–940.
Li et al. 2013. Multiplex and homologous recombination-mediated genome editing in Arabidopsis and Nicotiana benthamiana using guide RNA and Cas9. *Nature Biotechnology*, 31: 688–691.
Li et al. 2018. Expanding the scope of CRISPR/Cpf1-mediated genome editing in rice. *Molecular Plant*, 11: 995–998.
Limera et al. 2017. New biotechnological tools for the genetic improvement of major woody fruit species. *Frontiers in Plant Science* 8: 1418.
Liu et al. 2000. Molecular marker-facilitated pyramiding of different genes for powdery mildew resistance in wheat. *Plant Breeding*, 119 (1): 21–24.
Lockhart et al. 2000. Genomics, gene expression and DNA arrays. *Nature*, 405(6788): 827–836.
Lorenz, A.J. 2013. Resource allocation for maximizing prediction accuracy and genetic gain of genomic selection in plant breeding: a simulation experiment. G3: *Genes Genomes Genetics*, 3: 481–491.
Mackay et al. 2014. An eight-parent multiparent advanced generation inter-cross population for winter-sown wheat: creation, properties, and validation. *G3 (Bethesda)*. 18: 1603–1610.
Margulies et al. 2005. Genome sequencing in microfabricated high-density picolitre reactors. *Nature*, 437(-7057): 376–380.
Marroni et al. 2012. The quest for rare variants: pooled multiplexed next generation sequencing in plants. *Frontiers in Plant Science*, 3:133.
Mba, C. 2013. Induced mutations unleash the potentials of plant genetic resources for food and agriculture. *Agronomy*, 3: 200–231.
McLoughlin et al. 2018. Developing new RNA interference technologies to control fungal pathogens. *Canadian Journal of Plant Pathology,* 40: 325–335.
Meng et al. 2016. QTL mapping for agronomic traits using multi-parent advanced generation inter- cross (MAGIC) populations derived from diverse elite indica rice lines. *Field Crops Research*, 189:19–42.
Mezzetti et al. 2020. RNAi: What is its position in agriculture? *Journal of Pest Science,* 93: 1125–1130
Milos, P. 2008. Helicos BioSciences. *Pharmacogenomics*, pp 477–480.
Mitchell, P. L. and Sheehy, J. E. 2006. Supercharging rice photosynthesis to increase yield. *New Phytologist*, 171: 688–693.
Mochida, K. and Shinozaki, K. 2011. Advances in omics and bioinformatics tools for systems analyses of plant functions. *Plant and Cell Physiology*, 52(12): 2017–2038.
Molnar et al. 2011. Silencing signals in plants: a long journey for small RNAs. *Genome Biology*, 12(1): 1–8.
Moscou, M. and Bogdanove, A. 2009. Recognition by TAL Effectors. *Science*, 326:1501.
Moustafa, R.A.K. 2009. Development of salt_tolerant high_yielding barley lines via crossing between a mutant induced by EMS and a local cultivar. In: Shu QY, editor. *Induced Plant Mutations in the Genomics Era.* Proceedings of an International Joint FAO/IAEA Symposium.
MVD 2020 http://mvd.iaea.org/.
Nakagawa, H. 2009. Induced mutations in plant breeding and biological researches in Japan. In: Shu QY, editor. *Induced plant mutations in the genomics era*. Proceedings of an International Joint FAO/IAEA Symposium. Rome: Food and Agriculture Organization of the United Nations, pp. 48–54.
Nandi et al. 1997. Mapping QTLs for submergence tolerance in rice by AFLP analysis and selective genotyping. *Molecular Genetics and Genomics*, 255:1–8.

Narayanan et al. 2002. Molecular breeding for the development of Blast and Bacterial Blight resistance in Rice cv.IR50. *Crop Science,* 42:2072–2079.

Nat et al. 2007. Application of proteomics to investigate plant microbe interactions. *Current Proteomics*, 4: 28–43.

Nekrasov et al. 2013. Targeted mutagenesis in the model plant Nicotiana benthamiana using Cas9 RNA-guided endonuclease. *Nature Biotechnology*, 31: 691–693.

Nishimasu et al. 2014. Crystal structure of Cas9 in complex with guide RNA and target DNA. *Cell*, 156(5): 935–949.

Normile, D. 2006. Consortium aims to supercharge rice photosynthesis. *Science*, 313: 423.

Ogawa et al. 2018. Haplotype-based allele mining in the Japan-MAGIC rice population. *Science Report*, 8: 4379.

Ogita et al. 2003. Producing decaffeinated coffee plants. *Nature*, 423–823.

Oladosu et al. 2016. Principle and application of plant mutagenesis in crop improvement: a review. *Biotechnology & Biotechnological Equipment,* 30: 1–16.

Park et al. 2008. Overexpression of *Arabidopsis* ZEP enhances tolerance to osmotic stress. *Biochemical and Biophysical Research Communications*, 375: 80–85.

Pascual et al. 2015. Potential of a tomato MAGIC population to decipher the genetic control of quantitative traits and detect causal variants in the resequencing era. *Plant Biotechnology Journal*, 13:565–577.

Peterhansel et al. 2008. Metabolic engineering towards the enhancement of photosynthesis. *Photochemistry and Photobiology*, 84(6): 1317–1323.

Peterson et al. 2010. *Next Gen sequencing survey. North America equity research.* JP Morgan Chase & Co, New York.

Pidon et al. 2020. Allele mining unlocks the identification of RYMV resistance genes and alleles in African cultivated rice. *BMC Plant Biology*, 20(1): 222.

Pilcher et al. 2018. Genome-wide microarray analysis leads to identification of genes in response to herbicide, metribuzin in wheat leaves [published correction appears in PLoS ONE. 2018 Jun 19;13(6):e0199564]. *PLoS ONE*. 2017;12(12):e0189639.

Rafalski, A. 2002. Applications of single nucleotide polymorphisms in crop genetics. *Current Opinion in Plant Biology*, *5*(2): 94–100.

Raju et al. 2018. An epigenetic breeding system in soybean for increased yield and stability. *Plant Biotechnology Journal*, 16: 1836–1847.

Raliya et al. 2017. Nanofertilizer for precision and sustainable agriculture: current state and future perspectives. *Journal of Agricultural and Food Chemistry*, 66(26): 6487–6503.

Ray, S., and Satya, P. 2014. Next generation sequencing technologies for next generation plant breeding. *Frontiers in Plant Science*, 5:367. doi: 10.3389/fpls.2014.00367.

Reddy et al. 2009. ISSR marker based on GA and AG repeats reveal genetic relationship among rice varieties tolerant to drought, flood, salinity. *Journal of Zhejiang University Science B* 10:133–141.

Reddy et al. 2013. Optimization of callus induction and callus multiplication in rice (Oryza sativa L.) landraces. *Research in Plant Biology*, 3(5).

Renaut et al. 2006. Proteomics and low-temperature studies: bridging the gap between gene expression and metabolism. *Physiologia Plantarum*, 126(1): 97–109.

Ribaut et al. 2010. Molecular breeding in developing countries: challenges and perspectives. *Current Opinion in Plant Biology*, 13:1–6.

Roberts, J.K. 2002. Proteomics and a future generation of plant molecular biologists. *Plant Molecular Biology* 48:143–154.

Rodrigues, T.B. and Petrick, J.S. 2020. Safety considerations for humans and other vertebrates to agricultural uses of externally applied RNA molecules. *Frontiers in Plant Science* 11: 407.

Rothberg et al. 2011. An integrated semiconductor device enabling non-optical genome sequencing. *Nature*, 475(7356):348–352.

Ruel-Bergeron et al. 2015. Global update and trends of hidden hunger, 1995–2011: the hidden hunger index. *PLoS One*, *10*(12): e0143497.

Sallam, A. and Martsch, R. 2015. Association mapping for frost tolerance using multi-parent advanced generation inter-cross (MAGIC) population in faba bean (Vicia faba L.). *Genetica*, 143:501–14.

Salzberg, S. L. and J. A. Yorke. 2005. Beware of mis-assembled genomes. *Bioinformatics*, 21(24):4320–4321.

Sannemann et al. 2015. Multi-parent advanced generation inter-cross in barley: high-resolution quantitative trait locus mapping for flowering time as a proof of concept. *Molecular Breeding*, 35:1–16.

Sato et al. 2011. Sequence analysis of the genome of an oil-bearing tree, Jatropha curcas L. *DNA Research*, 18(1):65–76.
Schnable et al. 2009. The B73 maize genome: complexity, diversity, and dynamics. *Science*, 326(5956):1112–1115.
Schneider, A. 2002. Mapping of a nodulation loci sym9 and sym10 of pea. *Theoretical and Applied Genetics*, (104):1312–1316.
Sedlazeck et al. 2018. Piercing the dark matter: bioinformatics of long-range sequencing and mapping. *Nature Reviews Genetics*, 19(6):329–346.
Segal et al. 2003. A new opaque variant of maize by a single dominant RNAinterference- inducing transgene. *Genetics*, 165: 387–397.
Semagn et al. 2015. Quantitative trait loci mapping and molecular breeding for developing stress resilient maize for sub-Saharan Africa. *Crop Science*, 55:1.
Shan et al. 2013. Targeted genome modification of crop plants using a CRISPRCas system. *Nature Biotechnology*, 31: 686–688.
Sharma et al. 2019. Improvement of traditional varieties of rice through radiation induced mutation breeding. IGKV/ Pub/2019/ booklet/11, pp. 1–40.
Shi et al. 2009. Pyramiding multiple genes for resistance to soybean mosaic virus in soybean using molecular markers. *Molecular Breeding* 23(1): 113–124.
Shu et al. 2012. Principles and applications of plant mutation breeding. In: Shu QY, Forster BP, Nakagawa H, editors. *Plant Mutation Breeding and Biotechnology*. Wallingford: CABI, pp. 301–325.
Shukla et al. 2009. Precise genome modification in the crop species Zea mays using zinc-finger nucleases. *Nature*, 459: 437–441.
Shulaev et al. 2011. The genome of woodland strawberry (Fragaria vesca). *Nature Genetics*, 43(2): 109–116.
Singh et al. 2001. Pyramiding three bacterial blight resistance genes (xa5, xa13 and Xa21) using marker-assisted selection into indica rice cultivar PR106. *Theoretical and Applied Genetics*, 102:1011–1015.
Singh et al. 2011. Marker assisted selection: a paradigm shift in Basmati breeding. *Indian Journal of Genetics*, 71:120–128.
Singh et al. 2013. Marker-assisted simultaneous but stepwise backcross breeding for pyramiding blast resistance genes Piz5 and Pi54 into an elite Basmati rice restorer line 'PRR78.' *Plant Breeding*, 132:486–495.
Smith et al. 2000. Requirements for double-strand cleavage by chimeric restriction enzymes with zinc finger DNA-recognition domains. *Nucleic Acids Research*, 28: 3361–3369.
Song et al. 2017. Practical application of genomic selection in a doubled-haploid winter wheat breeding program. *Molecular Breeding*, 37(10):1–5.
Sprink et al. 2015. Plant genome editing by novel tools: TALEN and other sequence specific nucleases. *Current Opinion in Biotechnology*, 32: 47–53.
Srivastava et al. 2013. OnPLS integration of transcriptomic, proteomic and metabolomic data shows multi-level oxidative stress responses in the cambium of transgenic hipI-superoxide dismutase Populus plants. *BMC Genomics*, 14(1): 893.
Suh et al. 2013. Development of breeding lines with three pyramided resistance genes that confer broad-spectrum bacterial blight resistance and their molecular analysis in rice. *Rice*, 6:1–11.
Sun et al. 2015. DNA microarray revealed and RNAi plants confirmed key genes conferring low Cd accumulation in barley grains. *BMC Plant Biology*, 15(1): 259.
Sundaram et al. 2008. Marker assisted introgression of bacterial blight resistance in Samba Mahsuri., an elite indica rice variety. *Euphytica*. 160:411–422.
Suprasanna et al. 2009. Radiation induced in vitro mutagenesis, selection for salt tolerance and characterization in sugarcane. In *Induced plant mutations in the genomics era. Proceedings of an international joint FAO/IAEA symposium, International Atomic Energy Agency*, Vienna, pp. 145–147.
Tao et al. 2016. Genetic improvement of Japonica rice variety Wuyujing 3 for stripe disease resistance and eating quality by pyramiding Stv-bi and Wx-mq. *Rice Science*, 23:69–77.
Tilman, D. 2012. Biodiversity & environmental sustainability amid human domination of global ecosystems. *Daedalus*, 141(3): 108–120.
Tiwari et al. 2020. Phytohormones producing fungal communities: metabolic engineering for abiotic stress tolerance in crops. *Agriculturally Important Fungi for Sustainable Agriculture*. 171–197.
Toonen et al. 2013. ezRAD: a simplified method for genomic genotyping in non-model organisms. *PeerJ* 1:e203.
Townsend et al. 2009. High-frequency modification of plant genes using engineered zinc-finger nucleases. *Nature*, 459: 442–445.

Varshney et al. 2009. Next-generation sequencing technologies and their implications for crop genetics and breeding. *Trends in Biotechnology*, 27(9):522–530.

Vlk, D. and J. Řepková. 2017. Application of next-generation sequencing in plant breeding. *Czech Journal of Genetics and Plant Breeding*, 53(3): 89–96.

Wang et al. 2010. Assessment of FAE1 polymorphisms in three Brassica species using EcoTILLING and their association with differences in seed erucic acid contents. *BMC Plant Biology*, 10:137. doi: 10.1186/1471-2229-10-137.

Wang et al. 2011. The genome of the mesopolyploid crop species Brassica rapa. *Nature Genetics*, 43(10):1035–1039.

Wang et al. 2012. TILLING in extremis. *Plant Biotechnology Journal*, 10: 761–772.

Wang et al. 2016. Bidirectional cross kingdom RNAi and fungal uptake of external RNAs confer plant protection. *Nature Plants*, 2: 16151.

Wang et al. 2019. Transcriptome analysis of a rice cultivar reveals the differentially expressed genes in response to wild and mutant strains of *Xanthomonas oryzae* pv. *oryzae*. *Scientific Reports*, 9(1): 1–13.

Wendt et al. 2013. TAL effector nucleases induce mutations at a pre-selected location in the genome of primary barley transformants. *Plant Molecular Biology*, 83: 279–285.

Werner et al. 2005. Strategies for pyramiding resistance genes against the barley yellow mosaic virus complex (BaMMV, BaYMV, BaYMV-2). *Molecular Breeding*, 16:45–55.

Worrall et al. 2019. Exogenous application of RNAi-inducing double-stranded RNA inhibits aphid-mediated transmission of a plant virus. *Frontiers in Plant Science*, 10: 265.

Xia et al. 2012. Allelic variations of a light harvesting chlorophyll a/b-binding protein gene (Lhcb1) associated with agronomic traits in barley. *PLoS ONE*, 7:e37573.

Xiao et al. 1996. Genes from wild rice improve yield. *Nature*, 384:223–224.

Xiao et al. 1998. Identification of trait- improving quantitative trait loci alleles from a wild rice relative Oryza rufipogon. *Genetics*, 150:899–909.

Xu et al. 2011. Genome sequence and analysis of the tuber crop potato. *Nature*, 475: 189–195.

Yadav et al. 2019. Metabolic engineering to synthetic biology of secondary metabolites production. In: Gupta VK, Pandey A. editors. *New and Future Developments in Microbial Biotechnology and Bioengineering*. Amsterdam: Elsevier, pp. 279–320.

Yamamoto et al. 2010. Fine definition of the pedigree haplotypes of closely related rice cultivars by means of genome-wide discovery of single-nucleotide polymorphisms. *BMC Genomics*, 11(1):1–4.

Yang et al. 2012. Application of next-generation sequencing for rapid marker development in molecular plant breeding: a case study on anthracnose disease resistance in Lupinus angustifolius L. *BMC Genomics*, 13(1):318.

Yazdani et al. 2019. Ectopic expression of ORANGE promotes carotenoid accumulation and fruit development in tomato. *Plant Biotechnology Journal*, 17: 33–49.

Zaidi et al. 2017. CRISPR-Cpf1: A New Tool for Plant Genome Editing. *Trends Plant Science*, 22: 550–553.

Zhang et al. 2014. Comparative transcriptome analysis to reveal genes involved in wheat hybrid necrosis. *International Journal of Molecular Sciences*, 15(12):23332–23344.

Zhao, D. and Song, G.Q. 2014. Rootstock-to-scion transfer of transgene-derived small interfering RNA s and their effect on virus resistance in nontransgenic sweet cherry. *Plant Biotechnology Journal*, 12: 1319–1328.

Zhu, L. and Qian, Q. 2017. Development of "Purple Endosperm Rice" by engineering anthocyanin biosynthesis in endosperm: significant breakthrough in Transgene Stacking System, new progress in rice biofortification (in Chinese). *Chinese Bulletin of Botany*, 52: 539–542.

Zhu et al. 2018. From golden rice to aSTARice: bioengineering astaxanthin biosynthesis in rice endosperm. *Molecular Plant*, 11: 1440–1448.

Zhu et al. 2020. Plant synthetic metabolic engineering for enhancing crop nutritional quality. *Plant Communications*, 1(1): 100017.

Zotti et al. 2018. RNAi technology in crop protection against arthropod pests, pathogens and nematodes. *Pest Management Science*, 74: 1239–2125.

2

The Chemical Dialogue during Plant–Microbe Interaction: Implications in Sustainable Agriculture

Gunajit Goswami, Sudipta Sankar Bora, Jugabrata Das, Trishna Sarmah, and Madhumita Barooah*
Assam Agricultural University

CONTENTS

2.1 Introduction

In nature, plants interact with a plethora of active microorganisms, of which many are beneficial, some are commensal, and others are detrimental or pathogenic to the plants. The plant root system secretes a vast array of metabolic compounds to its immediate environment, viz. the rhizosphere that facilitates the creation of a niche for diverse microbial communities and dynamic interactions [1, 2]. Most of the interactions between microorganisms and plants occur in the rhizosphere. Lorenz Hitner, for the first time in 1904, defined the term "rhizosphere" as the volume of soil surrounded and influenced by plant roots [3, 4]. Later, Pinton and his co-workers redefined "rhizosphere" as the soil closely attached to the roots along with the root tissues with microbial colonization [2, 5]. Due to the abundance of plant-synthesized metabolites, microbial activities are always observed to be higher in the rhizosphere as compared to the adjacent bulk soil. In addition, certain signaling compounds and metabolites arising from the edaphic system and the microbes themselves shape the nature and structure of microbial community. As such, the microbial community in the soil surrounding the plant roots is more diverse and interactive than the bulk soil [6, 7]. The dimension and magnitude of plant–microbe interaction depend on the environmental condition, intrinsic microbial load, and developmental stage and genotype of the host plant. The interactions mainly occur in the rhizosphere, endosphere, and phylloplane. Such interactions mediated by the rhizobacteria, endophytes, mycorrhizae, and epiphytes benefit the host plants through growth improvement, biotic and abiotic stress management, and nutrient provisioning. In return for the shelter and nutritionally important metabolites received from the plant, the symbiotic microbes help in solubilizing and acquiring nutrients from the soil and delivering them to the hosts. In the classical example of plant–mycorrhizal symbiotic association, the symbiotic relationship helps the host plants in initial terrestrialization through nullifying environmental constraints such as nutrient and water deficit problems

* Corresponding author's email: m.barooah@aau.ac.in

DOI: 10.1201/9781003171416-2

(reviewed in [2]). Another important and classical leguminous plant–*Rhizobium* symbiotic relationship demonstrates the efficacy of the bacteria in fixation and mobilization of atmospheric nitrogen toward the host plant. Both these exemplary relationships share a common symbiotic signaling (Sym) pathway inducible in plant cells by mycorrhizal (Myc) and rhizobial nodulation (Nod) factors, respectively [8–11]. Apart from such symbiotic microorganisms, plants' rhizosphere also harbors non-symbiotic plant growth-promoting rhizobia (both bacteria and fungi), which are instrumental in disease suppression (elicitation of induced systemic resistance), pollutant degradation (heavy metals and recalcitrant compounds), and phytostimulation (phytohormones) (reviewed in [1]). It appears that few components of the Sym pathway cascades are also induced in such non-symbiotic relationships, indicating the evolutionary divergence of such signaling mechanisms among the beneficial microbes [12]. Irrespective of the type of interaction (i.e., neutral, beneficial, or detrimental), it is generally held that the initial stages of microbial colonization process are more or less similar among all the plant–microbe interactions. The colonization process starts with recognition (chemotaxis) followed by adherence and invasion (endophytes and pathogens) and finally colonization. The plant roots secrete chemical compounds, which attract the microbes toward the rhizosphere. Once the microbes reach the proximity of the plant roots, the next step, i.e., adhesion, begins, which establishes a direct contact with the host plants. The formation of biofilms through extracellular polysaccharide matrix plays a key role in the aggregation and adhesion process [2].

2.2 Role of Plant Root Exudates in Microbial Colonization

Plants produce root exudates at a significant carbon cost, and various microorganisms utilize these exudates as substrates for growth [13, 14]. Growing evidence suggests that the host plant through selective secretion of root exudates can actively modulate the rhizosphere microbial community. Soluble components of root exudates also function as chemical signaling agents (attractant or repellent) for soil microbial populations. Competing for the exudates, most of the beneficial and pathogenic microorganisms take part in a tripartite interaction with the plant host. Root exudates contain both high and low molecular weight compounds secreted by active and passive mechanisms. Low molecular weight compounds, which include amino acids, sugars, organic acids, hormones, and other secondary metabolites, are more diverse as compared to high molecular weight compounds. The later type mostly as less diverse are characterized as high-yield proteins, polysaccharides, and few secondary metabolites [2, 13–17]. The low molecular weight compounds such as amino acids and sugars are secreted by passive diffusion mechanism. Diffusions of these compounds depend on polarity, cytosolic pH, and membrane permeability. These compounds generally act as chemoattractant or chemorepellent for microbes. High molecular weight compounds consist of exopolysaccharides, secondary metabolites, etc., which are secreted by active mechanisms [6, 18]. Membrane-bound transporters such as the ATP-binding cassette (ABC) transporters, the multidrug and toxic compound extrusion (MATE) family exporters, major facilitator superfamily (MFS), and the aluminum-activated malate transporter (ALMT) family are known to actively secrete various root exudates [18–27]. High molecular weight components bear specific functions central to soil and plant health, microbial adhesion and invasion, etc. Lectins belonging to this group are involved in both symbiotic and plant defense interactions; arabinogalactan protein prevalent in the exopolysaccharide attracts PGPR while repelling plant pathogens. Plants also produce flavonoids such as rutin, which help in establishing mutualistic plant–fungal interaction [28–34]. Some plant hormones and secondary metabolites act as attractant of beneficial soil microorganisms and also provide defense against plants pathogens [35]. Plant hormones such as salicylic acid (SA) and jasmonic acid (JA) integral to plant defense mechanisms can help in recruiting antipathogenic microbes. It is assumed that such an active recruitment may occur as an outcome of the chemical crosstalk between the incoming microbes and the plant [36–38]. The microbes can also influence the composition and rate of root exudation. Microbe-originated compounds such as 2,4-diacetylphloroglucinol (DAPG), phenazine, and zearalenone can modulate amino acid concentration in the exudates. It is noteworthy that amino acids and their derivatives such as glutamic acid, proline, and betaines play a major role in bacterial osmoregulation, for which it is imperative that the rhizosphere microbial population constantly manipulate the plant root metabolome to its own advantage [39–45].

2.3 Chemotaxis Motility and Colonization by Rhizosphere Microbiota

Root colonization by rhizobacteria is a complex multistage process. Initially, a prospective bacterial population actively (flagellar movement) or passively (water flux) moves to the vicinity of plant roots in search of nutrients and shelter and firmly adheres to the roots through nonspecific adsorption. At this stage, the target plant also exerts a complex yet specific response to this bacterial adherence, including the expression of several metabolites through root exudation, which in turn induces bacterial gene expression as adaptive measures. The mobilization of soil bacteria from the soil and sediment to the rhizosphere occurs essentially through the mechanism of chemotaxis. The drivers of chemotaxis are certain plant metabolites that function as chemical cues recognizable through specific bacterial receptors [2, 6, 15]. Path finding research by Julius Adler in the 1960s first deciphered the mechanism of chemotaxis signal transduction in *Escherichia coli* [46, 47]. So far, eighteen classes of such mechanism have been discovered, consisting majorly of flagellar-based motility and minorly of type IV pili motility [48]. L-Malic acid secreted in the root exudate by certain plant species also serves as a chemotaxis mobility activator for beneficial *Bacillus subtilis* [49]. Similarly, organic acids, *viz.* oxalic acid, fumaric acid, and malic acid, also function as elicitors for chemotaxis movement and biofilm production during the colonization of the *Bacillus amyloliquefaciens* [50]. Bacterial phosphotransferase systems and/or periplasmic binding proteins are shown to be involved in detecting the presence of the ligand (chemical cues) that further facilitates the chemoreceptor binding [51–53]. In case of pathogenic fungi, another mechanism known as electrotaxis mediates the recruitment of the pathogenic oomycetes to the rhizosphere. The proton motive force generated due to the transports of protons and ions across the plant roots functions as external electric signals for motile zoospores [13, 54, 55]. While interacting with a new microorganism, the plant always considers it as foreign body and readily activates its own immune system against the incoming agent. Special types of molecules known as pathogen-/microbe-associated molecular patterns (PAMPs/MAMPs) produced by the microbes induce and activate the plant defense response. These highly conserved molecules among both pathogenic and beneficial microbes trigger the plant host's PAMP-/MAMP-triggered immunity (PTI or MTI). In order to establish any interaction, all types of symbiotic, non-symbiotic, or pathogenic microbes must employ specific strategies for the suppression of host PTI or MTI. During the process of colonization, the bacterium (MAMPs) needs to suppress the production of glucosinolate phytoanticipins, which would otherwise trigger ISR, including reduced expression of JA and production of SA. Reversibly changing colony morphology (phenotypic variation) also enables a bacterial population to escape initial host immune response. For instance, during initial colonization, PGPR *Pseudomonas brassicacearum* changes its morphology by lowering numbers of flagella by the action of alkaline protease (AprA) (reviewed in [1, 2, 15]). Salicylic acid (SA) is another important plant defense signaling regulator that acts against a vast array of plant pathogens (bio- or hemibiotrophic). Once induced, the master regulator NPR1 protein translocates to the nucleus and activates the transcription of SA-responsive genes [56–58]. Intensive SA expression (indicating increased NPR1 production) negatively correlates to rhizobia colonization (infection and nodulation) in plants. *Medicago truncatula* plants with overexpressed NPR1 in response to *Sinorhizobium meliloti* have been observed to suffer from deformed root hair development, a condition which can be ameliorated with RNAi-mediated NPR1 depletion. For the establishment of a mutualistic relationship with the leguminous plant host, rhizobia suppress this SA-dependent defense signaling through the onset of Nod signaling pathway [59–62]. Rhizobia have evolved sophisticated mechanisms for tackling plant defense response through modulating SA levels. Plants have evolved selection criteria for adopting compatible and incompatible *Rhizobium* strains using SA-mediated plant defense response; alfalfa plants have been found to overproduce SA levels in response to incompatible *Rhizobium* strains and vice versa. In addition, *Rhizobium* mutants for Nod factors were unable to reduce SA levels, indicating the necessity of these factors in infection and root nodulation [59]. Another *S. meliloti* mutant with a defective Nod factor gene was unable to downregulate β-glucanase MtBGLU1 gene in *M. truncatula* roots 24 h post-inoculation. This MtBGLU1 is a homologous gene to pathogenesis-related (PR) PR2 gene and plays an important role in plant defense related to Sym pathways. From these findings, it may be hypothesized that the Nod factors not only are necessary for the nodulation process in leguminous plants, but also play a quintessential role in modulating the host defense response [1, 63].

2.4 Role of Extracellular Polysaccharides in PMI

Lipopolysaccharides (LPS) and exopolysaccharides (EPS) have been the most widely known MAMPs in plants; however, recent reports also indicate the presence of other MAMPs, which include the cell wall components of fungi and bacteria such as peptidoglycans, glycoproteins [36, 64–67], and secondary metabolites such as AHLs [68]. In addition, siderophores [69], biosurfactants [70, 71] as well as antibiotics [36, 72–74] produced by different microorganisms are also regarded as MAMPs. The LPS are structural constituents of the outer membrane of gram-negative bacteria with the potential to elicit host immune responses [75]. Depending on the host and non-host situation, LPS can elicit differential host immune responses. The purified compound from *Sinorhizobium meliloti* evokes a very weak immune response in its natural leguminous plant host, *Medicago sativa*; however, the same compound has been observed to activate tremendous immune response in non-host tobacco cell cultures [76]. Application of a concoction of *S. meliloti* LPS and a yeast elicitor to alfalfa plants was unable to elicit the yeast elicitor-related transcriptional reprogramming and oxidative burst in the treated plants which indicated the possibility of LPS in host immune suppression [76–78]. This host-dependent response toward LPS elucidates the evolutionary significance of leguminous LPS perception systems and ability of the plants to differentiate between an incoming beneficial microbe and a pathogenic agent [1]. The exopolysaccharides (EPS) are high molecular weight sugar polymers secreted by bacteria having important roles in bacterial physiology [79] and also play a profound role (especially acidic EPS) in establishing mutualistic relationships. The EPS-deficient *Rhizobium* mutants evoked leguminous plant (host) immune response, including activation of host antimicrobial activity, confirming the usefulness of EPS in host defense suppression and consequently in the *Rhizobium* symbiosis process [80, 81]. The EPS are known to efficiently chelate the calcium anions, which blocks their cytosolic influx that leads to the inhibition of MAMP-triggered responses [82].

2.5 Quorum Sensing and Biofilm Production during Microbial Colonization

Quorum sensing (QS) is a special cell-to-cell communication between bacteria that occurs via diffusible chemical signals [83, 84]. A number of bacterial genes are regulated by QS, including those involved in biofilm formation. Biofilms are organized multilayer aggregate of bacterial cells embedded in a complex extracellular matrix of exopolysaccharides and proteins. Biofilm formation helps the bacterial population to survive under unfavorable environmental conditions. The formation of biofilms aids bacterial cells in adhering to environmental surfaces, including plant tissues, and is an intrinsic component of plant–microbe interactions [85]. Some bacteria trigger plant growth and provide protection against pathogens by forming thick biofilm on the root surface [86, 87]. Microbe–microbe interactions involving inter- or intra-species crosstalks always have a direct impact on plant–microbe interactions. During intra-species communications, bacteria use distinct QS signal compounds that serve as indicators for the presence of primary and secondary plant metabolites, competitor population, stress conditions, etc. During initial chemotaxis and adherence to the plant roots, QS compounds play a vital role in establishing microbial colonization. These QS compounds work as cell-dependent holistic stimuli that determine colonization pattern in bacteria, such as metabolic rate, reproduction, and pathogenicity in a synchronized manner. Such behaviors are triggered by the secretion of bacterial low molecular weight autoinducers (AIs) [2, 88–90]. Interspecies microbial crosstalks between competing microbes result in interference and disruption of AIs essential for QS signaling in plant pathogens. Beneficial microbes use various lytic enzymes (lactonase and acylase), volatile organic compounds (2,3-butanediol and acetoin), etc., which consequently disrupt the pathogenic bacteria-origin AHL production [91, 92]. In addition, the diketopiperazines [cyclo(L-Pro-L-Val/L-Phe/L-Tyr)] produced by *P. aeruginosa* can modulate auxin signaling and promote the growth of *A. thaliana* [93]. Such synchronized efforts of beneficial microbes protect the host plants from the colonization of soilborne pathogens and also help improve the growth of the host plant. The N-acyl homoserine lactones (AHLs) and N-(3-oxo-hexanoyl)-homoserine lactone (AI-1) secreted by gram-negative bacteria (though not exclusive to such bacteria) are the most well-known AIs till date. These molecules comprise of a hydrophilic lactone ring and a carbon chain (4–18 carbons) that decides the affinity of the entire molecule for water [90, 94]. In addition to AIs, cell-to-cell communication-related diverse molecules and peptides have also been reported from other microorganisms such as gram-positive bacteria and yeasts [94–103]. In *Bacillus subtilis*, oligopeptide-based ComX pheromone [104], gamma-butyrolactone-based A-factor; methylenomycin furans (reported from *Streptomyces* spp. [105, 106]); tyrosol and farnesol (from yeast [95, 97]); and ubiquitous AI furanosyl borate diester (AI-2) from a wide variety of bacteria [103, 107] play an important role in cell-to-cell communication. Few AHLs such as N-(butanoyl)-L-homoserine lactone (C4-HSL) and N-(hexanoyl)-L-homoserine lactone (C6-HSL) when used to treat plants can alter auxin/cytokinin ratio, thereby influencing the overall root architecture [90, 108]. Similarly, N-(3-oxo-tetradecanoyl)-L-homoserine lactone produced by *Sinorhizobium meliloti* promoted nodulation in *Medicago truncatula* [90]. In addition, N-(3-oxo-decanoyl)-L-homoserine lactone induced the formation of adventitious roots in *Vigna radiata* (mung bean) [90]. In an interesting phenomenon, short acyl chain AHLs have been reported to elongate primary root, while long acyl chain AHLs [e.g., N-(dodecanoyl) or N-(tetradecanoyl)-L-homoserine lactones] induce a plant defense system in *Arabidopsis thaliana* [90, 109]. The plant can more effectively translocate short-chain AHLs (e.g., C6-HSL) into leaves as compared to long-chain AHLs, indicating the presence of negative correlation between the size of AHLs and translocation efficiency [90]. In an effort to reduce QS signals and pathogenicity, the plants employ a number of strategies, which include inhibition of AI biosynthesis, enzymatic degradation of intermediate molecules/compounds, and interference in receptor binding (quorum quenching, QQ) [88, 91, 110–112]. Examples include the D-(+)-catechin produced by plants, which negatively regulates the binding of C4-HSL to RhlR of *P. aeruginosa* [113]. In contrast, plants also promote QS signaling for beneficial microbes; for example, secretion of IAA and cytokinins, the type III secretion system, flavonoids, etc., influence QS in soil bacterial populations [114–116].

2.6 Modulation of the Plant Immune System by Microorganisms

Recognition of MAMPs by the plant activates mitogen-activated protein kinases (MAPKs) and calcium-dependent protein kinases (CDPKs). These master signaling modules transduce the incoming early signals to a cascade of defensive measures such as formation of physical barriers (callose deposition and stomatal closure), production of antipathogenic agent (reactive oxygen species and secondary metabolites), and induction of phytohormone (ethylene) [117–120]. At this stage, microbes or pathogens play an elusive game by producing and delivering effectors as a part of effector-triggered susceptibility (ETS) [121–124]. These effectors (e.g., coronatine and syringolin A) are intended for masking the pathogen's presence so as to avoid PTI signals [125–127]. However, plants have also adopted another defense strategy called effector-triggered immunity (ETI), which readily gets activated in the presence of any pathogen effectors. This specific set of resistance genes, which help plants in the recognition of pathogen effectors, is known as R genes. In fact, R genes such as *NB-LRR* (proteins with nucleotide-binding and leucine-rich repeat domains) and *Xa21* (transmembrane protein rich in extracellular leucine repeats) impart resistance against an array of plant pathogens (bacteria, fungi, insects, nematodes, viruses, etc.) [2, 123, 128, 129]. The ETI is accompanied by SA-dependent plant defense mechanisms and programmed cell death, which prevents the proliferation of biotrophic pathogens. These biotrophic pathogens draw nutrients from living cells, through feeding structures. As a part of hypersensitive reaction or response (HR), programmed cell death inhibits the entry of pathogens into living tissues [130, 131]. PTI and ETI also safeguard distal and undamaged tissues through transducing long-distance signals and participate in memory-based defense capacity building in these regions for future attack. This SA-dependent immune preparedness is known as systemic acquired resistance (SAR), which confers resistance against various plant pathogens and herbivores too [2, 131–134]. It is to be emphasized that SAR is acquired upon infection by a pathogen, while induced systemic resistance (ISR) is triggered upon exposure to a beneficial microbe. While the former is SA-dependent, the latter is MAMPs-induced and SA-independent [2, 36, 128, 130, 131, 133–135].

ISR is integral to a plant's innate immunity that helps the host to react more responsibly (through priming) to the incoming beneficial microbe [36, 110]. As a part of ISR activity, plants produce and secrete secondary metabolites to block pathogenic attack. These secondary metabolites with antimicrobial properties are secreted either constitutively (phytoanticipins), or in response to certain pathogen attacks (phytoalexins) [136, 137]. In natural state, phytoanticipins remain in glycosylated form, which upon disturbance of tissue damage caused by pathogen attack gets converted into bioactive form aglycone by the action of β-glucosidases [138–140]. During mycorrhization of maize plant, the inactive glycosylated 2,4-dihydroxy-7-methoxy-1,4-benzoxazin-3-one (DIMBOA-Glc) is converted to 2,4-dihydroxy-7-methoxy-1,4-benzoxazin-3-one (DIMBOA) phytoanticipin, which also acts as a chemotaxis molecule for the recruitment of PGP *Pseudomonas putida* [35, 141]. HR during plant–microbe or plant–insect interactions leads to the production of phytoalexins and biosynthesis of phytohormones associated with defense signaling [142] such as that of the benzoxazinoids produced by grasses in response to insects and pathogens [143–145]; production of capsidiol and sesquiterpene by *Nicotiana benthamiana* is observed upon *Phytophthora infestans* infection and a number of glucosinolates, *viz.* sinigrin, glucotropaeolin, and gluconasturtiin, are produced due to damage in plants [140]. In addition to this distinct constitutive and inducible secondary metabolite-based plant defense response, it has also been observed that the expression and accumulation of few constitutive defense compounds increases with positive or negative interactions. For instance, rice infected with blast fungus *Magnaporthe grisea* increases the accumulation of antifungal momilactone A (a diterpene constitutive compound) in the leaves [2, 142, 146]. Endophytes, which are a specialized subset of the rhizosphere microbiota [147], enter the host plant through sites of opening in the roots (damaged tissues and cracks at epidermal conjunction due to natural growth), but can also enter plant tissue through stomata and hydathodes, or through flowers and fruits to a lesser extent [148, 149]. Inside the host plant, endophytes can participate in various direct or indirect growth-promoting activities, including nutrient acquisition, secretion and modulation of phytohormones (IAA, GA, and cytokinins) and antimicrobial agents, inhibition of ethylene (through the action of ACC deaminase), sugar metabolism, and disease suppression through outcompeting pathogens [150]. A comparative genomics study of poplar endophyte *Stenotrophomonas maltophilia* R551-3 and pathogenic *S. maltophilia* K279 revealed the existence of host insertion hotspots, antibiotic biosynthesis, resistance gene clusters, etc., in the former strain's core genome, which indicated that these bacteria have evolutionarily developed a host-specific cellular mechanism for plant penetration and better adaptation [151, 152]. Similarly, bacteria such as *Stenotrophomonas rhizophila* and *Burkholderia phytofirmans* have developed a molecular sensing mechanism for various abiotic stresses and accordingly rearrange gene expression pattern to the changing and challenged environment [151, 153].

2.7 Conclusions

Root exudates and associated microbial metabolites are the key players in determining the type and function of plant–microbe interaction, be it positive or negative. Once attached to the root surface, the rhizospheric microbes can modulate the phytohormone-based signaling pathways and the host plant defense system. The chemical signaling involved in plant–microbe interactions significantly determines the nature and magnitude of a plant's immune response toward symbionts and plant pathogens. Understanding the mechanisms by which plants distinguish and favor plant growth-promoting microbes over a pathogen remains an active area of research, particularly with respect to how the host immune system differentially senses the incoming microbial communities and favors the colonization of beneficial microbes over the pathogenic ones. Additionally, it will be interesting to know how perturbations in the root exudates impact the plant–microbe chemical dialogues and signaling. Regardless of complexity, studies on signals and molecules as the determinants of beneficial and pathogenic microbe-specific plant host immune response can pave the way for developing improved multispectral bioinoculum, including biofertilizers and biocontrol agents. Currently, most of the available reports in this arena encircle around genomics and transcriptomics with minimal studies on metabolomics. With recent advances in metabolomics tools and technologies, including mass spectrometry, it is expected that omics approaches integrating genomics, proteomics, and metabolomics will unravel rhizosphere signalomics and elucidate the cryptic signaling processes.

REFERENCES

1. Zamioudis C and Pieterse CMJ (2012) Modulation of host immunity by beneficial microbes. *Mol Plant Microbe Interact* 25:139–150.
2. Chagas FO et al. (2018) Chemical signaling involved in plant–microbe interactions. *Chem Soc Rev* 47:1652–1704.
3. Bakker PAHM et al. (2013) The rhizosphere revisited: root microbiomics. *Front Plant Sci* 4:165.
4. Hartmann et al. (2009) Plant-driven selection of microbes. *Plant Soil* 321:235–257.
5. Nannipieri et al. (2007) The rhizosphere. In: Pinton R, Varanini Z, Nannipieri, P (eds) *Biochemistry and Organic Substances at the Soil-Plant Interface*, 2nd ed. CRC Press, Taylor and Francis Group, New York.
6. Badri D V. and Vivanco JM (2009) Regulation and function of root exudates. *Plant Cell Environ* 32:666–681.
7. Mendes et al. (2011) Deciphering the rhizosphere microbiome for disease-suppressive bacteria. *Science* 332:1097–1100.
8. Bonfante P and Genre A (2010) Mechanisms underlying beneficial plant–fungus interactions in mycorrhizal symbiosis. *Nat Commun* 1:48.
9. Maillet et al. (2011) Fungal lipochitooligosaccharide symbiotic signals in arbuscular mycorrhiza. *Nature* 469:58–63.
10. Oldroyd GED and Downie JA (2008) Coordinating nodule morphogenesis with rhizobial infection in legumes. *Annu Rev Plant Biol* 59:519–546.
11. Oldroyd et al. (2009) Reprogramming plant cells for endosymbiosis. *Science* 324:753–754.
12. Sanchez et al. (2005) Pseudomonas fluorescens and Glomus mosseae trigger DMI3-dependent activation of genes related to a signal transduction pathway in roots of Medicago truncatula. *Plant Physiol* 139:1065–1077.
13. Bais et al. (2006) The role of root exudates in rhizosphere interactions with plants and other organisms. *Annu Rev Plant Biol* 57:233–266.
14. Deveau et al. (2012) Communication between plant, ectomycorrhizal fungi and helper bacteria. In: Witzany G (ed) *Biocommunication of Fungi*. Springer, Dordrecht, Netherlands, pp. 229–247.
15. Badri et al. (2009) Rhizosphere chemical dialogues: plant-microbe interactions. *Curr Opin Biotechnol* 20:642–650.
16. Chaparro et al. (2014) Rhizosphere microbiome assemblage is affected by plant development. *ISME J* 8:790–803.
17. Walker et al. (2003) Root Exudation and Rhizosphere Biology. *Plant Physiol* 132:44–51.
18. Weston et al. (2012) Mechanisms for cellular transport and release of allelochemicals from plant roots into the rhizosphere. *J Exp Bot* 63:3445–3454.
19. Wilkens S (2015) Structure and mechanism of ABC transporters. *F1000Prime Rep* 7:14.
20. Badri et al. (2009) An ABC transporter mutation alters root exudation of phytochemicals that provoke an overhaul of natural soil microbiota. *Plant Physiol* 151:2006–2017.
21. Badri et al. (2008) Altered profile of secondary metabolites in the root exudates of Arabidopsis ATP-binding cassette transporter mutants. *Plant Physiol* 146:762–771.
22. Kang et al. (2011) Plant ABC Transporters. *Arab B* 9:e0153–e0153.
23. Kretzschmar et al. (2011) Functions of ABC transporters in plants. *Essays Biochem* 50:145–160.
24. Liu et al. (2009) Aluminum-activated citrate and malate transporters from the MATE and ALMT families function independently to confer Arabidopsis aluminum tolerance. *Plant J* 57:389–399.
25. Takanashi et al. (2014) The multidrug and toxic compound extrusion (MATE) family in plants. *Plant Biotechnol* 31:417–430.
26. Reddy et al. (2012) The major facilitator superfamily (MFS) revisited. *FEBS J* 279:2022–2035.
27. Hwang et al. (2016) Plant ABC transporters enable many unique aspects of a terrestrial plant's lifestyle. *Mol Plant* 9:338–355.

28. De Hoff et al. (2009) Plant lectins: the ties that bind in root symbiosis and plant defense. *Mol Genet Genomics* 282:1–15.
29. Nguema-Ona et al. (2013) Arabinogalactan proteins in root-microbe interactions. *Trends Plant Sci* 18:440–449.
30. Cannesan et al. (2012) Effect of arabinogalactan proteins from the root caps of pea and brassica napus on Aphanomyces euteiches zoospore chemotaxis and germination. *Plant Physiol* 159:1658–1670.
31. Xie et al. (2012) A plant arabinogalactan-like glycoprotein promotes a novel type of polar surface attachment by Rhizobium leguminosarum. *Mol Plant Microbe Interact* 25:250–258.
32. Vicré et al. (2005) Root Border-Like Cells of Arabidopsis. Microscopical Characterization and Role in the Interaction with Rhizobacteria. *Plant Physiol* 138:998–1008.
33. Lagrange et al. (2001) Rutin, the phenolglycoside from eucalyptus root exudates, stimulates Pisolithus hyphal growth at picomolar concentrations. *New Phytol* 149:349–355.
34. Kalinova J and Radova S (2009) Effects of rutin on the growth of Botrytis cinerea, Alternaria alternata and Fusarium solani. *Acta Phytopathol Entomol Hungarica* 44:39–47.
35. Neal et al. (2012) Benzoxazinoids in Root Exudates of Maize Attract Pseudomonas putida to the Rhizosphere. *PLoS One* 7:e35498.
36. Doornbos et al. (2012) Impact of root exudates and plant defense signaling on bacterial communities in the rhizosphere. A review. *Agron Sustain Dev* 32:227–243.
37. Carvalhais et al. (2013) Activation of the jasmonic acid plant defence pathway alters the composition of rhizosphere bacterial communities. *PLoS One* 8:e56457.
38. Lebeis SL et al. (2015) Salicylic acid modulates colonization of the root microbiome by specific bacterial taxa. *Science* 349:860–864.
39. Huang et al. (2014) Rhizosphere interactions: root exudates, microbes, and microbial communities. *Botany* 92:267–275.
40. Moe LA (2013) Amino acids in the rhizosphere: from plants to microbes. *Am J Bot* 100:1692–1705.
41. Meier et al. (2013) Fungal communities influence root exudation rates in pine seedlings. *FEMS Microbiol Ecol* 83:585–595.
42. Talboys et al. (2014) Auxin secretion by Bacillus amyloliquefaciens FZB42 both stimulates root exudation and limits phosphorus uptake in Triticum aestivium. *BMC Plant Biol* 14:51.
43. Rocha et al. (2007) Signal transduction-related responses to phytohormones and environmental challenges in sugarcane. *BMC Genomics* 8:71.
44. Phillips et al. (2004) Microbial products trigger amino acid exudation from plant roots. *Plant Physiol* 136:2887–2894.
45. Wood JM (2011) Bacterial osmoregulation: A paradigm for the study of cellular homeostasis. *Annu Rev Microbiol* 65:215–238.
46. Hazelbauer GL (2012) Bacterial chemotaxis: the early years of molecular studies. *Annu Rev Microbiol* 66:285–303.
47. Parkinson et al. (2015) Signaling and sensory adaptation in Escherichia coli chemoreceptors: 2015 update. *Trends Microbiol* 23:257–266.
48. Wuichet K and Zhulin IB (2010) Origins and diversification of a complex signal transduction system in prokaryotes. *Sci Signal* 3:ra50.
49. Rudrappa et al. (2008) Root-secreted malic acid recruits beneficial soil bacteria. *Plant Physiol* 148:1547–1556.
50. Yuan et al. (2015) Organic acids from root exudates of banana help root colonization of PGPR strain Bacillus amyloliquefaciens NJN-6. *Sci Rep* 5:13438.
51. Hazelbauer GL and Lai W-C (2010) Bacterial chemoreceptors: providing enhanced features to two-component signaling. *Curr Opin Microbiol* 13:124–132.
52. Hazelbauer et al. (2008) Bacterial chemoreceptors: high-performance signaling in networked arrays. *Trends Biochem Sci* 33:9–19.
53. Neumann et al. (2012) Chemotactic signaling via carbohydrate phosphotransferase systems in Escherichia coli. *Proc Natl Acad Sci* 109:12159–12164.
54. van West et al. (2002) Oomycete plant pathogens use electric fields to target roots. *Mol Plant Microbe Interact* 15:790–798.
55. Hua et al. (2008) A, Phytophthora sojae G-protein α subunit is involved in chemotaxis to soybean isoflavones. *Eukaryot Cell* 7:2133–2140.

56. Dong X (2004) NPR1, all things considered. *Curr Opin Plant Biol* 7:547–552.
57. Leon-Reyes et al. (2009) Ethylene modulates the role of Nonexpressor of pathogenesis-related genes1 in cross talk between salicylate and jasmonate signaling. *Plant Physiol* 149:1797–1809.
58. Pieterse et al. (2009) Networking by small-molecule hormones in plant immunity. *Nat Chem Biol* 5:308–316.
59. Martínez-Abarca et al. (1998) Involvement of Salicylic Acid in the Establishment of the Rhizobium meliloti -Alfalfa Symbiosis. *Mol Plant-Microbe Interact* 11:153–155.
60. Stacey et al. (2006) Effects of endogenous salicylic acid on nodulation in the model legumes lotus japonicus and medicago truncatula. *Plant Physiol* 141:1473–1481.
61. van Spronsen et al. (2003) Salicylic acid inhibits indeterminate-type nodulation but not determinate-type nodulation. *Mol Plant Microbe Interact* 16:83–91.
62. Peleg-Grossman et al. (2009) NPR1 protein regulates pathogenic and symbiotic interactions between rhizobium and legumes and non-legumes. *PLoS One* 4:e8399.
63. Mitra RM and Long SR (2004) Plant and bacterial symbiotic mutants define three transcriptionally distinct stages in the development of the medicago truncatula/sinorhizobium meliloti symbiosis. *Plant Physiol* 134:595–604.
64. Felix et al. (1993) Specific perception of subnanomolar concentrations of chitin fragments by tomato cells: induction of extracellular alkalinization, changes in protein phosphorylation, and establishment of a refractory state. *Plant J* 4:307–316.
65. Erbs et al. (2008) Peptidoglycan and muropeptides from pathogens agrobacterium and xanthomonas elicit plant innate immunity: structure and activity. *Chem Biol* 15:438–448.
66. Gust et al. (2007) Bacteria-derived peptidoglycans constitute pathogen-associated molecular patterns triggering innate immunity in Arabidopsis. *J Biol Chem* 282:32338–32348.
67. Newman et al. (2007) Invited review: priming, induction and modulation of plant defence responses by bacterial lipopolysaccharides. *J Endotoxin Res* 13:69–84.
68. Schuhegger et al. (2006) Induction of systemic resistance in tomato by N-acyl-L-homoserine lactone-producing rhizosphere bacteria. *Plant Cell Environ* 29:909–918.
69. Meziane et al. (2005) Determinants of Pseudomonas putida WCS358 involved in inducing systemic resistance in plants. *Mol Plant Pathol* 6:177–185.
70. Ongena et al. (2007) Surfactin and fengycin lipopeptides of Bacillus subtilis as elicitors of induced systemic resistance in plants. *Environ Microbiol* 9:1084–1090.
71. Tran et al. (2007) Role of the cyclic lipopeptide massetolide A in biological control of Phytophthora infestans and in colonization of tomato plants by Pseudomonas fluorescens. *New Phytol* 175:731–742.
72. Iavicoli et al. (2003) Induced systemic resistance in Arabidopsis thaliana in response to root inoculation with Pseudomonas fluorescens CHA0. *Mol Plant Microbe Interact* 16:851–858.
73. Audenaert et al. (2002) Induction of systemic resistance to Botrytis cinerea in tomato by Pseudomonas aeruginosa 7NSK2: role of salicylic acid, pyochelin, and pyocyanin. *Mol Plant Microbe Interact* 15:1147–1156.
74. Boller T and Felix G (2009) A renaissance of elicitors: perception of microbe-associated molecular patterns and danger signals by pattern-recognition receptors. *Annu Rev Plant Biol* 60:379–406.
75. Zeidler et al. (2004) Innate immunity in Arabidopsis thaliana: lipopolysaccharides activate nitric oxide synthase (NOS) and induce defense genes. *Proc Natl Acad Sci U S A* 101:15811–15816.
76. Albus et al. (2001) Suppression of an elicitor-induced oxidative burst reaction in Medicago sativa cell cultures by Sinorhizobium meliloti lipopolysaccharides. *New Phytol* 151:597–606.
77. Scheidle et al. (2005) The Lipid A substructure of the Sinorhizobium meliloti lipopolysaccharides is sufficient to suppress the oxidative burst in host plants. *New Phytol* 165:559–565.
78. Tellström et al. (2007) The lipopolysaccharide of Sinorhizobium meliloti suppresses defense-associated gene expression in cell cultures of the host plant medicago truncatula. *Plant Physiol* 143:825–837.
79. Deka et al. (2019) Bacterial exopolysaccharide promotes acid tolerance in Bacillus amyloliquefaciens and improves soil aggregation. *Mol Biol Rep* 46:1079–1091.
80. Niehaus et al. (1993) Plant defence and delayed infection of alfalfa pseudonodules induced by an exopolysaccharide (EPS I)-deficient Rhizobium meliloti mutant. *Planta* 190:415–425.
81. Jones et al. (2008) Differential response of the plant Medicago truncatula to its symbiont Sinorhizobium meliloti or an exopolysaccharide-deficient mutant. *Proc Natl Acad Sci U S A* 105:704–709.
82. Aslam et al. (2008) Bacterial polysaccharides suppress induced innate immunity by calcium chelation. *Curr Biol* 18:1078–1083.

83. Abisado et al. (2018) Bacterial quorum sensing and microbial community interactions. *MBio* 9:e02331–17.
84. Miller MB and Bassler BL (2001) Quorum sensing in bacteria. *Annu Rev Microbiol* 55:165–199.
85. Ramey et al. (2004) Biofilm formation in plant–microbe associations. *Curr Opin Microbiol* 7:602–609.
86. Assmus et al. (1995) In situ localization of Azospirillum brasilense in the rhizosphere of wheat with fluorescently labeled, rRNA-targeted oligonucleotide probes and scanning confocal laser microscopy. *Appl Environ Microbiol* 61:1013–1019.
87. Bais et al. (2004) Biocontrol of Bacillus subtilis against infection of Arabidopsis roots by Pseudomonas syringae is facilitated by biofilm formation and surfactin production. *Plant Physiol* 134:307–319.
88. Atkinson S and Williams P (2009) Quorum sensing and social networking in the microbial world. *J R Soc Interface* 6:959–978.
89. An et al. (2014) Bacterial quorum sensing and metabolic slowing in a cooperative population. *Proc Natl Acad Sci U S A* 111:14912–14917.
90. Schikora et al. (2016) Beneficial effects of bacteria-plant communication based on quorum sensing molecules of the N-acyl homoserine lactone group. *Plant Mol Biol* 90:605–612.
91. Hong et al. (2012) Quorum quenching revisited--from signal decays to signalling confusion. *Sensors (Basel)* 12:4661–4696.
92. Chernin et al. (2011) Quorum-sensing quenching by rhizobacterial volatiles. *Environ Microbiol Rep* 3:698–704.
93. Ortíz-Castro et al. (2008) Plant growth promotion by Bacillus megaterium involves cytokinin signaling. *Plant Signal Behav* 3:263–265.
94. Biswa P and Doble M (2013) Production of acylated homoserine lactone by Gram-positive bacteria isolated from marine water. *FEMS Microbiol Lett* 343:34–41.
95. Yajima A (2014) Recent progress in the chemistry and chemical biology of microbial signaling molecules: quorum-sensing pheromones and microbial hormones. *Tetrahedron Lett* 55:2773–2780.
96. Fuqua C and Greenberg EP (2002) Listening in on bacteria: acyl-homoserine lactone signalling. *Nat Rev Mol Cell Biol* 3:685–695.
97. Chen et al. (2004) Tyrosol is a quorum-sensing molecule in Candida albicans. *Proc Natl Acad Sci U S A* 101:5048–5052.
98. Enjalbert B and Whiteway M (2005) Release from quorum-sensing molecules triggers hyphal formation during Candida albicans resumption of growth. *Eukaryot Cell* 4:1203–1210.
99. Chen H and Fink GR (2006) Feedback control of morphogenesis in fungi by aromatic alcohols. *Genes Dev* 20:1150–1161.
100. Albuquerque P and Casadevall A (2012) Quorum sensing in fungi – a review. *Med Mycol* 50:337–345.
101. Kugler et al. (2000) Phenotypic variation and intracellular parasitism by Histoplasma capsulatum. *Proc Natl Acad Sci* 97:8794–8798.
102. Nazzaro et al. (2013) Quorum Sensing and Phytochemicals. *Int J Mol Sci* 14:12607–12619.
103. Hense BA and Schuster M (2015) Core principles of bacterial autoinducer systems. *Microbiol Mol Biol Rev* 79:153–169.
104. Thoendel M and Horswill AR (2010) Biosynthesis of peptide signals in gram-positive bacteria. *Adv Appl Microbiol* 71:91–112.
105. Aigle B and Corre C (2012) Chapter seventeen - waking up streptomyces secondary metabolism by constitutive expression of activators or genetic disruption of repressors. In: Hopwood DABT-M (ed) *Natural Product Biosynthesis by Microorganisms and Plants, Part C*. Academic Press, pp 343–366.
106. Willey JM and Gaskell AA (2011) Morphogenetic signaling molecules of the streptomycetes. *Chem Rev* 111:174–187.
107. Rezzonico et al. (2012) Detection of AI-2 receptors in genomes of Enterobacteriaceae suggests a role of type-2 quorum sensing in closed ecosystems. *Sensors* 12(5):6645–6665.
108. von Rad et al. (2008) Response of Arabidopsis thaliana to N-hexanoyl-DL-homoserine-lactone, a bacterial quorum sensing molecule produced in the rhizosphere. *Planta* 229:73–85.
109. Schenk ST and Schikora A (2015) AHL-priming functions via oxylipin and salicylic acid. *Front Plant Sci* 5:1–7. doi: 10.3389/fpls.2014.00784
110. Lareen et al. (2016) Plant root-microbe communication in shaping root microbiomes. *Plant Mol Biol* 90:575–587.
111. Teplitski et al. (2011) Perception and degradation of N-acyl homoserine lactone quorum sensing signals by mammalian and plant cells. *Chem Rev* 111:100–116.

112. Koh et al. (2013) Plant-derived natural products as sources of anti-quorum sensing compounds. *Sensors (Basel)* 13:6217–6228.
113. Vandeputte et al. (2010) Identification of catechin as one of the flavonoids from Combretum albiflorum bark extract that reduces the production of quorum-sensing-controlled virulence factors in Pseudomonas aeruginosa PAO1. *Appl Environ Microbiol* 76:243–253.
114. Hartmann et al. (2014) Bacterial quorum sensing compounds are important modulators of microbe-plant interactions. *Front Plant Sci* 5:131.
115. Chalupowicz et al. (2009) Regulatory interactions between quorum-sensing, auxin, cytokinin, and the Hrp regulon in relation to gall formation and epiphytic fitness of pantoea agglomerans pv. gypsophilae. *Mol Plant-Microbe Interact* 22:849–856.
116. Pérez-Montaño et al. (2011) Nodulation-gene-inducing flavonoids increase overall production of autoinducers and expression of N-acyl homoserine lactone synthesis genes in rhizobia. *Res Microbiol* 162:715–723.
117. Zipfel C and Oldroyd GED (2017) Plant signalling in symbiosis and immunity. *Nature* 543:328–336.
118. Li et al. (2014) Transcriptome analysis of Acidovorax avenae subsp. avenae cultivated in vivo and co-culture with Burkholderia seminalis. *Sci Rep* 4:1–8.
119. Tena et al. (2011) Protein kinase signaling networks in plant innate immunity. *Curr Opin Plant Biol* 14:519–529.
120. Clay et al. (2009) Glucosinolate metabolites required for an Arabidopsis innate immune response. *Science* 323:95–101.
121. Bardoel et al. (2011) Pseudomonas evades immune recognition of flagellin in both mammals and plants. *PLOS Pathog* 7:e1002206.
122. de Jonge et al. (2010) Conserved fungal LysM effector Ecp6 prevents chitin-triggered immunity in plants. *Science* 329:953–955.
123. Jones JDG and Dangl JL (2006) The plant immune system. *Nature* 444:323–329.
124. Pel MJC and Pieterse CMJ (2013) Microbial recognition and evasion of host immunity. *J Exp Bot* 64:1237–1248.
125. Melotto et al. (2006) Plant stomata function in innate immunity against bacterial invasion. *Cell* 126:969–980.
126. Geng et al. (2014) The phytotoxin coronatine is a multifunctional component of the virulence armament of Pseudomonas syringae. *Planta* 240:1149–1165.
127. Groll et al. (2008) A plant pathogen virulence factor inhibits the eukaryotic proteasome by a novel mechanism. *Nature* 452:755–758.
128. Boller T and He SY (2009) Innate immunity in plants: an arms race between pattern recognition receptors in plants and effectors in microbial pathogens. *Science* 324:742–744.
129. Dodds PN and Rathjen JP (2010) Plant immunity: towards an integrated view of plant-pathogen interactions. *Nat Rev Genet* 11:539–548.
130. Lowe-Power et al. (2016) Degradation of the plant defense signal salicylic acid protects ralstonia solanacearum from toxicity and enhances virulence on tobacco. *MBio* 7:e00656–16.
131. Pieterse et al. (2014) Induced systemic resistance by beneficial microbes. *Annu Rev Phytopathol* 52:347–375.
132. Dempsey DA and Klessig DF (2012) SOS - too many signals for systemic acquired resistance? *Trends Plant Sci* 17:538–545.
133. Spoel SH and Dong X (2012) How do plants achieve immunity? Defence without specialized immune cells. *Nat Rev Immunol* 12:89–100.
134. Vlot et al. (2009) Salicylic acid, a multifaceted hormone to combat disease. *Annu Rev Phytopathol* 47:177–206.
135. Pieterse et al. (2012) Hormonal modulation of plant immunity. *Annu Rev Cell Dev Biol* 28:489–521.
136. Baetz U and Martinoia E (2014) Root exudates: the hidden part of plant defense. *Trends Plant Sci* 19:90–98.
137. VanEtten et al. (1994) Two classes of plant antibiotics: phytoalexins versus "Phytoanticipins." *Plant Cell* 6:1191–1192.
138. Schlaeppi K and Mauch F (2010) Indolic secondary metabolites protect Arabidopsis from the oomycete pathogen Phytophthora brassicae. *Plant Signal Behav* 5:1099–1101.
139. Schlaeppi et al. (2010) Disease resistance of Arabidopsis to Phytophthora brassicae is established by the sequential action of indole glucosinolates and camalexin. *Plant J* 62:840–851.

140. Morant et al. (2008) beta-Glucosidases as detonators of plant chemical defense. *Phytochemistry* 69:1795–1813.
141. Song et al. (2011) Induction of DIMBOA accumulation and systemic defense responses as a mechanism of enhanced resistance of mycorrhizal corn (Zea mays L.) to sheath blight. *Mycorrhiza* 21:721–731.
142. Hasegawa et al. (2010) Phytoalexin accumulation in the interaction between rice and the blast fungus. *Mol Plant Microbe Interact* 23:1000–1011.
143. Niemeyer HM (2009) Hydroxamic Acids Derived from 2-Hydroxy-2H-1,4-Benzoxazin-3(4H)-one: Key Defense Chemicals of Cereals. *J Agric Food Chem* 57:1677–1696.
144. Ahmad et al. (2011) Benzoxazinoid metabolites regulate innate immunity against aphids and fungi in maize. *Plant Physiol* 157:317–327.
145. Wouters et al. (2016) Benzoxazinoids: reactivity and modes of action of a versatile class of plant chemical defenses. *J Braz Chem Soc* 27:1379–1397.
146. Kato-Noguchi et al. (2008) Secretion of momilactone A from rice roots to the rhizosphere. *J Plant Physiol* 165:691–696.
147. Reinhold-Hurek B and Hurek T (1998) Life in grasses: diazotrophic endophytes. *Trends Microbiol* 6:139–144.
148. Hardoim et al. (2015) The hidden world within plants: ecological and evolutionary considerations for defining functioning of microbial endophytes. *Microbiol Mol Biol Rev* 79:293–320.
149. Compant et al. (2010) Plant growth-promoting bacteria in the rhizo- and endosphere of plants: Their role, colonization, mechanisms involved and prospects for utilization. *Soil Biol Biochem* 42:669–678.
150. Afzal et al. (2019) Plant beneficial endophytic bacteria: mechanisms, diversity, host range and genetic determinants. *Microbiol Res* 221:36–49.
151. Alavi et al. (2014) Stenotrophomonas comparative genomics reveals genes and functions that differentiate beneficial and pathogenic bacteria. *BMC Genomics* 15:482.
152. Ryan et al. (2009) The versatility and adaptation of bacteria from the genus Stenotrophomonas. *Nat Rev Microbiol* 7:514–525.
153. Sheibani-Tezerji et al. (2015) Transcriptome profiling of the endophyte burkholderia phytofirmans PsJN indicates sensing of the plant environment and drought stress. *MBio* 6:e00621.

3

Implication of Microbial Signals: Plant Communication

Manmeet Kaur* **and H.S. Sodhi**
Punjab Agricultural University

CONTENTS

3.1 Introduction

Plants are rooted in their environment and thus cannot escape stressful situations, unlike animals and insects. Plants therefore use a range of methods to recognize or reduce stresses. It has become obvious that plant-based microbiota plays a crucial role in preventing the adverse effects of stress on plants (Bakker et al., 2018). Plants invest a significant amount of their fibrous carbon ("20%") in the preservation

* Corresponding author's email: manmeet1-mb@pau.edu

DOI: 10.1201/9781003171416-3

of their rhizosphere- and phyllosphere-related microbiota through exudation of "sugars," "amino acids," and "organic acids." In exchange, "beneficial microbiota (PGPB)" and "fungi (PGPF)" generally provide plants with significant advantages, including "improved mineral consumption," "nitrogen fixation," and "biocontrol," making them crucial in plant growth and health (Philippot et al., 2018). Setting up intimate, beneficial interactions between plant and microbe requires an exchange and colonization of signaling molecules. An enormous number of defensive features have been developed by the plants, which effectively reduce their foes. Defenses, however, are seldom faultless, so plants as static, non-interactive organisms do not exist. Plants require a "language," and volatile organic compounds (VOCs) are "words" in plants' vocabulary in order to communicate without interaction with them. A very well cognitive appraisal of a nitrogen-binding symbiosis of "Rhizobials" and "Leguminosae" plants is the formation of nodules. The formation of nodule requires coordinated development of organogenetic bacterial infection and root nodule and is initiated by biosynthesis on plant flavonoid perception of rhizobial nodulation factors (Oldroyd et al 2011). Plant receptor nodulation variables actually recognize and cause "transcriptional" and "physiological" changes that result in "root hair curling" and "bacterial traps"; "creation of an appropriate nitrogen fixation environment"; and "bacterial growth initiation."

Besides these mutualistic relations, many other microbes naturally free to live can boost plant growth. By associating these microbes with their plants host, nutrient uptake and root architectural changes can be stimulated or plant health can be promoted (Zamioudis et al., 2014). For instance, selected microbes may benefit from plant health, which suppress pathogens through antibiotics or nutrient competition, or cause a host immune response that is known as induced systemic resistance (ISR) (Venturi et al., 2016).

"PGPB" and "PGPF" are being enormously used to increase the health of cultures, without improved inputs of fertilizer and pesticides, as "biocontrol" agents and "biostimulants." Apart from well-studied symbioses of nitrogen-fixing bacteria, the molecular signals and mechanisms governing their action in plants are, however, relatively little known. By interfering with root development, PGPB encourage plant growth and health. Selected *Pseudomonas* spp. strains, for example, induce change in root architecture through the activation of development programs that prevent primary root elongation and encourage lateral root formation and hair growth (Zamioudis et al., 2013). Even without direct connections with the PGPB, developmental changes in microbe-induced components can be observed, demonstrating that gaseous rhizobacterial volatile organic compounds (VOCs) are potent inducers of plant growth and are drivers of root architectural changes (Blom et al., 2011). VOCs play important roles, including the inter-plant communications and the associated microbiota. All through various stages of plant development, extensive communication takes place between plants and microorganisms, in which both partner signaling molecules contribute.

Fungal and bacterial species can discern the host plant and initiate its rhizosphere colonization strategy by generating canonical plant-regulating agents such as auxins or cytokinins. But on the other hand, plants are able to identify and modify the defense and growth response of the microorganisms they derive accordingly. This molecular dialogue, generally through highly coordinated cellular processes, produces the desired consequence of the relationship ranging from pathogenesis to symbiosis (Lugtenberg et al., 2009). As such, communication between shoot and root can give survival to the plant, potentially restricting or preventing diseases. Bacterial and fungal phytopathogens cannot be limited to infecting only aerial or root tissues. For example, beneficial soil and fungi can provide immunity to a wide range of foliar conditions by activating plant defenses and thus decrease the susceptibility of a plant to a disease attack (Pieterse et al., 2014). Signals in plant pathogens have, for many years, been a key theme in phytopathology, while more recent efforts to discover signals in plant communications with non-pathogens are being made.

3.2 Plants: Contribution to Microbes

Secondary metabolites released through various parts of the root system create the unique soil environment known as rhizosphere. These compounds belong to the following three main classes and are collectively called root exudates:

1. low molecular weight compounds
2. high molecular weight compounds
3. volatile organic compounds (VOCs).

The main portion of exudates are low molecular weight compounds consisting of "sugars," "amino acids," "organic acids," "phenolics," and "vitamins." High molecular weight compounds include "mucilage" and "proteins"; secondary metabolites include "alcohols" and "aldehydes"; and volatiles constitute "carbon dioxide" (Schulz et al., 2007). These substances can act as transmissions for microbial attractions; for example, malate can be used in microbial nutrition as a carbon source. The physical, biochemical, and ecological attributes of the rhizosphere are determined by the interplay between the distinct compounds released, the time of release, and any uniquely constitutive or over-expression substances. Bacteria, fungi, actinomycetes, and algae are the forms of microorganisms within a rhizosphere. Rhizosphere is a dynamic process in which root and microorganism interactions and communication contribute significantly to sustaining plant development and productivity. Rudrappa and associates have recently investigated the impacts of root exudates as signaling molecules, showing that root-secreted malic acid recruits new beneficial soil bacteria *Bacillus subtilis* and this connection plays a role in the plant protection from foliar pathogen *Pseudomonas syringae* (Rudrappa et al., 2008).

Elicitors are molecules that participate in plant defense; most of them are derived directly from beneficial or pathogenic microbes (Mackey et al., 2006). The ampleness of a range of secondary metabolites, which include "indole glucosinolates," "phytoalexins," and "alkamides," which can have their effect on communication with microbial populations, results in exogenous applications of defense, such as "salicylic acid," "methyl jasmonate," and "nitrogen oxides."

3.3 Chemical Signals in the Rhizosphere and Phyllosphere

Plants use chemical signals to perceive and modulate their environment as stationary organisms. Plants transmit and receive across each ground signal and maintain aerial and root associations with microbial organisms. The area above ground around the plant is known as the phyllosphere. There are plants where chemical messages can be sent and received with microbes and adjacent plants. However, extensive environmental changes lead to "high levels of ultraviolet radiation," "water scarcity," "minimum nutrient availability," and low "microbial density," and variety in the phyllosphere (Andrew et al., 2000). The rhizosphere environment, on the contrary, is rich in nutrients, is protected from severe environmental perturbations, is generally a prime location for microbial growth, and surrounds plant roots directly. The rhizosphere therefore also encourages competition among microbes in this prime location for a spot. Much more intriguingly, plants amplify their root secretions for the manipulation of rhizospheric interactions and communications. The interesting facet of the new microbial recruits is that plants can delight precise microbes preferentially (Whipps, 2001).

3.4 Characterized Chemical Compounds with Role in Plant–Microbe Interactions

To date, a number of microbial molecules that play a role in the development of complex plant–microbe interactions have been identified. These molecules are generally classified into three groups: "plant hormones (microbial phytohormones)," "toxins," and "stimulatory compounds." In the scenario of microbial phytohormones, here we largely focus on microbial molecules that imitate or affect endogenous plant hormones or their activism and on microbial enzymes that change the abundance and/or distribution of plant hormones.

3.4.1 Microbial Phytohormones

Phytohormones play a critical role in regulating the growth and development of plants. Plant hormones, derived from the Greek word "hormon," which emanate from photosynthetic activity, constrain how energy is channeled into "plant growth," "development," and "reproduction." Except perhaps animals, plants have no specialized phytohormonal organ and normally all cells are capable of biosynthesis (Davies et al., 2010). Phytohormones go through a lot of significant cellular and developmental processes in the entire plant through processes of active and passive diffusion, often at a concentrational and situational level. The appropriate response is directed by the local integration of the hormonal signaling networks through crosstalk. Pathogenic and beneficial microbes produce a variety of molecules, which may imitate or interfere with host transcription and enzymatic pathways for producing plant hormones such as "auxins," "cytokinins," and "gibberellins" (Muller et al., 2015). The production of such phytohormones by microorganisms does not play an immediate role in the development of microbes, but it contributes to their interaction with the environment, especially plants, just as a secondary metabolite.

3.4.2 Cytokinins

The "cytokininine (CK)" family includes numerous essential regulatory agencies of intricate universal development interactions for plants, such as "cell division," "leaf senescence," "mobility of nutrients," and "germination of seeds" (Hwang et al., 2012). CK biosynthesis is monitored in plants by means of the genes "isopentenyltransferase (IPT)," which encode enzymes that limit the effect of the initial CK biosynthesis reaction. Accumulating findings of modalities of CK signaling also divulged complicated interactions with certain other sequence signals such as "auxin" and "salicylic acid," and at the same time, CKs also make a significant contribution to plant immunity (Argueso et al., 2012). Such strongly believed examples also include accumulation of salicylic acid in "tobacco" and *Arabidopsis* increased biosynthesis of "phytoalexins" and a reduction in tobacco "abscisic acid (ABA)." CKs have proved to delay senescence through a reduction in the oxidative plant explosion and photosynthesis. Given their effect on plant immunity, it may be no wonder that CKs are produced not only by plants, but also by their microbes with different intentions (Grobkinsky et al., 2014).

3.4.3 Indole-3-Acetic Acid

The main auxin in plants is indole-3-acetic acid (IAA), which is critical to "plant growth," "development," and "protection." For the growth promotion of many beneficial microbes, microbial production of IAA is essential (Duca et al., 2014). Indolic compounds, such as "IAA," occur in large amounts in the rhizosphere and are capable of acting as signaling molecules intra- and inter-species. Indoles modify a range of bacteria mechanisms, including "biofilm formation," "virulence," and "antibiotic resistance." Oddly enough, bacteria that were unable synthesize indoles may modify or deteriorate such compounds, thus leading to an increasing number of indole derivative products in nature (Lee et al., 2015). Auxins play an important role in coordinating and regulating many development- and growth-related processes in both the above-ground and below-ground tissues over the entire plant's life cycle (Davies et al., 2010). The auxins may act antagonistically and synergistically, respectively, on the defense-associated compounds "salicylic acid (SA)" and "jasmonic acid (JA)" (Naseem et al., 2015). Plant pathogens figured out ways of using the plant's auxin machinery to alter their own development programs. The soil bacterium *Agrobacterium tumefaciens* is responsible for the proliferation and growth of tumors in plants through their integration into the genome of the host plant of auxin and cytokinin biosynthesis genes. Furthermore, a successful plant pathogen infection frequently tends to depend on an incursion of the tissue of the plants by means of the so-called natural or lateral breaks or leaf stomata. Changes in the opening and closing of stomatal auxins by deregulation could thus have a significant impact on plant resistance to such pathogens (Acharya et al., 2009). Most rhizobacteria can synthesize "de novo IAA" via different biosynthetic pathways, for instance, the pathway to "indole-3-pyruvic acid (IPA)" and the recently reviewed pathway to "indole-3-acetamide (IAM)." Throughout this chapter, the IPA pathway is taken when "L-tryptophan" is deaminated to form IPA, which is converted in a decarboxylasis and ultimately oxidized by oxidases to generate "indole-3-acetaldehyde (IAAld)" (Duca et al., 2014). IAA biosynthesis by IAM pathway was mostly defined for plant pathogens. It combines multiple enzymatic steps, i.e., the conversion by "tryptophan 2-monooxygenase" to a "tryptophan intermediate" with "IAA hydrolase" followed by hydrolysis. Two bacterial genes were identified, which eventuate in innumerable few bacterial genera, including *Pseudomonas*, which are coding monooxygenase and hydrolase, respectively, *iaAM* and *iaaH* (Patten et al., 2013).

3.4.4 Gibberellins

Gibberellins are omnipresent plant hormones that incorporate multiple metabolic functions needed for the development and growth of a plant, including "germination of seeds" and "senescence." Similar to CKs, "gibberellic acid (GA)," formerly known as "Gibberella fujikuroi," was first isolated from a plant pathogen of rice, *F. fujikuroi* (Hedden et al., 2015). Gibberellins were isolated from a rice leaf by filtration and were therefore subsequently characteristic of active fungal compounds that could develop pathogenic symptoms, such as seedling elongation and infertility. Endophytic fungi can also produce GAs for plants to cope with certain stresses.

3.4.5 Defense Hormones: Mediators of Plant–Microbe Interactions

Plant hormones were first found to have their roles in plant development and normal work in the plant, but they also help to communicate outside the plant to influence the functioning and resistance of diseases. "Salicylic acid (SA)," "jasmonic acid (JA)," and "ethylene (ET)" are quintessentially renowned for their functions in plant protection through stimulation of systemic resistance and induced systemic resistance. "Systemic Acquired Resistance (SAR)" reacts to pathogens, while "induced systemic resistance (ISR)" is stimulated by beneficial bacteria and encourages a less dangerous protection reaction to prepare the plant for a pathogenic attack that could potentially be prompting. Many trials have been conducted on the interaction between a beneficial soil microbe and ISR in plants, attempting to make them more resistant against a foliar pathogen. In recent years, plant hormones that were originally studied in other plant processes have been found to play an important role in plant defense. In addition, certain of these signals can even affect other organisms, including related fungi and bacteria (Spence et al., 2014). In many plant–microbe interactions, the defense-related plant hormones "jasmonic acid," "salicylic acid," and "ethylene" play a major role (Bari et al., 2009). "JA" and "SA" are the specific control bodies for plant growth and defense, and they act abrasively to combat necrotrophic and biotrophic pathogens, respectively. In contrast, biotrophic pathogens activate SA-dependent defenses via "NPR1," the transcriptional coactivator of a broad range of defense genes. Increased "SA signals" concurrently suppress the "JA pathway" (Pieterse et al., 2012). Different microbial mechanisms are used to suppress this interaction between the hormonal networks of the factory to suppress defense strategy. The plants in turn have the ability to trigger the appropriate reaction by modulation of hormonal balances, both beneficial and adverse plant-associated microbes.

A diversity of "beneficial plant-associated microbes" are often reported to produce "SA" in substantial amounts. SA production is often attributed to the production of siderophores. These are compounds with iron-chelating properties, which help to obtain bacterial iron when there is little iron available, and have a salicylate movement. It is thus not clear to what extent microbial SA is just an in vitro artifact, or whether it continues to play an effective role in plant–microbe interactions and modulation of host immunity (Bakker et al., 2014).

ET is produced as a major modulated signal of plant immunity through numerous plant–pathogen interactions (Glick et al., 2014). In the interaction with positive microbes, necrotrophenetic pathogens, and insects, it also has an important role together with the JA pathway. ET acts primarily as a pathway crosstalk modulator of these hormones (Broekgaarden et al., 2015). The accumulation of ET is believed to occur in two phases under stress conditions: first, a rapid conversion of the endogenous pool (ACC) to the ET, and second, a de novo synthesis of the ACC. Typically, this second phase deals with "senescence," "chlorosis," and "leaf abscission" and is therefore harmful to plant growth and development. Here, beneficial PGPB can be used to reduce ACC levels via the activity of the ACC deaminase enzyme under conditions of stress. ACC deaminase cleaves ACC into "ammonia" and "α-ketobutyrate," reducing the amount of ET that is produced in the plant (Glick et al., 2014).

3.4.6 Stimulatory Compounds

PGPB (plant growth-promoting bacteria) and PGPF (plant growth-promoting fungi) promote plant growth in a number of ways, including microbial phytohormone production. By inducing systemic defense initiation, also known as ISR, some PGPB enhance plant immunity (Pieterse et al., 2014). A number of causal factors, such as "lipopolysaccharides (LPS)" and "cellular envelopes"; secretory compounds such as "siderophores," "lipopeptides (LPs)," and "volatiles"; "antibiotics" such as "2,4-diacetylphloroglucinol (DAPG)"; sensing molecules such as "phenazine" and "pyocyanin"; "N-acyl homoserine lactones"; and flagella, have been identified as the elicitors of ISR. In the absence of bacteria, ISR could be reproduced by the use of several purified bacterial determinants. However, there were no impediments in their ability to elicit ISR from single knockout mutants no longer producing the respective determinants, suggesting the redundancy between bacterial determinants for elicitation of ISR (Meziane et al., 2005).

3.4.6.1 Siderophores

Siderophores are low molecular weight compounds produced in the environment by sequestration of ferric ions (Fe^{3+}) typically by microbes under low iron availability conditions. In most soil profiles, iron supplies are low, as the iron pool is dominated by poorly soluble ferric hydroxides. Through their iron-sequestering activities, producer microbes have relevant biological control traits to compete with their neighbors, including soilborne pathogens. The *P. protegens* strain Pf-5 showed that its siderophores make a significant contribution to "mycotoxin fusaric acid resistance," caused by several pathogens of the genus *Fusarium* by their iron sequestration activity (Denance et al., 2013). Besides siderophore biosynthesis, there are two main pathways. One relies on "non-ribosomal peptide synthetase (NRPS)," and another on siderophore syntheses, which belong to the *IucA/IucC* protein family (Lorenzo et al., 1986). Many *Pseudomonas* spp. produce "pyoverdines," and biosynthesis entails seven genomically dispersed BGCs within one strain. Siderophores generated on the second track are less frequently investigated in detail, also known as the NRPS-independent track or NIS route (Visca, 2007).

3.4.6.2 Lipopeptides

LPs are a lipid tail attributed to short oligopeptides, whether linear or cyclic (CLPs). They serve as microbial surfactants that work at lower surface or interfacial stress. Numerous organisms, including bacteria and fungi, produce biosurfactants and can affect "cell differentiation," "signaling," "biofilm development," and "motility" through their activity. The so-called weather resistance and the ability to reduce surface tension in water have been combined in plant-related environments. The wetting of leaf tissue is intended to encourage –microorganisms' motility and may also serve as a stage for the interchange of signals and nutrients (Xin et al., 2016). LPs are quite well known for their broad spectrum of antimicrobial activity and, in particular, act as a major defense against protozoa in addition to their function as biosurfactants. The principal action mode is the creation of pores in the membrane, which leads to an imbalance in transmembrane ion fluxes and cell death. The duration and concentration of the fluid movement, and the type, total amount and specification of the amino acids in peptide chain, differ greatly in their structural appearance. Much like most siderophores, CLPs are formulated by means of large multi-domain NRPS that are part of even larger BGCs with different additional proteins related to both transport and transcription. LPs improve both useful and harmful microbes with regard to their potential for colonization in their own communities through their biosurfactant activities (Raaijmakers et al., 2010).

3.4.7 Toxins

In addition, due to the toxin contamination of food and animal food and their virulence effect in several plant–pathogen interactions, the role of secondary metabolites has been closely studied in agriculture research. The maize hybrids with Texas male sterile cytoplasm were particularly sensitive to a secondary metabolite called T-toxin produced by an unknown race of the fungal pathogen, the *Cochliobolus heterostrophus* (Ullstrup et al., 1972). Despite their huge chemical complexity and diversity, a tiny handful of primary precursors produce all secondary metabolites. As a result, fungal secondary metabolites are usually classified into four canonical chemical classes based on the class of enzymes involved in the first intermediate biosynthesis. Polyketides such as "aflatoxin," "T-toxins," and "perylenequinone toxins" are included in these products. Non-ribosomal peptides, such as "HC-toxins," and syrodesmic siderophores, such as "ferricrocin," are present. The core enzymes "PKSs," "NRPSs," "terpene synthases," and "dimethylallyltryptophan synthases (DMATS)" are the respective governing elements of biosynthesis in each secondary metabolite scenario.

3.4.7.1 Polyketides

Polyketides are the richest in secondary microorganisms that represent a variety of small molecules structurally and functionally from environmental toxins such as aflatoxin B1 to pharmaceuticals such as tetracycline or cholesterol-lowering medicines such as lovastatin. Fungal polyketides are assembled linearly with large proteins known as PKS type I that contain a multi-domain module required to extend and modify the chain one-round (Newma et al., 2014). Consequently, the diversity of the structures of fungal polyketides results partly from the number of iterations and other changes in the PKS enzyme. "Perylenequinones" are among the most fascinating polyketides. Perylenequinones distribute a characteristic pentacyclic conjugated chromophore that enables the production of reactive oxygen species in the presence of oxygen and ambient light. *Cercosporin* is one of the most successfully studied "perylenequinones" produced by most members of the genus *Cercospora*, a fungal genus made of many well-known and destructive plant pathogens found worldwide (Daub et al., 2000).

3.4.7.2 NRPs

Multi-domain, multi-modular enzymes known as "nRPs" are derived from amino acids. Amino acids that are connected in a linear or cyclic manner can either be proteinogenic or non-proteinogenic. The difference between NRPs is due to the length of the peptide chain, whether cyclic, and the domain function variations. Like polyketides, further changes can occur through the tailoring of enzymes encoded in the NRPS gene cluster. "Victor C," a cyclic pentapeptide synthesized in oat Victoria Blight, is one of the most infamous NRPs in the world (Wolpert et al., 2016). The bacterial lipopeptide toxin classes, including "syringomycin," "syringopeptin," and "corpeptin," are known to cause direct cellular damage to plants through their ability to form pores and thus contribute to the virulence. The pathogens are contributed by the phytotoxic properties of cichopeptin from *P. cichorii* (Trantas et al., 2015).

3.4.7.3 Terpenes

Terpenes are synthesized from "terpene cyclases," the enzymes necessary for the biosynthesis of various terpene substrates, including "geranyl," "farnesyl," and "geranyl diphosphates." "Sesqui-," "di-," and "triterpenoids" belong to terpene classes. Apart from terpene synthesis and cyclases-generating underpinnings, tailoring enzymes are often combined to produce bioactive toxins. *Fusarium* head blight is a serious disease that can kill highly productive crops of wheat and barley within a few weeks (McMullen et al., 1997).

3.4.7.4 Indole Alkaloids

Indole alkaloids are mainly extracted from "tryptophan" and "dimethylallyl pyrophosphate" by DMAT. The ergot alkaloids, such as "ergotamine" and "ergovaline," produced by *Clavicipitaceae* species are undoubtedly the best alkaloids (Florea et al., 2017). *Claviceps* spp. makes resting structures known as ergots, which eventually saturate food and feed with ergot alkaloids. Fortunately, the evidence that ergot alkaloids play an ecological role implies that they protect ergot-producing fungi against insects and herbivores. The concentrations of alkaloids have been shown to increase with plant age, with seasonal timing, which suggests that climate change might impact endophyte/host dynamics (Fuchs et al., 2017).

3.5 Plant Flavonoid Signals

A different class of natural products generated from the secondary metabolism of plants is flavonoids. They are polyaromatic compounds with a 15-carbon skeleton and can be classified according to the structure. Flavonoids are a phenylpropanoid pathway product; a chalcone synthase that catalyzes 4-coumaroyl-CoA condensation and three malonyl-CoA molecules to form a chalcone flavonoid precursor provides the first devoted step in the flavonoid branch. The flavonoid pathway is one of the best researched specialist metabolite biosynthesis pathways. To date, more than 10,000 plant flavonoids have been found and their synthesis appears to be omnipresent in plants (Ferrer et al., 2008). Their diversity comes from the production of several basal flavonoid structures, which include "flavones," "flavonols," "flavan-3-ols," and "flavanones." The skeleton flavonoid can be changed to diverse end products through "glycosylation," "malonylation," "methylation," "hydroxylation," "acylation," "prenylation," or "polymerization" (Winkel-Shirley, 2001). These substitutes have significant consequences on the "function," "solubility," "mobility," and "degradation of flavonoids."

Flavonoid synthesis is generally well known, and the majority of enzymes, often of multiple species, have been identified (Du et al., 2010). In the cytosolic side of the endoplasmic reticulum, flavonoid synthesis begins in enzyme complexes (Jorgensen et al., 2005). Certain enzyme complexes lie in the tonoplastic where flavonoid intermediates can be used for the reaction and storage of glycosylated substances in the vacuole (Aoki et al., 2000). In higher plants, the generic flavonoid class is omnipresent, but certain plant families may have specific structures (Dixon et al., 2002). Flavonoids are a variety of polyphenolic compounds produced through the secondary metabolism of plants. They perform a multifunctional role in the communication of rhizospheric plant microbes and plants. It may also be possible to transport flavonoids inside and between cells and tissues. Flavonoids within cells are likely to be moved through "vesicle-mediated transport," "ABC (ATP-binding cassette)," "MATE (multidrug and toxin extrusion)," or "membrane-bound transporter of families with ATP" (Zhao and Dixon, 2009). Conjugation of glutathione to cytoplasm flavonoids and ATP-driven transport via glutathione S-transferase pumps allow transporting flavonoids into vacuoles (Goodman et al., 2004).

3.6 Impact of Rhizosphere Community Structure on Flavonoid-Mediated Communications

The existence of microorganisms in rhizosphere, without a doubt, has repercussions for rhizosphere flavonoid quality and quantity (Cooper, 2004). It can either be done by modifying root exudation regularities, or by using the exuded flavonoid microbial catabolism. Flavonoids in legumes are vital signals for the initiation of nitrogen symptoms by acting as injectors of nodulation and genes related to the nodulation (Broughton et al., 2000). Flavonoids are perpetually exuded by legume hosts, but concentration levels in the rhizosphere raise considerably when rhizobial strains are interoperable (Schmidt et al., 1994). Flavonoid structures engage with rhizobial "Nod proteins" in order to trigger transcripts of nodulatory genes that encode lipo-chitooligosaccharide biosynthesis Nod factors that deform root-cutting hair and help rhizobial entry through infection fibers. The successful development of infectious threads probably depends on the rhizobial production of extracellular polysaccharides and proteins, with the secretion of flavonoid structures (Broughton et al., 2000).

These rhizobia can reveal their host leguminous crops in soils due to the chemotaxis of rhizobia to the root exudates. "Nod D," a "LysR-type" regulator acting as a transcription activator for other node genes, and "Nod A," "Nod B," and "Nod C," which combine the collared structure of the node factor lipo-chitooligosaccharide, which induces symbiotic signals from the hosts, have been identified to be responsible for the recognition of plant signal molecules (Sugiyama and Yazaki, 2012). The signals by flavonoids secreted by legume plants, such as "luteolin" in alfalfa, "7,4′-dihydroxyflavone" and "geraldone" in white clove, and "daidzein" and "genistein" in soybean, have been identified using a reporter system consisting of a node promoter, which was fused into "LacZ" (Redmond et al., 1986). "Alfalfa-S" has been studied inducing node gene expression by flavonoids. As a model, meliloti's interaction shows the expression of nodal genes only by host-specific "flavonoids," such as "luteolin," while the non-competing flavonoids such as "naringenin" compete with "Nod D1 luteolin binding."

When Nod factors are perceived, leguminous plants quickly react to signals, which contributes to the emergence of root nodules. Flavonoid biosynthesis genes have proved to be stimulated during the initial nodulation stage; for instance, *B. japonicum* induces the expression of "Phenylalanine ammonia-lyase (PAL)" (Estabrook and Sengupta-Gopalan, 1991). The presence of flavonoids in the roots has been shown to be indispensable during Nod factor synthesis in stimulation (Zhang et al., 2009). During infection, the concentration of kaempferol increases, inhibits the transport of auxin, and causes an accumulation of auxin at sites of the beginning of nodules (Wasson et al., 2006).

3.6.1 Modification of Patterns of Exudation

The presence of microorganisms or microbial components (e.g., Nods and cell walls) causes changes in plant phenylpropanoid biosynthesis expression of enzymes. Enriched plant biosynthesis gene expression is sometimes linked to qualitative and quantitative changes in the exudation or accumulation of end products in the root of plants. Root exudation is a complex phenomenon involving processes leading to transportation of C from roots to soil and exudation. The long-distance transport of carbon in sources usually occurs in the phloem via the generally recognized phloem flow mechanism of Munch. Under this mechanism, phloem metabolites transport the turgor difference between sink and source organs, which are ascertained by source–sink activities through the concentration gradients (De Schepper et al., 2013). Roots also generate high molecular weight compounds to the soil by deposition of rhizome, mainly root cells and mucilage, with the addition of a wide range of secondary metabolites. Such rhizodeposits can serve soil physical functions. For illustration, the root–microbe and the root–symbiont relationships are an essential macronutrient for rhizospheric microbes.

It is well recognized that the interactions between root exudates and microbial diversity are strong (Eisenhauer et al., 2017), but the extent to which they influence one another remains unknown. Elicited primary metabolites affect microbial diversity (Steinauer et al., 2016). Simultaneously, the preferential use of exudates by plants affects specific concentration degrees outside roots. In addition, soil microbes can release compounds such as "2,4-diacetylphloroglucinol" and "zearalenone" that stimulate primary metabolites' root exudation, such as "amino acids" (Phillips et al., 2004). Curring models depicting "C pools/fluxes" and dynamics of vegetation were mainly suggested as "source-driven" models. It was asserted that the monitoring of environmental evidence (e.g., "temperature," "water," and "available nutrients") for meristematic activity (declining capability) is essential since these are mostly limited to plant growth before photosynthesis responds to the source capability (Korner, 2015).

Although source–sink dynamics are widely recognized as essential for comprehending the appropriation of plant carbon, microorganisms should also be seen as a strong photosynthesis plant, thus encouraging root to exude (Savage et al., 2016). Root exudates are used for sensing and interacting with the abiotic and biotic components and for helping to acquire nutrients (Martin et al., 2017). Nevertheless, interaction studies often only addressed specific compounds (e.g., secondary metabolites) and thus left hidden potentials of the greater proportion of root exudates, i.e., primary metabolites. The architecture of the root system provides an essential factor for determining the plant's ability to adapt to hot spots of nutrients, maximizing soil resources (Khan et al., 2016).

3.6.2 Microbial Catabolism

In liquid culture, the flavonoid catabolism of the rhizosphere-known microbial species has been asserted. Even so, the fate of flavonoids, which are subject to the rhizosphere concert of catabolism, is hardly investigated. The microbial catabolism of rhizosphere soil flavonoids will affect the power of the original plant root signal. Initially, hydrolysis of exuded flavonoid glycosides into a stronger Nod gene occurs, leading to rhizobial activity (Hartwig and Phillips, 1991). When present in the aglyconous form, the biodegradation of a parent flavonoid can produce new flavonoid structures (e.g., "naringenin" and "chalcone" intermediates are produced during "quercetin biodegradation" before C ring cleavage destroys the flavonoid motif altogether). Thus, flavonoid signals could be microbial-attenuating or altering a crucial aspect of rhizosphere ecology and symbiotic development.

3.7 Role of Volatile Organic Compounds

Plants are in ongoing dialog with organisms in their environment through VOCs. It is a very important communication, as it enables the plants and organisms to interact to adapt their "growth," "development," "defense," "spread," and "cycles of life to maximum fitness." VOCs belong to various chemical classes, such as "terpenoids," "benzene," "phenylpropanoids," and "fatty acidic molecules," including "green leaf volatiles" and minor classes such as the "nitriles," "oximes," and "sulfides," in plants themselves as well as in organisms in their environments (Figure 3.1). The interactions between "plants and mutualists" and "pesticides and pathogens," and though not much has yet been dealt with experimentally, multi-trophic interactions mediate the interactions between the organic compounds. The semiochemicals functioning in these interactions can be produced either in the form of a component or as a response to external interactions and stimuli (Massalha et al., 2017). While semiochemicals may travel long distances, plant–plant and plant–microbe communications usually occur at relatively short distances and plant volatiles with a role in the interaction of plant insects are perceived at distances of several hundred meters. The resulting extremely wide dilution of the volatiles and precursors and their large variability for the chemical structures and properties of the VOCs present challenges for analysis (Fu et al., 2017). Plants are a source of volatile (and non-volatile) compounds that can affect "physiological" and "metabolistic" plant repairs

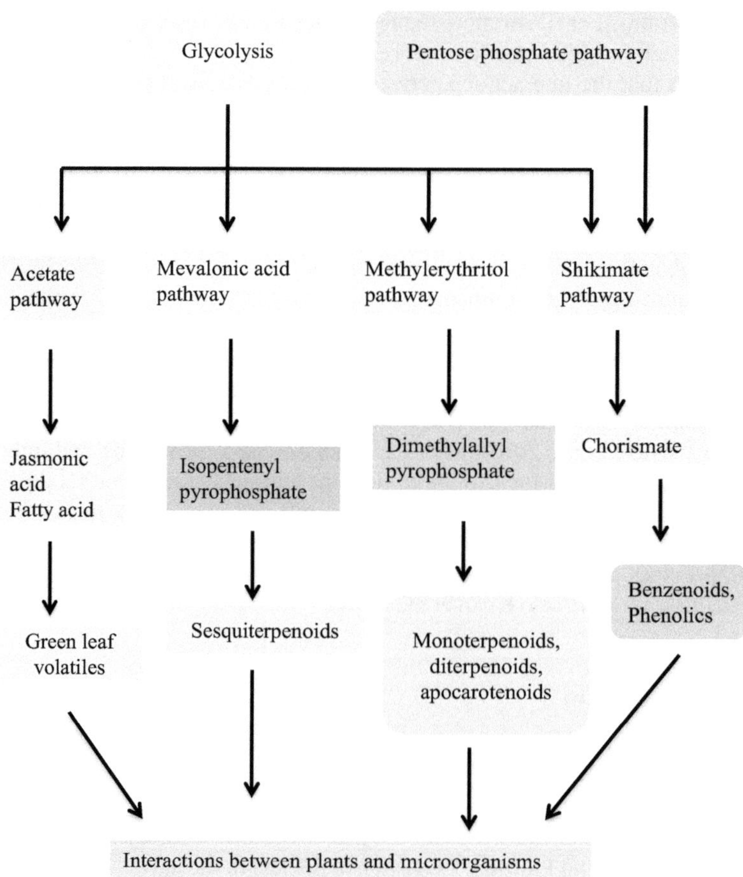

FIGURE 3.1 Biosynthetic pathway for major volatile compounds.

and are the source of volatile interaction not only between species, but also with plant-related microorganisms (Wenke et al., 2019).

Volatile compounds including "3-hydroxy-2-butanone,""2,3-butanediol," and "2-pentylfuran" produced by microorganisms such as *Bacillus*, *Pseudomonas*, *Arthrobium*, *Fusarium*, and *Alternaria rheizobactia* can stimulate root growth (Fincheira and Quiroz, 2018). This indicates that plants not only produce volatile products, but also perceive them and react to them by changing growth or defense (Schenkel et al., 2018). Volatiles induced by microbes can inhibit the growth of microbes and act as local bacteria agents (Sharifi et al., 2018).

Not only do plants activate their defense responses in response to a pathogen attack, but also the infection can induce volatile emissions into headspace and can result in an immune response in neighboring plants (Riedlmeier et al., 2017). But the immune (or defense) response can also be reduced if multiple volatile signals have to be integrated. Hormonal crosstalk has a role to play in balancing trade-offs that can also reduce volatile emissions (Erb, 2018). Plants can therefore produce certain volatiles on microbial interactions, but in return, microbial biosynthesis with VOC effectors has been demonstrated to be able to suppress (Sharifi et al., 2018). Microbes can also arbitrate tritrophic interactions in conjunction with their host plant (Shikano et al., 2017). Viruses that are dependent on a vector, for example, can change plant volatile biosynthesis for the recruitment of a vector insect. In the volatile communication of plants, the role of microorganisms also reaches flower volatility. Microbes can invade and occupy a nectar and alter a flower's volatile amalgamation, which can modify the visitation and thus the reproduction of plants (Rering et al., 2018).

3.8 Molecular Mechanisms Underlying Volatile Perception

The signaling role of volatiles within and between plants and how plants can perceive these signals, in particular, is an intriguing function (Cofer et al., 2018b). Plants often show remarkably specific reactions to various volatiles. Therefore, one may need to examine alternative scenarios, one example being the perception of carbon dioxide in guard cells in plants. These cells in *Arabidopsis* have an S-type anion channel called "slow anion channel 1 (SLAC1)" (Zhang et al., 2018a). There are also interaction sites of other ion channels for specific volatility that can alter their activity just like CO_2/HCO_3 can change the activity of "SLAC1" and its stomatal moment (Zebelo et al., 2012). The transcriptional "TOPLESS-like (TPL)" protein, which is a caryophyllene analogue, has been identified in the very recent past as a binding of volatile sesquiterpenes (Nagashima et al., 2018). TPL proteins are known to regulate different hormonal reactions such as jasmonic acid. Enzymes have also been shown to play a role in transportation and sensing volatility within or among plants. "Methyl salicylate (MeSA)" is transported through a tobacco phloem (*Nicotiana tabadefenccum*) to a non-infected, systemically resistant leaf with a tobacco mosaic virus infection (Park et al., 2007).

3.9 Chemistry of Plant–Plant Signaling

Numerous chemical compounds have been implicated in herbivorous, parasitoid, and predatory signals, but we will focus on examining the compounds involved in inter-plant and transmission signals. In plants-to-plants signals, typically many of those volatiles that work are de novo synthesized when herbivores are attacked. Several compounds, including "(E)-2-hexenal" (Arimura et al., 2001); green leaf volatiles "(Z)-3-hexen-1-ol" (Ruther et al., 2005) and "cis-3-hexenyl acetate" (Frost et al., 2008); "terpenes"; "myrcenic" and "ocimenal" volatiles (Godard et al., 2008); and "phytohormone methyl jasmonate" (Godard et al., 2008), are reported to function between and within plant signals. A range of C6 compounds including "aldehydes," "alcohols," and "esters" are present in green leaf volatiles. These compounds, shaped as a result of the "lipoxygenase pathway," are quickly emitted during plant disruption by mechanical and herbivorous damage (Fall et al., 1999). Through these compounds, any mechanical damage can be shown and plants could be receiving early signals. However, emissions that are highly correlated to herbivores are not equally reliable than emissions such as "DMNT" and "TMTT" (Dicke et al., 1994).

Of the majority of secondary compounds, comprising approximately 40,000 compounds, 43 are "terpenoids," which include a minimum of 1,000 and 6,500 "monoterpenes" and "sesquiterpenes," respectively. All terpenes are derived from "isopentenyl diphosphate (IPP)" and its allylic isomer "dimethylallyl diphosphate (DMAPP)." IPP is synthesized by cytosol through the mevalonate (MVA) pathway in plants, while it is synthesized by "2-C-methyl-D-erythritol 4-phosphate (MEP) pathway" in plastids. Some terpenoids are essential oils and resins and are produced and stored in specialized structures such as glands or resin pipes. These structures will be broken, and the compounds will be released after damage by herbivores. Herbivore feeding can induce the de novo biosynthesis of terpenoids locally and systemically. As a group, therefore, terpenoids are able to provide herbivore-damage-related signals to receiving plants. Methyl jasmonate is an integral component of plant defense response to insect feeding, volatile derivatives of jasmonic acid.

Methyl jasmonate was shown to increase the production of proteinase inhibitors under controlled conditions in tomato plant leaves (Farmer et al., 1990). MeSA is a phenolic compound that is synthesized of salicylic acid and is an important component in plant protection. In response to aphid feed damage, it is released by considerable quantities from plants and emitted by tobacco as a response to infection with "tobacco mosaic virus." An increased resistance to tobacco mosaic virus has been shown in plants that are exposed to "methyl salicylate" (Shulaev et al., 1997).

3.10 Microbial Signals Involved in Plant Growth and Development

The rhizosphere is able to produce a wide variety of microorganisms that restrict plant growth and development. Bacterial and fungal phytohormone production, for instance auxins and cytokinins, may affect the proliferation of the cell by sprinkling the tumor as it does for *Agrobacterium tumefaciens* and *Ustilago maydis*, or modify the root system architecture with lateral roots and root hair overproduction with a subsequent increased nutrient and water intake. Therefore, it is possible to identify whether a microbial interaction can be beneficial or damaging to the balance between auxin and cytokinin and to the site of hormone accumulation. Additional microbial signaling has been shown to play a role in plant morphology, including "N-acyl-L-homoserine lactones (AHLs)." Gram-negative bacteria such as *Pseudomonas* belong to a group of bacterial quorum sensing signals. These compounds allow gene expression depending on population density to be regulated by bacterial cells. Most recently, AHLs have been found to be able to detect plants, alter "gene expression" in roots and shoots, and modulate "cell growth" and "defense reactions."

a. **Auxins and Cytokinins:** "Auxins" and "cytokinins" interact in the control of many important developmental processes in plants, particularly in "apical dominance" and "root and shoot development." The balance between auxin and cytokinin is a key regulator of in vitro organogenesis. Exposing callus cultures to a high auxin-to-cytokinin ratio results in root formation, whereas a low ratio of these hormones promotes shoot development. Although both auxins and cytokinins can be produced in roots and shoots, the production of these signals does not occur randomly, but is regulated by the location of the producing tissues, the developmental stage of the plant, and environmental growth conditions such as light and temperature. Young shoot organs such as the first true leaves and developing primary and lateral roots are important sites of IAA production (Ljung et al., 2005). From the tissues involved in hormone production, the signals move through specific transport systems and different mechanisms to regulate plant growth and development. IAA can be transported from shoot to root through the vascular tissue. In addition to long-distance auxin transport, local transport of IAA along and across tissues is important for auxin localization in small groups of cells, for example in an emerging lateral root primordium or in the root cap during gravitropism. This can be achieved by the action of specific influx and efflux transporter systems. Auxin importers include members of the amino acid permease family AUX1 (auxin resistant 1) and LAX (likeaux1), and PGP4, which belongs to the MDR/PGP (multidrug resistance/P-glycoprotein) family (Wu et al 2007). Conversely, the root cap produces cytokinins, which appear to regulate primary root growth and gravitropism. The positive effect of cytokinins on growth at the whole plant level has been demonstrated by the identification of genes involved in cytokinin perception and signaling. Three sensor histidine kinases, CRE1/AHK4/WOL, AHK2, and AHK3, have been shown to act as cytokinin receptors (Kakimoto et al., 2003).

b. **N-Acyl-L-Homoserine-Lactones:** Many bacteria regulate diverse cellular processes in concert with their population size, a process commonly referred to as quorum sensing (QS) (Reading et al., 2006). Bacterial cell-to-cell communication utilizes small diffusible signals, which the bacteria both produce and perceive. The bacteria couple gene expression to population density by eliciting a response only when the signaling reaches a critical threshold. The population as a whole is thus able to modify its behavior as a single unit. In gram-negative bacteria, the QS signals most commonly used are AHLs. It is now apparent that AHLs are used for regulating diverse behaviors in rhizosphere-inhabiting bacteria and that plants may produce their own metabolites, which may interfere with QS signaling (Figure 3.2).

AHLs consist of a lactone residue attributed to an acyl side chain, that is homoserine. The specificity is derived from acyl chain length (4–18 carbon atoms), C3 position substitution, and acyl chain saturation (Raffa et al., 2004). The long, medium, or short category of AHL may be broadly determined, depending on whether its "acyl" content consists of more than eight, between eight and twelve, or below twelve carbon atoms. These molecules freely distribute within rhizosphere through the bacterial membrane. AHLs are excellent candidates in the rhizosphere for the mediation of this communication (Gantner et al., 2006).

Mathesius and coworkers (2003) published their first work showing that plants can perceive AHLs and found that nanomolar concentrations of two different AHL types, "3-oxo-C12-HL" and "3-oxo-C16:1-HL," grown axenically, caused significant changes in the accumulation of more than 150 proteins in *Medicago truncatula*. These proteins are found to have features in "plant protection," "stress responses," "energetic and metabolic activities," "transcriptional regulation," and "protein processing"

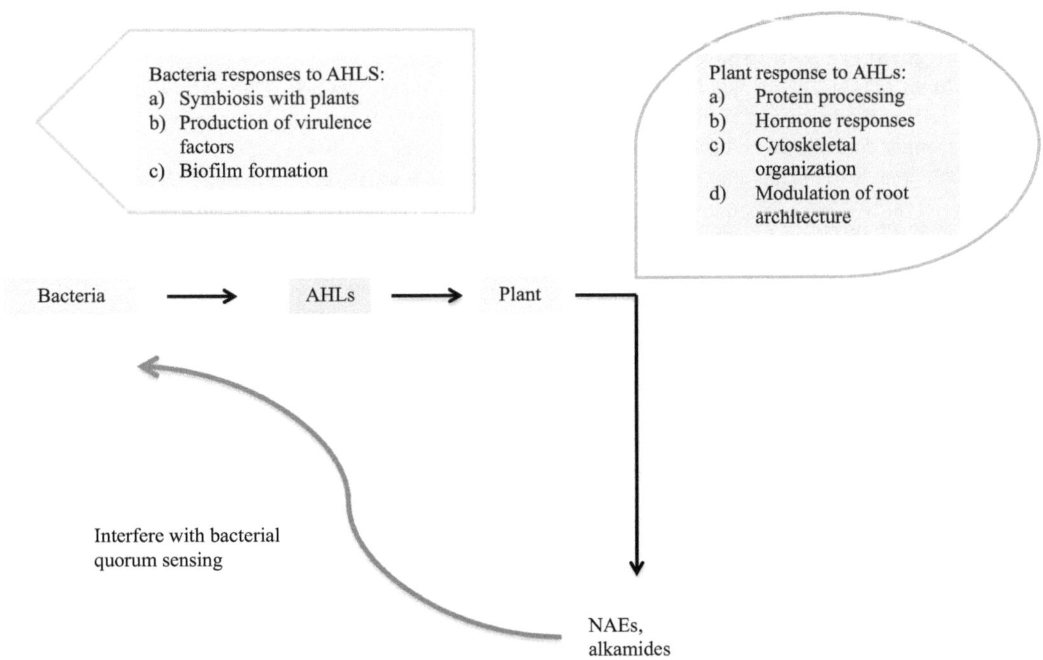

FIGURE 3.2 Communications between bacteria and plants mediated by AHLs and plant-produced signals.

(Mathesius et al., 2003). These results have also been confirmed in *Arabidopsis thaliana* by microarray analysis. The presence of AHL-producing bacteria in the rhizosphere of tomatoes induced "salicylic acid"- and "ethylene"-reliant protection measures that are important to stimulate systemic resistance in plants and provide protection from *Alternaria alternata*, the fungal pathogen. In addition, plants in a process dependent on the length of the acyl chain were found to have AHLs. Increasing the conduct of the stomata and transpiration in shoots leads to the application of the homoserine lactone, an AHL product breached through soil bacteria in the bean root. The plant and the bacteria benefit from an increasing intake of mineral nutrients (Joseph et al., 2003).

3.11 Conclusions and Future Aspects

The intricacy and immensity of chemical signals in the rhizosphere make the multi-transmission study an intimate task. However, a very complex system diverged between the plants and their microbial neighbors, each of which attempts to deceive their own endurance to maximize the fitness of their plants. In conclusion, there is still a small amount of knowledge of organic phytogenic composites and their role is not fully understood in interspecific communication between plants. A better understanding of functional plant flavonoids and volatile organic receptors may enhance our ability to assess the environmental importance of molecular communication over and beneath the earth in plant communities.

REFERENCES

Acharya BR and Assmann SM (2009) Hormone interactions in stomatal function. *Plant Mol. Biol*, 69: 451–462.

Andrews JH and Harris RF (2000) The ecology and biogeography of microorganisms of plant surfaces. *Annu Rev Phytopathol*, 38:145–180.

Aoki et al. (2000) Flavonoids of leguminous plants: structure, biological activity and biosynthesis. *J Plant Res*, 113: 475–488.

Argueso et al. (2012) Two component elements mediate interactions between cytokinin and salicylic acid in plant immunity. *PLoS Genet*, 8, e1002448.

Arimura et al. (2001) Plant-plant interactions mediated by volatiles emitted from plants infested by spider mites. *Biochem Syst Ecol*, 29:049–1061.

Bakker et al. (2018) The Soil-borne legacy. *Cell*, 172, 1178–1180.

Bari R and Jones J D G (2009) Role of plant hormones in plant defence responses. *Plant Mol Biol*, 69, 473–488.

Blom et al. (2011) Production of plant growth modulating volatiles is widespread among rhizosphere bacteria and strongly depends on culture conditions. *Environ Microbiol*, 13: 3047–3058.

Broekgaarden et al. (2015) Ethylene: traffic controller on hormonal crossroads to defense. *Plant Physiol*, 169, 2371–2379.

Broughton et al. (2000) Keys to symbiotic harmony. *J Bacteriol* 182: 5641–5652.

Cofer et al. (2018b) From acetoin to (Z)-3-hexen-1-ol: the diversity of volatile organic compounds that induce plant responses. *J Agric Food Chem*, 66: 11197–11208.

Cooper JE (2004) Multiple responses of rhizobia to flavonoids during legume root infection. *Adv Bot Res* 41,1–62.

Daub ME and Ehrensha M (2000) The photoactivated cercospora toxin cercosporin: contributions to plant disease and fundamental biology. *Annu Rev Phytopathol*, 38: 461–490.

Davies PJ (2010) *Plant Hormones – Biosynthesis, Signal Transduction, Action!* Springer, Dordrecht, 3rd edn.

de Lorenzo V and Neilands JB (1986) Characterization of iucA and iucC genes of the aerobactin system of plasmid ColV-K30 in Escherichia coli. *J Bacteriol*, 167: 350–355.

Dicke M (1994) Local and systemic production of volatile herbivore-induced terpenoids: their role in plant-carnivore mutualism. *J Plant Physiol*, 143:465–472.

Dixon et al. (2002) The phenylpropanoid pathway and plant defence – a genomics perspective. *Mol Plant Pathol* 3: 371–390.

Du et al. (2010) Genetic and metabolic engineering of isoflavonoid biosynthesis. *Appl Microbiol Biotechnol*, 86: 1293–1312.

Duca et al. (2014) Indole-3-acetic acid in plant-microbe interaction. *Eupub*, 106:85–125.

Eisenhauer et al. (2017) Root biomass and exudates link plant diversity with soil bacterial and fungal biomass. *Sci Rep*, 7:44641.
Erb M (2018) Volatiles as inducers and suppressors of plant defense and immunity-origins, specificity, perception and signaling. *Curr Opin Plant Biol*, 44: 117–121.
Estabrook EM and Sengupta-Gopalan C (1991) Differential expression of phenylalanine ammonia-lyase and chalcone synthase during soyabean nodule development. *Plant Cell* 3: 299–308.
Farmer EE and Ryan CA (1990) Interplant communication—airborne methyl jasmonate induces synthesis of proteinase-inhibitors in plant leaves. *Proc Nat Acad Sci USA*; 87:7713–7716.
Ferrer et al. (2008) Structure and function of enzymes involved in the biosynthesis of phenylpropanoids. *Plant Physiol Biochem*, 46 :356–370.
FincheiraP and Quiroz A (2018) Microbial volatiles as plant growth inducers. *Microbiol Res*, 208: 63–75.
Florea et al. (2017) Ergot Alkaloids of the Family Clavicipitaceae. *Phytopathology*, 107: 504–518.
Frost et al. (2008) Priming defense genes and metabolites in hybrid poplar by the green volatile cis-3-hexenyl acetate. *New Phytol*, 180:722–734.
Fu et al. (2017) Analytical method for metabolites involved in biosynthesis of plant volatile compounds. *RSC Adv*, 7: 19363–19372.
Fuchs et al. (2017) Plant age and seasonal timing determine endophyte growth and alkaloid biosynthesis. *Fungal Ecol*, 29: 52–58.
Gantner et al. (2006) In situ quantitation of the spatial scale of calling distances and population density-independent N-acylhomoserine lactone-mediated communication by rhizobacteria colonized on plant roots. *FEMS Microb Ecol*, 56:188–94.
Glick BR (2014) Bacteria with ACC deaminase can promote plant growth and help to feed the world Microbiol. *Research*, 169: 30–39.
Großkinsky et al. (2014) Abscisic acid-cytokinin antagonism modulates resistance against Pseudomonas syringae in tobacco. *Phytopathology*, 104, 1283–1288.
Hartwig U and Phillips DA (1991) Release and modification of nod gene inducing flavonoids from alfalfa seeds. *Plant Physiol*, 95: 804–807.
Hedden P and Sponsel V (2015) A century of gibberellin research. *Plant Growth Regul*, 34: 740–760.
Jorgensen et al. (2005) Metabolon formation and metabolic channeling in the biosynthesis of plant natural products. *Curr Opin Plant Biol*, 8: 280–291.
Joseph CM and Phillips DA (2003) Metabolites from soil bacteria affect plant water relations. *Plant Physiol Biochem*, 41:189–92.
Joshi et al. (2016) Plant phenolic volatiles inhibit quorum sensing in pectobacteria and reduce their virulence by potential binding to ExpI and ExpR proteins. *Sci Rep*, 6: 38126.
Khan et al. (2016). Root system architecture and abiotic stress tolerance: current knowledge in root and tuber crops. *Front Plant Sci*, 7:1584.
Korner C (2015) Paradigm shift in plant growth control. *Curr Opin Plant Biol*, 25: 107–114.
Lee et al. (2015) Roles of indole as an interspecies and interkingdom signaling molecule. *Trends Microbiol*, 23: 707–718.
Ljung et al. (2005) Sites and regulation of auxin biosynthesis in Arabidopsis roots. *Plant Cell*, 17:1090–104.
Lugtenberg B and Kamilova F (2009) Plant growth promoting bacteria. *Annu Rev Microbiol*, 63:541–556.
Mackey D and McFall AJ (2006) MAMPs and MIMPs: proposed classifications for inducers of innate immunity. *Mol Microbiol*, 61:1365–1371.
Martin et al. (2017). Ancestral alliances: plant mutualistic symbioses with fungi and bacteria. *Science*, 356:eaad4501.
Massalha et al. (2017) Small molecules below-ground: the role of specialized metabolites in the rhizosphere. *Plant J*, 90: 788–790.
Mathesius et al. (2003) Extensive and specific response of a eukaryote to bacterial quorum-sensing signals. *Proc Natl Acad Sci USA*; 100:1444–1449.
McMullen et al. (1997) Scab of Wheat and Barley: A Re-emerging Disease of Devastating Impact. *Plant Dis*, 81: 1340–1348.
Meziane et al. (2005) Determinants of Pseudomonas putida WCS358 involved in inducing systemic resistance in plants. *Mol Plant Pathol*, 6: 177–185.
Muller LJ (2015) Bacteria and fungi controlling plant growth by manipulating auxin: balance between development and defense. *J Plant Physiol*, 172, 4–12.

Newman et al. (2014) Systematic domain swaps of iterative, nonreducing polyketide synthases provide a mechanistic understanding and rationale for catalytic reprogramming. *J Am Chem Soc*, 136: 7348–7362.

Oldroyd et al. (2011) The rules of engagement in the legume-rhizobial symbiosis. *Annu Rev Genet*, 45: 119–144.

Park et al. (2007) Methyl salicylate is a critical mobile signal for plant systemic acquired resistance. *Sci*, 318: 113–116.

Patten et al. (2013) Activity, distribution and function of indole-3-acetic acid biosynthetic pathways in bacteria. *Crit Rev Microbiol*, 39:395–415.

Philippot et al. (2013) Going back to the roots: the microbial ecology of the rhizosphere. *Nat Rev Microbiol*, 11: 789–799.

Phillips et al. (2004) Microbial products trigger amino acid exudation from plant roots. *Plant Physiol*, 136: 2887–2894.

Pieterse et al. (2012) Hormonal modulation of plant immunity. *Annu Rev Cell Dev Biol*, 28: 489–521.

Pieterse et al. (2014) Induced systematic resistance by beneficial microbes. *Annu Rev Phytopathol*, 52: 347–375.

Raaijmakers et al. (2010) Natural functions of lipopeptides from Bacillus and Pseudomonas: more than surfactants and antibiotics. *FEMS Microbiol Rev*, 34: 1037–1062.

Reading NC and Sperandio V (2006) Quorum sensing: the many languages of bacteria. *FEMS Microbiol Letts*, 254:1–11.

Redmond et al. (1986) Flavones induce expression of nodulation genes in Rhizobium. *Nature*, 323:632–635.

Rering et al. (2018) Nectar-inhabiting microorganisms influence nectar volatile composition and attractiveness to a generalist pollinator. *New Phytol*, 220: 750–759.

Riedlmeier et al. (2017) Monoterpenes support systemic acquired resistance within and between plants. *Plant Cell*, 29: 1440–1459.

Ruther J and Kleier S (2005) Plant-plant signalling: ethylene synergizes volatile emission in Zea mays induced by exposure to (Z)-3-hexen-1-ol. *J Chem Ecol*, 31:2217–2222.

Savage et al. (2016). Allocation, stress tolerance and carbon transport in plants: how does phloem physiology affect plant ecology? *Plant Cell Environ*, 39: 709–725.

Schenkel et al. (2018) Fungi indirectly affect plant root architecture by modulating soil volatile organic compounds. *Front Microbiol*, 9: 1847.

Schepper et al. (2013). Phloem transport: a review of mechanisms and controls. *J Exp Bot*, 64: 4839–4850.

Schmidt et al. (1994) Nod factors of Bradyrhizobium Japonicum and Rhizobium Sp NGR234 induce flavonoid accumulation in soybean root exudate. *Mol Plant Microbe In* 7: 384–390.

Schulz S and Dickschat JS (2007) Bacterial volatiles: the smell of small organisms. *Nat Prod Rep*; 24:814–842.

Sharifi R and Ryu CM (2018) Revisiting bacterial volatile-mediated plant growth promotion: lessons from the past and objectives for the future. *Ann Bot*, 122: 349–358.

Shikano et al. (2017) Tritrophic interactions: microbe-mediated plant effects on insect herbivores. *Annu Rev Phytopathol* 55, 313–331.

Shulaev et al. (1997) Airborne signalling by methyl salicylate in plant pathogen resistance. *Nature*, 385:718–721.

Spence et al. (2014) Natural rice rhizospheric microbes suppress rice blast infections. *BMC Plant Biol*, 14:130.

Steinauer et al. (2016). Root exudate cocktails: the link between plant diversity and soil microorganisms? *Ecol Evol*, 6: 7387–7396.

Sugiyama A and Yazaki K (2012) Root exudates of legume plants and their involvement in interactions with soil microbes. In: Vivanco JM, Baluska F (eds) *Secretions and Exudates in Biological Systems*. Springer, pp. 27–48.

Trantas et al. (2015) Comparative genomic analysis of multiple strains of two unusual plant pathogens: *Pseudomonas corrugata* and *Pseudomonas mediterranea*. *Front. Microbiol*, 6: 811.

Ullstrup J (1972) The Impacts of the Southern Corn Leaf Blight Epidemics of 1970–1971. *Annu Rev Phytopathol*, 10: 37–50.

Venturi V and Keel C (2016) Signaling in the Rhizosphere. *Trends Plant Sci*, 21: 187–198.

Visca et al. (2007) Pyoverdine siderophores: from biogenesis to biosignificance. *Trends Microbiol*, 2007, 15, 22–30.

Wenke et al. (2019) Volatiles of rhizobacteria Serratia and Stenotrophomonas alter growth and metabolite composition of Arabidopsis thaliana. *Plant Biol*, 21: 109–119.

Whipps JM (2001) Microbial interactions and biocontrol in the rhizosphere. *J Exp Bot*, 52:487–511.

Winkel-Shirley B (2001) Flavonoid biosynthesis. A colorful model for genetics, biochemistry, cell biology, and biotechnology. *Plant Physiol*, 126: 485–493.

Wolpert TJ and Lorang JM, (2016) Victoria Blight, defense turned upside down. *Physiol Mol Plant Pathol*, 95: 8–13.

Wu et al. (2007) Mutations in Arabidopsis multidrug resistance-like ABC transporters separate the roles of acropetal and basipetal auxin transport in lateral root development. *Plant Cell,* 19:1826–1837.

Xin et al. (2016) Bacteria establish an aqueous living space in plants crucial for virulence. *Nature*, 524–539.

Zamioudis et al. (2014) β-Glucosidase BGLU42 is a MYB72-dependent key regulator of rhizobacteria-induced systemic resistance and modulates iron deficiency responses in Arabidopsis roots. *New Phytol*, 204: 368–379.

Zamioudis et al. (2013) Unraveling root developmental programs initiated by beneficial *Pseudomonas* spp. bacteria. *Plant Physiol*, 162: 304–318.

Zebelo et al. (2012) Plasma membrane potential depolarization and cytosolic calcium flux are early events involved in tomato (Solanum lycopersicon) plant-to-plant communication. *Plant Sci*, 196: 93–100.

Zhang et al. (2009) Flavones and flavonols play distinct critical roles during nodulation of Medicago truncatula by Sinorrhizobium meliloti. *Plant J*, 57:171–183.

Zhang et al. (2018a) Identification of SLAC1 anion channel residues required for CO2/bicarbonate sensing and regulation of stomatal movements. *Proc Natl Acad Sci USA*, 115: 11129–11137.

4

Molecular Aspects of Host–Pathogen Interaction

Ashim Debnath
A.N.D.U.A & T, Kumarganj

Ratul Moni Ram*
KVK Saran

Asha Kumari
ICAR-IARI

Paromita Saikia and Dharmendra Singh Lagoriya
DBT-AAU Centre, A.A.U.

CONTENTS

4.1 Introduction

Plants are one of the primary sources of food and shelter for humans, animals, birds, and a wide range of microbes, including bacteria, fungi, viruses, insects, and even other parasitic plants. Plants, because of their quiescent nature, have evolved to a wide range of constitutive or inducible biochemical and molecular factors, developing exceptional approaches. Plant immune response has resulted in a strong defensive system capable of resisting possible attack by invaders due to continuous co-evolution between

* Corresponding author's email: ratulbhu006@gmail.com

DOI: 10.1201/9781003171416-4

them and their host (Allard et al. 2008). They provide both long-term and short-term defense to upcoming threats such as pathogen assault (Nurnberger et al. 2004). The gene-for-gene hypothesis suggests that for every avirulence (*Avr*) gene in the pathogen, there is a corresponding resistant (*R*) gene in the host and vice versa (Flor 1955). The interaction between *R* genes and the *Avr* genes will decide the fate of the interaction, whether a plant is prone or resistant to pathogen invasion. Several studies have demonstrated that gene-for-gene interactions caused by pathogenic assaults and the initiation of acquired resistance in plants entail a variety of hormonal signaling molecules (van Loon 2006). It should be typical for plants to actively synthesize phytohormones such as ethylene, jasmonic acid, salicylic acid, and reactive oxygen species (ROS) before *R* gene upregulation. The biological significance of these compounds in imparting disease resistance has captivated the attention of researchers (Montesano et al. 2003). In recent years, the use of transgenic plants, as well as the generation of mutants with modified *R* genes, has revealed unique perceptions in phenomenon involved in pathogen recognition, signal transduction, and consequent disease resistance in plants. With an evolved hierarchy of biological order, the specificity of interactions between plants and diseases remains an inexplicable development. The investigation of this process is a critical component of modern plant pathology. However, great potential for crop development appears to be on the horizon, since the entire sequencing of *Arabidopsis* plant order, which might be a molecular analysis model, has become a reality. Numerous *R* genes have been discovered that bestow resistance to a wide range of pathogens in a wide range of plant species.

Plants defend themselves against pathogens in general through two divisions of immunity. One is the innate immune response, triggered through MAMP/PAMP (microbe-/pathogen-associated molecular pattern). The plant–pathogen communication can take different forms and is facilitated by plant- and pathogen-derived components such as proteins, carbohydrates, and lipopolysaccharides (Boyd et al. 2013). The pathogen-generated molecules are the crucial components that confirm their pathogenicity and allow them to thrive within the host. Plant-derived components, on the other hand, are preoccupied with identifying pathogens to induce a defensive response. The major contact between plants and microorganisms occurs in the apoplast and is facilitated by recognition of specific microbial elicitors by plant receptor proteins (Dodds and Rathjen 2010). Plant membrane-localized pattern recognition receptors (PRRs) identify these microbial elicitors, also known as pathogen-associated molecular patterns (PAMPs) (Boyd et al. 2013; Zipfel 2014). This offers security against a wide variety of diseases by utilizing cell surface plant pattern recognition receptors (PRRs) that recognize extracellular PAMPs. The second immune response is adaptive immunity, also known as effector-triggered immunity (ETI). This delivers total host resistance and is triggered by specific interaction between the *R* gene and the *Avr* gene (Abramovitch et al. 2006; Dodds et al. 2010; Wilkinson et al. 2019). As explained in the PTI/ETI zigzag model of the plant immune system, the close association between the host and the pathogen is the result of a long-term co-evolutionary process in which the fungal pathogen and the host plant each strive to keep ahead by evolving ways to overcome resistance/pathogenicity (Jones et al. 2006; Lo Presti et al. 2015). Other models have also been presented, including "effector-triggered defense" (ETD) (Stotz et al. 2014), the "invasion model" (Cook et al. 2015), the "spatial immunity model" (Van der Burgh et al. 2019), and the "spatial invasion model" (Kanyuka et al. 2019). These models facilitate pathogen detection on a broader scale and thus increase our domain of knowledge.

The revelation of the molecular theory of host–pathogen interaction has changed our perspective about pathogenesis, and this helps in designing better management tools. An inclusive approach based on the combined understanding of the molecular basis of the plant's "defensive systems" and pathogens' "invasion systems" will certainly enable us to develop innovative techniques for plant disease management.

4.2 Different Stages of Host–Pathogen Interaction

A host–pathogen interaction is crucial for a better understanding of the infectious disease, in addition to its treatment and prevention. Through different studies and analyses of different stages of infection, the molecular mechanism of pathogen invasion, and the proliferation of pathogen in the host system, the components of host–pathogen interaction may be mainly categorized into four phases: (i) host invasion by pathogen, (ii) evasion of host defense system, (iii) replication of pathogen inside the host, and (iv) host

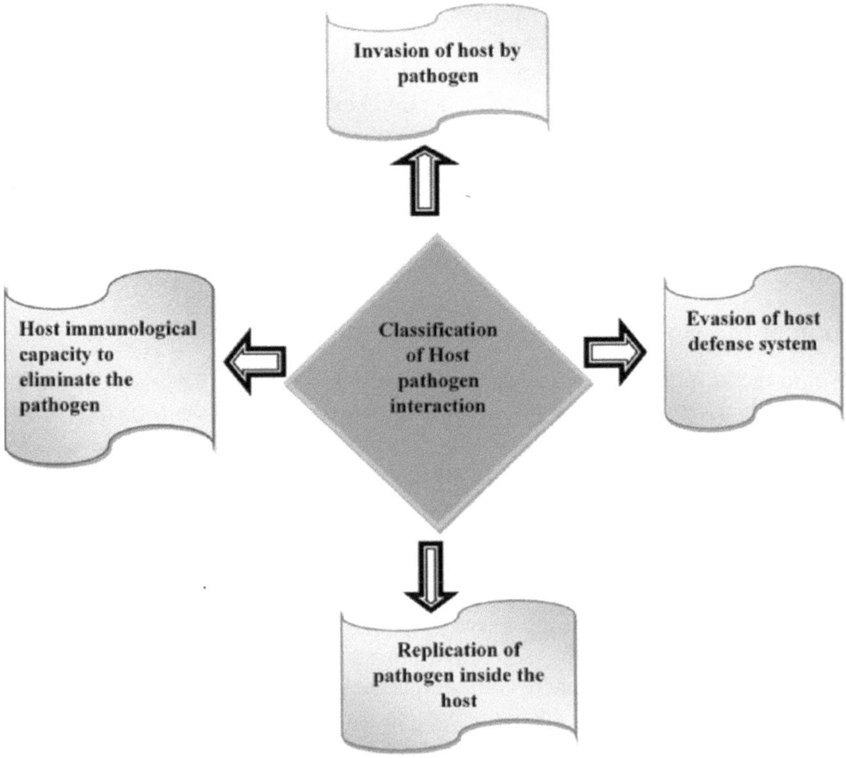

FIGURE 4.1 Classification of host–pathogen interactions.

immunological capacity to eliminate the pathogen (Sen et al. 2016). There are several virulence factors found in pathogens that can cause disease in the host system. Therefore, the above classification has been done based on these stages, which are depicted in Figure 4.1.

4.2.1 Invasion of the Host by a Pathogen

The key event in systemic infection is the penetration of the pathogen to one or multiple host barriers. When a pathogen comes in contact with any organism, it first recognizes its host. Therefore, the first step of infection is usually adherence or attachment of the pathogen to any host surface if only the pathogen is not inserted directly into the tissue (as by a wound, insect vectors, or other similar means). However, for most pathogens, the exact means of adherence are still not well resolved, particularly for protozoa and fungi. One of the methods of pathogen invasion is by secreting toxins or effector proteins in the host body; for example, gram-negative bacteria consist of secretion systems that cause virulence in the host body (Costa et al. 2015). Till now, eight general types of protein secretion system have been reported (i.e., Type I–Type VIII) that are significantly diverse from each other. However, gram-positive bacteria consist of the same type of protein secretion system, but with one system restricted to this group; that is, the Type VII system (Tseng et al. 2009) and Type VIII secretion system refer to the Curli biogenesis pathway (Depluverez 2016). Most of these proteins secreted by the pathogen can penetrate the host cells and induce physiological changes and promote colonization, viz. multivalent adhesion molecules, toxins, and urease enzyme.

The multivalent adhesion molecules (MAMs) help the pathogen to hold onto the host cell during the initial stages of infection. MAM7 facilitates the bacterial attachment to the host body via protein–protein (fibronectin) interactions and/or protein–lipid (phosphatidic acid) interactions. They are mostly present in the outer membrane of gram-negative bacteria (Krachler et al. 2011). Toxins are chemically diverse compounds that include polypeptides of variable lengths, cytotoxic proteins, and alkaloids capable of initiating infection when adhering to or absorbed by the host cell. Many organisms produce a toxin, which is harmful to both plant and animals. Reduced expression of toxin in a pathogen has less effect on the induction of host TCR signaling pathway than higher toxin expression at the time of the invasion. Urease (Ure) is a cytosolic enzyme that catalyzes the hydrolysis of urea to form ammonia and carbon dioxide and is present in most of the species of *Mycobacterium*. In several pathogenic microorganisms, urease is considered to have a virulence factor. It has been observed that toxicity in the renal epithelium is caused by the ammonia released from the urease of the urinary pathogen such as *P. mirabilis*, which causes kidney stone and complete inactivation of the renal epithelium by producing urease enzyme (Breitenbach et al. 1988).

4.2.2 Evasion of the Host Defense System

To survive within the host, a pathogen needs to overcome or inactivate the attack of the host immune system by different evasion strategies. It is seen that *Mycobacterium tuberculosis* (*Mtb*) induces several gene transcriptions that are associated with the fortification and evasion from a host system (Rachman et al. 2006). To prevent extreme inflammatory response by the host defense system, *Mycobacterium* generates signals for anti-inflammation through glycolipids to adapt with the host system (Torrelles et al. 2010). Macrophages are another important element of the host defense system. This scavenging mechanism of the host cell secures the access and elimination of pathogenic bacteria present in phagolysosomes and also activates the immune system by presenting the bacterial antigen. However, once internalized within the host, the pathogen tries to resist entry into a phagolysosome by developing various strategies to overcome the damages caused by the microenvironment of this organelle by exploiting normal host cell function (Pieters et al. 2001). The activation of macrophages occurs due to multiple factors such as signal transduction molecules, gene(s) encoding receptors, transcription factors, and bacterial constituents that turn on Toll-like receptors, viz. muramyl dipeptide, lipoteichoic acids, lipopolysaccharides, and heat shock proteins.

Nairz et al. (2010) reported that the enzyme ornithine decarboxylase is responsible for the inhibition of macrophage's ability to engulf and ingest killed pathogens. It was observed that *Mtb* present inside the macrophage can adapt to anaerobic conditions to invade the host defense system for a longer duration (Schnappinger et al. 2003). Pathogens usually act by affecting the hosts' biological pathways in the process of invading their defense system. Following the theory, targeting of hub proteins provides a more efficient system for viruses to change the cellular pathways of the host for their benefit; in humans, EBV (Epstein–Barr virus) targets the protein that tends to reside in the human interactome (Calderwood et al. 2007). The interaction between a host and a pathogen is often seen to be controlled by RNAi molecules such as miRNA, siRNA, and shRNA (Scaria et al. 2006, 2007; Ghosh et al. 2009). Konig et al. (2009) analyzed the role of siRNA in host–pathogen interaction by comparing the genome-wide siRNA study with the available online databases. Short linear motifs (SLiMs) present in the pathogen show resemblance to the SLiMs present in the host system. In eukaryotes, pathogens also use motif mimicry to manipulate the host signaling pathways by accepting SLiM-induced protein interaction (Via et al. 2015).

4.2.3 Replication of Pathogen Inside the Host

Once inside the cytosol, a pathogen colonizes the host cell in multiple ways to enhance its growth and number. First, the pathogen needs to protect few important genes to sustain in the body of the host; however, many other genes transcribing their definite protein are required to survive in the outside environment. In context to this, a research analysis revealed that about 1,087 metabolic pathways and transport responses are being catalyzed by 1,083 genes that are essential for *Salmonella typhimurium* to survive outside the host cell, while only a few inducing specific metabolic pathways are required for its survival inside the host system (Raghunathan et al. 2009). Similarly, the erythrocytic malaria parasite requires the gene encoding for the protease enzyme, which is important for its cellular process inside its host to replicate effectively (Pandey et al. 2006). Another strategy of pathogen, mostly viruses, to maintain its number of replicons in the host cell is the integration of own DNA in the genome of the host (Rappoport and Linial 2012).

In the case of some pathogens, they can hold up to more than one microenvironment for their proliferation. The genes that express Snm (Secretion in *mycobacterium*) protein secretion in a mutated *Mtb* (*Mycobacterium smegmatis*) are homologs of their *Mtb* analogues. Some hosts unconsciously help the pathogen to survive in them due to few factors. These are factors known as host factors and are responsible for facilitating replication, transcription, translation, and ultimately the growth of the pathogen. For example, the replication of the influenza virus is assisted by 295 cellular cofactors of its host in the initial stages (Konig et al. 2010). Amid all of the cofactors, 181 play a key role in host–pathogen interactions, 219 aid in effective growth of the influenza virus, 23 facilitate entry of the virus, and 10 are needed for the steps involved in virus replication inside the host. However, small molecule inhibitors, such as ATPase and CAMK2B, which have multiple functions, are also capable of resisting the replication of the influenza virus. The process of apoptosis is also found to help increase bacterial growth in the host cell. The host cell deletion method is influenced by apoptosis, which initiates inflammation and triggers defense system.

4.2.4 Host-Inherent Capacity to Eradicate the Pathogen

A host constantly deals with numerous pathogens in its entire life time, but only a few of them able to cause disease. An infectious disease occurs when a pathogen successfully evades the innate host defense mechanism. So to resist the invasion and to eliminate the pathogen, hosts launch their immune responses concerning their pathogen by triggering the expression of certain defense-related genes, by autophagy, the role of glycoconjugates and dendritic cells (Rescigno and Borrow 2001; Talat et al. 2004; Singh et al. 2006; Vergne et al. 2006; Mishra et al. 2011; Mege 2016).

The complement system is one of the most effective host defense mechanisms that can mediate lytic destruction of many kinds of pathogens, including bacteria and viruses having a lipid membrane (Cooper 1984). The complement system of the host is triggered by direct recognition of the pathogen or the absence of a complement control mechanism on the surface of the pathogen. If complement control proteins are absent, then it allows amplification of complement by the alternative pathway on any unprotected surface (Pangburn et al. 2008). On the other hand, a complement system can also direct the effect on pathogen by interacting with various effector cells, including monocytes, neutrophils, basophils, neutrophils, mast cells, and lymphocytes (Cooper 1984). Therefore, all these different factors prove to be essential in predicting the host–pathogen interaction.

4.3 Genetic and Molecular Basis of Host–Pathogen Interaction

The induction of plant defense gets initiated through recognition of particular effectors protein by host gene product, also known as *R* genes. Several R genes have been reported and characterized in various crops and are efficiently utilized in crop improvement programs (Bent and Mackey 2007). The use of the *R* gene provides several benefits as compared with chemical management practices; the prime benefit includes the maximum reduction in inoculum density with minimum impairment to crops and ecosystem in a sustainable manner (Agrios 2005). The initial concept regarding pathogen virulence and host resistance was explained by the gene-for-gene concept in flax rust, and later, the same concept was found applicable in various other diseases caused by bacteria, fungi, viruses, nematodes, parasitic higher plants, etc. The inheritance studies of virulence versus avirulence reveal that a specific gene controls virulence and the absence of this gene leads to avirulence. R protein-induced resistance is effective only when the cognate effector protein (Avr) of a strain of pathogen is recognized by the host's R protein (Gururani et al. 2012; Petit-Houdenot and Fudal 2017). To elucidate the interaction between a host and a pathogen, a thorough study on the structural and functional characteristics of the *R* gene is needed.

4.3.1 Disease Resistance in Plants through Innate Immunity

Innate immunity is a characteristic feature that mediates resistance to a plant against the aggressors (pathogen, insect, herbivores, etc.). It comprises basal resistance and *R* gene-mediated resistance.

4.3.1.1 Basal Defense

Basal resistance is the initial state of defense governing both host and non-host resistances. Physical as well as chemical barriers play an active role against a wide range of plant pathogens. The physical barrier includes the formation of a thick waxy cell wall, trichomes, and thorns, which are most effective against non-specialized opportunistic aggressors, but prove to be inefficient against more specialized attackers. Moreover, the antimicrobial substances, i.e., phenolic compounds, tannins, and fatty acid derivatives, form the chemical barrier (Chisholm et al. 2006; Gururani et al. 2012; Gill et al. 2015).

Pattern-triggered immunity (PTI) defends plants against most of the aggressors and contains numerous defense layers being induced after the recognition of specific molecular patterns. The initial interaction between the plants and pathogens is facilitated by the recognition of microbial elicitors (MAMPs/PAMPs) by the pattern recognition receptors (PRRs) of plants (Jones and Dangle 2006; Spoel and Dong 2012). These receptors are usually receptor kinases and receptor-like proteins that are present on plant cells surface (Zipfel 2014; Boutrot and Zipfel 2017). The PRRs detect invading pathogens via molecular patterns. Few PRRs detect patterns, i.e., PAMPs and MAMPs, which show the occurrence of chemical signals that are not released from the host plant itself (Zipfel et al. 2004; Basu et al. 2018). Even the damaged or wounded cells also produce chemical cues that the PRR can perceive (Heil and Land 2014; Boutrot and Zipfel 2017). The regulation of PTI is controlled by a complex signaling network that occurs in between the host plant and perceived patterns (Bigeard et al. 2015; Couto and Zipfel 2016). Two types of defensive pathways usually occur in plants, i.e., jasmonic acid (JA)- and salicylic acid (SA)-dependent defenses. The former is generally more effective against herbivores and necrotrophic pathogens, while the latter imparts defense against biotrophic pathogens (Glazebrook 2005). PTI imparts defenses such as cell wall strengthening, formation of pathogenesis-related (PR) proteins, and accumulation of volatile organic compounds (VOCs) such as terpenes and phenylpropanoids (van Loon et al. 2006). However, to counteract this defense, a group of invaders have developed few strategies. The virulent pathogen races can surpass this PTI through metabolites, effector proteins, and small RNAs (sRNAs), which they can inject into host cells (Weiberg et al. 2013; Turuno et al. 2016).

4.3.1.2 R Gene-Mediated Defense

During initial infection, a phytopathogen secretes avirulence (Avr) or effector proteins directly into the plant cells. These effector proteins interfere with the host defense activities by changing the physiological state of the host plant and also by helping pathogen colonization (Hammond-Kosack et al. 2007). In turn, the host plant also shows significant resistance with a resistance protein encoded by *R* genes against a specific race of pathogen. So, the immunity, which has been initiated against the pathogen effector activity, is known as effector-triggered immunity (ETI). For recognizing a plant pathogen, R proteins are involved in R–Avr interaction in a gene-for-gene relationship, where the host plant with *R* gene resists pathogen races with the corresponding *Avr* genes (Mundt 2014;). The *Avr* genes present in different plant pathogens, i.e., fungi, bacteria, viruses, and oomycetes, elicit a resistance response (Table 4.1). This resistance initiates a defensive mechanism in the host plant, which is known as hypersensitive reaction (HR) (Gururani et al. 2012; Pandolfi et al. 2017). The HR reaction is active against biotrophic pathogens, where pathogen growth is being ceased by restricting their mineral and nutrient access and also by localized, programmed cell death of the infected area (Jones and Dangl 2006; Hiruma et al. 2013). Post-pathogen attack within an hour of infection, reactive oxygen intermediates (ROI) are formed, which cause local cell death. The formation of ROIs is the most important aspect of the HR mechanism. Major ROIs include hydrogen peroxide and superoxide, which cross cellular membrane to cause cell damage and convert into ROS (Thakur and Sohal 2013). However, pathogens such as nematodes can influence the ROS pathway to change the compatible reaction between the host and pathogen (Ali et al. 2015). HR also induces the synthesis of salicylic acid, a signaling molecule that initiates another defense mechanism called systemic acquired resistance (SAR) (Rasul et al. 2019). A proper explanation of the interaction of R–Avr protein has been summarized in the "guard hypothesis," according to which an indirect interaction occurs between R protein and Avr effector protein. Initially, Avr protein interacts with another protein and changes its conformation, which allows R protein to bind Avr protein and activate resistance

TABLE 4.1
Different Pathogen *Avr* Genes Interacting with Corresponding *R* Genes

Category of Pathogen	Host	*R* Gene	Pathogen	*Avr* Gene	Reference
Fungi	*Linum usitatissimum*	*L*	*Melampsora lini*	*AyrL*	Dodds et al. (2006)
	Oryza sativa	*Pi-ta*	*Magnaporthe grisea*	*Avr-Pita*	Khang et al. (2008)
	Hordeum vulgare	*Rpg1*	*Puccinia graminis* f. sp. *tritici*	*Avr-Rpg1*	Kleinhofs et al. (2009)
	Zea mays	*Rp1*	*Puccinia sorghi*	*AvrRP-1-D*	Collins et al. (1999)
	Hordeum vulgare	*Mla*	*Blumeria graminis*	*AvrMla*	Zhou et al. (2001)
Bacteria	*Oryza sativa*	*Xa1, Xa21*	*Xanthomonas oryzae*	*Avr-Xa1, Avr-Xa21*	Song et al. (1995); Yoshimura et al. (1998)
	Lycopersicon esculentum		*Pseudomonas syringae* pv. *tomato*	*Avr-Pto, Avr-PtoB*	Kim et al. (2002); Abramovitch et al. (2003)
	Capsicum annuum	*Bs2*	*Xanthomonas campestris*	*Avr-Bs2*	Minsavage et al. (1990); Tai et al. (1999)
Viruses	*Oryza sativa*	*elF(iso)4G1*	Rice yellow mottle virus	*Vpg*	Hebrard et al. (2010)
	Brassica napus	*CI*	Turnip mosaic virus	*TuRBO1*	Jenner et al. (2003)
	Phaseolus vulgaris	*BV1 protein*	Bean dwarf mosaic virus	*Bdm*	Garrido-Ramirez et al. (2000)
	Pisum sativum	*sbm1*	Pea seed-borne mosaic virus	*Vpg*	Gao et al. (2004)
	Capsicum annuum	*pvr1*	Potato virus Y	*VPg*	Kang et al. (2005)
Oomycetes	*Arabidopsis thaliana*	*RPP1-Nd/WsB*	*Hyaloperonospora arabidopsidis*	*ATR1*	Rehmany et al. (2005)
	Lactuca sativa	*Dm3*	*Bremia lactucae*	*Avr3*	Michelmore et al. (2008)
	Glycine max	*Rps1a, Rps3a, Rps3c*	*Phytophthora sojae*	*Avr1a, Avr3a, and Avr3c*	Mao et al. (1996); Qutob et al. (2009); Dong et al. (2009)
	Solanum tuberosum	*R1*	*Phytophthora infestans*	*Avr1*	Ballvora et al. (2002)
Nematodes	*Capsicum annuum*	*CaMi*	*Meloidogyne incognita*	-	Chen et al. (2007)
	Lycopersicon esculentum	*Mi*	*Meloidogyne incognita*	-	Milligan et al. (1998)
	Triticum sp.	*Cre3*	*Heterodera avenae*	-	Lagudah et al. (1997)
	Solanum tuberosum	*Hero, Grol e4*	*Globodera rostochiensis*	-	Williamson and Kumar (2006)

response (Glazebrook 2005; Gururani et al. 2012; Rasul et al. 2019). In *Arabidopsis thaliana, R* gene *RPS5* interacts with *Avr* gene AvrPphB from pathogen *Pseudomonas syringae* and demonstrates the most accurate explanation of the guard hypothesis (Kaloshian 2004).

4.3.2 Major Classes of *R* Gene

The *R* gene provides resistance against diverse plant pathogens in a particular environment. Based on the types of domains and amino acid motifs, the *R* gene is classified into eight classes (Figure 4.2). Among these different classes, the majority of the *R* genes contain a nucleotide-binding site (NBS) and a leucine-rich repeat receptor (LRR) motif (Wei et al. 2016; Pandolfi et al. 2017).

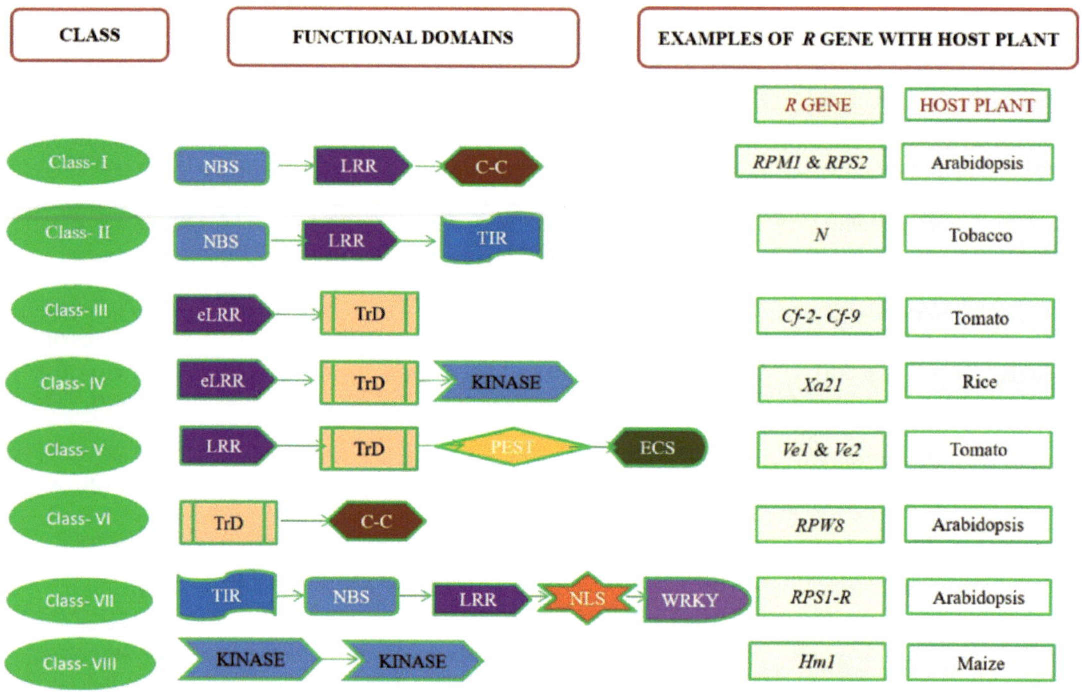

FIGURE 4.2 Major classes of plant resistance (*R*) genes with their functional domains arrangement. Abbreviations: NBS, nucleotide-binding site; LRRs, leucine-rich repeats; CC, coiled coil; TIR, Toll-interleukin-1 receptor; eLRRs, extracellular leucine-rich repeats; TrD, transmembrane domain; PEST, protein degradation domain (proline–glycine–serine–threonine); ECS, endocytosis cell signaling domain; NLS, nuclear localization signal; WRKY, amino acid domain. (From Gururani et al., *Physiol. Molecular Plant Pathol.*, 78, 51–65, 2012. With permission.)

The first major class of resistance genes consists of membrane-associated cytoplasmic proteins, which contain LRR and NBS motifs along with a putative coiled coil (CC) domain at the N-terminus. The RPM1 and RPS2 resistance genes of *Arabidopsis* are examples of this class. The second class comprises the genes encoding for cytoplasm proteins with a C-terminal LRR, an NBS, and an N-terminal Toll-interleukin-1 receptor (TIR) domain, e.g., *N* and L^6 genes conferring resistance to TMV and *M. lini*, respectively. The third major class consists of extracytoplasmic leucine-rich repeats (eLRRs) with a transmembrane domain (TrD), and this family is devoid of NBS motif. The resistance genes (*Cf-2–Cf-9*) against *Cladosporium fulvum* are examples of this class. The fourth class comprises of an extracellular LRR domain and a transmembrane domain (TrD) with an intracellular serine–threonine kinase domain, e.g., *Xa21* gene governing resistance against rice pathogen *Xanthomonas oryzae* (Yang et al. 2012).

In the fifth class, the putative extracellular LRRs motif is attached with the transmembrane domain (TrD). PEST (Pro–Glu–Ser–Thr) domain is also attached for protein degradation along with short proteins motifs (ECS) that is used for endocytosis. The examples include *Ve1* and *Ve2* genes found in tomato plants with the same kind of functional domain. The sixth major class contains an extracellular LRR domain along with a putative coiled coil (CC) domain mediating broad-spectrum resistance. An example is found in *Arabidopsis*, where *RPW8* gene confers resistance against powdery mildew disease. In the case of the seventh class, two extra functional domains (NLS and WRKY) are attached with TIR-NBS-LRR domain. For example, *RRS1R* gene in *Arabidopsis* gives resistance against bacterial wilt pathogen *Ralstonia solanacearum*. The eighth major class of the R gene is devoid of both LRR and NBS domains; however, it uses the interaction between enzymes (Takken and Goverse 2012). *Hm1* gene present in maize plant shows protection against fungal pathogen *Cochliobolus carbonum*, which is a typical member of this class. Another example, *P. syringae* Pto protein contains a serine–threonine protein kinase domain devoid of LRRs; however, in barley, *Rpg1* gene conferring resistance to *P. graminis* f. sp. *tritici* encodes a receptor kinase-like protein and is devoid of any other domains.

4.4 Metabolomics in Studying Host–Pathogen Interaction

Plant disease contributes to significant crop losses worldwide along with those incurred by other biotic and abiotic stresses (Singh 2014; Pandey et al. 2017). Plant pest and diseases account for 31%–42% of the annual crop losses globally (Agrios 2005). The prime pathogens responsible for causing plant diseases are fungi, bacteria, mycoplasmas, viruses and subviral particles, and parasitic nematodes. In history, we have witnessed few important diseases that marked a huge social and economic impact on the lives of people, viz. Irish potato famine (1845), coffee rust in Ceylon (1870), grape downy mildew in France (1878), southern corn blight in the USA (1940), and Great Bengal famine (1943) (Yoshida et al. 2013; Ram et al. 2016). These losses drove the attention of researchers toward understanding the mechanism of disease development by different pathogens and designing new methods for disease prevention/-management. One of the safest methods to combat pathogen attack is the use of a resistant gene against a virulent gene of the pathogen (Bent and Mackey 2007). So, the attempt to create disease-resistant cultivars is mainly dependent upon breeding practices to incorporate resistant alleles to selected lines (Falk 2010; Vanderplank 2012). However, the possible limitation of this technique was that it doesn't explicate the underlying mechanism of disease development for any host–pathogen interaction. To accomplish this agreement, researchers have rather depended on studies at the phenotype and molecular levels to evaluate symptomatology, gene and protein profiling, and so forth (Putham 1995; Wise 2007). Our comprehension of many plant diseases has been progressed by the usage of various model plants such as *Arabidopsis*, maize, and tomato (Piquerez et al. 2014). Finally, more studies on host–pathogen interactions using the omics-based approach have significantly contributed to a better realization of plant disease mechanisms (Etalo et al. 2013; Asselin et al. 2015; Ram et al. 2018).

4.4.1 Concept of Metabolomics

Metabolomics deals with the comprehensive profiling of all minor molecules within an organism and captures the result of information cascade initiated through the genome and progressing through transcriptome and proteome (Liu and Locasale 2017). Metabolomics has gained widespread attention in the last two decades and has appeared as an important tool in the study of plant–pathogen enigma (Summer et al. 2003). Metabolomics was initiated in the 1970s with medical profiling of human metabolites, which aided in diagnostics and drug development. However, its applicability in plants started in the 1990s, with investigations on the mode of action of herbicides. Oliver et al. (1998) proposed the theory of "metabolome," and later on, various scientists performed researches in this regard. However, Nicholson et al. (1999) attempted to define metabolomics as "the quantitative measurement of the dynamic multiparametric metabolic response of living systems to pathophysiological stimuli or genetic modification." Following this, metabolomics paved the way for numerous experiments on disease diagnostics and drug screening. The term "metabolomics" was coined by "Fiehn" and was defined as "a comprehensive and quantitative analysis of all metabolites in a biological system" (Fiehn 2001). The current metabolomics technology involves the use of sophisticated instruments such as high-pressure liquid chromatography linked with high-resolution mass spectrometry (MS) and nuclear magnetic resonance (NMR) spectroscopy. This facilitates high resolution into the chemical phenotypes of study organisms and thus the ability to capture and generate profile information of thousands of metabolites (Allwood et al. 2008; Crandall et al. 2020). A list of different online databases for metabolomics studies is presented in Table 4.2.

4.4.2 Role of Metabolomics in Plant Pathology

Preceding the advent of metabolomics, the advancement of other omics-based studies such as genomics, proteomics, transcriptomics, and secretomics has added significantly to our comprehension of plant diseases and elucidated the possible phenomenon behind pathogen establishment and disease development. Genomics examines the hereditary architecture of the two organisms (plant and pathogen) and has been valuable to screen how they adjust to infection pressure (Bergelson et al. 2001; Moller and Stukenbrock 2017). Transcriptomic studies facilitated a thorough knowledge about various genes that are concerned

TABLE 4.2
List of Different Databases for Metabolomic Studies

S. No.	Name	Website address
1.	PM: Plant Metabolomics	http://plantMetabolomics.org/
2.	HMDB: Human Metabolome Database	https://hmdb.ca/
3.	NMD: National Microbiological Database	https://www.foodsafety.govt.nz/industry/general/nmd/
4.	KEGG: Kyoto Encyclopedia of Genes and Genomes	http://www.genome.jp/kegg/
5.	YMDB: The Yeast Metabolome Database	http://www.ymdb.ca/
6.	NIST: National Institute of Standards and Technology	http://www.NIST.gov/srd/
7.	HMP: The Human Microbiome Project	http://www.hmpdacc.org/
8.	PathDB: Pathogen Database	http://www.ncgr.org/pathdb/
9.	MetaCyc Encyclopedia of Metabolic Pathways	http://metacyc.org/
10.	MNPD: Microbial Natural Products Database	http://naturalprod.ucsd.edu/
11.	SYSTOMONAS Database	http://systomonas.tu-bs.de/
12.	BioCyc Pathway	http://biocyc.org/
13.	HumanCyc	http://bicyc.org
14.	Lipidomics: SphinGOMAP	http://sphingomap.org/
15.	New drug and its metabolite database	http://www.ualberta.ca/_gjones/mslib.htm
16.	PubChem Database	http://www.pubmed.gov
17.	ChemSpider Database	http://www.chemspider.com
18.	ChEBI: Chemical Entities of Biological Interest	http://www.ebi.ac.uk/chebi/
19.	IIMDB: In Vivo/In Silico Metabolites Database	http://metabolomics.pharm.uconn.edu/iimdb/
20.	METLIN Metabolite Database	http://metlin.scripps.edu/
21.	ARM: Atomic Reconstruction of Metabolism	http://www.metabolome.jp
22.	Fiehn GC-MS Database	http://fiehnlab.ucdavis.edu/
23.	BMRB: BioMagResBank	http://www.bmrb.wisc.edu/metabolomics/
24.	Metagene	http://www.metagene.de/program/a.prg
25.	MMCD: Madison Metabolomics Consortium Database	http://mmcd.nmrfam.wisc.edu/

with disease development or reprogrammed for conferring resistance to the pathogen. However, proteomic studies deliver key information on host–pathogen interaction, mode of infection, and various regulatory elements, viz. transcription factors or mediators involved in the downstream signaling pathway for the establishment of the disease (Kav et al. 2007; Rampitsch and Bykova 2012; Kalita and Ram 2018). Metabolomics as far as we are concerned at present times is a new endeavor in molecular plant pathology. Although specialists have since decades ago perceived the significance of a single metabolite during the infection process, only in recent times have scientists started admiring global analysis accessible through metabolomics. Conventional strategies to examine the diseases depended on phenotypic examinations, for example, correlations of manifestation improvement among resistant and susceptible cultivars, and different molecular analysis of plant defenses (Chen and Kim 2009; Gonzalez-Lamothe et al. 2009). After the recognition of resistance and virulence targets of the pathogens, the necessity to perform complete research on phenotypes related to these interactions has emerged. Metabolomics bestows the depiction of plants' metabolism during advancement and in light of a wide scope of abiotic and biotic boosts, including various types of stresses (Mechin et al. 2014; Hong et al. 2016; Leary et al. 2016).

Metabolomics is especially valuable in discovering new compounds that might be related to the bioactivity of plant extracts utilized in disease management (Mumtaz et al. 2017). The same concept can be utilized for elucidating host and pathogen responses, whether the reaction will be compatible or not. In case of compatible reaction, disease development would take place, while in incompatible one, the pathogen would fail to harm the host. For example, a metabolomic analysis on root and stem rot of soybean revealed the presence of many secondary metabolites and sugars and in resistant cultivars signified the possible role of these compounds in plant defense (Zhu et al. 2018). NMR analysis studies on citrus canker displayed the expression of an antimicrobial compound "sarcotoxin," which led to a decrease in pathogen-induced metabolites accumulated in infected plant cells (do Prado Apparecido 2017). Owing to the comprehensiveness of the essential structure of metabolites and their extensively conserved structures among specific ones, metabolomics is an especially viable approach to contemplate host–pathogen associations across plant cultivars and pathogen races (Castro-Moretti et al. 2020)

4.4.3 Role in Studying Host–Pathogen Interactions

Metabolomics also provides its benefits in understanding interactions between a particular host and pathogen. It usually identifies the metabolic breakdown caused by the disease-causing factor and its producers. It is known that a pathogen produces few effector molecules during the infection process, which agitate host processes to slow down its defense machinery or divert the readiness of essential nutrients for the pathogen growth. So, these elicitors can be well tracked during the metabolome analysis. Moreover, in addition to elicitors, pathogens also produce various toxins, viz. Victorin, HC toxin, and T-toxin, which act as the main disease inciting factor. Metabolomics has the property to characterize and identify them too. For example, rubrofusarin and few other toxins were identified along with secondary metabolites using GC–MS and UPLC–MS/MS technique (Schmidt et al. 2018). Metabolomic studies on host–pathogen interactions have been explicated in various important plant diseases such as rice blast, bacterial wilt of solanaceous crops, and soybean cyst nematode. Table 4.3 depicts different metabolites formed and their effect on different plant species.

4.5 Online Repositories for Host–Pathogen Interaction

With the development of modern tools and technologies, system biology is enjoying a great prospect made possible by the vast online resources available. Biological data analytics is regarded as an imperative means to contribute to a better understanding of the subject through the extraction of latent features, relationships, and associated mechanisms. Moreover, it provides a clear perspective of data analysis from both computational and biological points of view. The host–pathogen interaction is the process of developing relationships under favorable environments that can lead to the development of immune or infectious disease depending on the host's resistance or susceptibility status. To have a smooth running for experiments related to the host–pathogen mechanism, several online repositories have been developed to track host and pathogen developmental processes by examining host–pathogen nature or using the omics approaches. Nevertheless, a lot of information is required to understand and develop new strategies before an outbreak of a pathogen occurs. However, the various online repositories available are listed in Table 4.4.

4.6 Conclusions

In nature, there is always a battle between the host and pathogen to overpower each other. Both the host and pathogen develop evolutionary mechanisms to fulfill their objectives. There is a saying that "disease is an exception while resistance is a rule." Thus, to challenge this general perception, the pathogens undergo mutations or alterations in themselves to incite diseases. The interaction between a host and a pathogen is a crucial step that ultimately decides their fate. The interaction of a pathogen with its host is a unique mechanism, and it may vary in case of different pathogens. So, the study of host–pathogen interactions describes to us the different steps involved in pathogenesis and also the measures adopted by the host plant against the invaders. The classification of host–pathogen interactions provides a clear picture of how disease development occurs and the genes involved in conferring resistance and virulence. To counterattack the invading pathogen, plants have got innate immunity, which comprises PTI and ETI. There are around eight different classes of the *R* gene having different domains and governing resistance to plants against the virulent races of the pathogen. The interaction of the *R* gene of plants and the *Avr* gene of pathogens leads to a cascade of biochemical reactions and activation of defense responses in

TABLE 4.3

A Representative List of Few Metabolites Playing a Crucial Role in Plant–Pathogen Interactions

S. No.	Molecule	Host	Function	Reference
1.	Methyl jasmonate, salicylic acid	*Oryza sativa*	Cell signaling against rice blast	Yang et al. (2017); Tezuka et al. (2019)
2.	R-linalool	*Zea mays*	Defense against insects	Tolosa et al. (2019); Huff and Pitts (2019)
3.	Benzothiadiazole	*Beet vulgaris*	Providing resistance against tobacco necrosis virus	Burketova et al. (1999)
4.	Sarcotoxin	Transgenic *Citrus* spp.	Defense against citrus canker	do Prado Apparecido et al. (2017)
5.	Methyl jasmonate	*Brassica napus*	Accumulation of glucosinolates	Doughty et al. (1995)
6.	Camalexin	*Arabidopsis thaliana*	Defense against *Phytophthora brassicae*	Schlaeppi and Mauch (2010)
7.	(Z)-3-hexenyl propionate, (Z)-3-hexenyl butyrate	*Lycopersicon esculentum*	Defense against *Pseudomonas syringae*	Lopez-Gresa et al. (2018)
8.	Salicylic acid	*Lycopersicon esculentum*	Chilling tolerance and resistance to pathogens	Garcia- Magallon et al. (2002)
9.	Benzothiadiazole	*Helianthus annuus*	Prevents infestation of parasitic weed *Orobanche cumana*	Sauerborn et al. (2002)
10.	Chitosan	*Raphanus sativus, Arachis hypogaea, Nicotiana tabacum, Pisum sativum*	Induces hypersensitive reaction and lignification; promotes defense activation against pathogens	Maksimov et al. (2003)
11.	4-Methoxyxyclobrassinin	*Brassica napus*	Defense against *Plasmodiophora brassicae*	Pedras et al. (2008)
12.	Salicylic acid and 4-aminobutyric acid	*Pisum sativum*	Provide resistance against *Erysiphe polygoni*	Katoch (2005)
13.	*β*-Amino butyric acid	*Citrus sinensis*	Inhibits spore germination and germ tube formation of *Penicillium italicum*	Tavallali et al. (2008)
14.	Benzothiadiazole and humic acid	*Glycine max*	Reduce wilt and damping-off incidence and enhance growth	Abdel-Monaim et al. (2011)
15.	Viral coat protein Harpin from TMV	*Lycopersicon esculentum, Nicotiana tabacum*	Activation of hypersensitive response	Montesano et al. (2003)

plants. The omics-based studies have facilitated the expansion of our knowledge behind the fundamental issues in disease ecology, such as plant defense, stress response, and potential for disease suppression. Metabolomics emerged as a powerful tool in understanding host–pathogen interactions. Numerous databases are available for the smooth conduction of such studies. Lastly, when these approaches (genomics, proteomics, transcriptomics, metabolomics, etc.) are integrated correctly at an accurate scale and for the exact research query, it promises to disclose a multidimensional view of plant disease.

4.7 Future Prospects

The continual threat of yield and quality loss from plant diseases is an utmost disruptive concern in agriculture presently. At the moment, this is mostly being addressed by the use of agrochemicals. The identification and utilization of genes has opened up new avenues in the development of disease-resistant plant cultivars. The genetic foundation of disease resistance to a wide range of phytopathogens, as well as the methods by which the R gene product recognizes microorganism elicitors and the plant defense

TABLE 4.4

List of Host–Pathogen Interaction Repositories Serving as an Online Database

S. No.	Repository Name	Repository URL	Description
1.	Host–Pathogen Interaction Database	https://hpidb.igbb.msstate.edu/	The HPIDB is a public resource dedicated to understanding the molecular interactions between major organisms and pathogens to which they are susceptible.
2.	PHI-base	http://www.phi-base.org/aboutUs.htm	PHI is a database of verified pathogens that infect animals, plants, fungi, and insect hosts.
3.	HoPaCI-db: The Host-Pathogen and Coxiella Interaction database	http://mips.helmholtz-muenchen.de/HoPaCI/	It is a manually created database to act as a knowledge base resource for applications in the bioinformatics field. This database includes host–pathogen interaction elements related to biological and cellular components, *etc*.
4.	Pathogen Host Interaction Network Analysis (PHINA)	http://bioinf.modares.ac.ir/-software/PHINA/VirusFamilies.php	The database describes human and virus pathogen-associated host–pathogen protein–protein interaction networks.
5.	The pathogen interaction gateway (PIG)	https://www.hsls.pitt.edu/obrc/-index.php?page=URL1233163392	PIG includes functional annotation of host plant–protein interactions and information about protein interactions.
6.	DisGeNET database	https://www.disgenet.org/home/	DisGeNET is the largest publicly available collection of genes and variants associated with human diseases and helps prioritize genotype–phenotype relationships.
7.	HPIDB	https://hpidb.igbb.msstate.edu/about.html	It is a molecular interactive database for an agricultural host–pathogen system that helps in the development of new strategies for crop plants against infectious diseases.
8.	VirHostNet 3.0: towards systems biology of virus/host interactions	https://virhostnet.prabi.fr/	VirHostNet is an open gold standard knowledge base of host protein–protein interaction networks.
9.	GPS-Prot: Data Visualization for Protein-Protein Interactions	http://gpsprot.org/index.php	GPS-Prot is the fastest and easiest way to start searching for queries as follows: "With whom does my protein of interest bind?" "What are the publications and data supporting this conversation?"
10.	PATRIC, Bacterial Bioinformatics Resource Center	https://www.patricbrc.org/	It serves as a resource center for bioinformatics for all bacteria. It provides information related to all pathosystem integration and data analysis tools for bacterial infectious diseases for the biomedical field.
11.	MPD: a pathogen genome and metagenome database	http://data.mypathogen.org/pgdb/	Mypathogen database (MPD) is the first database for genomes of microbes and metagenomics. MPD acts as a data storage center and management system for statistical analysis.
12.	Database of virulence factors in fungal pathogen	http://sysbio.unl.edu/DFVF/	Fungal pathogens have caused various diseases from plants to animals and have resulted in death and disability in humans, devastated crops, and led to global wildlife extinction or population decline.

(*Continued*)

TABLE 4.4 (*Continued*)

List of Host–Pathogen Interaction Repositories Serving as an Online Database

S. No.	Repository Name	Repository URL	Description
13.	The virulence factor database (VFDB)	http://www.mgc.ac.cn/VFs/main.htm	It is an online resource center curated for virulence factors (VFs) related to bacterial pathogens. The main objective of this database is to include in-depth knowledge and characterization of bacterial VFs for the treatment and prevention of infectious diseases.
14.	PathoPlant	http://www.pathoplant.de/	PathoPlant is a database on components of the signal transduction pathway related to plant–pathogen interactions and plant pathogenesis.
15.	OmicsDB::Pathogens—a database for exploring functional networks of plant pathogens	https://pathogens.omicsdb.org/about/	The database allows comparison and transfer of information across species and helps in the visualization of genomics, transcriptomics, co-expression networks, protein interaction networks.
16.	PHIDIAS	http://www.phidias.us/	PHIDIAS is a database of pathogen–host interactions that cause various infectious diseases to humans and animals.
17.	PLEXdb	http://www.plantgdb.org/prj/PLEXdb/	The Plant Expression Database (PLEXdb) is a community partnership database for plant and pathogen gene expression profiles with phenotypic and genotypic data. It helps in the formulation of hypotheses based on genotypic and phenotypic information.
18.	IMEx data	http://www.imexconsortium.org/	This matrix is a centralized database for biological domain-specific resources such as DB and HPIDB.
19.	CCSB interactome mapping and ORFeome	http://interactome.dfci.harvard.edu/A_thaliana/index.php?page=networklist	CCSB interactome database includes the details of *H. sapiens, Virhostome, A. thaliana, C. elegans, S. cerevisiae*; Fragmentome includes proteome-wide binary protein–protein interaction.
20.	PlaD	http://zzdlab.com/plad/	PlaD is a transcriptomics database for plant defense responses to pathogens.

prevents pathogen development, is future research objectives in the field of plant pathology. The insights gathered from future research will undoubtedly aid in providing strong disease resistance and in reducing the use of ecologically harmful agrochemicals. In a nutshell, adornment of the plant world with the best molecular weapons, which will defend themselves from pathogen attacks, is feasible solely by dissecting the molecular basis of host–pathogen interactions.

REFERENCES

Abramovitch et al. (2003). *Pseudomonas* type III effector AvrPtoB induces plant disease susceptibility by inhibition of host programmed cell death. *The EMBO Journal*, *22*(1), 60–69.

Agrios, G. N. (2005). *Plant Pathology*. 5th (ed), Academic Press, New York.

Ali et al. (2015). Plant-nematode interactions: from genomics to metabolomics. *International Journal of Agriculture and Biology*, *17*(6).

Allard et al. (2008). Strategies of attack and defense in plant–oomycete interactions, accentuated for *Phytophthora parasitica* Dastur (syn. *P. nicotianae* Breda de Haan). *Journal of Plant Physiology*, *165*, 83–94.

Allwood et al. (2008). Metabolomic technologies and their application to the study of plants and plant–host interactions. *Physiologia Plantarum*, *132*(2), 117–135.

Asselin et al. (2015). Perturbation of maize phenylpropanoid metabolism by an AvrE family type III effector from *Pantoea stewartii*. *Plant Physiology*, *167*(3), 1117–1135.

Ballvora et al. (2002). The R1 gene for potato resistance to late blight (Phytophthora infestans) belongs to the leucine zipper/NBS/LRR class of plant resistance genes. *The Plant Journal*, *30*(3), 361–371.

Basu et al. (2018). Altering plant defenses: herbivore-associated molecular patterns and effector arsenal of chewing herbivores. *Molecular Plant-Microbe Interactions*, *31*(1), 13–21.

Bent, A. F. and Mackey, D. (2007). Elicitors, effectors, and R genes: the new paradigm and a lifetime supply of questions. *Annual Review of Phytopathology*, *45*, 399–436.

Bergelson et al. (2001). Evolutionary dynamics of plant R-genes. *Science*, *292*(5525), 2281–2285.

Bigeard et al. (2015). Signaling mechanisms in pattern-triggered immunity (PTI). *Molecular Plant*, *8*(4), 521–539.

Boutrot, F. and Zipfel, C. (2017). Function, discovery, and exploitation of plant pattern recognition receptors for broad-spectrum disease resistance. *Annual Review of Phytopathology*, *55*, 257–286.

Boyd et al. (2013). Plant–pathogen interactions: disease resistance in modern agriculture. *Trends in Genetics*, *29*(4), 233–240.

Breitenbach, J. M. and Hausinger, R. P. (1988). Proteus mirabilis urease. Partial purification and inhibition by boric acid and boronic acids. *Biochemical Journal*, *250*(3), 917–920.

Burketova et al. (1999). Benzothiadiazole as an inducer of β-1, 3-glucanase and chitinase isozymes in sugar beet. *Biologia Plantarum*, *42*(2), 279–287.

Calderwood et al. (2007). Epstein–Barr virus and virus human protein interaction maps. *Proceedings of the National Academy of Sciences*, *104*(18), 7606–7611.

Castro-Moretti et al. (2020). Metabolomics as an emerging tool for the study of plant-pathogen interactions. *Metabolites*, *10*(2), 52.

Chen, X. Y. and Kim, J. Y. (2009). Callose synthesis in higher plants. *Plant Signaling & Behavior*, *4*(6), 489–492.

Chen et al. (2007). CaMi, a root-knot nematode resistance gene from hot pepper (*Capsium annuum L.*) confers nematode resistance in tomato. *Plant Cell Reports*, *26*(7), 895–905.

Chisholm et al. (2006). Host-microbe interactions: shaping the evolution of the plant immune response. *Cell*, *124*(4), 803–814.

Collins et al. (1999). Molecular characterization of the maize Rp1-D rust resistance haplotype and its mutants. *The Plant Cell*, *11*(7), 1365–1376.

Converse, S. E. and Cox, J. S. (2005). A protein secretion pathway critical for Mycobacterium tuberculosis virulence is conserved and functional in Mycobacterium smegmatis. *Journal of Bacteriology*, *187*(4), 1238.

Cook et al. (2015). Understanding plant immunity as a surveillance system to detect invasion. *Annual Review of Phytopathology*, *53*, 541–563.

Cooper, N. R. (1984). The role of the complement system in host defense against virus diseases. In: Notkins AL, Oldstone MBA (eds) *Concepts in Viral Pathogenesis* (pp. 20–25). Springer, New York.

Costa et al. (2015). Secretion systems in Gram-negative bacteria: structural and mechanistic insights. *Nature Reviews Microbiology*, *13*(6), 343–359.

Couto, D. and Zipfel, C. (2016). Regulation of pattern recognition receptor signalling in plants. *Nature Reviews Immunology*, *16*(9), 537.

Crandall et al. (2020). A multi-omics approach to solving problems in plant disease ecology. *Plos One*, *15*(9), e0237975.

Depluverez et al. (2016). The role of bacterial secretion systems in the virulence of Gram-negative airway pathogens associated with cystic fibrosis. *Frontiers in Microbiology*, *7*, 1336.

Ding et al. (2002). Jasmonate and salicylate induce the expression of pathogenesis-related-protein genes and increase resistance to chilling injury in tomato fruit. *Planta*, *214*(6), 895–901.

do Prado Apparecido et al. (2017). NMR-based metabolomics of transgenic and non-transgenic sweet orange reveals different responses in primary metabolism during citrus canker development. *Metabolomics*, *13*(2), 20.

Dodds, P. N. and Rathjen, J. P. (2010). Plant immunity: towards an integrated view of plant-pathogen interactions. *Nature Reviews Genetics,* 11, 539–548.

Dodds et al. (2006). Direct protein interaction underlies gene-for-gene specificity and coevolution of the flax resistance genes and flax rust avirulence genes. *Proceedings of the National Academy of Sciences*, *103*(23), 8888–8893.

Dong et al. (2009). The *Phytophthora sojae* avirulence locus Avr3c encodes a multi-copy RXLR effector with sequence polymorphisms among pathogen strains. *PLoS One*, 4(5), e5556.

Doughty et al. (1995). Selective induction of glucosinolates in oilseed rape leaves by methyl jasmonate. *Phytochemistry*, *38*(2), 347–350.

Etalo et al. (2013). System-wide hypersensitive response-associated transcriptome and metabolome reprogramming in tomato. *Plant Physiology*, *162*(3), 1599–1617.

Falk, D.E. (2010). Generating and maintaining diversity at the elite level in crop breeding. *Genome*, *53*, 982– 991.

Fiehn, O. (2001). Combining genomics, metabolome analysis, and biochemical modelling to understand metabolic networks. *Comparative and Functional Genomics*, *2*(3), 155–168.

Flor, H. H. (1955). Host-parasite interactions in flax rust-its genetics and other implications. *Phytopathology*, 45, 680–685

Gao et al. (2004). The potyvirus recessive resistance gene, sbm1, identifies a novel role for translation initiation factor eIF4E in cell-to-cell trafficking. *The Plant Journal*, *40*(3), 376–385.

Garcia-Magallon et al. (2002). Aplicación del ácido benzoico en forma foliar al cultivo de *Lilium* cv. Dreamland. In: Garcia Magallon E (ed) *Memoria del XIX Congreso Nacional de Fitogenética*, p. 72, Sociedad Mexicana de Fitogenética, Saltillo, Mexico.

Garrido-Ramirez et al. (2000). Bean dwarf mosaic virus BV1 protein is a determinant of the hypersensitive response and avirulence in *Phaseolus vulgaris*. *Molecular Plant-Microbe Interactions*, *13*(11), 1184–1194.

Ghosh et al. (2009). Cellular versus viral microRNAs in host–virus interaction. *Nucleic Acids Research*, *37*(4), 1035–1048.

Gill et al. (2015). Host versus nonhost resistance: distinct wars with similar arsenals. *Phytopathology*, *105*(5), 580–587.

Glazebrook, J. (2005). Contrasting mechanisms of defense against biotrophic and necrotrophic pathogens. *Annual Review of Phytopathology*, *43*, 205–227.

Gonzalez-Lamothe et al. (2009). Plant antimicrobial agents and their effects on plant and human pathogens. *International Journal of Molecular Sciences*, *10*(8), 3400–3419.

Gururani et al. (2012). Plant disease resistance genes: current status and future directions. *Physiological and Molecular Plant Pathology*, *78*, 51–65.

Hammond-Kosack, K. E. and Kanyuka, K. (2007). *Resistance Genes (R genes) in Plants*. John Wiley and Sons, Ltd., Chichester, UK.

Hébrard et al. (2010). Direct interaction between the Rice yellow mottle virus (RYMV) VPg and the central domain of the rice eIF (iso) 4G1 factor correlates with rice susceptibility and RYMV virulence. *Molecular Plant-Microbe Interactions*, *23*(11), 1506–1513.

Heil, M. and Land, W. G. (2014). Danger signals–damaged-self recognition across the tree of life. *Frontiers in Plant Science*, *5*, 578.

Hiruma et al. (2013). Glutathione and tryptophan metabolism are required for Arabidopsis immunity during the hypersensitive response to hemibiotrophs. *Proceedings of the National Academy of Sciences*, *110* (23), 9589–9594.

Hong et al. (2016). Plant metabolomics: an indispensable system biology tool for plant science. *International Journal of Molecular Sciences*, *17*(6), 767.

Huff, R.M. and Pitts, R.J. (2019). An odorant receptor from *Anopheles gambiae* that demonstrates enantioselectivity to the plant volatile, linalool. *PLoS One*, *14*, e0225637.

Jenner et al. (2003). The dual role of the potyvirus P3 protein of Turnip mosaic virus as a symptom and avirulence determinant in brassicas. *Molecular Plant-Microbe Interactions*, *16*(9), 777–784.

Jones, J. D. G. and Dangl, J. L. (2006). The plant immune system. *Nature*, *444*, 323–329.

Jung et al. (2009). Priming in systemic plant immunity. *Science*, *324*(5923), 89–91.

Kalita, P. J. and Ram, R. M. (2018). Proteomics in plant pathology: a tool for disease management. *International Journal of Chemical Studies*, *6*(4), 1804–1813.

Kaloshian, I. (2004). Gene-for-gene disease resistance: bridging insect pest and pathogen defense. *Journal of Chemical Ecology*, *30*(12), 2419–2438.

Kang et al. (2005). The pvr1 locus in Capsicum encodes a translation initiation factor eIF4E that interacts with Tobacco etch virus VPg. *The Plant Journal*, *42*(3), 392–405.

Kanyuka, K. and Rudd, J. J. (2019). Cell surface immune receptors: the guardians of the plant's extracellular spaces. *Current Opinion in Plant Biology*, *50*, 1–8.

Katoch, R. (2005). Effect of elicitors and *E. polygoni* inoculation on the activity of phenol metabolizing enzymes in garden pea (*Pisum sativum* L.). *Indian Journal of Agricultural Biochemistry, 18*(2), 87–91.

Kav et al. (2007). Application of proteomics to investigate plant-microbe interactions. *Current Proteomics, 4*(1), 28–43.

Khang et al. (2008). Genome organization and evolution of the AVR-Pita avirulence gene family in the *Magnaporthe grisea* species complex. *Molecular Plant-Microbe Interactions, 21*(5), 658–670.

Kim et al. (2002). Two distinct Pseudomonas effector proteins interact with the Pto kinase and activate plant immunity. *Cell, 109*(5), 589–598.

Kleinhofs et al. (2009). Barley stem rust resistance genes: structure and function. *The Plant Genome, 2*(2), 109–120.

Konig et al. (2010). Human host factors required for influenza virus replication. *Nature, 463*(7282), 813–817.

König et al. (2008). Global analysis of host-pathogen interactions that regulate early-stage HIV-1 replication. *Cell, 135*(1), 49–60.

Krachler et al. (2011). Outer membrane adhesion factor multivalent adhesion molecule 7 initiates host cell binding during infection by gram-negative pathogens. *Proceedings of the National Academy of Sciences, 108*(28), 11614–11619.

Lagudah et al. (1997). Map-based cloning of a gene sequence encoding a nucleotide-binding domain and a leucine-rich region at the Cre3 nematode resistance locus of wheat. *Genome, 40*(5), 659–665.

Liu, X. and Locasale, J. W. (2017). Metabolomics: a primer. *Trends in Biochemical Sciences, 42*(4), 274–284.

Lo Presti et al. (2015). Fungal effectors and plant susceptibility. *Annual Review of Plant biology, 66*, 513–545.

Lopez-Gresa et al. (2018). A new role for green leaf volatile esters in tomato stomatal defense against *Pseudomonas syringe* pv. tomato. *Frontiers in Plant Science, 9*, 1855.

Luna et al. (2011). Callose deposition: a multifaceted plant defense response. *Molecular Plant-Microbe Interactions, 24*(2), 183–193.

Maksimov et al. (2003). "Chitin-specific peroxidases in plants," *Biochemistry, 68*(1), 111–115.

Mao, Y. and Tyler, B. M. (1996). The *Phytophthora sojae* genome contains tandem repeat sequences which vary from strain to strain. *Fungal Genetics and Biology, 20*(1), 43–51.

Mechin et al. (2014). Impact of the brown-midrib bm 5 mutations on maize lignins. *Journal of Agricultural and Food Chemistry, 62*(22), 5102–5107.

Mege, J. L. (2016). Dendritic cell subtypes: a new way to study host-pathogen interaction. *Virulence, 7*(1), 5.

Michelmore, R. and Wong, J. (2008). Classical and molecular genetics of *Bremia lactucae*, cause of lettuce downy mildew. *European Journal of Plant Pathology, 122*(1), 19–30.

Milligan et al. (1998). The root knot nematode resistance gene Mi from tomato is a member of the leucine zipper, nucleotide binding, leucine-rich repeat family of plant genes. *The Plant Cell, 10*(8), 1307–1319.

Minsavage et al. (1990). Gene-for-gene relationships specifying disease resistance in *Xanthomonas campestris* pv. *vesicatoria* - pepper interactions. *Molecular Plant-Microbe Interactions, 3*(1), 41–47.

Mishra et al. (2011). Lipoarabinomannan and related glycoconjugates: structure, biogenesis and role in *Mycobacterium tuberculosis* physiology and host–pathogen interaction. *FEMS Microbiology Reviews, 35*(6), 1126–1157.

Moller, M. and Stukenbrock, E. H. (2017). Evolution and genome architecture in fungal plant pathogens. *Nature Reviews Microbiology, 15*(12), 756.

Montesano et al. (2003). Pathogen derived elicitors: searching for receptors in plants. *Molecular Plant Pathology, 4*(1), 73–79.

Mumtaz et al. (2017). An overview of recent developments in metabolomics and proteomics–phytotherapic research perspectives. *Frontiers in Life Science, 10*(1), 1–37.

Mundt, C. C. (2014). Durable resistance: a key to sustainable management of pathogens and pests. *Infection, Genetics and Evolution, 27*, 446–455.

Nairz et al. (2010). The struggle for iron–a metal at the host–pathogen interface. *Cellular Microbiology, 12* (12), 1691–1702.

Nicholson et al. (1999). 'Metabonomics': understanding the metabolic responses of living systems to patho-physiological stimuli via multivariate statistical analysis of biological NMR spectroscopic data. *Xenobiotica, 29*(11), 1181–1189.

Nurnberger et al. (2004). Innate immunity in plants and animals: striking similarities and obvious differences. *Immunological Reviews, 198*(1), 249–266.

O'Leary et al. (2016). Early changes in apoplast composition associated with defence and disease in interactions between Phaseolus vulgaris and the halo blight pathogen *Pseudomonas syringae* pv. *phaseolicola. Plant, Cell & Environment, 39*(10), 2172–2184.

Oliver et al. (1998). Systematic functional analysis of the yeast genome. *Trends in Biotechnology, 16*, 373–378.

Pandey et al. (2006). Falstatin, a cysteine protease inhibitor of *Plasmodium falciparum*, facilitates erythrocyte invasion. *PLoS Pathogens, 2*(11), e117.

Pandey et al. (2017). Impact of combined abiotic and biotic stresses on plant growth and avenues for crop improvement by exploiting physio-morphological traits. *Frontiers in Plant Science, 8*, 537.

Pandolfi et al. (2017). Resistance (R) genes: applications and prospects for plant biotechnology and breeding. *Current Protein and Peptide Science, 18*(4), 323–334.

Pangburn et al. (2008). Discrimination between host and pathogens by the complement system. *Vaccine, 26*, I15–I21.

Pedras et al. (2008). Metabolic changes in roots of the oilseed canola infected with the biotroph *Plasmodiophora brassicae*: phytoalexins and phytoanticipins. *Journal of Agricultural and Food Chemistry, 56*(21), 9949–9961.

Petit-Houdenot, Y. and Fudal, I. (2017). Complex interactions between fungal avirulence genes and their corresponding plant resistance genes and consequences for disease resistance management. *Frontiers in Plant Science, 8*, 1072.

Piasecka et al. (2015). Secondary metabolites in plant innate immunity: conserved function of divergent chemicals. *New Phytologist, 206*(3), 948–964.

Pieters, J. (2001). Evasion of host cell defense mechanisms by pathogenic bacteria. *Current Opinion in Immunology, 13*(1), 37–44.

Piquerez et al. (2014). Improving crop disease resistance: lessons from research on Arabidopsis and tomato. *Frontiers in Plant Science, 5*, 671.

Putnam, M. L. (1995). Evaluation of selected methods of plant disease diagnosis. *Crop Protection, 14*, 517–525.

Qutob et al. (2009). Copy number variation and transcriptional polymorphisms of Phytophthora sojae RXLR effector genes Avr1a and Avr3a. *PloS One, 4*(4), e5066.

Rachman et al. (2006). Unique transcriptome signature of *Mycobacterium tuberculosis* in pulmonary tuberculosis. *Infection and Immunity, 74*(2), 1233–1242.

Raghunathan et al. (2009). Constraint-based analysis of metabolic capacity of *Salmonella typhimurium* during host-pathogen interaction. *BMC Systems Biology, 3*(1), 1–16.

Ram et al. (2016). Comprehensive approaches in plant disease management. In:Sarma BK, Singh A (eds) *Microbial Empowerment in Agriculture: A key to Sustainability and Crop Productivity.* Biotech Books, New Delhi, pp. 417–443.

Ram et al. (2018). Biocontrol technology: eco-friendly approaches for sustainable agriculture. In: Brah D, Azevedo V (eds) *Omics Technologies and Bio-Engineering: Towards Improving Quality of Life Volume II Microbial, Plant, Environmental and Industrial Technologies.* Academic, London, pp. 177–190.

Rampitsch, C. and Bykova, N. V. (2012). Proteomics and plant disease: advances in combating a major threat to the global food supply. *Proteomics, 12*, 673–690.

Rappoport, N. and Linial, M. (2012). Viral proteins acquired from a host converge to simplified domain architectures. *PLoS Computational Biology, 8*(2), e1002364.

Rasul et al. (2019). Genetic basis for biotic stress resistance in plants from Solanaceae family: a review. *International Journal of Agriculture and Biology, 22*, 178–194.

Rehmany et al. (2005). Differential recognition of highly divergent downy mildew avirulence gene alleles by RPP1 resistance genes from two Arabidopsis lines. *The Plant Cell, 17*(6), 1839–1850.

Rescigno, M. and Borrow, P. (2001). The host-pathogen interaction: new themes from dendritic cell biology. *Cell, 106*(3), 267–270.

Sauerborn et al. (2002). Benzothiadiazole activates resistance in sunflower (*Helianthus annuus*) to the root-parasitic weed *Orobanche cuman. Phytopathology, 92*(1), 59–64.

Scaria et al. (2006). Host-virus interaction: a new role for microRNAs. *Retrovirology, 3*(1), 1–9.

Scaria et al. (2007). Host–virus genome interactions: macro roles for microRNAs. *Cellular Microbiology, 9*(12), 2784–2794.

Schlaeppi, K. and Mauch, F. (2010). Indolic secondary metabolites protect Arabidopsis from the oomycete pathogen *Phytophthora brassicae. Plant Signalling Behaviour, 5*, 1099–1101.

Schmidt et al. (2018). Deciphering the genome and secondary metabolome of the plant pathogen Fusarium culmorum. *FEMS Microbiology Ecology*, *94*(6), fiy078.

Schnappinger et al. (2003). Transcriptional adaptation of *Mycobacterium tuberculosis* within macrophages: insights into the phagosomal environment. *The Journal of Experimental Medicine*, *198*(5), 693–704.

Sen et al. (2016). A review on host–pathogen interactions: classification and prediction. *European Journal of Clinical Microbiology & Infectious Diseases*, *35*(10), 1581–1599.

Singh, H. B. (2014). Management of plant pathogens with microorganisms. *Proceedings of the Indian National Science Academy*, *80*(2), 443–454.

Singh et al. (2006). Human IRGM induces autophagy to eliminate intracellular mycobacteria. *Science*, *313* (5792), 1438–1441.

Singh et al. (2010). Comparison of virus interactions with human signal transduction pathways. *Proceedings of the First ACM International Conference on Bioinformatics and Computational Biology*, 17–24.

Song et al. (1995). A receptor kinase-like protein encoded by the rice disease resistance gene, Xa21. *Science*, *270*(5243), 1804–1806.

Spoel, S. H. and Dong, X. (2012). How do plants achieve immunity? Defence without specialized immune cells. *Nature Reviews Immunology*, *12*(2), 89–100.

Stotz et al. (2014). Effector-triggered defence against apoplastic fungal pathogens. *Trends in Plant Science*, *19*, 491–500.

Sumner et al. (2003). Plant metabolomics: large-scale phytochemistry in the functional genomics era. *Phytochemistry*, 62(6), 817–836.

Tai et al. (1999). Expression of the Bs2 pepper gene confers resistance to bacterial spot disease in tomato. *Proceedings of the National Academy of Sciences*, *96*(24), 14153–14158.

Talaat et al. (2004). The temporal expression profile of *Mycobacterium tuberculosis* infection in mice. *Proceedings of the National Academy of Sciences*, *101*(13), 4602–4607.

Takken, F. L. and Goverse, A. (2012). How to build a pathogen detector: structural basis of NB-LRR function. *Current Opinion in Plant Biology*, 15, 375–384.

Tavallali et al. (2008). Effects of 5-aminobutyric acid on the induction of resistance to *Penicillium italicum*. *World Applied Sciences Journal*, *5*(3), 345–351.

Tezuka et al. (2019). The rice ethylene response factor OsERF83 positively regulates disease resistance to *Magnaporthe oryzae*. *Plant Physiology and Biochemistry*, *135*, 263–271.

Thakur, M. and Sohal, B. S. (2013). Role of elicitors in inducing resistance in plants against pathogen infection: a review. *International Scholarly Research Notices*, *2013*.

Tolosa et al. (2019). Molasses grass induces direct and indirect defense responses in neighbouring maize plants. *Journal of Chemical Ecology*, *45*(11), 982–992.

Torrelles, J. B. and Schlesinger, L. S. (2010). Diversity in Mycobacterium tuberculosis mannosylated cell wall determinants impact adaptation to the host. *Tuberculosis*, *90*(2), 84–93.

Torres et al. (2006). Reactive oxygen species signaling in response to pathogens. *Plant Physiology*, *141*(2), 373–378.

Toruno et al. (2016). Plant-pathogen effectors: cellular probes interfering with plant defenses in spatial and temporal manners. *Annual Review of Phytopathology*, *54*, 419–441.

Tseng et al. (2009). Protein secretion systems in bacterial-host associations, and their description in the Gene Ontology. *BMC Microbiology*, *9*(1), 1–9.

Van der Burgh, A. M. and Joosten, M. H. A. J. (2019). Plant immunity: Thinking outside and inside the box. *Trends in Plant Science*, *24*, 587–601.

van Loon et al. (2006). Significance of inducible defense-related proteins in infected plants. *Annual Review of Phytopathology*, *44*, 135–162.

Vanderplank, J. E. (2012). *Disease Resistance in Plants*. Elsevier, Orlando, FL.

Vergne et al. (2006). Autophagy in immune defense against Mycobacterium tuberculosis. *Autophagy*, *2*(3), 175–178.

Via et al. (2015). How pathogens use linear motifs to perturb host cell networks. *Trends in Biochemical Sciences*, *40*(1), 36–48.

Wei et al. (2016). Dramatic number variation of R genes in Solanaceae species accounted for by a few R gene subfamilies. *PLoS One*, *11*(2), e0148708.

Weiberg et al. (2013). Fungal small RNAs suppress plant immunity by hijacking host RNA interference pathways. *Science*, *342*(6154), 118–123.

Wilkinson et al. (2019). Surviving in a hostile world: plant strategies to resist pests and diseases. *Annual Review of Phytopathology*, *57*, 505–529.
Williamson, V. M. and Kumar, A. (2006). Nematode resistance in plants: the battle underground. *Trends in Genetics*, *22*(7), 396–403.
Wise et al. (2007). Transcript profiling in host–pathogen interactions. *Annual Review of Phytopathology*, *45*, 329–369.
Wu et al. (2018). Mapping the Arabidopsis metabolic landscape by untargeted metabolomics at different environmental conditions. *Molecular Plant*, *11*(1), 118–134.
Yang et al. (2012). Receptor-like kinases and receptor like proteins: keys to pathogen recognition and defense signaling in plant innate immunity. *Frontiers in Biology*, *7*, 155–166.
Yang et al. (2017). Activation of ethylene signaling pathways enhances disease resistance by regulating ROS and phytoalexin production in rice. *The Plant Journal*, *89*(2), 338–353.
Yoshimura et al. (1998). Expression of Xa1, a bacterial blight-resistance gene in rice, is induced by bacterial inoculation. *Proceedings of the National Academy of Sciences*, *95*(4), 1663–1668.
Zhou et al. (2001). Cell-autonomous expression of barley Mla1 confers race-specific resistance to the powdery mildew fungus via a Rar1-independent signaling pathway. *The Plant Cell*, *13*(2), 337–350.
Zhu et al. (2018). Metabolomics analysis of soybean hypocotyls in response to *Phytophthora sojae* infection. *Frontiers in Plant Science*, *9*, 1530.
Zipfel, C. (2014). Plant pattern-recognition receptors. *Trends in Immunology*. 35, 345–351.
Zipfel et al. (2004). Bacterial disease resistance in *Arabidopsis* through flagellin perception. *Nature*, *428* (6984), 764–767.

5

Omics: A Potential Tool to Delineate the Mechanism of Biocontrol Agents against Plant Pathogens

S. Harish*
Tamil Nadu Agricultural University

S. Parthasarathy
Amrita School of Agricultural Sciences, Amrita Vishwa Vidyapeetham

S.R. Prabhukarthikeyan, Arabinda Mahanty, Basavaraj Teli, and P.C. Rath
ICAR – National Rice Research Institute

K. Anandhi
Tamil Nadu Agricultural University

CONTENTS

* Corresponding author's email: harish.s@tnau.ac.in

DOI: 10.1201/9781003171416-5

5.1 Introduction

Plant diseases are a major threat to global food security (Ons et al. 2020). In the International Year of Plant Health, our focus should be to combat microbial pathogens using disease-suppressive beneficial microbes. The biocontrol of plant disease involves the use of a microorganism to suppress the phytopathogen, which in turn reduces the disease. Various microbial biocontrol agents recognized in several studies are commercially available as of date for effective utilization. The biological management of plant diseases has been considered a prospective approach in recent years, because chemical applications result in the settlement of detrimental remains in the soil and plants, which may lead to serious ecological problems. Thus, discovering sustainable methods of plant disease management is of utmost prominence (Eljounaidi et al. 2016). However, the method of reduction of plant diseases by biocontrol agents and their interaction with the introduced environment in the complex situations is less understood (Smith et al. 1999). Consequently, the molecular interaction that occurs between the biocontrol agent and pathogen in the rhizosphere needs to be understood. This may allow us to improve the biocontrol strategies by manipulating the soil environment, which helps the beneficial organism to establish itself in the rhizosphere (Chaur, 1998). However, very little information on the function of plant and microbial genes expressed during such associations is known. Omics technologies had a marvelous influence on the investigation of the interaction of biocontrol agents with plants and their diseases, which enabled researchers to unveil the biological control mechanism in a more precise way. The multidimensional strategies of omics tools make it authoritative that they involve various biological sciences. The integrated multi-omics involves computational integration of genomics, proteomics, transcriptomics, metabolomics, ionomics, lipidomics, glycomics, and culturomics for interactive analysis, expression studies, molecular phylogenetics, microarray informatics, and systems biology. To some extent, it also deals with synthetic biology, mainly big data analysis and computer-aided drug design and discovery. Here we briefly narrate the exploitation of omics technology during the interaction of microbial bioagents with their targeted pathogens at the site of action. The effectiveness of biocontrol agents depends on the interactions among the host plant, beneficial microbes, pathogenic microbes, and the environment. Recent tools in omics and computational biology have provided effective strategies to light up the mechanisms that strengthen these interactions. The efficient interaction of biocontrol agents dealing with disease suppression results from competition for nutrients and environment, parasitism, lysis, antibiosis, induced

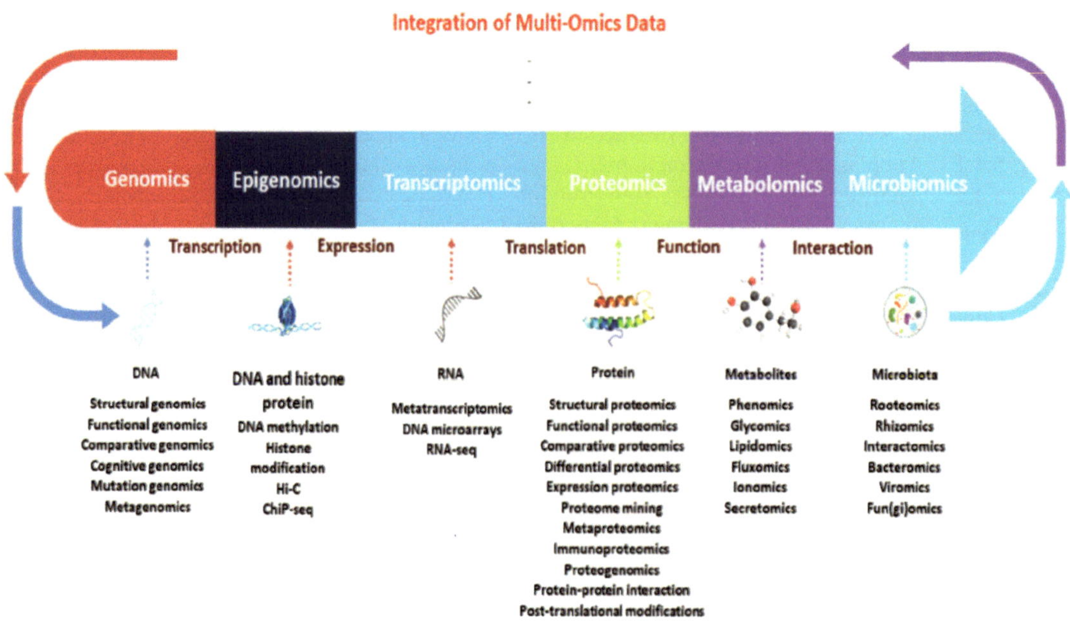

resistance, etc. (Thomashow and Weller 1996; Verma et al. 2018). Moreover, biocontrol agents induce cell surface and cytoplasmic modifications, leading to a physical barrier at the site of infection by the pathogens (Benhamou et al. 2000). Plant surface colonization of biocontrol agents tends to increase the expression of extensins, which also strengthens the plant cell wall against pathogenic microbes (Pinski et al. 2019). With this background, this chapter reviews the various omics technologies utilized in plant–pathogen–biocontrol agent interaction.

5.2 Genomics

In recent years, the biological control of plant diseases has attained a lot of interest due to its extensive practice in agriculture as biofungicides and in industry as a source of lytic enzymes. Diversified methods, including phenotypic and genotypic characterization, have been employed to understand the mechanisms regulating the antagonism and survival of potential biocontrol agents. Multi-site action in biocontrol agents requires versatile computational approaches to elucidate the mechanisms of microbial functions. In this context, gaining insights into the genomics of biocontrol organisms can enable the development of effective formulations and strategies that can manage plant diseases in diverse agroecological conditions. Comparison of gene ontology, organization, function, and regulation in biocontrol agents has enabled a deeper understanding of biology, lifestyle, evolution, biocontrol efficacy, mode of action, macromolecular proteins up- (or) downregulated, and metabolites/enzymes involved in biosynthesis and degradation, which are significantly well conserved in the microbial world. The genomics of the important biocontrol agents has been dealt with hereunder.

5.2.1 *Trichoderma* sp.

Trichoderma spp. are beneficial microorganisms present in the soil and terrestrial environment, colonize the rhizosphere, maintain soil health, and are present in almost all agro-climatic zones (Mukherjee et al. 2013). They can penetrate the living cells and can also dwell as saprophytes. The inherent ability of the fungus to antagonize the harmful phytopathogens in the soil has made them potential commercial biopesticides. In the biopesticide market, approximately 60% of the products registered are based on *Trichoderma*, which has a significant impact on crop protection. The genus *Trichoderma* was first reported by Persoon (1794), but came to attention only after Weindling displayed that *Trichoderma* spp. can destroy other fungi and minimize diseases. The research on the discovery of antimicrobial genes on these fungi has continued unabated since 1934. Within the *Trichoderma* spp., *T. reesei*, which is a known model organism for industrial production of cellulolytic enzymes, is the first organism to be sequenced (Martinez et al. 2008). Two hundred different *Trichoderma* spp. have recently been identified by molecular characterization of their genomic and mitochondrial DNA (Bissett et al. 2015). The evolution of mycoparasitism and ecological fitness of *Trichoderma* spp. is understood by the sequence analysis of the first three genomes, *viz. T. reesei*, *T. virens*, and *T. atroviride* (Druzhinina et al. 2011; Kubicek et al. 2011). Many genome sequencing projects have embattled the identification of few species such as *T. reesei*, *T. virens*, *T. atroviride*, *T. harzianum*, *T. asperellum*, *T. longibrachiatum*, and *T. citrinoviride* (http://genome.jgi-psf.org/programs/fungi) to reveal their mechanisms behind the action as biocontrol agents (Martinez et al. 2008; Kubicek et al. 2011; Mukherjee et al. 2013). Promoters are central genetic elements that are required to drive the transcription of an introduced or native gene. *Trichoderma* enzyme systems have most intensively been studied, and their genes, such as *cbh1*, *cbh2*, *xyn1*, *xyn2*, and *egl1*, and the respective promoters are well characterized for further use (Steiger et al. 2011). The most often used promoter is the *cbh1*, which enables high expression levels and is therefore often used in heterologous gene expression studies, and it can be induced by sophorose, lactose, or cellulose and repressed by glucose (Qin et al. 2012). Zou et al. (2012) replaced the CRE1 elements in the cbh1 promoter by ACE2 and HAP2/HAPC3/HAP5 elements. *T. harzianum* T6776 genome consists of 1,573 sequence scaffolds of 39.73 Mbp and a GC content of 48.50%, which was assessed using CEGMA (Baroncelli et al. 2015). *T. gamsii* T6085 is another promising isolate which controls *Fusarium* head blight, reduces mycotoxin accumulation, competes with the pathogen, and thus reduces the disease in field conditions. Its genome was completely sequenced using Illumina mate-paired sequencing technology and assembled using Velvet 1.2.08. *T. gamsii* draft genome consists of 381 scaffolds, with a length of 37.97 Mbp and GC content of 49.00%, which was evaluated using CEGMA, version 2.4 (Baroncelli et al. 2016). The whole-genome shotgun approach was used to sequence the genome of *T. atrobrunneum* (*T. harzianum* species complex) ITEM 908. The sequence was assembled using the Spades, version 5.0, software, which consists of 804 contigs, GC% of 49.18, and 8,649 genes, involved in mycoparasitism and antagonistic activities (Fanelli et al. 2018). The sequence of *T. harzianum* B97 whole genome revealed genes responsible for the alleviation of abiotic stresses and plant growth promotion in agriculture (Compant et al. 2017). The mitochondrial genome of *T. atroviride* ATCC 26799 consists of 14 core protein-coding genes and four *orf* genes, through a length of 32,758 bp (314.81x; GenBank accession no. MN125601). A highly conservative gene order was noticed among the core genes and *rns* gene, thus providing an opportunity for genome improvement (Kwak et al. 2020).

5.2.2 *Pseudomonas* sp.

Members of the *Pseudomonas fluorescens* species complex consists of diversified functional species, notable as an antagonistic microorganism, naturally occurs in soil, water, plant tissues as endophyte, plant root (rhizosphere) and leaf (phyllosphere) as epiphyte colonization, which currently consists of more than fifty named species. Owing to their varied lifestyles is their capability to produce array of secondary metabolites that include mutual relationship with plants and other microbial populations of their bionetworks (Mark et al. 2011). Plant-associated isolates of *P. fluorescens* produce antimicrobials and plant growth-promoting hormones to combat pathogens in plants and soil (Haas and Defago, 2005). *Pseudomonas* spp. antagonize phytopathogens through synthesis of antibiotics, *viz.* 2,4-DAPG, pyrrolnitrin, pyoluteorin, phenazines, other secondary metabolites, hydrogen cyanide (HCN), siderophores, plant regulating hormones, such as auxins, cytokinins, and gibberellins, as well as macro- and micronutrient solubilization, which directly influences plant growth and promotions. Loper et al. (2008) used the sequence-guided targeted mutagenesis to functionally analyze a Pf-5 gene cluster (*rzx*). The first full-length genome of antagonistic *P. fluorescens* Pf-5 is composed of one circular chromosome of 7,074,893 bp with 6,144 open reading frames (ORFs), possessing 3,822 genes with known functions and secondary metabolite and siderophore gene clusters (Paulsen et al. 2005). Silby et al. (2009) reported that *P. fluorescens* SBW25 (6,722,539 bp) genome encodes 6,009 coding sequences (CDSs), whereas *P. fluorescens* Pf0-1 (6,438,405 bp) possesses 5,741 CDSs. Draft genome sequence of *P. fluorescens* WH6 using a hybrid next-generation sequencing approach revealed 6.27 megabases, with predicted 5,876 genes which showed synteny with orthologous genes of *P. fluorescens* SBW25 (Kimbrel et al. 2010). Similarly, whole genome sequence of *P. protegens* Cab57 against cucumber–*Pythium ultimum* pathosystem, using paired-end and whole-genome shotgun sequencing, revealed a size of 6,827,892 bp, GC content of 63.3%, and 6,186 predicted protein-coding sequences. Comparative genomics of *P. protegens* CHA0 and Pf-5 using the JSpecies program revealed four gene clusters (*phl, prn, plt*, and *hcn*), which were fully conserved in the Cab57 genome (Takeuchi et al. 2014). *P. fluorescens* UM270 genome consists of 6,047,974 bp, with a GC content of 62.66% and 5,509 genes, including communication genes, *viz. acdS* and *iaaMH*, involved in growth promotion and antagonistic activities (Santoyo et al. 2012; Glick, 2014). Besides, Type II to VI secretion systems were identified, which are important for survival, competence, and colonization in the rhizosphere and root systems (Annette et al. 2013). *P. fluorescens* strain PICF7 chromosome consists of 6,136,735 bp with 5,567 protein-coding genes, which act against the *Verticillium dahliae*, the incitant of olive wilt (Martínez-García et al. 2015). Genome mining of *P. fluorescens* BRZ63, an antagonist against soil-borne pathogens of oil seed rape, revealed a genome size of 6,335,040 bp and GC content of 64% with 6,120 genes related to the biocontrol and secondary metabolites (Chlebek et al. 2020). Whole genome sequences of the bacterial isolates were aligned using the LASTZ program, and unique genomic regions were identified using the IslandViewer program (Rodriguez-R and Konstantinidis 2014). antiSMASH program was used for the analysis of secondary metabolite production clusters (Takeuchi et al. 2014). Transposon mutagenesis is a spine of functional genomic research in many bacterial species. A total of 9,797 genome data with 631 complete sequences of *Pseudomonas* sp. are available at the Web site: http://www.pseudomonas.com.

5.2.3 *Bacillus* sp.

Bacillus sp. have widely been used as biofungicides due to their ability to synthesize antimicrobial lipopeptides, antagonize phytopathogens, produce endospores, and tolerate unfavorable environmental conditions (Olishevska et al. 2019). *Bacillus subtilis*, a gram-positive bacterium, comprises a genome of 4,214,810 base pairs with 4,100 protein-coding genes (Kunst et al. 1997). The 16S rRNA gene sequences of several *Bacillus* species (*B. velezensis* FZB42, *B. velezensis* FJAT-45028, *B. velezensis* CAU B946, *B. subtilis* H1, *B. subtilis* 168, *B. licheniformis* SRCM103583, *B. licheniformis* ATCC 14580, *B. altitudinis* CHB19, *B. altitudinis* GQYP101, *B. pumilus* SF-4, and *B. pumilus* ZB201701) were identified from the complete genome sequences (GenBank IDs: CP000560.2, CP047157.1, HE617159.1, CP026662.1, NC_000964.3, CP035404.1, CP034569.1, CP043559.1, CP040514.1, CP047089.1, and CP029464.1, respectively). *Bacillus cereus* UW85 was able to control damping off, caused by *Phytophthora megasperma* f. sp. *medicaginis* on alfalfa. The genome consists of 5,522,108 bp, with 23 contigs and an N50 contig size of 240,092 bp (Lozano et al. 2016). *Bacillus atrophaeus* GQJK17 was effective against pathogenic fungi. Its genome sequences consist of 4,325,818 bp with 4,181 coding DNA sequences with a GC content of 43.3%. Eight candidate gene clusters responsible for producing antimicrobial secondary metabolites were also identified (Ma et al. 2018). Olishevska et al. (2019) reported that *Bacillus velezensis* UFLA258 possesses a single chromosome of 3.95 Mb in length, with 3,949 genes and a GC content of 46.69%. The sequence of *B. velezensis* PG12 contains 3,990,845 bp, 3,884 CDSs, and an average GC content of 46.45% (Zeng et al. 2019). The genome sequence of *B. velezensis* AL7, exhibiting antagonistic activity to *Verticillium dahlia* in cotton, possesses one chromosome with 3,894,709 bp, with a GC content of 46.64% (Liu et al. 2020). PacBio sequencing of *B. velezensis* strain YB-130, showing a broad antifungal activity against various phytopathogens, revealed 12 gene clusters, which code for secondary metabolites with special reference to lanthipeptide (Xu et al. 2020). Leal et al. (2021) reported that the *Bacillus subtilis* PTA-271 genome possesses 4,001,755 bp, with 43.78% GC content and 3,945 protein-coding genes. Similarly, *B. velezensis* AK-0 genome sequence has a 3,969,429-bp circular chromosome possessing 3,808 genes (Kim et al. 2021) (Table 5.1).

5.3 Proteomics

Understanding the molecular mechanism of plant defense reactions against biotic stresses provides a vital foundation for the development of stress-resistant varieties. Pathogens and biocontrol agents induce an array of molecular responses that are vital for the health of the plant. This response can be studied using various omics platforms such as genomics, transcriptomics, proteomics, and metabolomics. Among these, proteomics is one of the most powerful tools for understanding the molecular mechanisms of disease and pest resistance. Proteomics is the study of the whole set of proteins expressed under a given condition in a particular cell. Unlike the genome, which is static, the proteome is highly dynamic. This dynamic nature of the proteome provides a scope to identify proteins that are synthesized in response to pathogenic infestations and are responsible for host immunity. Identification of such proteins could be backtraced to the respective genes and could be used for the development of disease-/pest-resistant varieties.

Significant developments have been achieved in proteomics technology in recent years. Gel-based methods have long been used to identify differentially expressed proteins. While it still retains its relevance, gel-free methods have also been developed, which have many added advantages. Irrespective of the methodology used, proteomics technology provides a scope for understanding the changes in plants that crop up during stress responses. It allows us to detect the protein changes that arise during the host–pathogen interaction as well as identify genes involved in defense signaling pathways. According to the previous reports, proteins associated with the synthesis of major plant metabolites and defense proteins are regulated during the host–pathogen–PGPR interactions. A study demonstrated that, in *Arabidopsis*, *Pseudomonas fluorescens* FPT9601-T5 either upregulated or downregulated the ethylene-responsive genes (Wang et al. 2005). In another experiment, *Bacillus cereus* NMSL88 induced proteins linked to growth and defense pathways in rice (Wang et al. 2013). *Bacillus amyloliquefaciens* strain BS5 either

TABLE 5.1

Recent Complete Genome Sequences of Potential Biocontrol Agents

S. No.	Biocontrol Agents	Accession Number	References
1.	*T. harzianum* T6776	JOKZ00000000	Baroncelli et al. (2015)
2.	*T. virens* FT-333	JTGJ00000000	Kuo et al. (2015)
3.	*T. gamsii* T6085	JPDN00000000	Baroncelli et al. (2016)
4.	*T. harzianum* B97	MRYK00000000	Compant et al. (2017)
5.	*Trichoderma* sp. ITEM908	PNRQ00000000	Fanelli et al. (2018)
6.	*P. fluorescens* Pf-5	CP000076	Paulsen et al. (2005)
7.	*P. fluorescens* SBW25	AM181176	Silby et al. (2009)
8.	*P. fluorescens* Pf0-1	CP000094	Silby et al. (2009)
9.	*P. fluorescens* WH6	AEAZ00000000	Kimbrel et al. (2010)
10	*P. fluorescens* UM270	JXNZ00000000	Santoyo et al. (2012)
11.	*P. protegens* Cab57	AP014522	Takeuchi et al. (2014)
12.	*P. fluorescens* BRZ63	PRJNA529642	Chlebek et al. (2020)
13.	*B. cereus* UW85	LYVD01000000	Lozano et al. (2016)
14.	*B. atrophaeus* GQJK17	CP022653	Ma et al. (2018)
15.	*B. velezensis* UFLA258	NZ_CP039297	Olishevska et al. (2019)
16.	*B. velezensis* PG12	PIWI00000000	Zeng et al. (2019)
17.	*B. velezensis* AL7	CP045926	Liu et al. (2020)
18.	*B. velezensis* YB-130	CP054562	Xu et al. (2020)
19.	*B. subtilis* PTA-271	JACERQ000000000	Leal et al. (2021)
20.	*B. velezensis* AK-0	CP047119	Kim et al. (2021)

upregulated or downregulated proteins, *viz.* ATP synthase, 2-cys peroxiredoxin, ribosomal protein (50s), serine/threonine protein kinase, and trehalose-6-phosphate phosphatase, during the challenge inoculation with *Bipolaris oryzae* (Prabhukarthikeyan et al. 2019).

5.3.1 Protein Identification

Proteomics studies identify the proteins that are differentially regulated in an organism under contrasting conditions (e.g., control vs. treated/infected). Identification of proteins with relative alterations in abundance in different samples in combination with pathway analysis tools provides an understanding of the mechanism involved in a biological phenomenon. Since the development of the proteomics technology, there has been constant progress in protein identification techniques. During the last decade of the 20th century and the first few years of the 21st century, the identification of differentially expressed proteins was mostly carried out by 2D gel electrophoresis in combination with MALDI-TOF-MS or LC-MS (O'Farrell 1975; Lilley et al. 2002). However, advancements in hyphenated technologies such as LC-MS have revolutionized the field of proteomics research. Thus, the techniques used for protein identification could be broadly categorized into gel-based and gel-free (mainly the chromatography–mass spectrometry) techniques.

5.3.1.1 Gel-Based Techniques

Gel-based proteomics includes one-dimensional (1D) gel electrophoresis and two-dimensional (2D) gel electrophoresis. In 1D gel electrophoresis, the mixture of proteins is separated by their molecular weight; all the proteins with the same molecular weight in a protein extract from a cell accumulate in the same space (forming a band) in a polyacrylamide gel. In 2D gel electrophoresis, initially in the first dimension, the proteins are resolved on the basis of their isoelectric points (pI) on a polyacrylamide gel containing ampholytes of different pH ranges. (These polyacrylamide gels are available as ready-made strips known as immobilized pH gradient or IPG strips.) Then, in the second dimension, the IPG strips are placed over

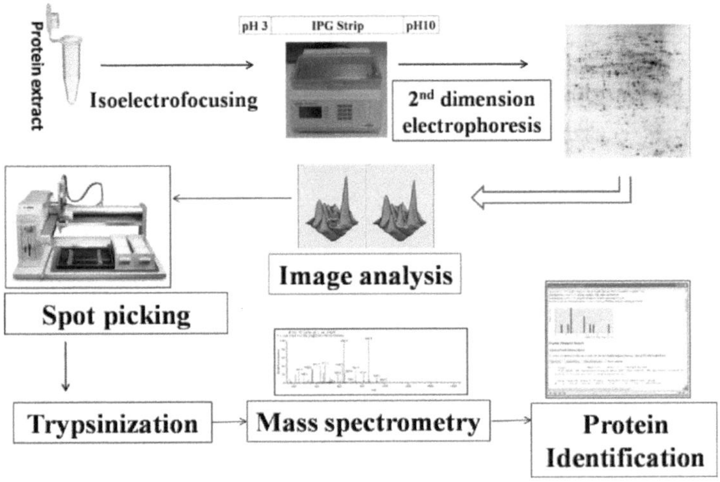

FIGURE 5.1 The workflow of 2D gel electrophoresis.

a SDS–PAGE gel and voltage is applied, to separate the proteins on the basis of their molecular weight. Unlike in 1DE, in 2DE, different proteins with the same molecular weight occupy different spaces, forming protein spots in a polyacrylamide gel. Each spot in a 2D gel represents a single protein subunit or a polypeptide. These protein spots are removed from the gel, are digested with a protease enzyme (trypsin/chymotrypsin), and then can be recognized with the help of mass spectrometric techniques (Figure 5.1).

5.3.1.1.1 Differential Gel Electrophoresis (DIGE)—An Advanced Variant of Gel Electrophoresis

Typically, Coomassie brilliant blue (CBB) or silver stain are used to stain the conventional 2D gels. However, these stains have a narrow dynamic range and linearity. Moreover, separate gels are resolved for different groups of samples (control/treated) and are compared with an image analysis process. Any small inadvertent variation in the experimentation process leads to differences in the gels of different groups, making it difficult for analysis. To overcome these issues, an advanced version of gel electrophoresis, where three samples can be separated simultaneously on a single gel, has been developed. Three different fluorescent dyes (Cy2, Cy3, and Cy5) are mixed with three samples prior to electrophoresis. Due to the differential fluorescent properties (excitation and emission spectra) of the dyes overlaid, multichannel images can be generated using a fluorescent scanner without compromising the mobility of proteins in different samples (Unlu et al. 1997; Alban et al. 2003). This technique provides researchers with a number of advantages, such as the exclusion of technical replicates, as variation is kept to a minimum between different gels and there is precise quantification of induced biological change among samples (https://2d-gel-analysis.com/starters-guides/dige-guide/).

Even though gel-based proteomics is a highly sensitive and useful technique, it has got its own set of limitations such as quantitative reproducibility and the ability to study certain classes of proteins. This has led to the development of a parallel gel-free approach. In gel-based proteomics, the change in abundance of proteins is deduced from the densitometric analysis of the gel images, and the mass spectrometry is used only for the identification of the peptides, whereas in gel-free proteomics, the mass spectrometry is used for both identification and quantification of the peptides.

5.3.1.2 Gel-Free Techniques

As mentioned earlier, in gel-free proteomics, both identification and quantification of proteins are done with a mass spectrometer. However, the complex mixture of proteins cannot be directly introduced into it, and thus, the proteins are fractionated with chromatographic or OFFGEL electrophoresis techniques.

Chromatographic techniques are more popular and have many advantages. The chromatographic techniques used for protein fractionation are described below (Abdallah et al. 2012).

5.3.1.2.1 Chromatographic Techniques for Protein Fractionation

5.3.1.2.1.1 Ion Exchange Chromatography In this technique, the column for fractionation is coated with ion exchangers, which attract proteins based on their charge. Although both cation exchangers (CX) and anion exchangers (AX) can be used for protein fractionation, the use of strong cation exchangers (SCX) is more common. The peptide mixture is loaded on the SCX columns, which under acidic conditions enable the column to bind positively charged peptides. The peptides are then eluted using different pH gradients.

5.3.1.2.1.2 Reversed-Phase Chromatography This is one of the most widely used chromatographic techniques in proteomics research. The analyte partition coefficient between the hydrophobic stationary phase and the polar mobile phase forms the basis for separation. The stationary phase coated over the inner surface of the column is made up of hydrophobic material. Thus, the hydrophobic peptides get trapped in the column, and thus, the hydrophilic peptides are eluted faster than the hydrophobic ones. The mobile phases generally used are water and acetonitrile. Ion pair reagents such as formic acid or trifluoroacetic acid (TFA) are also added (Manadas et al. 2010).

5.3.1.2.1.3 Two-Dimensional Liquid Chromatography (2D-LC) and Multidimensional Protein Identification Tool (MudPIT) In two-dimensional liquid chromatography, more than one separation technique, such as anion exchange, size exclusion, affinity, and reversed-phase chromatography, is used to get a better exposure to the proteome. In most cases, reversed-phase chromatography is used as the second dimension due to the compatibility of solvents used in this method with mass spectrometry. In MudPIT technology, the microcapillary column consists of both the SCX and RP phases (Fournier et al. 2007; Elschenbroich et al. 2009).

5.3.1.2.1.4 Mass Spectrometry for Identification and Quantification of Peptides After the fractionation of peptides, there are two approaches through which the identification and quantification of the peptides can be accomplished: the label-based approach and the label-free approach. The very concept of the approach is that if different sample mixtures are labeled with reagents of different molecular masses, then the labeled peptides in the samples can be distinguished from each other on the basis of their mass shift. A number of chemistries are available for the labeling of proteins, such as SILAC (stable isotope labeling by/with amino acids in cell culture), iTRAQ (isobaric tag for relative and absolute quantitation), ICAT (*isotope-coded affinity tag*), tandem mass tag (TMT), and metabolic labeling. Out of these, iTRAQ and SILAC are the most popular and widely used methods and are thus being discussed here.

5.3.1.2.1.5 Stable Isotope Labeling by/with Amino Acids in Cell Culture (SILAC) Here, two or more groups to be compared are grown in media containing different isotopes of elements such as C^{13} or N^{15}. These elements get incorporated into the proteins during cell growth and create a mass difference in proteins without making any changes to their other properties. This mass difference, which is observed in MS spectra as peak intensities, is used for deduction of the changes in protein abundance in the cells (Schneider and Hall, 2005). Even though this technique is very useful, its application is confined to systems where the growth of an organism/cell is being maintained in a controlled environment with media containing specified isotopes.

5.3.1.2.1.6 Isobaric Tag for Relative and Absolute Quantitation (iTRAQ) iTRAQ is one of the frequently used label-based, gel-free proteomics techniques. In this technique, the peptides are tagged with different iTRAQ reagents after extraction and digestion. The iTRAQ reagents contain different isotopes and are presently available in two modules such as 4-plex and 8-plex. The 4-plex iTRAQ module contains four reagents such as 114, 115, 116, and 117, which can be used for labeling four different samples, whereas the 8-plex module can be used for the analysis of eight samples in a single experiment (Ross et al. 2004). After labeling, the samples are pooled and fractionated by liquid chromatography. The fractionated peptides can be analyzed with MALDI-TOF/TOF-MS or

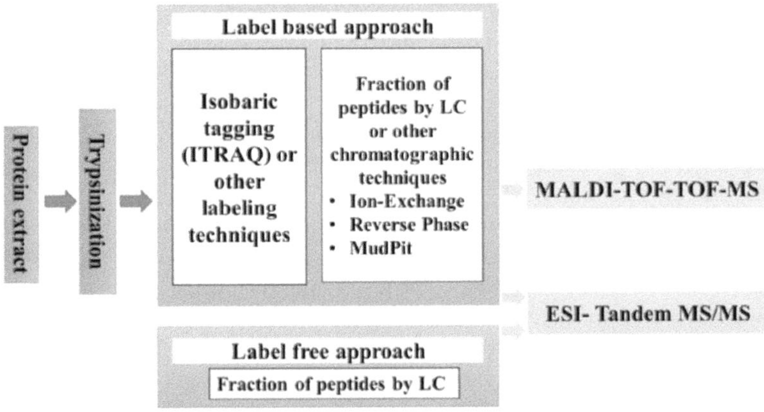

FIGURE 5.2 Workflow of gel-free proteomics.

with ESI-MS. If it is analyzed with an MALDI-TOF-TOF-MS, then the fractionated peptides are spotted on the MALDI plate, which is then transferred to the MALDI instrument in off-line mode. However, when analyzed through ESI-MS, the LC system is directly connected with the MS system and the transfer of peptides occurs in online mode. The peptides in different samples are labeled with different isobars, and labeled peptides do not illustrate any change in MS; rather, the signal from the same peptide from all samples gets clubbed up. However, different tags generate tag-specific reporter ion, and the signal intensities ratio from these tags is used to quantify peptides in a given sample (Figure 5.2).

5.3.1.3 *Label-Free Techniques*

The label-based techniques use labeling reagents that considerably increase the cost of analysis. Moreover, multiple sample preparation steps are also involved in this method. These factors encouraged the development of label-free techniques that are easier to perform, reproducible, and cost-effective. Here, the correlation among protein abundance or the number of MS/MS spectra is used for the quantification of peptides. There are two different label-free approaches followed for quantification of the peptides: one through spectral counting and the other through spectral peak intensity measurement. The spectral counting approach is based on the detail that peptides from more abundant proteins produce a higher number of peptides. The protein abundance in the sample is calculated based on the number of MS/MS spectra of the peptides (Chelius and Bondarenko 2002; Liu et al. 2004). Silva et al. (2005) reported that precursor ion signal intensity in consecutive LC/MS analysis is used to detect the same peptide in diverse samples through a spectral peak intensity approach.

It needs a high level of technical expertise to ensure reproducibility in gel-based proteomics, whereas the gel-free techniques are technically less demanding and thus more reproducible. The cost of analysis in label-free proteomics is way less than in both gel-based and label-based, gel-free techniques. The coverage of proteins in gel-based proteomics is far less than in gel-free techniques. The label-based approach can also be used in gel-free techniques to analyze only tagged peptides. However, in the label-free approach, all the peptide constituents in the sample can be quantitatively compared.

5.3.2 Fundamentals of Plant–Microbe (PM) Interactions

Interactions between plants and microbes can be harmful or beneficial depending on the nature of the interaction. In beneficial PM interactions, the host plants and the microbes develop mutual and beneficial associations that lead to enhanced resistance against biotic stresses (Reid, 2011). The best examples of beneficial PM interactions are nitrogen-fixing bacteria and arbuscular mycorrhizal fungi (AMF) for nitrogen fixation and phosphate absorption, respectively (Oldroyd et al. 2011; Smith and Smith, 2011).

The plant–microbe interaction is initiated by the process of quorum sensing and leads to microbial biofilm formation on plant tissues (Mathesius et al. 2003; Ramey et al. 2004). The formation of siderophore is another biologically important event that provides the plant growth-promoting bacteria to colonize the plant root (Haas and Defago, 2005; Crowley, 2006). Besides, beneficial microbes produce various phytohormones, including auxins and cytokinins, to promote plant growth in crop plants. PGPR strains protect the plant from pathogens by producing antibiotic substances or imparting induced systemic resistance.

In harmful PM interactions, many phytopathogens such as fungi, oomycetes, bacteria, viruses, and mycoplasmas are involved and cause severe diseases in crop plants. The phytopathogens invade through the natural opening of the plants and trigger the immune system. The outcome of plant–pathogen interactions is influenced by a variety of factors such as host susceptibility, pathogen virulence, and climatic factors (Brader et al., 2017). Both the plant and the pathogen are co-evolving with each other. Nowadays, new pathogens are evolving and causing outbreaks of new emerging diseases. The plant–pathogen interaction is always a complex process, and introducing a new pathogen has been a difficult task. An understanding of plant–pathogen interactions will help to develop better management strategies to solve the problems.

5.3.3 Records of Plant, Pathogen, and PGPR Interactions through Proteomic Approaches

Understanding the molecular responses of the plant to biotic stresses provides the foundation for the development of biotic stress-resistant plants. Genome-wide identification of genes during the interaction of pathogens and biocontrol agents is one of the important steps to delineate the mechanism of disease resistance. A number of strategies are employed for the recognition of genes associated with the defense response of a plant, and these are collectively known as "functional genomics" strategies. Functional genomics studies could employ the large mass of information generated through high-throughput genome and transcriptome sequencing projects and proteomics studies.

Proteomics has been utilized to examine the following interactions: plant–fungi, plant–bacteria, plant–virus, plant–PGPR, and recently, plant–pathogen–PGPR. Several studies have been reported on two-way interactions, i.e., the interaction between the plant and the pathogen (Colditz et al., 2005; Zhou et al. 2006). The *Xanthomonas oryzae–Oryza sativa* interaction through 2DE and MS analysis showed that differential proteins are involved in the plant–bacteria interaction (Mahmood et al., 2006).

A proteomic study was carried out to investigate the interaction between *Triticum aestivum* and *Fusarium graminearum* in which the proteins associated with the signaling pathways, PR proteins, and nitrogen metabolism were found to be downregulated (Zhou et al., 2006). Similarly, proteomic studies carried out to identify the differentially expressed proteins from rice leaves upon exposure to *M. oryzae* have shown that defense protein levels were high in the incompatible interaction compared to the compatible interaction, thus depicting the role of these proteins in resistance (Kim et al., 2004). A comparative proteome study explained that heat shock proteins and pathogenicity-related protein (PR10) are induced in *Medicago truncatula* upon inoculation with *Aphanomyces euteiches* (Colditz et al, 2004). Schenkluhn et al. (2010) reported that, during the plant–virus interaction, DIGE is utilized as a tool to recognize the differential proteins. DIGE and nano-ESI-LC-MS/MS were used by Di Carli et al. (2010) to characterize the differentially expressed protein in wild and transgenic resistant tomato plants. Manikandan et al. (2018) investigated the comparative proteomic analysis of virulent and avirulent strains of *F. oxysporum* f. sp. *lycopersici* and reported the role of proteins in the pathogenicity and disease development.

On the other hand, studying the three-way interaction between host, pathogen, and PGPR could provide an understanding of the PGPR-mediated defense response. Saveetha et al. (2009) demonstrated the protein expression in rice challenged with the sheath blight pathogen in response to *P. fluorescens* treatments using 2D gel electrophoresis and identified 23 differentially expressed proteins. Similarly, Senthil (2013) identified differentially expressed proteins during the interaction of *Chaetomium globosum* with *Pythium aphanidermatum* in chilli. Archana (2014) identified 22 differentially expressed proteins in mango upon challenge inoculation with *Colletotrichum gloeosporioides*. Through protein profiling, Prabhukarthikeyan et al. (2017) demonstrated the molecular mechanism behind the tripartite interaction between *P. fluorescens*, *P. aphanidermatum*, and turmeric plants. Similarly, several authors have documented the two-way and three-way interactions using proteomic approaches (Tables 5.2 and 5.3).

TABLE 5.2
Two-Way Interaction

S. No.	Plant–PGPR Interaction	Techniques	Results	Reference
1.	*Glomus mosseae* on *Pteris vittata*	2DE, LC-Q-TOF-MS/MS	Arsenic tolerance activity observed in AMF-treated plants.	Bona et al. (2001)
2.	*Piriformospora indica* on Barley	LC-MS/MS	Increased the plant growth and attenuated the NaCl-induced lipid peroxidation.	Baltruschat et al. (2008)
3.	*Pseudomonas fluorescens strain KH-1* on rice plants	2DE-MS analysis	Induced 23 important proteins in rice that are involved in plant growth promotion.	Saveetha et al. (2009)
4.	*Glomus irregulare* colonized on *Medicago truncatula*	2DE/MALDI/TOF	Alleviated the heavy metal stress tolerance, including cadmium (Cd).	Aloui et al. (2011)
5.	*Piriformospora indica* on Barley	2DE/MALDI	Forty-five differentially abundant proteins related to growth promotion and plant defense responses were noticed.	Ghabooli et al. (2013)
6.	*Medicago truncatula–Rhizophagus irregularis* (formerly *Glomus intraradices*)	GeLC-MS/MS	Symbiosis elicited changes in membrane trafficking- and nutrient uptake processes.	Abdallah et al. (2014)
7.	*Glomus mosseae* and *Amorpha fruticosa*	iTRAQ-LC-MS/MS	The expression of seven *A. fruticosa* mycorrhizal proteins and eleven different categories of plant proteins assigned to energy related, membrane transporter, stress and defense responsive, etc. were investigated	Song et al. (2015)
8.	*Paenibacillus polymyxa* E681 on *Arabidopsis thaliana*	2DE-MALDI-TOF/TOF	Strain E681 promoted plant growth and defense proteins.	Kwon et al. (2016)
9.	*Candidatus Glomeribacter gigasporarum–Gigaspora margarita*	2DE-iTRAQ-LC/-MS-MS	*Ca. G. gigasporarum* promoted the fungal oxidative phosphorylation and increased the respiratory activity.	Vannini et al. (2016)
10.	*Paenibacillus polymyxa* SQR-21 on watermelon	LC-MS	Strain SQR-21 treatment induced the proteins involved in photosynthesis growth and other physiological activities.	Yaoyao, et al. (2017)
11.	PGPR strains (*Stenotrophomonas maltophilia* and *Bacillus* sp.) on rice cultivar MR219-9	2DE-MS analysis	Induced the proteins responsible for plant growth promotion in aerobic rice cultivar MR219-9.	Naher et al. (2018)
12.	*Piriformospora indica* on *Brassica napus*	LC-MS/MS	46 (biological process), 23 (primary metabolic process), and 20 (cellular metabolic process) proteins attributed to enhancing growth, yield, and nutritional quality in *B. napus*.	Srivastava et al. (2018)
13.	*Bacillus velezensis* 5113 on wheat	2DE-MALDI-TOF/TOF analysis	Induced the metabolic and regulatory functions, which promoted both growth and abiotic stress tolerance in wheat plants.	Abd El-Daim et al. (2019)
14.	*Pseudomonas aeruginosa* (RP2) on groundnut	2DE/MALDI-TOF	Increased root morphogenesis, with significant change in metabolites.	Ankati et al. (2019)
15.	*Funneliformis mosseae* on soybean	iTRAQ-LC-MS/MS	Growth and disease resistance promotion through increased activity of oxidative phosphorylation, glycolysis, and amino acid metabolism.	Bai et al. (2019)

TABLE 5.3
Three-Way Interaction

S. No.	Plant–Pathogen–PGPR Interaction	Techniques	Results	Reference
1.	Maize–*Colletotrichum graminicola* and *Pythium ultimum–Trichoderma harzianum* strain T22	Total protein assay	β-1,3-Glucanase, exochitinase, and endochitinase increased in both roots and shoots.	Harman et al. (2004)
2.	Banana–banana bunchy top virus–endophytic bacteria	Native gel electrophoresis	More isoforms of PR proteins, *viz.* peroxidase, chitinase, and β-1,3-glucanase, in the banana plants challenged with mixtures of plant growth-promoting endophytic bacteria and BBTV viruliferous aphids were noticed.	Harish et al. (2009)
3.	*Pennisetum glaucum–Sclerospora graminicola–Pseudomonas fluorescens*	2DE-MS/MS	Sixty-three differentially abundant proteins associated with energy and metabolism, stress, and defense category were identified.	Anup C.P. et al. (2015)
4.	*Cucumis sativus–Fusarium oxysporum* f. sp. *cucumerinum (FOC)–Paenibacillus polymyxa NSY50*	2DE-MALDI-TOF/TOF	Enhanced the abundance of proteins involved in defense mechanisms.	Du et al. (2016)
5.	*Curcuma longa–Pythium aphanidermatum–Pseudomonas fluorescens*	2DE-MALDI-TOF-PMF	Identification of 12 differentially expressed proteins involved in disease resistance.	Prabhukarthikeyan et al. (2017)
6.	*Morinda citrifolia–Meloidogyne incognita–Bacillus subtilis*	2DE-MALDI-TOF-MS/MS	A total of 15 proteins were found to be differentially expressed, which are functionally important, defense-related proteins *against M. incognita*.	Kavitha et al. (2017)
7.	Rice–*Bipolaris oryzae–Bacillus amyloliquefaciens*	2DE-MALDI-TOF-MS/MS	Tripartite interaction expressed the proteins that are mainly associated with plant metabolism, defense response, and disease resistance.	Prabhukarthikeyan et al. (2019)
8.	Groundnut–*Sclerotium rolfsii–Pseudomonas aeruginosa (RP2)*	2DE/MALDI-TOF/MS-MS	Lower incidence of stem rot disease and increased seed survival rate due to infection.	Ankati et al. (2019)
9.	Rice–*Rhizoctonia solani–Bacillus subtilis* EPB24	2DE/MALDI-TOF/MS-MS	Upregulation of putative disease resistance protein RGA1, NBS-LRR proteins, serine–threonine protein kinase, chitinase, β-1,3-glucanase, ascorbate peroxidases, hydroxymethyl CoA ligase, PAL, and iron superoxide dismutase.	Durgadevi et al. (2021)

5.4 Secretomics in Plant–Pathogen Interaction

Molecules secreted by the pathogens are the major attributes that are involved in pathogenesis, leading to successful colonization inside the host (Gupta et al., 2015). Pathogen-derived molecules such as proteins, sugars, and lipopolysaccharides are recognized by plant-derived molecules, leading to a defense response in the host. The recognition of the pathogen-associated molecular patterns (PAMPs) with the pattern recognition receptors (PRRs) occurs in the apoplast, which in turn triggers the defense response in the host (Zipfel, 2014). Some of the PAMP molecules involved in PAMP-triggered immunity (PTI) include bacterial flagellin, flg22, elongation factor (EF)-Tu peptide elf18, and chitin, which are recognized by the plant PRRs, *viz.* flagellin-sensing 2 (FLS2), EF-Tu receptor (EFR), and chitin elicitor receptor kinase 1 (CERK1) (Liu et al., 2013). Similarly, pathogens convey various effector proteins, *viz.* avirulence (Avr) protein molecules, which are recognized by the resistance (R) proteins of plants to induce effector-triggered immunity (ETI) (Jones and Dangl, 2006). In this context, analyzing the changes of apoplastic proteins through a proteomics approach is highly essential, since the primary interface between plants and phytopathogens occurs in the apoplast. Many studies have commenced to understand the role of apoplast proteins in plant–pathogen interactions (Tanveer et al., 2014). Initially, secreted proteins from the apoplast were isolated by suspension-cultured cells (Agrawal et al., 2010). However, in the in planta system, we could get different proteins than in the *in vitro* system (Jung et al., 2008). The apoplastic protein extraction method is described hereunder.

5.4.1 Apoplastic Protein Extraction

The proteins can be extracted from the apoplast directly and analyzed using gel-based systems. Apoplastic proteins are normally extracted by vacuum-infiltration-centrifugation (VIC) and gravity extraction methods (Agrawal et al., 2010). In tobacco and *Arabidopsis*, the apoplastic proteins are extracted from the leaves by VIC method (De-la-Pena et al., 2008). In short, the leaves are cut into pieces and washed extensively to remove the cytoplasmic proteins, and the extraction buffer is infiltrated into the leaves through vacuum. Finally, low-speed centrifugation is used to extract the apoplastic proteins. However, a lower amount of apoplastic proteins is obtained by this method, and it is more complex to isolate from waxy coated leaves. To overcome the limitations of the VIC method, calcium based VIC was used (Floerl et al., 2008). Here, the leaves are reciprocally shaken in calcium buffer for 1h on ice. It is then followed by vacuum infiltration, centrifugation, and phenol precipitation. The purity of the proteins was assessed using cytoplasmic marker enzyme activity assays and western blot analysis.

5.4.2 Analysis of *in planta* Secreted Proteins

The apoplastic proteins expression has been recognized through various researches during plant–microbe interactions. Agrawal et al. (2010) reported that apoplastic proteins were exuded through the "Golgi endoplasmic reticulum pathway," but non-classical protein secretion pathways were also involved in the process. Floerl et al. (2008) reported that the interaction of *Verticillium longisporum* with *Arabidopsis* and *Brassica* plants led to the upregulation of PR proteins, which are involved in plant defense. Shenton et al. (2012) used a gel-based proteomics approach to identify proteins during the interaction of *M. oryzae* in rice. Similarly, Kim et al. (2013) identified 732 secreted proteins during the rice–*M. oryzae* interaction. Wang et al. (2013) identified 109 apoplastic proteins from rice, of which only 6 proteins were secreted from rice and the remaining proteins were secreted by the bacterial blight pathogen. Besides, Kim et al. (2014) also identified 501 apoplastic proteins during the rice–*Cochliobolus miyabeanus* interaction. Gupta et al. (2015) reported that mannitol dehydrogenase (MTD) and superoxide dismutase were also secreted in the apoplast during the plant–pathogen interaction.

5.5 Transcriptomics in Plant–Pathogen–Antagonist Interaction

Plants are constantly interacting with a wide variety of potential pathogens and beneficial microbes within their environment. During compatible interactions, the pathogen and/or biocontrol agents manipulate the immune system of the plant, leading to susceptibility, whereas in incompatible interactions, the plant prevents the invasion of the pathogen by inducing various defense mechanisms. Northern blots, RNase protection assays, qPCR, differential display RT-PCR, and serial/cap analysis of gene expression were used to evaluate the expression changes of a substantial number of gene transcripts. Of late, microarray and next-generation RNA sequencing have been used to analyze gene expression under specific conditions. Transcriptomics tools are mainly expression sequence tag (EST), microarray,

RNA-Seq, SAGE, etc., which are powerful to executing profiling, gene function annotation, and understanding ecology and evolution. It is also referred to as "integromics," whereby merging information from different omics tools results in crisp information. Transcriptome profiling plays a vital role in deciphering the mechanisms behind altered gene expression during plant–pathogen–biocontrol agent interactions. Transcriptome data provide numerous opportunities and trials to investigate the molecular mechanisms of plant immunity (Qi et al., 2018). Hundreds of positive transcriptome profiling in the areas of plant–phytopathogen and plant–beneficial microbe interactions have been performed to date, establishing high-throughput transcriptomics as a mature platform for unraveling the molecular mechanism of such interactions (Perazzolli et al., 2016). Transcriptomics is an assembly of several techniques, and each of these techniques has its own applications, utilities, inherent advantages, and limitations. Transcriptomics is widely applied to analyze gene expression and identify pathways in response to abiotic and biotic stresses in plants. Potential biocontrol agents have extensively been studied using whole genome sequencing, in combination with functional annotation using mutagenesis as a tool for the documentation of key characters behind biocontrol activity (Hennessy et al., 2017). However, very scanty studies have been undertaken to study biocontrol agent–pathogen interactions. Hence, the mechanism of resistance induced by biocontrol agents against phytopathogens can be unraveled through transcriptomic approaches, which can help in designing suitable management options.

The main aims of the transcriptomics are the following:

i. To determine the changes in the expression of transcripts, including mRNAs (coding RNA), small RNAs, and noncoding RNAs.
ii. To find out the transcriptional structure of the genes, i.e., their start site, 5′–3′ ends, differential splicing, and RNA editing.
iii. To quantify gene expression level during different experimental conditions.

5.5.1 Application of Transcriptomics

Depending upon the different methods to detect multiple gene transcriptional changes, transcriptomics can be categorized into three main types: polymerase chain reaction (PCR)-, hybridization-, and sequencing-based methods.

New methods, *viz.* ESTs and DDRT-PCR, provide a beneficial response of microbes to experimental plants and conditions (Moustafa et al. 2016). Differential-display reverse transcription PCR (DDRT-PCR) is another powerful fingerprinting technique that allows extensive analysis of gene expression among several cell populations and helps in identifying the differentially expressed cDNAs from two or more samples. Briefly, this method consists of five steps: (i) total RNA isolation from the experimental samples, (ii) RNA converted to single-stranded cDNA by reverse transcription, (iii) PCR amplification of cDNAs using the 3′ anchored primer and an arbitrary primer, (iv) denatured polyacrylamide gel electrophoresis (PAGE) separation of PCR products, and (v) cloning, sequencing, and characterization of differentially expressed bands. This technique has numerous advantages: rapidity, simplicity, sensitivity, the ability to identify differentially expressed genes (DEGs) in more than one sample, and the requirement of minute quantities of base material. It is one of the most primitive but valuable transcript profiling techniques because it does not require any prior knowledge of the genome, ESTs, or cDNA libraries. The progress of EST libraries linked with "differential gene expression" (DGE) tools (Green et al., 2001) offers a broad view of voluminous biological processes (Qi et al., 2018). The inability to identify the rare transcripts, generation of false-positive signals during band elution, and a probability of overlapping expression patterns due to contamination from the adjacent bands are a few of the disadvantages of this technique. Bozkurt et al. (2008) carried out a DDRT-PCR study to identify the resistance mechanism of wheat during wheat yellow rust disease. Out of 60 differentially expressed bands, 39 bands were randomly cloned and sequenced, which showed similarity to resistance-related and pathogenesis-related responses. These genes were further confirmed by quantitative real-time PCR (qRT-PCR). Seeds of tomato colonized by *Trichoderma harzianum* revealed hypothetical proteins such as 3-hydroxybutyryl-CoA dehydrogenase, 40S ribosomal protein S3a, fatty acid desaturase, phospholipase A2, secretion-related small

GTPase, and ubiquitin-activating enzyme through differential display (Mehrabi-Koushki et al., 2012). DDRT-PCR analysis on isolated antagonistic factors of *Bacillus endophyticus* revealed the upregulation of a gene with a well-known antimicrobial serine protease-like protein mostly produced by plants, animals, and insects (Moustafa et al., 2016). Comparative transcriptomic analyses based on ESTs and SAGE revealed different strategies of *Trichoderma* mycoparasitism (Atanasova et al., 2013). Later, such gene expression analyses moved to the use of microarrays (Herrera-Estrella, 2014). Nowadays, high-throughput next-generation sequencing (NGS) technologies have replaced microarrays to identify an array of differentially expressed antagonistic genes under diverse conditions. The availability of Next-seq technologies facilitated transcriptome sequencing with greater accuracy, less time, and big data generation. Forty-six million paired-end reads were attained during the plant–pathogen (*Armillaria mellea*)–biocontrol agent (*Trichoderma atroviride*) using the Illumina HiSeq 2000 at Fasteris, and 28,309 DEGs were identified using the DESeq2 package (Perazzolli et al., 2016). Subsequently, colonization of maize roots by *Trichoderma virens* induced expression of a wide range of genes (Malinich et al., 2019). Yuan et al. (2019) reported that 2,029 DEGs were noticed in *T. harzianum* Tr-92 by RNA sequencing. *Trichoderma*-treated tomato plants overexpressed transcripts coding for defense-related transcription factors (AP2-ERF, bZIP, MYB, NAC, and WRKY) (Coppola et al., 2019), elucidating the mechanisms of pest and disease resistance. Guo et al. (2020) identified 16,723 functional genes from *T. harzianum* ACCC30371 using transcriptomics analysis, whereas 402 biocontrol genes were identified. These upregulated genes illustrate an integrated biocontrol mechanism, among which mycoparasitism is the most dominant. Using another fungal biocontrol agent, the three-way transcriptomic analysis during interaction showed that *Helminthosporium solani* gene expression was highly reduced in chickpea when co-inoculated with *Clonostachys rosea* (Lysøe et al., 2017). The transcriptomic analysis of *C. rosea* in response to deoxynivalenol and *Fusarium graminearum* secretome revealed 24,112 unigenes with secondary metabolism-related genetic repertoire (Demissie et al., 2018). During the interaction of *Pseudomonas fluorescens* In5 with *Rhizoctonia solani*, genes involved in the synthesis of non-ribosomal synthetases and hydrogen cyanide were expressed, substantiating antagonism in dual culture (Hennessy et al., 2017). Transcriptome profiling of *Chaetomium globosum* strain Cg2 with *Bipolaris sorokiniana* BS112 from wheat using RNA-Seq revealed transcripts involved in catalytic activity, hydrolytic activity, and metabolic activity (Darshan et al., 2020).

5.6 Culturomics

The culturome is defined as the high-throughput cultivation, identification, and description of extensive microbial species or strains from human and plant microbiota or from the environment (Greub, 2012). The modern advent of comprehensive culturing methods has revealed that these prokaryotes can be cultured. The culturomics approach was introduced by the team of Didier Raoult and Jean-Christophe Lagier, as an alternative to metagenomics, which mainly relies on the existence of homologous sequences to identify new species. The abundant prokaryotes identified by culturomics were unnoticed in metagenomic approaches by pyrosequencing (Lagier et al., 2012). In order to culture the "uncultivable microbes," and since these unknown species are rather "uncultured yet" (Lagier et al., 2015), all microbes are culturable using the precise situation (Bilen et al., 2018). For plant microbiome culturomics, host plant-based culture media are used as a unique source. Culturome is the most effective approach for finding bacterial stocks in *Arabidopsis thaliana* (Bai et al., 2015) and rice (Zhang et al., 2019), but it is expensive and laborious. It involves various selective enrichment conditions combined with metabolomics and molecular approaches for the identification of bacterial colonies. Thus, culturomics complements metagenomics by overwhelming the bias in metagenomic methods. Obviously, uncultivable mycoparasites cannot be used for biocontrol programs. But culturomic strategies can be used to identify their biocontrol potential (Gdanetz and Trail, 2017). Thus far, the applied biocontrol elucidations have typically used a single strain. This culturomics approach may be thought-provoking by developing microbiome-based strategies based on consortia such as "synthetic communities" (Sergaki et al., 2018). The engineering of ecosystems using microbial consortia with mycoparasites can lead to the development of hopeful strategies for sustainable disease management programs.

5.7 Concluding Remarks

Omics tools enable the study and analysis of complex cellular mechanisms involved in the interaction of plants, microbes, and biocontrol agents. These approaches have increased our understanding of two-way and three-way interactions between plants, pathogens, and biocontrol agents, particularly in disease management. In various crop systems, focused research studies are conducted to elucidate the mechanism and effects of triadic interaction and to enhance the host defense response. Earlier research focused exclusively on a single omics tool for manipulating the antagonistic mechanism of biocontrol agents. To complete the mechanism mapping, integration is being performed to examine the potential determinants active during the process. Additionally, omics has aided in the discovery of previously unknown genes, proteins, enzymes, and compounds involved in the biocontrol process. Culturomics has enabled the study of all mechanisms concurrently and in a simplified fashion. Additionally, the evolution of novel strains can be studied. Thus, by utilizing integrated omics concepts, we have made significant advancements in our understanding of plant–microbe–biocontrol agent interactions, which may aid in developing strategies for sustainable agriculture.

Acknowledgment

The authors acknowledge the DST-SERB (No. CRG/2020/004589), New Delhi, India for granting fellowship to carry out the above research program

REFERENCES

Abd El-Daim, I. A., Bejai, S., and Meijer, J. (2019). *Bacillus velezensis* 5113 induced metabolic and molecular reprogramming during abiotic stress tolerance in wheat. *Science Reports*, 9, 16282. DOI: 10.1038/s41598-019-52567-x.

Abdallah, C., Dumas-Gaudot, E., Renaut, J., Sergeant, K. (2012). Gel-based and gel-free quantitative proteomics approaches at a glance. *International Journal of Plant Genomics*, 2012, 1–17.

Abdallah, C., Valot, B., Guillier, C., Mounier, A., Balliau, T., Zivy, M., van Tuinen, D., Renaut, J., Wipf, D., Dumas-Gaudot, E. and Recorbet, G. (2014). The membrane proteome of *Medicago truncatula* roots displays qualitative and quantitative changes in response to arbuscular mycorrhizal symbiosis. *Journal of Proteomics*, 108, 354–368.

Agrawal, G.K., Jwa, N.S., Lebrun, M.H., Job, D., and Rakwal, R. (2010). Plant secretome: unlocking secrets of the secreted proteins. *Proteomics* 10, 1–29. DOI: 10.1002/pmic.200900514.

Alban, A., David, S.O., Bjorkesten, L. et al. (2003) A novel experimental design for comparative two-dimensional gel analysis: two-dimensional difference gel electrophoresis incorporating a pooled internal standard. *Proteomics*, 3(1), 36–44.

Aloui, A., Recorbet, G., Robert, F., Schoefs, B., Bertrand, M., Henry, C., Gianinazzi-Pearson, V., Dumas-Gaudot, E. and Aschi-Smiti, S. (2011). Arbuscular mycorrhizal symbiosis elicits shoot proteome changes that are modified during cadmium stress alleviation in *Medicago truncatula. BMC Plant Biology*, 11(1), 1–17.

Ankati, S., Rani, T.S. and Podile, A.R. (2019). Changes in root exudates and root proteins in groundnut–*Pseudomonas* sp. interaction contribute to root colonization by bacteria and defense response of the host. *Journal of Plant Growth Regulation*, 38(2), 523–538.

Annette, A.A. and Ann, M.H. (2013). Biofilm formation in the rhizosphere. In: de Bruijn FJ (ed.), *Multispecies Interactions and Implications for Plant Growth. Molecular Microbial Ecology of the Rhizosphere*, Volume 2, First Edition, USA: John Wiley & Sons, Inc.

Anup, C. P., Melvin, P., Shilpa, N., Gandhi, M.N., Jadhav, M., Ali, H. and Kini, K.R. (2015). Proteomic analysis of elicitation of downy mildew disease resistance in pearl millet by seed priming with β-aminobutyric acid and *Pseudomonas fluorescens. Journal of Proteomics,* 120, 58–74.

Archana, S. (2014). Interactive proteomics and molecular characterization of PGPE mediated defense response in mango against *Colletotrichum gloeosporioides*, an incitant of anthracnose disease. *Ph.D. (Ag.) Thesis*, Tamil Nadu Agricultural University, Coimbatore, India. p. 155.

Atanasova, L., Le, C.S., Gruber, S., Coulpier, F., Seidl-Seiboth, V., Kubicek, C.P., Druzhinina, I.S. (2013) Comparative transcriptomics reveals different strategies of *Trichoderma* mycoparasitism. *BMC Genomics*, 14, 121–136.

Bai, L., Sun, H.B., Liang, R.T. and Cai, B.Y. (2019). ITRAQ proteomic analysis of continuously cropped soybean root inoculated with *funneliformis mosseae*. *Frontiers in Microbiology*, 10, 61.

Baltruschat, H., Fodor, J., Harrach, B.D., Niemczyk, E., Barna, B., Gullner, G., Janeczko, A., Kogel, K.H., Schäfer, P., Schwarczinger, I. and Zuccaro, A. (2008). Salt tolerance of barley induced by the root endophyte *Piriformospora indica* is associated with a strong increase in antioxidants. *New Phytologist*, 180(2), 501–510.

Baroncelli, R., Piaggeschi, G., Fiorini, L., Bertolini, E., Zapparata, A., Pè, M.E., Sarrocco, S., Vannacci, G. (2015). Draft whole-genome sequence of the biocontrol agent *Trichoderma harzianum* T6776. *Genome Announcements*, 3(3), e00647-15.

Baroncelli, R., Zapparata, A., Piaggeschi, G., Sarrocco, S., & Vannacci, G. (2016). Draft whole-genome sequence of *Trichoderma gamsii* T6085, a promising biocontrol agent of *Fusarium* head blight on wheat. *Genome announcements*, 4(1), e01747-15.

Benhamou, N., Gagné, S., Quéré, D.L., Dehbi, L. (2000). Bacterial mediated induced resistance in cucumber: beneficial effect of the endophytic bacterium *Serratia plymuthica* on the protection against infection by *Pythium ultimum*. *Biochemistry and Cell Biology*, 90, 45–56.

Bilen, M., Dufour, J.C., Lagier, J.C., Cadoret, F., Daoud, Z., Dubourg, G., Raoult, D. (2018). The contribution of culturomics to the repertoire of isolated human bacterial and archaeal species. *Microbiome*, 6(1), 94.

Bissett, J., Gams, W., Jaklitsch, W., Samuels, G.J. (2015). Accepted *Trichoderma* names in the year 2015. *IMA Fungus*, 6(2), 263–295.

Bona, E., Marsano, F., Massa, N., Cattaneo, C., Cesaro, P., Argese, E., di Toppi, L.S., Cavaletto, M. and Berta, G. (2011). Proteomic analysis as a tool for investigating arsenic stress in *Pteris vittata* roots colonized or not by arbuscular mycorrhizal symbiosis. *Journal of Proteomics*, 74(8), 1338–1350.

Bozkurt, O., Unver, T., and Akkaya, M. (2008). Genes associated with resistance to wheat yellow rust disease identified by differential display analysis. *Physiological and Molecular Plant Pathology*, 71, 251. DOI: 10.1016/j.pmpp.2008.03.002.

Brader, G., Compant, S., Vescio, K., Mitter, B., Trognitz, F., Ma, L. J., & Sessitsch, A. (2017). Ecology and genomic insights into plant-pathogenic and plant-nonpathogenic endophytes. *Annual Review of Phytopathology*, 55, 61–83.

Chaur, T. (1998). General mechanisms of action of microbial biocontrol agents. *Plant Pathology Bulletin*, 7, 155–166.

Chelius, D. and Bondarenko, P.V. (2002). Quantitative profiling of proteins in complex mixtures using liquid chromatography and mass spectrometry. *Journal of Proteome Research*, 1(4), 317–323.

Chelius, D. and Bondarenko, P.V. (2020). Genome Mining and Evaluation of the Biocontrol Potential of *Pseudomonas fluorescens* BRZ63, a New Endophyte of Oilseed Rape (*Brassica napus* L.) against Fungal Pathogens. *International Journal of Molecular Sciences*, 21(22), 8740. DOI: 10.3390/ijms21228740.

Chlebek, D., Pinski, A., Żur, J., Michalska, J., Hupert-Kocurek, K. (2004). Proteomic approach: identification of *Medicago truncatula* proteins induced in roots after infection with the pathogenic oomycete *Aphanomyces euteiches*. *Plant Molecular Biology*, 55, 109–120.

Colditz, F., Braun, H.P., Jacquet, H., Niehaus, K. and Krajinski, F. (2005). Proteomic profiling unravels insights into the molecular background underlying increased *Aphanomyces uteiches* tolerance of *Medicago truncatula*. *Plant Molecular Biology*, 59, 387–406.

Compant, S., Gerbore, J., Antonielli, L., Brutel, A., Schmoll, M. (2017). Draft genome sequence of the root-colonizing fungus *Trichoderma harzianum* B97. *Genome Announcements*, 5, e00137-17. DOI: 10.1128/genomeA.00137-17.

Coppola, M., Diretto, G., Digilio, M.C., Woo, S.L., Giuliano, G., Molisso, D., Pennacchio, F., Lorito, M., Rao, R. (2019). Transcriptome and metabolome reprogramming in tomato plants by *Trichoderma harzianum* strain T22 primes and enhances defense responses against aphids. *Frontiers in Physiology*, 10, 745.

Crowley, D.E. (2006). Microbial siderophores in the plant rhizospheric. In: Barton, LL and Abadia, J (eds.), *Iron Nutrition in Plants and Rhizospheric Microorganisms*, Dordrecht: Springer, pp. 169–198.

Darshan, K., Aggarwal, R., Bashyal, B.M., Singh, J., Shanmugam, V., Gurjar, M.S., Solanke, AU. (2020). Transcriptome profiling provides insights into potential antagonistic mechanisms involved in *Chaetomium globosum* against *Bipolaris sorokiniana*. *Frontiers in Microbiology*, 11, 578115. DOI: 10.3389/fmicb.2020.578115.

De-la-Pena, C., Lei, Z., Watson, B.S., Sumner, L.W., and Vivanco, J.M. (2008). Root-microbe communication through protein secretion. *Journal of Biological Chemistry*, 283, 25247–25255. DOI: 10.1074/jbc.m801967200.

Demissie, Z.A., Foote, S.J., Tan, Y., Loewen, M.C. (2018). Profiling of the Transcriptomic Responses of *Clonostachys rosea* upon treatment with *Fusarium graminearum* secretome. *Frontiers in Microbiology*, 9, 1061. DOI: 10.3389/fmicb.2018.01061.

Di Carli, M., Villani, M. E., Bianco, L., Lombardi, R., Perrotta, G., Benvenuto, E., Donini, M. (2010). Proteomic analysis of the plant-virus interaction in cucumber mosaic virus (CMV) resistant transgenic tomato. *Journal of Proteome Research*, 9, 5684–5697. DOI: 10.1021/pr100487x.

Druzhinina, I.S., Seidl-Seiboth, V., Herrera-Estrella, A., Horwitz, B.A., Kenerley, C.M., Monte, E., Mukherjee, P.K., Zeilinger, S., Grigoriev, I.V., Kubicek, C.P. (2011). *Trichoderma*: the genomics of opportunistic success. *Nature Reviews Microbiology*, 9(10), 749–759.

Du, N., Shi, L., Yuan, Y., Li, B., Shu, S., Sun, J. and Guo, S. (2016). Proteomic Analysis reveals the positive roles of the plant-growth-promoting rhizobacterium NSY50 in the response of cucumber roots to *Fusarium oxysporum* f. sp. *cucumerinum* inoculation. *Frontiers in Plant Science*, 7, 1859. DOI: 10.3389/fpls.2016.01859.

Durgadevi, D., Harish, S., Manikandan, R., Prabhukarthikeyan, S.R., Alice, D., and Raguchander, T. (2021). Proteomic profiling of defense/resistant genes induced during the tripartite interaction of *Oryza sativa*, *Rhizoctonia solani* AG1-1A, and *Bacillus subtilis* against rice sheath blight. *Physiological and Molecular Plant Pathology*, DOI: 10.1016/j.pmpp.2021.101669.

Eljounaidi, K., Kyu, S., Bae, H. (2016). Bacterial endophytes as potential biocontrol agents of vascular wilt diseases—review and future prospects. *Biological Control*, 103, 62–68.

Elschenbroich, S., Ignatchenko, V., Sharma, P., Schmitt-Ulms, G., Gramolini, A.O., and Kislinger, T. (2009). Peptide separations by on-line MudPIT compared to isoelectric focusing in an off-gel format: application to a membrane-enriched fraction from C2C12mouse skeletal muscle cells. *Journal of Proteome Research*, 8(10), 4860–4869.

Fanelli, F., Liuzzi, V.C., Logrieco, A.F., Altomare, C. (2018). Genomic characterization of *Trichoderma atrobrunneum* (*T. harzianum* species complex) ITEM 908: insight into the genetic endowment of a multitarget biocontrol strain. *BMC genomics*, 19(1), 1–8.

Floerl, S., Druebert, C., Majcherczyk, A., Karlovsky, P., Kües, U., and Polle, A. (2008). Defense reactions in the apoplastic proteome of oilseed rape (*Brassica napus* var. *napus)* attenuate *Verticillium longisporum* growth but not disease symptoms. *BMC Plant Biology*, 8, 129. DOI: 10.1186/1471-222-9-8-129.

Fournier, M.L., Gilmore, J.M., Martin-Brown, S.A., Washburn, M.P. (2007) Multidimensional separations-based shotgun proteomics. *Chemical Reviews*, 107(8), 3654–3686.

Gdanetz, K., and Trail, F. (2017). The wheat microbiome under four management strategies, and potential for endophytes in disease protection. *Phytobiomes Journal*, 1, 158–168. DOI: 10.1094/PBIOMES-05-17-0023-R.

Ghabooli, M., Khatabi, B., Ahmadi, F.S., Sepehri, M., Mirzaei, M., Amirkhani, A., Jorrín-Novo, J.V. and Salekdeh, G.H. (2013). Proteomics study reveals the molecular mechanisms underlying water stress tolerance induced by *Piriformospora indica* in barley. *Journal of Proteomics*, 94, 289–301.

Glick, B.R. (2014). Bacteria with ACC deaminase can promote plant growth and help to feed the world. *Microbiology Research*, 169(1), 30–39. DOI: 10.1016/j.micres.2013.09.009.

Green, C. D., Simons, J. F., Taillon, B. E., & Lewin, D. A. (2001). Open systems: panoramic views of gene expression. *Journal of immunological methods*, 250(1–2), 67–79.

Greub, G. (2012). Culturomics: a new approach to study the human microbiome. *Clinical Microbiology and Infection*, 18(12), 1157–1159.

Guo, R., Wang, Z., Zhou, C., Liu, Z., Zhang, P., Fan, H. (2020). Transcriptomic analysis reveals biocontrol mechanisms of *Trichoderma harzianum* ACCC30371 under eight culture conditions. *Journal of Forestry Research*, 31(5), 1863–1873.

Gupta, R., Lee, S.E., Agrawal, G.K., Rakwal, R., Park, S., Wang, Y. and Kim, S.T. (2015). Understanding the plant-pathogen interactions in the context of proteomics-generated apoplastic proteins inventory. *Frontiers in Plant Science*, 6, 352. DOI: 10.3389/fpls.2015.00352.

Haas, D. and Defago, G. (2005). Biological control of soil-borne pathogens by fluorescent pseudomonads. *Nature Reviews Microbiology*, 3(4), 307–319. DOI: 10.1038/nrmicro1129.

Harish, S., Kavino, M., Kumar, N., Balasubramanian, P., & Samiyappan, R. (2009). Induction of defense-related proteins by mixtures of plant growth promoting endophytic bacteria against Banana bunchy top virus. *Biological Control*, 51(1), 16–25.

Harman, G.E., Petzoldt, R., Comis, A., and Chen, J. (2004). Interactions between *Trichoderma harzianum* strain T22 and maize inbred line Mo17 and effects of these interactions on diseases caused by *Pythium ultimum* and *Colletotrichum graminicola*. *Phytopathology*, 94(2), 147–153.

Hennessy, R.C., Glaring, M.A., Olsson, S., Stougaard, P. (2017). Transcriptomic profiling of microbe–microbe interactions reveals the specific response of the biocontrol strain *P. fluorescens* In5 to the phytopathogen *Rhizoctonia solani*. *BMC Research Notes*, 10(1), 1–9.

Herrera-Estrella A. (2014). Genome-wide approaches toward understanding mycotrophic *Trichoderma* species. In: Gupta, VK, Schmoll, M, Herrera-Estrella, A, Upadhyay, RS, Druzhinina, I, and Tuohy, M, (eds.), *Biotechnology and Biology of Trichoderma*, Netherlands: Elsevier, pp. 455–464.

Jones, J.D. and Dangl, J.L. (2006). The plant immune system. *Nature*, 444, 323–329. DOI: 10.1038/nature05286.

Jung, Y.H., Jeong, S.H., Kim, S.H., Singh, R., Lee, J.E., Cho, Y.S., et al. (2008). Systematic secretome analyses of rice leaf and seed callus suspension cultured cells: workflow development and establishment of high-density two-dimensional gel reference maps. *Journal of Proteome Research*, 7, 5187–5210. DOI: 10.1021/pr8005149.

Kavitha, G., Nakkeeran, S. and Ahila, D. (2017). A proteomic approach to identify *Bacillus subtilis* induced defense related proteins in Noni challenged with *Meloidogyne incognita*. DOI: 10.21276/ijpaes.

Kim, S.T., Kim, S.G., Hwang, D.H., Kang, S.Y., Kim, H.J., Lee, B.H., Lee, J.J. and Kang, K.Y. (2004). Proteomic analysis of pathogen-responsive proteins from rice leaves induced by rice blast fungus, *Magnaporthe grisea*. *Proteomics*, 4: 3569–3578.

Kim, S.G., Wang, Y., Lee, K.H., Park, Z.Y., Park, J., Wu, J., et al. (2013). In- depth in-sight into *in vivo* apoplastic secretome of rice-*Magnaporthe oryzae* interaction. *Journal of Proteomics*, 78, 58–71. DOI: 10.1016/j.jprot.2012.10.029.

Kim, J.Y., Wu, J., Kwon, S.J., Oh, H., Lee, S.E., Kim, S.G., et al. (2014). Proteomics of rice and *Cochliobolus miyabeanus* fungal interaction: insight into proteins at intracellular and extracellular spaces. *Proteomics*, 14, 2307–2318. DOI: 10.1002/pmic.201400066.

Kim, Y.S., Lee, Y., Cheon, W., Park, J., Kwon, H.T., Balaraju, K., Kim, J., Yoon, Y.J., Jeon, Y. (2021). Characterization of *Bacillus velezensis* AK-0 as a biocontrol agent against apple bitter rot caused by *Colletotrichum gloeosporioides*. *Scientific Reports*, 11(1), 1–4.

Kimbrel, J.A., Givan, S.A., Halgren, A.B., et al. (2010). An improved, high-quality draft genome sequence of the Germination-Arrest Factor-producing *Pseudomonas fluorescens* WH6. *BMC Genomics*, 11, 522 https://doi.org/10.1186/1471-2164-11-522.

Kubicek, C.P., Herrera-Estrella, A., Seidl-Seiboth, V., Martinez, D.A., Druzhinina, I.S., Thon, M., Zeilinger, S., Casas-Flores, S., Horwitz, B.A., Mukherjee, P.K., Mukherjee, M. (2011). Comparative genome sequence analysis underscores mycoparasitism as the ancestral life style of *Trichoderma*. *Genome Biology*, 12(4), 1 5.

Kunst, F., Ogasawara, N., Moszer, I., Albertini, A.M., Alloni, G.O., Azevedo, V., Bertero, M.G., Bessières, P., Bolotin, A., Borchert, S., Borriss, R. (1997). The complete genome sequence of the gram-positive bacterium *Bacillus subtilis*. *Nature*, 390(6657), 249–256. DOI: 10.1038/36786.

Kuo, H.C., Wang, T.Y., Chen, P.P., Chen, R.S., Chen, T.Y. (2015). Genome sequence of *Trichoderma virens* FT-333 from tropical marine climate. *FEMS Microbiology Letters*, 362(7), fnv036. DOI: 10.1093/femsle/fnv036.

Kwak, Y. (2020). Complete mitochondrial genome of the fungal biocontrol agent *Trichoderma atroviride*: genomic features, comparative analysis and insight into the mitochondrial evolution in *Trichoderma*. *Frontiers in Microbiology*, 11, 785.

Kwon, Y. S., Lee, D. Y., Rakwal, R., Baek, S., Lee, J.H., Kwak, Y., Seo, J., Chung, W.S., Bae, D., and Kim, S.G. (2016). Proteomic analyses of the interaction between the plant-growth promoting rhizobacterium *Paenibacillus polymyxa* E681 and *Arabidopsis thaliana*. *Proteomics*, 16, 122–135. DOI: 10.1002/pmic.201500196.

Lagier, J.C., Armougom, F., Million, M., Hugon, P., Pagnier, I., Robert, C., Bittar, F., Fournous, G., Gimenez, G., Maraninchi, M., Trape, J.F., Koonin, E.V., La Scola, B., Raoult, D. (2012). Microbial culturomics: paradigm shift in the human gut microbiome study. *Clinical Microbiology and Infection*, 18(12), 1185–1193.

Lagier, J. C., Hugon, P., Khelaifia, S., Fournier, P. E., La Scola, B., & Raoult, D. (2015). The rebirth of culture in microbiology through the example of culturomics to study human gut microbiota. *Clinical microbiology reviews*, 28(1), 237–264.

Leal, C., Fontaine, F., Aziz, A., Egas, C., Clément, C., Trotel-Aziz, P. (2021). Genome sequence analysis of the beneficial *Bacillus subtilis* PTA-271 isolated from a *Vitis vinifera* (cv. Chardonnay) rhizospheric soil: assets for sustainable biocontrol. *Environmental Microbiome*, 16(1), 1–4.

Lilley, K.S., Razzaq, A., Dupree, P. (2002). Two-dimensional gel electrophoresis: recent advances in sample preparation, detection and quantitation. *Current Opinion in Chemical Biology*, 6(1), 46–50.

Liu, H., Sadygov, R.G., Yates, J.R. (2004). A model for random sampling and estimation of relative protein abundance in shotgun proteomics. *Analytical Chemistry*, 76(14), 4193–4201.

Liu, J., Ding, P., Sun, T., Nitta, Y., Dong, O., Huang, X., et al. (2013). Heterotrimeric G proteins serve as a converging point in plant defense signaling activated by multiple receptor-like kinases. *Plant Physiology*, 161, 2146–2158. DOI: 10.1104/pp.112.212431.

Liu, H., Zeng, Q., Wang, W., Zhang, R., Yao, J. (2020). Complete genome sequence of *Bacillus velezensis* strain AL7, a biocontrol agent for suppression of cotton *Verticillium wilt. Microbiology Resource Announcements*, 9, e01595-19. DOI: 10.1128/MRA.01595-19.

Loper, J.E., Henkels, M.D., Shaffer, B.T., Valeriote, F.A., Gross, H. (2008). Isolation and identification of rhizoxin analogs from *Pseudomonas fluorescens* Pf-5 by using a genomic mining strategy. *Applied and Environmental Microbiology*, 74, 3085–3093.

Lozano, G.L., Holt, J., Ravel, J., Rasko, D.A., Thomas, M.G., Handelsman, J. (2016). Draft genome sequence of biocontrol agent *Bacillus cereus* UW85. *Genome Announcements*, 4(5), e00910-16.

Lysøe, E., Dees, M.W., Brurberg, M.B. (2017). A Three-Way Transcriptomic Interaction Study of a Biocontrol Agent (*Clonostachys rosea*), a Fungal Pathogen (*Helminthosporium solani*), and a Potato Host (*Solanum tuberosum*). *Molecular Plant-Microbe Interactions*, 30(8), 646–655. DOI: 10.1094/MPMI-03-17-0062-R.

Ma, J., Wang, C., Wang, H., Liu, K., Zhang, T., Yao, L., Zhao, Z., Du, B., Ding, Y. (2018). Analysis of the complete genome sequence of *Bacillus atrophaeus* GQJK17 reveals its biocontrol characteristics as a plant growth-promoting rhizobacterium. *BioMed Research International*, 2018, 9473542.

Mahmood, T., Jan, A., Kakishima, M. and Komatsu, S. (2006). Proteomic analysis of bacterial-blight defense responsive proteins in rice leaf blades. *Proteomics* 6, 6053–6065.

Malinich, E.A., Wang, K., Mukherjee, P.K., Kolomiets, M., Kenerley, C.M. (2019). Differential expression analysis of *Trichoderma virens* RNA reveals a dynamic transcriptome during colonization of *Zea mays* roots. *BMC Genomics*, 20(1), 1–9.

Manadas, B., Mendes, V.M., English, J., Dunn, M.J. (2010). Peptide fractionation in proteomics approaches. *Expert Review of Proteomics*, 7(5), 655–663.

Manikandan, R., Harish, S., Karthikeyan, G. and Raguchander, T. (2018). Comparative proteomic analysis of different isolates of *Fusarium oxysporum* f. sp. *lycopersici* to exploit the differentially expressed proteins responsible for virulence on tomato plants. *Frontiers in Microbiology*, 9, 420. DOI: 10.3389/fmicb.2018.00420.

Mark, W.S., Winstanley, C., Godfrey, S.A.C., Levy, S.B., Jackson, R.W. (2011). *Pseudomonas* genomes: diverse and adaptable. *FEMS Microbiology Reviews*, 35(4), 652–680. DOI: 10.1111/j.1574-6976.2011.00269.x.

Martinez, D., Berka, R.M., Henrissat, B., Saloheimo, M., Arvas, M., Baker, S.E., Chapman, J., Chertkov, O., Coutinho, P.M., Cullen, D., Danchin, E.G. (2008). Genome sequencing and analysis of the biomass-degrading fungus *Trichoderma reesei* (syn. *Hypocrea jecorina*). *Nature Biotechnology*, 26(5), 553–560.

Martínez-García, P.M., Ruano-Rosa, D., Schilirò, E., Prieto, P., Ramos, C., Rodríguez-Palenzuela, P., Mercado-Blanco, J. (2015). Complete genome sequence of *Pseudomonas fluorescens* strain PICF7, an indigenous root endophyte from olive (*Olea europaea* L.) and effective biocontrol agent against *Verticillium dahliae*. *Standards in Genomic Sciences*, 10:10. DOI: 10.1186/1944-3277-10-10.

Mathesius, U., Mulders, S., Gao, M., Teplitski, M., Caetano-Anolles, G., Rolfe, B. G., et al. (2003). Extensive and specific responses of a eukaryote to bacterial quorum-sensing signals. *Proceedings of the National Academy of Sciences of the United States of America*, 100, 1444–1449.

Mehrabi-Koushki, M., Rouhani, H., Mahdikhani-Moghaddam, E. (2012). Differential display of abundantly expressed genes of *Trichoderma harzianum* during colonization of tomato-germinating seeds and roots. *Current Microbiology*, 65(5),524–533. DOI: 10.1007/s00284-012-0189-1.

Moustafa, M.F., Taha, T.H., Helal, M., Alrumman, S.A. (2016). Differential-display reverse transcription-PCR (DDRT-PCR): a new technology for molecular detection and studying one of the antagonistic factors of *Bacillus endophyticus* strain SA against *Staphylococcus aureus* (MRSA). *3 Biotech*, 6(2), 121. DOI: 10.1007/s13205-016-0439-1.

Mukherjee, P.K., Horwitz, B.A., Herrera-Estrella, A., Schmoll, M., Kenerley, C.M. (2013). *Trichoderma* research in the genome era. *Annual Review of Phytopathology*, 51, 105–129.

Naher, U.A., Panhwar, Q.A., Othman, R., Shamshuddin, J., Ismail, M.R. and Zhou, E. (2018). Proteomic study on growth promotion of PGPR inoculated aerobic rice (*Oryza sativa* L.) cultivar MR219-9. *Pakistan Journal of Botany*, 50(5), 1843–1852.

O'Farrell, P.H. (1975). High resolution two dimensional electrophoresis of proteins. *Journal of Biological Chemistry*, 250(10), 4007–4021.

Oldroyd, E.D.G., Murray, J.D., Poole, P.S., and Downie, J.A. (2011). The rules of engagement in the legume-rhizobial symbiosis. *Annual Review of Genetics*, 45, 119–144. DOI: 10.1146/annurev-genet-110410-132549.

Olishevska, S., Nickzad, A., Déziel, E. (2019). *Bacillus* and *Paenibacillus* secreted polyketides and peptides involved in controlling human and plant pathogens. *Applied Microbiology and Biotechnology*, 103(3), 1189–1215.

Ons, L., Bylemans, D., Thevissen, K., Cammue, B.P.A. (2020). Combining biocontrol agents with chemical fungicides for integrated plant fungal disease control. *Microorganisms*, 8(12), 1930. DOI: 10.3390/microorganisms8121930.

Paulsen, I.T., Press, C.M., Ravel, J., Kobayashi, D.Y., Myers, G.S., Mavrodi, D.V., DeBoy, R.T., Seshadri, R., Ren, Q., Madupu, R., Dodson, R.J. (2005). Complete genome sequence of the plant commensal *Pseudomonas fluorescens* Pf-5. *Nature Biotechnology*, 23(7), 873–878.

Perazzolli, M., Herrero, N., Sterck, L., et al. (2016). Transcriptomic responses of a simplified soil microcosm to a plant pathogen and its biocontrol agent reveal a complex reaction to harsh habitat. *BMC Genomics*, 17, 838. DOI: 10.1186/s12864-016-3174-4.

Pinski, A., Betekhtin, A., Hupert-Kocurek, K., Mur, L.A.J., Hasterok, R. (2019). Defining the genetic basis of plant–endophytic bacteria interactions. *International Journal of Molecular Sciences*, 20, 1947.

Prabhukarthikeyan, S.R., Manikandan, R., Durgadevi, D., Keerthana, U., Harish, S., Karthikeyan, G., Raguchander, T. (2017). Bio-suppression of turmeric rhizome rot disease and understanding the molecular basis of tripartite interaction among *Curcuma longa*, *Pythium aphanidermatum* and *Pseudomonas fluorescens*. *Biological Control*, 111, 23–31.

Prabhukarthikeyan, S.R., Yadav, M. K., Anandan, A., Aravindan, S., Keerthana, U., Raghu S., Baite, M.S., Parameswaran, C., Panneerselvam, P., Rath, P.C. (2019). Bio-protection of brown spot disease of rice and insight into the molecular basis of interaction between *Oryza sativa*, *Bipolaris oryzae* and *Bacillus amyloliquefaciens*. *Biological Control*, 137, 104018.

Qi, H., Jiang, Z., Zhang, K., Yang, S., He, F., Zhang, Z. (2018). PlaD: a transcriptomics database for plant defense responses to pathogens, providing new insights into plant immune system. *Genomics, Proteomics & Bioinformatics*, 16(4), 283–293.

Ramey, B.E., Koutsoudis, M., von Bodman, S.B., and Fuqua, C. (2004). Biofilm formation in plant–microbe associations. *Current Opinion in Microbiology*, 7, 602–609. DOI: 10.1016/j.mib.2004.10.014.

Reid, A. (2011). Microbes helping to improve crop productivity. *Microbe*, 6, 435–439.

Rodriguez-R, L.M. and Konstantinidis, K.T. (2014). Bypassing cultivation to identify bacterial species. *Microbe*, 9, 111–118.

Ross, P.L., Huang, Y.N., Marchese, J.N., et al. (2004). Multiplexed protein quantitation in Saccharomyces cerevisiae using amine-reactive isobaric tagging reagents. *Molecular and Cellular Proteomics*, 3(12), 1154–1169.

Santoyo, G., Orozco-Mosqueda Ma Del, C., Govindappa, M. (2012). Mechanisms of biocontrol and plant growth-promoting activity in soil bacterial species of Bacillus and Pseudomonas: a review. *Biocontrol Science and Technology*, 22, 855–872. DOI: 10.1016/j.mib.2004.10.014.

Saveetha, K., Karthiba, L., Raveendran, M., Saravanakumar, D., Suresh, S., Raguchander, T., Balasubramanian, P. and Samiyappan, R. (2009). Understanding the molecular basis of plant growth promotional effect of *Pseudomonas fluorescens* on rice through protein profiling. *Proteome Science*, 7, 47.

Schenkluhn, L., Hohnjec, N., Niehaus, K., Schmitz, U., Colditz, F. (2010). Differential gel electrophoresis (DIGE) to quantitatively monitor early symbiosis- and pathogenesis-induced changes of the *Medicago truncatula* root proteome. *Journal of Proteomics*, 73(4), 753–768.

Schneider, L.V. and Hall, M.P. (2005). Stable isotope methods for high-precision proteomics. *Drug Discovery Today*, 10(5), 353–363.

Senthil, R. (2013). Interactive proteomic approach to identify the differentially expressed proteins involved in the management of damping off and anthracnose diseases using liquid formulation of *Chaetomium globosum* in chilli. *Ph. D. Thesis,* Tamil Nadu Agricultural University, Coimbatore, India. p. 174.

Sergaki, C., Lagunas, B., Lidbury, I., Gifford, M. L., and Schäfer, P. (2018). Challenges and approaches in microbiome research: from fundamental to applied. *Frontiers in Plant Science*, 9, 1205. DOI: 10.3389/fpls.2018.01205.

Shenton, M.R., Berberich, T., Kamo, M., Yamashita, T., Taira, H., and Terauchi, R. (2012). Use of intercellular washing fluid to investigate the secreted proteome of the rice–*Magnaporthe* interaction. *Journal of Plant Research*, 125, 311–316. DOI: 10.1007/s10265-012-0473-y.

Shrivastava, N., Jiang, L., Li, P., Sharma, A.K., Luo, X., Wu, S., Pandey, R., Gao, Q. and Lou, B. (2018). Proteomic approach to understand the molecular physiology of symbiotic interaction between *Piriformospora indica* and *Brassica napus*. *Scientific Reports*, 8(1), 1–13.

Silby, M.W., Cerdeño-Tárraga, A.M., Vernikos, G.S., Giddens, S.R., Jackson, R.W., Preston, G.M., Zhang, X.X., Moon, C.D., Gehrig, S.M., Godfrey, S.A., Knight, C.G. (2009). Genomic and genetic analyses of diversity and plant interactions of *Pseudomonas fluorescens*. *Genome Biology*, 10(5), 1–6.

Silva, J.C., Denny, R., Dorschel, C.A., et al. (2005). Quantitative proteomic analysis by accurate mass retention time pairs. *Analytical Chemistry*, 77(7), 2187–2200.

Smith, S.E. and Smith, F.A. (2011). Roles of arbuscular mycorrhizas in plant nutrition and growth: new paradigms from cellular to ecosystem scales. *Annual Review of Plant Biology*, 62, 227–250. DOI: 10.1146/annurev-arplant-042110-103846.

Smith, K.P., Handelsman, J., Goodman, R.M. (1999). Genetic basis in plants for interactions with disease-suppressive bacteria. *Proceedings of the National Academy of Sciences*, 96(9), 4786–4790.

Song, F., Qi, D., Liu, X., Kong, X., Gao, Y., Zhou, Z. and Wu, Q. (2015). Proteomic analysis of symbiotic proteins of *Glomus mosseae* and *Amorpha fruticosa*. *Scientific Reports*, 5(1), 1–10.

Steiger, M.G., Vitikainen, M., Uskonen, P., Brunner, K., Adam, G., Pakula, T., Penttila, M., Saloheimo, M., Mach, R.L. and Mach-Aigner, A.R. (2011). Transformation system for *Hypocrea jecorina* (*Trichoderma reesei*) that favors homologous integration and employs reusable bidirectionally selectable markers. *Applied and Environmental Microbiology*, 77, 114–121.

Takeuchi, K., Noda, N., Someya, N. (2014). Complete genome sequence of the biocontrol strain *Pseudomonas protegens* Cab57 discovered in Japan reveals strain-specific diversity of this species. *PLoS ONE*, 9(4), e93683. DOI: 10.1371/journal.pone.0093683.

Tanveer, T., Shaheen, K., Parveen, S., Kazi, A.G., and Ahmad, P. (2014). Plant secretomics: identification, isolation, and biological significance under environmental stress. *Plant Signaling & Behavior*, 9(8), e29426. DOI: 10.4161/psb.29426.

Thomashow, L.S. and Weller, D.M. (1996). Current Concepts in the Use of Introduced Bacteria for Biological Disease Control: Mechanisms and Antifungal Metabolites. *InPlant-Microbe Interactions*, Boston, MA: Springer, pp. 187–235.

Unlu, M., Morgan, M.E., and Minden, J.S. (1997). Difference gel electrophoresis: a single gel method for detecting changes in protein extracts. *Electrophoresis*, 18(11), 2071–2077.

Vannini, C., Carpentieri, A., Salvioli, A., Novero, M., Marsoni, M., Testa, L., de Pinto, M.C., Amoresano, A., Ortolani, F., Bracale, M. and Bonfante, P. (2016). An interdomain network: the endobacterium of a mycorrhizal fungus promotes antioxidative responses in both fungal and plant hosts. *New Phytologist*, 211(1), 265–275.

Verma, S.K., Kingsley, K.L., Bergen, M.S., Kowalski, K.P., White, J.F. (2018). Fungal disease prevention in seedlings of rice (*Oryza sativa*) and other grasses by growth-promoting seed-associated endophytic bacteria from invasive *Phragmites australis*. *Microorganisms*, 6, 21.

Wang, Y., Ohara, Y., Nakayashiki, H., Tosa, Y., Mayama, S. (2005). Microarray analysis of the gene expression profile induced by the endophytic plant growth-promoting rhizobacteria, *Pseudomonas fluorescens* FPT9601-T5 in *Arabidopsis*. *Molecular Plant-Microbe Interactions*, 18, 385–396.

Wang, W., Chen, L. N., Wu, H., Zang, H. et al. (2013a). Secretome analysis of the rice bacterium *Xanthomonas oryzae* (*Xoo*) using *in vitro* and *in planta* systems. *Proteomics*, 13, 1901–1912. DOI: 10.1002/pmic.201200454.

Wang, Y., Kim, S.G., Wu, J., Huh, H.H., Lee, S.J., Rakwal, R., et al. (2013b). Comparative proteomic analysis of rice seedlings in response to inoculation with *Bacillus cereus*. *Letters in Applied Microbiology*, 56(208), 215.

Xu, W., Zhang, L., Goodwin, P.H., Xia, M., Zhang, J., Wang, Q., Liang, J., Sun, R., Wu, C., Yang, L. (2020). Isolation, identification, and complete genome assembly of an endophytic *Bacillus velezensis* YB-130, potential biocontrol agent against *Fusarium graminearum*. *Frontiers in Microbiology*, 2020, 11.

Yaoyao, E., Yuan, J., Yang, F., Wang, L., Ma, J., Li, J., Pu, X., Raza, W., Huang, Q. and She Q. (2017). PGPR strain *Paenibacillus polymyxa* SQR-21 potentially benefits watermelon growth by reshaping root protein expression. *AMB Express*, 7, 104. DOI: 10.1186/s13568-017-0403-4.

Yuan, M., Huang, Y., Jia, Z., Ge, W., Zhang, L., Zhao, Q., Song, S., Huang, Y. (2019). Whole RNA-sequencing and gene expression analysis of *Trichoderma harzianum* Tr-92 under chlamydospore-producing condition. *Genes Genomics*, 41(6), 689–699. DOI: 10.1007/s13258-019-00812-y.

Zeng, Q., Xie, J., Li, Y., Chen, X., Wang, Q. (2019). Draft genome sequence of an endophytic biocontrol bacterium, *Bacillus velezensis* PG12, isolated from apple fruit. *Microbiology Resource Announcements*, 8(41).

Zhang, J., Liu, Y. X., Zhang, N., Hu, B., Jin, T., Xu, H., Qin, Y., Yan, P., Zhang, X., Guo, X. and Hui, J. (2019). NRT1. 1B is associated with root microbiota composition and nitrogen use in field-grown rice. *Nature biotechnology*, 37(6), 676–684.

Zhou, W., Eudes, F. and Laroche, A. (2006). Identification of differentially regulated proteins in response to a compatible interaction between the pathogen *fusarium graminearum* and its host, *Triticum aestivum*. *Proteomics*, 6, 4599–609.

Zipfel, C. (2014). Plant pattern-recognition receptors. *Trends in Immunology*, 35, 345–351. DOI: 10.1016/j.it.2014.05.004.

Zou, G., Shi, S., Jiang, Y., van den Brink, J., de Vries, R. P., Chen, L., Zhang, J., Ma, L., Wang, C., & Zhou, Z. (2012). Construction of a cellulase hyper-expression system in Trichoderma reesei by promoter and enzyme engineering. *Microbial Cell Factories*, 11(1), 1–12.

6

Bioinformatics Approaches to Improve and Enhance the Understanding of Plant–Microbe Interaction: A Review

Surojit Sen*
Mariani College

Sunayana Rathi*
Assam Agricultural University

CONTENTS

6.1 Introduction

Increasing the crop yield and productivity is the primary goal of all agricultural activities. In the present scenario of global warming, optimization of plant production system for better yield in areas of limited fertility is targeted. The growth and productivity of plants depend much on the interaction with the microbes present in their immediate environment. Plants share their habitat with complex microbiota that include bacteria, oomycetes, fungi, archaea, and viruses (Agler et al., 2016). The complexity is determined by the shared environment and the biotic and abiotic interactions involved at different levels. The outcome of host–parasite interaction depending on the resources available may be positive, neutral,

* Corresponding author's email: surojit0368@yahoo.co.in & sunayana.rathi@aau.ac.in

DOI: 10.1201/9781003171416-6

or negative. Microbes can thus be considered as mutualistic, commensal, or pathogenic and require a well-balanced interaction for the sustainable productivity of plants. Hence, it is very important to understand the interactions involved and modes of control at the molecular level. A knowledge base developed in this direction will help treat and prevent infection and also reduce crop loss.

In host–pathogen relationship prediction, biochemistry-based approaches play a major role, which may be further supported by bioinformatics. Bioinformatics approach may play an important role by utilizing extremely large data sets generated in the post-genomic era. Further, the use of modern techniques such as machine learning and network analysis will provide a better insight into the interaction between host and pathogen and develop new strategies. Generally, two strategies are adopted for the management of host–pathogen interactions. The first one is to reduce or disable the virulence of the pathogen by targeting the machinery it uses. The other one is to target the host machinery so that the host immune system can be boosted to shield it from the pathogen attack. Therefore, it is highly significant to predict the key host–pathogen interactions in order to get the outcome as desired. Whichever may be the strategy, in both the cases, a thorough understanding of the interaction network is needed and bioinformatics may help a lot to understand the mechanism involved and decipher the knowledge of plant–microbe interactions.

Bioinformatics approaches to study host–pathogen interaction can be broadly divided into two categories: biological and computational. The biological category is inspired by the traditional biological knowledge of structure and homology, while computational methods are data-driven and need high-throughput computational tools such as network analysis and machine learning.

6.2 Genomics

With the improvement of sequencing techniques, modern genomics has produced large amounts of publicly available DNA sequence data and subsequently, a huge amount of data have been produced in other fields of omics too. The development of computational science and internet has helped biologist to submit and archive these data in retrievable databases/repositories. Host–pathogen interaction data are not an exception to that. These data can be easily retrieved from these open-source portals and analyzed to gather knowledge. The metadata developed may help in understanding the host–pathogen interaction more precisely and develop new strategies in this direction.

Novel techniques in genomic field have transformed the identification and detection strategy of host–pathogen interactions. These techniques can also provide new insights to understand their underlying dynamics. The availability of genomic data has helped biologists to study genomic signatures of host–pathogen interactions by searching for the association of single gene to genome-wide scans. The genomic sequences available can be used for phylogenetic and comparative analysis of host and pathogen. By thorough genome scans, mutations causing resistance in host can be easily detected and further comparative population genetic studies of the host can help in presuming the impact of pathogen. It is also seen that comparative sequence analysis of resistant and susceptible host can identify the differences in size orientation and location of the genes involved. Genes involved in known pathosystem can be targeted to study in unknown systems of our interest. In this regard, whole-genome comparisons can help and this has become feasible only due to the gradual reduction in the costs of high-throughput sequencing recently. Since a host's response during infection by a particular pathogen most often involves multiple genes, whole-genome approaches have high potentiality of unfolding polygenic responses (Daub et al., 2013).

In order to understand the underlying genetics of the interactions between hosts and pathogens, genetic variation can be studied at different levels such as within species, across species, within population, or across population. Genotype–phenotype association studies can also be used to understand the genetic architecture more precisely. Hence, genome-wide association study (GWAS) is successfully used in unrevealing the host's responses to pathogen exposure. Although the whole-genome approach has been established as a benchmark for many host–pathogen studies, there are many limitations such as non-availability of reference genome of many non-model organisms and poor annotation of reference genome. This may lead to low rate of discovery of important regions of host genome responding to pathogen.

Combining selection scans with association studies can reveal the differences in infectious disease susceptibilities and identify specific protective genes and alleles. Once the resistant genes/QTLs are identified, they can be introgressed and pyramided by marker-assisted selection or through genetic engineering.

TABLE 6.1

Bioinformatics Databases/Repositories of Host–Pathogen Interactions

Name	URL	Description
VFDB	http://www.mgc.ac.cn/VFs/main.htm	Virulence factor database of bacterial pathogens.
PATRIC	http://www.patricbrc.org/	Provides integrated data and analysis tools for bacterial infectious diseases.
ViPR	https://www.viprbrc.org/brc/home.spg?decorator=vipr	Virus pathogen database and analysis resource.
Expasy	https://www.expasy.org/	Swiss bioinformatics resource portal.
ViralZone	https://viralzone.expasy.org/	Resource for viral data.
V-pipe	https://www.expasy.org/resources/v-pipe	Bioinformatics pipeline for assessing viral genetic diversity.
HPIDB 3.0	https://hpidb.igbb.msstate.edu/	Host–pathogen interaction database.
GPS-Prot	http://gpsprot.org/	Data Visualization for Protein-Protein Interactions.
PHI-base	http://www.phi-base.org/	Host–pathogen interaction database.
DIP	https://dip.doe-mbi.ucla.edu/dip/Main.cgi	Database of Interacting Proteins.
BioGRID	https://thebiogrid.org/	Database of Protein, Genetic, and Chemical Interactions.
IntAct	https://www.ebi.ac.uk/intact/	Molecular Interaction Database.
PID	http://www.ndexbio.org/#/user/301a91c6-a37b-11e4-bda0-000c29202374	Pathway interaction database.
PHIDIAS	http://www.phidias.us/	Pathogen-Host Interaction Data Integration and Analysis.
PHISTO	https://www.phisto.org/	Pathogen-Host Interaction Search Tool.
HoPaCI-db	http://mips.helmholtz-muenchen.de/HoPaCI	Host–Pathogen Interaction database.
mentha	http://mentha.uniroma2.it/index.php	Interactome database.
MINT	https://mint.bio.uniroma2.it/	Molecular Interaction database.
SIGNOR 2.0	https://signor.uniroma2.it/	Signaling network database
MatrixDB	http://matrixdb.univ-lyon1.fr/	Extracellular matrix proteins, proteoglycans, and polysaccharides interaction database.
IMEx	http://www.imexconsortium.org/	Molecular interaction data.

There are many toolkits and repositories of scripted pipelines available for genomic data analysis, such as https://github.com/pditommaso/awesome-pipeline, core R packages (https://cran.r-project.org/web/packages/GenomicTools/index.html), R Bioconductor, R Markdown (https://rmarkdown.rstudio.com), or Jupyter (https://jupyter.org); for graphical user interface-guided data integration analysis, "Galaxy" can be used. Databases such as NCBI, EMBL, DDBJ, and Stanford genomic resource (http://genome-www4.stanford.edu/) also provide many tools for visualization and analysis of genomic data (Table 6.1).

6.3 Gene expression Data

6.3.1 Expressed Sequence Tags

ESTs are obtained from cDNA libraries by partial random sequencing. They are single-read mRNA sequences of approximately 300–500 nucleotides in length. ESTs represent expressed genes of organs or tissues at a specific developmental stage. An enormous number of ESTs have been produced from thousands of species in the past few years and are available freely in databases such as dbEST of NCBI, DDBJ, and EMBL. In case of non-model organisms, where whole-genome sequencing data are not available, EST data sets are utilized as an alternative for providing valuable resources to develop gene-associated markers such as SSR and SNP.

TABLE 6.2
Widely Used Databases and Tools for Gene Expression Data

Databases/Tools	Description
GEO (Gene Expression Omnibus) (https://www.ncbi.nlm.nih.gov/geo/)	Database for gene expression profiling and RNA methylation profiling derived from microarray and RNA-Seq experiments.
ArrayExpress (http://www.ebi.ac.uk/arrayexpress/)	Repository of functional genomics data.
AMAD software package	Provides basic microarray data storage and retrieval capabilities.
MGOS database	Contains data obtained from *O. sativa* and *M. grisea.*
OryzaExpress (http://bioinf.mind.meiji.ac.jp/OryzaExpress/)	Gene expression database for rice.
NASCArrays (http://arabidopsis.info/affy)	Repository of microarray data of *Arabidopsis thaliana* with data mining tools.
PathoPlant (http://www.pathoplant.de)	Microarray expression data of co-regulated genes involved in plant defense responses.
PLEXdb (http://www.plexdb.org)	Plant and plant–pathogen microarrays.
OmicsDB::Pathogens (http://pathogens.omicsdb.org)	A database for exploring functional networks of plant pathogens.
PlaD (http://systbio.cau.edu.cn/plad/index.php or http://zzdlab.com/plad/index.php)	Transcriptomics database for plant defense responses to pathogens.
PHI-base (www4.rothamsted.bbsrc.ac.uk/phibase/)	Database for pathogen–host interactions.

6.3.2 Microarrays

Microarray is a laboratory technique used to detect the expression profile of thousands of genes at the same instance. Microarray tool can be used to analyze RNA expression profile of both pathogens and hosts by the help of microarray chips to ensure gene expression and identify regulatory mechanisms involved in the pathogenic state. It can assist in hypothesizing functions of uncharacterized resistant genes of host and also in identifying virulence genes that promote colonization or those that cause damage to host tissue. It is also used to identify the genetic polymorphism of specific loci associated with a particular trait. Hence, by this technique, genes involved in pathogenicity can be identified in the study of host–pathogen interactions. This can be achieved by measuring and comparing the gene expression of host cells before and after infection. The gene expression pattern analysis can provide insight into the gene regulatory network for host during all stages of infection. Many microarray studies have been performed in the past decades, leading to accumulation of enormous amount of expression data. The need to store and analyze these data has led to the creation of many new expression databases. Some of these databases/tools of gene expression data are listed in Table 6.2.

6.3.3 RNA-Seq

In contrast to microarrays, genes with low abundance, sequence variation, and even novel transcripts can be easily identified by RNA-Seq. Moreover, since the expression analysis for non-model organisms can be performed by RNA-Seq, the expensive step of producing species-specific arrays can be avoided. Because of these advantages, recently, RNA-Seq technology has become popular for studying genome-wide expression profile. Entire RNA molecules are sequenced to measure the expression levels of all transcripts in order to harness knowledge of known as well as novel unidentified defense genes of host and effector genes of pathogen. Using the RNA-Seq method, the total transcriptional activity of both the host and pathogen can be studied before and after infection. Data can be analyzed for identifying the differentially expressed genes during infection. Plant–pathogen mixed RNA-Seq databases are available, which can be accessed and analyzed using bioinformatics tools.

More recently, high-throughput RNA sequencing has been developed, which paved the way for capturing all classes of coding and noncoding transcripts in both the pathogen and the host. This technique, called dual RNA-Seq technique, not only allows understanding the physiological changes in pathogen and host during infection, but also reveals hidden molecular phenotypes of virulence associated with small noncoding RNAs that were not visible in standard assays. The assay pipeline involves the following steps: RNA extraction → rRNA depletion → deep sequencing → parallel read mapping with host and pathogen genome → cross-mapping → aligned reads → normalization.

The normalized reads are then subjected to downstream analyses such as quantification, differential expression, pathway analysis, and network inference.

Differential expression analysis is generally done using popular tools such as edgeR, DESeq2, and limma/voom, available through Bioconductor of R statistical programming language. Various algorithms are also available for this purpose of downstream analysis, among which pipelines such as Tuxedo suite are standard.

The list of genes (both pathogen and host) produced as a result of differential expression analysis can further be interpreted in terms of gene function to hypothesize new tests. Databases such as Gene Ontology (GO) and the Kyoto Encyclopedia of Genes and Genomes (KEGG) provide software suites for this purpose. Further, the metabolic network can be reconstructed by specialized knowledge bases such as BioCyc. Tools are also available for the reconstruction of molecular signature and gene set enrichment analysis from RNA-Seq data. The link between the identified genes can be inferred by network analysis called network inference (NI), and global regulatory networks can be constructed from the expression data.

6.3.4 MicroRNAs

MicroRNAs (miRNAs) are small noncoding RNAs that are endogenously found in organisms for the regulation of gene expression. They are derived from single-stranded RNA precursors that can form stem-loop structures. Generally, they are 18–24 nucleotides long and depending on the extent of base-pairing with the target mRNAs, miRNAs can silence gene expression. It is found that host miRNAs target pathogen virulence genes, while pathogen's miRNAs target plant resistance genes. Thus, miRNA can mediate trans-kingdom gene regulation and can play a great role in host–pathogen interactions.

Experimentally, miRNAs are discovered by cloning (Long and Chen, 2009), microarray screening (Barad et al., 2004), *in situ* hybridization (Yao et al., 2012), or next-generation sequencing of small RNAs (Landgraf et al., 2007), while computationally, miRNAs and their targets are identified by *in silico* genomic or EST sequence analysis. miRBase (http://www.mirbase.org/) is a searchable database for published miRNA sequences and annotation. Several computational prediction downloadable programs are available, such as miRPlant (http://sourceforge.net/projects/mirplant/), miRNA EMBL (http://www.russelllab.org/miRNAs/), MIREAP (https://sourceforge.net/projects/mireap), miRA (https://github.com/mhuttner/miRA), C-mii (http://www.biotec.or.th/isl/c-mii), and Web servers such as microHARVESTER, miRU, DIANA Tools, miRanda, and EIMMo are also used. For target prediction, tools such as psRNATarget (http://plantgrn.noble.org/psRNATarget/) (Dai and Zhao, 2011), TAPIR (Bonnet et al., 2010) (http://bioinformatics.psb.ugent.be/webtools/tapir/), TargetScan (http://www.targetscan.org/vert_ 72/), and miRTour Web server are frequently used. miRDB is an online database for miRNA target prediction and functional annotation. Apart from these, various R packages are available at https://bioconductor.org for miRNA prediction.

The predicted miRNA can further be validated using quantitative reverse transcriptase PCR (qRT-PCR).

6.4 Protein–Protein Interaction (PPI) Prediction

Eukaryotic cells have thousands of gene products in their proteome, which undergoes complex interactions throughout life, forming functional pathways to provide signals from outside the cell and a proper cellular response to the signals. Proteins are the workhorses of host–pathogen interaction network too.

Protein–protein interaction is the most prominent way how a pathogen interacts with its host. Proteins are a sequence of amino acids bonded by peptide to form a string called primary structure. The sequence of amino acids in the primary structure determines the structure as well as the function of the protein. Local folding of the primary structure caused by interaction between the side chains of amino acids results in the secondary structure (alpha-helix and beta-sheet). Alpha-helices are responsible for structure and membrane spanning domains, while beta-sheets provide the docking site for enzymatic reactions. In tertiary structure, further folding of beta-sheets and alpha-helices occurs to form a complex three-dimensional entity. This structural entity is anchored by ionic interactions, disulfide bridges, hydrophobic interactions, and van der Waals forces. Even after folding, a number of post-translational modifications such as cleavage, phosphorylation, glycosylation, or ubiquitination take place. After all these complex modifications, protein attains a final shape and is ready for interactions. Hence, the determination of three-dimensional structure of proteins is of utmost importance in the study of protein–protein interaction between pathogen and host. But unfortunately, the determination of 3D structure of a protein is difficult and time-consuming. Traditionally, X-ray crystallography/NMR is used for 3D structure determination, but unfortunately, many proteins get distorted during crystallization. Till date, only a small fraction of 3D structures of host/pathogen proteome has been determined.

The more the primary sequence similarity, the more the chance of an interaction among the proteins (interologs). Protocols have been developed to map known sequences of interaction interface onto pairs of sequences (homologous or orthologous) in different organisms. At least 80% sequence similarity is required for this purpose, and hence, correct determination of PPI decreases as the evolutionary distance increases. Such homologous proteins with primary sequence similarity are searched for in the pathogen/or host, which may also interact with known annotated proteins (interologs). Interolog mapping has a high false-positive hit rate; hence, in order to improve the quality of the mapping results, further filtering based on cellular localization, biological functions, and temporal expression profile is required to significantly identify potential host–pathogen protein–protein interactions.

6.4.1 Homology-Based Prediction

Similar sequences usually have similar functions; homology-based prediction methods work based on this assumption. This has been found true in case of a large degree of similarity or evolutionary conservation of proteins under investigation, which are quite abundant. It is an anticipation that protein–protein interaction would be conserved across related species. In this prediction method, the genomic data of hosts and pathogens are analyzed to identify proteins homologous with the known interacting protein. These are then compared for the determination of likelihood of any protein–protein interactions occurring between the host and the pathogen. In this approach, interaction templates of host and pathogen genomic sequences are considered to find out the probable sets of PPIs. In order to filter out non-homologous sets, a homology detection algorithm is applied to these PPIs. Then, further filtration is done to the newly obtained sets through stage-specific and tissue-specific expression data of pathogen and host. Filtering is also done with the help of predicted localized data. Homology-based approaches are widely used for the prediction of host–pathogen PPI as this approach is considered to be simple and having a well-supported biological background. Simple data such as template PPIs and protein sequences are required for the purpose of prediction, and hence, they can be adapted and applied to other multiple host–pathogen systems.

A major drawback of homology-based prediction is the detection of high rate of false positives (Mariano and Wuchty, 2017). Further, protein pairs predicted by these *in silico* approaches may have differential temporal and spatial expressions and hence may rarely have the chance to interact. Successful homology-based prediction approaches, therefore, require filters that account for these criterions. The use of random forest classifiers can help in this regard (Figure 6.1).

6.4.2 Structure-Based Prediction

It is generally believed that when a pair of protein structures is similar to known interacting pair of proteins, it is more likely that they will be interacting in a similar pattern. The structural information of proteins can be used extensively for the prediction of host–pathogen interactions computationally by comparing with already known interactions with other proteins. In this approach, the host and pathogen genomes are first scanned for structural similarity with already known protein complexes to find out probable interactions using the structural similarity. The result is finally filtered by expression data of tissue-specific host proteins and stage-specific pathogen proteins. The set of proteins hence identified have a high interaction probability.

There are number of tools available for predicting PPI based on 3D structures of the interacting proteins, such as docking and MD (molecular dynamics) simulation. But unfortunately, till date only a small fraction of 3D structures of host/pathogen proteome has been determined. Hence, there is little focus on the prediction of host–pathogen PPIs through this technique. However, bioinformatics tools also provide an alternative to solve the 3D structures of the proteins whose X-ray structures are not available.

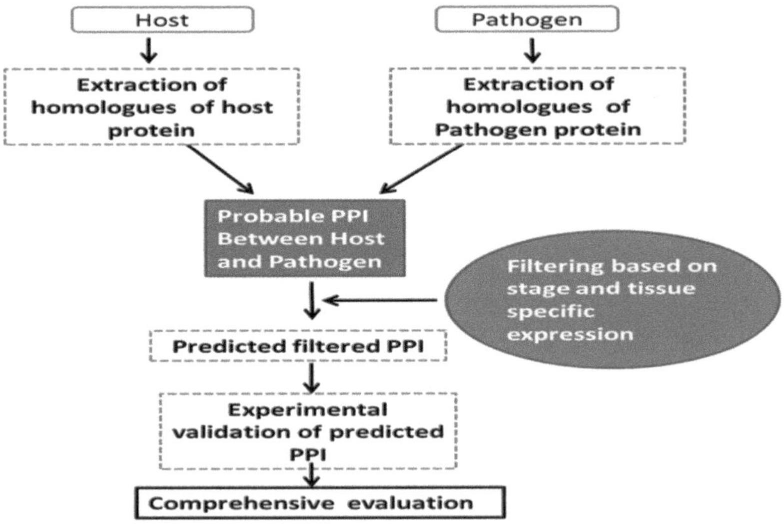

FIGURE 6.1 Predictions of host–pathogen interactions using homology-based approach.

Protein structure prediction is usually done by homology modeling. This method attempts to create a new structure based on a set of known structures with sequence similarity. In addition, molecular dynamics approach uses Newtonian mechanics to simulate the atom-scale interactions. There are reports that successfully used these approaches to provide insights into host–pathogen interactions or plants' defense mechanisms (Sarma et al., 2012; Dehury et al., 2013; Dehury et al., 2015).

For interspecies PPI prediction, the 3D structural homology is identified by scanning host and pathogen genomes. Proteins having similarity to known protein complexes are assessed for the detection of putative interactions using structural information. The left out interactions are filtered based on the biological context for several pathogens. This strategy was first adopted by Davis et al. (2007).

Although prediction based on structure is a powerful tool, it seems that pathogens, in order to achieve binding stability, evolve their protein interfaces without sequence or structural similarity to native interacting proteins. Pathogen protein interfaces sometimes overlap with and even compete with or mimic the endogenous host protein interfaces (Figure 6.2).

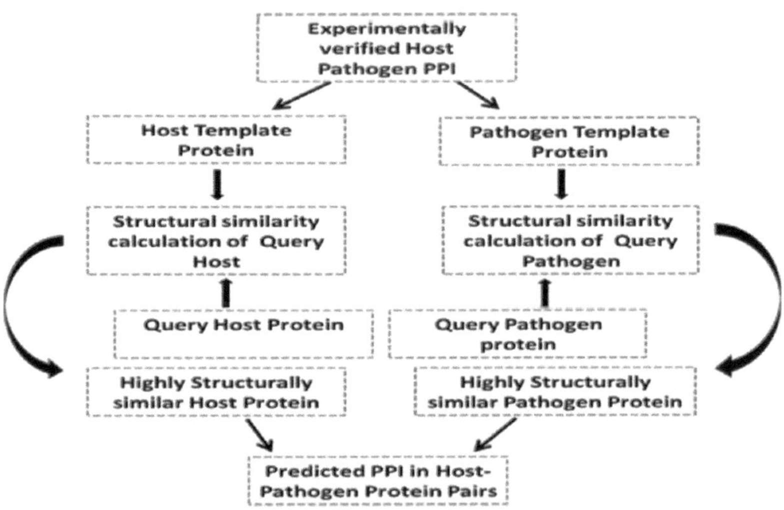

FIGURE 6.2 Predictions of host–pathogen interactions using structure-based approach.

6.4.3 Domain-Based Approaches

A protein domain is the conserved region of a protein's three-dimensional structure that is responsible for a specific biological function. Domains are created while the proteins get folded in nature, and they evolve independently (Hleap and Blouin, 2016). Domains are the regions of contact during PPIs, and hence, domain–domain interactions (DDIs) drive PPIs considerably. A number of studies have been carried out based on known intra-species DDIs for the prediction of HP-PPIs. Protein–domain association studies are highly predictive when machine learning algorithms such as support vector machine (SVM) and random forest (RF) are used (Barman et al., 2014). Based on the information of primary sequence of interacting domains, the large-scale detection of hypothetical interactions between proteins has been possible. In a study of human protein interaction network, Dyer and his associates combined DDIs with protein sequence k-mers and topological properties of protein using support vector machine algorithm to predict host–pathogen interactions (Dyer et al., 2011). Domain-based prediction method is helpful in the identification of common functionality features, which allow pathogens to interact with multiple hosts.

6.4.4 Motif- and Integration-Based Approaches

Motifs are small recognizable regions of protein having unique biological functions. Proteins interact through a reusable set of binding motifs with their partners. Motifs complement each other among the partners despite slight difference between individual proteins. Motifs always show a conserved pattern, and hence during analysis, once these patterns are recognized and validated, the remaining part of data can be discarded for reducing the computational cost. Biologists have exploited this in various studies involving protein–protein interaction networks and identification of transcription factor binding sites (Das and Dai, 2007), prediction of secondary structure (Bi et al., 2008) as well as enzymatic activity studies and identification of functional residues (Hulo et al., 2008). In all of these studies, a known data set is analyzed to detect statistical over-representation of patterns. These patterns are then applied to new proteins to draw inferences.

Motif identification can be carried out by various tools available in the open-source portals. Tools such as miniMotif and PSI-BLAST look for an over-represented set of amino acids for the identification of these sites. These *de novo* motif algorithms compare a set of proteins with a known function with a set of proteins without the desired function. However, these *de novo* algorithms are not fit for instances where the function is already well studied. In such cases, databases such as PROSITE, eukaryotic linear motif (ELM), and PFAM can be used to curate a large collection of linear sequence motifs. These functional motifs can be used to predict protein-protein interactions of pathogen and host.

6.4.5 Motif–Domain and Motif–Motif Interaction-Based Approaches

These approaches can also be used as foundations for host–pathogen PPI prediction and have gained importance recently. Motifs of one protein sometimes interact with domains or even motifs of another protein. This has been studied in HIV–human interactions by Evans and his coworkers (2009). They generated an HIV-1 and human interactome with the help of annotated ELMs in HIV-1 proteins that interacted with counterpart human protein domains. Integration of primary and secondary sequence information enhances *in silico* host–pathogen PPI predictions. However, other auxiliary data can also be used to reduce the impact of false positives. Currently, assimilation of features such as domain information, sequence features, ELM data, GO features, graph topological properties, and gene co-expression data are used to train the classifiers. This strategy was successfully used by Coelho and his coworkers to predict the human oral microbial interactome by incorporating domain information, protein sequence features, and GO annotations (Coelho et al., 2014).

6.4.6 Surface Electrostatics and Epitope Prediction

Epitopes are the antigenic determinants of pathogens and can be recognized by the host immune system. Interacting protein surfaces show electrostatic as well as non-covalent interactions. Antibodies, which bind to the epitopes, also show a number of electrostatic and non-covalent interactions. These interactions take place either through backbone carbon or through side chain carbon. This allows a number of host proteins to recognize pathogen antigens by shared physicochemical properties. Several computational protein interaction identification tools have already integrated these electrostatic attributes, thereby enabling epitope prediction on the basis of surface energetics.

6.4.7 Analysis of Dynamic Character of PPI

Three-dimensional structures obtained from X-ray crystallographic method are cumbersome, time-consuming, and expensive, and also many proteins' structure gets distorted during crystallization. So, recent studies have employed NMR to complement the static crystallographic data with dynamic functional data. This allows the proteins to be studied in their natural state, the way they actually fold in nature. Solution-state NMR can determine the interaction between pathogen protein and the domains of host protein by emphasizing how surface charge distribution, intrinsic disorder, and mimicry of host protein help in specific binding. NMR united with molecular dynamics simulations can enhance the prediction process if a preexisting structure is available. A blend of cryo-electron microscopy, MDs, and solid-state NMR can further help in building a model on how interaction takes place in natural environment.

6.5 Machine Learning-Based Predictions

Prediction methods based on machine learning are widely used in host–pathogen interactions. Figure 6.3 illustrates some of the machine learning methods that are being widely used to study the host–pathogen interactions.

Supervised learning has been used for the successful prediction of PPIs in the host–pathogen domain by considering more than 35 features of host and pathogen. The features considered can be sequence similarity, gene expression profiles, similarity in post-translational modifications, GO, tissue distribution, and various other features of host and pathogen interactome. After the initial analysis, top three or top six features of utmost importance are selected so that the data can be classified into interacting and non-interacting classes. In most of the cases, supervised learnings exploit RF classifiers for these kinds of classifications.

Naive Bayes algorithm is used for the classification of training samples based on similarity. The similarity here is measured with the help of Smith–Waterman local alignment algorithm. Input features such as amino acid composition, amino acid frequencies, and amino acid properties are used, and finally,

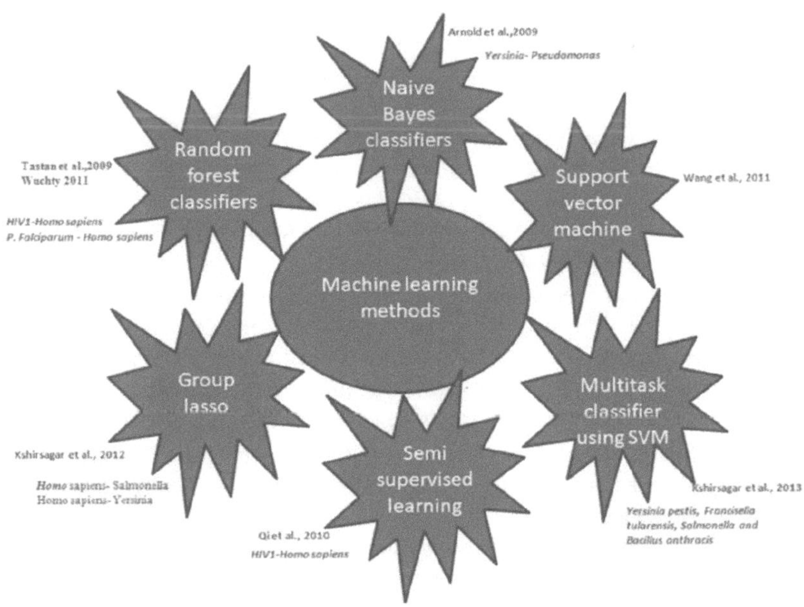

FIGURE 6.3 Various machine learning methods.

most important features are strategized using feature selection strategy. In some cases, features derived from the secondary structure are also used. PSIPRED software is used for structure prediction, and from the predicted structure, features are selected for input vector.

Machine learning approaches such as SVM algorithms are trained by carefully picking positive and negative training sets of protein interactions. All approaches of prediction by supervised machine learning need appropriate training for both positive and negative sets with sequence or higher-order information to robustly classify interacting proteins between host and pathogen. Inclusion of non-interacting set in the training data influences the accuracy of identification of interacting pairs from the non-interacting ones. In order to create a negative training set, highly dissimilar protein sequences from other organisms are selected. Dissimilarity here is compared with the interacting proteins of the pathogen in question. SVM with these training sets increases the prediction accuracy.

Multi-task classification frameworks can be used for establishing relationship between host and multiple pathogens. Based on the similarity of infection initiated by different pathogens, this machine learning technique is used for the classification of PPI into interacting and non-interacting classes. The classification is based upon the hypothesis that similar pathogen targets the same critical biological process of the host.

In semi-supervised multi-task method of prediction, the data set of host is processed through both supervised and semi-supervised learning. The supervised classifier works on labeled PPI data and trains the semi-supervised classifier with partially labeled PPIs. The supervised classifier shares network layers with the semi-supervised classifier. This entire framework is used to improve the prediction of interacting pairs.

Group lasso technique, on the other hand, can be used to improve the supervised learning-based prediction. In this technique, the missing data set values are replaced by the values generated from cross-species information. This has been successful in increasing the prediction accuracy by more than 70%.

6.6 Systems Biology Approach

Systems biology is a holistic approach to understanding the complex biological systems using mathematical modeling and analysis of high-throughput data. It focuses on single- or multi-level computational analysis and modeling of experimental data resulting from new hypotheses. It can be approached in two ways: bottom-up and top-down. In the bottom-up approach, sub-models are built and later integrated to find out the integration of cell components, which is followed by building of *in silico* models comprising all pathways of cell–pathogen or host–pathogen interaction.

In the top-down approach, a genome-wide analysis is performed with the help of data obtained from omics technologies (such as genomics, transcriptomics, proteomics, and metabolomics).

Identification of key molecules and their interaction is carried out in the following three steps:

1. **Identification of Key Molecules (Biomarkers)**: First, biomarkers such as DNA/RNA sequences, proteins, mutations (SNPs), transcripts of coding region (microarray or RNA-Seq data of differentially expressed genes), noncoding transcripts (miRNAs and piRNAs), or metabolites are identified. Sometimes machine learning approach is used for the prioritization of key biomarkers.
2. **Network Modeling of Regulatory Interactions:** The next step is the systems-level understanding of the molecular mechanisms of all the involved biological processes by means of mathematical modeling. The network is generally represented by nodes which denote the molecules such as proteins, DNA, RNA, or metabolites and edges representing interactions between the nodes. Based on the prior knowledge of omics data, interaction networks such as gene regulatory networks or even genome-wide networks can be generated and inferred.
3. **Identification of Disease Modules:** Then the group of molecules and interactions among them, which are linked to a phenotype of interest, is identified. The next step is the integrated analysis of interaction networks for the discovery of disease-associated modules. This integrated study can reveal disease modules with partially overlapping molecular mechanism. Proteins and their degree of overlap correlate biological similarity or disease symptoms. This can be used successfully for discovering the affected mechanism.

TABLE 6.3

List of Tools Related to Systems Biology

Name	Description
Cytoscape	Data integration, network visualization, and analysis.
MEGA	Phylogenetic analysis and creation of dendrograms.
GenMAPP	Visualization and analysis of genomic data in the context of pathways
BioTapestry	Interactive tool for building, visualizing, and simulating genetic regulatory networks.
PathVisio	Tool for displaying and editing of biological pathways.
PathView	Pathway-based data integration and visualization.
InCroMap	Integration of omics data and experimental data and their joint visualization in pathways.
CellDesigner	Diagram editor for gene regulatory networks.
Complex Pathway Simulator (COPASI)	Simulation and analysis of biochemical networks.
SBMLToolbox	Analysis of SBML model in MATLAB.

Systems biology and computational modeling can be employed to metabolic engineering to anticipate the effect of genetic engineering on metabolism. Recently, constraint-based modeling (examining the function of metabolic networks by relying on physicochemical constraints) has gained popularity and has been proven successful for large-scale microbial networks (Price et al., 2003). Once the gene network or metabolic network are identified or disease-associated modules are modeled, gene editing techniques can be used to control plant–pathogen interactions to obtain customized plants with enhanced yield. Highly efficient gene editing tools such as zinc finger nucleases (ZFNs), transcription activator-like effector nucleases (TALENs), and clustered regularly interspaced short palindromic repeats (CRISPRs) can help in achieving such a goal (Table 6.3).

6.7 Conclusions

Bioinformatics has played and will continue to play a significant role in exploiting the data available for exploring the host–pathogen interaction and enhancing our knowledge in this field. All the processes of prediction mentioned here has their own advantages and disadvantages. Hence, the use of these tools and selecting the right one, needs deeper exploration. Further, the non-availability of reference genome of non-model organism, proper annotation, and curation are the major challenges. Since the results and inference drawn depend much on the quality of the input data, these challenges are to be addressed properly. Although *in silico* techniques help to speed up the prediction process, time-to-time validation is required to conclusively decide on the causes and consequences of host–pathogen interactions and to combat them.

REFERENCES

Agler et al. (2016). Microbial hub taxa link host and abiotic factors to plant microbiome variation. *PLoS Biol.* **14**(1): e1002352.

Arnold et al. (2009). Sequence based prediction of type III secreted proteins. *PLoS Pathogen.* **5**(4):e1000376.

Barad et al. (2004). MicroRNA expression detected by oligonucleotide microarrays: system establishment and expression profiling in human tissues. *Genome Res.* **14**: 2486–2494.

Barman et al. (2014). Prediction of interactions between viral and host proteins using supervised machine learning methods. *PLoS One.* **9**: e112034.

Bi et al. (2008). A comparative study on computational two-block motif detection: algorithms and applications. *Mol. Pharm.* **5**(1): 3–16.

Bonnet et al. (2010). TAPIR, a web server for the prediction of plant microRNA targets, including target mimics. *Bioinformatics.* **26**: 1566–1568.

Carvalho Leite et al. (2017). Computational prediction of host-pathogen interactions through omics data analysis and machine learning. In: Rojas, I., Ortuño, F. (Eds.), *Bioinformatics and Biomedica Engineering.* Springer International Publishing, 5th International Work-Conference, IWBBIO 2017, Granada, April 26–28, 2017, Proceedings, Part II.

Coelho et al. (2014). Computational prediction of the human– microbial oral interactome. *BMC Syst. Biol.* **8**: 24.

Dai X and Zhao PX. (2011). psRNATarget: a plant small RNA target analysis server. *Nucl Acids Res.* **39**: W155–159.

Das M K and Dai H K. (2007). *A* survey of DNA motif finding algorithms. *BMC Bioinformatics.* **8** (Suppl 7): S21.

Daub et al. (2013). Evidence for polygenic adaptation to pathogens in the human genome. *Mol Biol Evol.* **30**: 1544–1558.

Davis et al. (2007). Host-pathogen protein interactions predicted by comparative modelling. *Protein Sci.* **16**: 2585–2596.

Dehury et al. (2013). Structural analysis and molecular dynamics simulations of novel δ-endotoxin Cry1Id from Bacillus thuringiensis to pave the way for development of novel fusion proteins against insect pests of crops. *J Mol Model.* **19**: 5301–5316.

Dehury et al. (2015). Molecular recognition of avirulence protein (avrxa5) by eukaryotic transcription factor xa5 of rice (Oryza sativa L.): insights from molecular dynamics simulations. *J Mol Graph Model.* **57**: 49–61.

Dix et al. (2016). Use of systems biology to decipher host-pathogen interaction networks and predict biomarkers. *Clin Microbiol Infect.* **22**(7): 600–606.

Dyer et al. (2011). Supervised learning and prediction of physical interactions between human and HIV proteins. *Infect Genet Evol.* **11**: 917–923.

Esvelt KM and Wang HH. (2013). Genome-scale engineering for systems and synthetic biology. *Mol Syst Biol.* **9**: 641.

Evans et al. (2009). Prediction of HIV-1 virus-host protein interactions using virus and host sequence motifs. *BMC Med Genom.* **2**: 27.

Hleap J and Blouin C. (2016). The semantics of the modular architecture of protein structures. *Curr Prot Pept Sci.* **17**: 62–71.

Hulo N et al. (2008). The 20 years of PROSITE. *Nucl Acids Res.* **36**: D245.

Kshirsagar et al. (2012). Techniques to cope with missing data in host–pathogen protein interaction prediction. *Bioinformatics.* **28**(18): i466–i472.

Kshirsagar et al. (2013). Multitask learning for host–pathogen protein interactions. *Bioinformatics.* **29**(13): i217–i226.

Landgraf et al. (2007). A mammalian microRNA expression atlas based on small RNA library sequencing. *Cell.* **129**: 1401–1414.

Long J-EE and Chen H-XX. (2009). Identification and characteristics of cattle microRNAs by homology searching and small RNA cloning. *Biochem Genet.* **47**: 329–343.

Mariano R and Wuchty S. (2017). Structure-based prediction of host pathogen protein interactions. *Curr Opin Struct Biol.* **44**: 119–124.

Mathur et al. (2020). MicroRNA-mediated trans-kingdom gene regulation in fungi and their host plants. *Genomics.* **112**(5): 3021–3035.

Price et al. (2003). Genome-scale microbial in silico models: the constraints-based approach. *Trends Biotechnol.* **21**: 162–169.

Qi et al. (2010). Semi-supervised multi-task learning for predicting interactions between HIV-1 and human proteins. *Bioinformatics.* **26**: i645–i652.

Sarma et al. (2012). A comparative proteomic approach to analyse structure, function and evolution of rice chitinases: a step towards increasing plant fungal resistance. *J Mol Model.* **18**: 4761–4780.

Tastan et al. (2009). Prediction of interactions between HIV-1 and human proteins by information integration. *Pac Symp Biocomput.* **51**: 6–527.

Wang et al. (2011). High-accuracy prediction of bacterial type III secreted effectors based on position-specific amino acid composition profiles. *Bioinformatics.* **27**(6): 777–784.

Westermann et al. (2017). Resolving host-pathogen interactions by dual RNA-seq. *PLoS Pathog.* **13**(2): e1006033.

Wuchty S. (2011). Computational prediction of host–parasite protein interactions between P. falciparum and H. sapiens. *PLoS One.* **6**: e26960.

Yao et al. (2012). In situ detection of mature miRNAs in plants using LNA modified DNA probes. *Methods Mol Biol.* **883**: 143–154.

Zhang Y. (2005). miRU: an automated plant miRNA target prediction server. *Nucl Acids Res.* **33**: W701–W704.

7

Plant–Microbe Interactions in the Age of Sequencing

Kuntal Kumar Dey*
Assam Agricultural University

Sebantee Ganguly*
University of Calcutta

CONTENTS

7.1 Introduction

Plants reside with various types of microorganisms during different stages of their life. Some of these are beneficial or detrimental, whereas some don't interfere with the plant's microbiome. Based on the localization of the microbial community, they are termed rhizospheric, phyllospheric, or endospheric microbiota. The root-associated microbial community inside the soil is termed rhizospheric, whereas phyllospheric microbiota cover the microbial community above the soil surface and endospheric microbiota refer to the community inside the plant body. The rhizosphere- and endosphere-oriented microbiomes interact with plants and are involved in plant maintenance. For example, nitrogen-fixing bacteria help in plant growth through nitrogen fixation. It has also been reported that microbial communities help in quorum sensing (QS) during interplant communication; a receptor called QS-LuxR is identified in several plants, which modulates root growth and responds to bacterial QS signaling molecules (Bai et al., 2012; Ortíz-Castro et al., 2008). Microbes can also help in controlling plant immunity; for example, *Xanthomonas campestris* pv. *campestris* produces a DSF (cis-11-methyl-2-dodecenoic acid) that suppresses plant innate immunity (Tran et al., 2020). The key questions in this context are the following: (i) Does the microbial community have any impact on plant growth regulation? (ii) If yes, is it organism-specific or community-dependent? (iii) How to identify the organism and the responsible mechanism? Addressing these questions helps in

* Corresponding author's email: kuntalk.dey@gmail.com & sebanteeganguly@gmail.com

DOI: 10.1201/9781003171416-7

understanding the plant–microbe interaction toward the use of microbial community for the improvement of plant health and yield.

Gradually, in search of an answer, the first step is to identify the microbial community, followed by the identification of key method through which microbes can interact and finally the identification of novel microorganism and their structural and functional annotation. Next-generation sequencing (NGS) technology helps in achieving those objectives with time efficiency. Metagenome or amplicon sequencing helps in the survey of microbial communities. In contrast, gene expression-dependent functional annotation can be done through metatranscriptomics and metaproteomics. Microbial metabolomics identifies the presence of versatile metabolites in the microbial community. The metabolomic study helps in getting a clear picture through estimation of the metabolome in hosts and microbes. Lastly, a combination of sequencing platforms can be used to build a novel microbial genome and its annotation to comprehend the interaction between the host and microbe.

All of these techniques can be utilized to decipher the biology behind plant–microbe interaction, but require quality analysis. Here comes the intervention of bioinformatics analysis pipeline and databases. Massive data generated through sequencing platforms are needed to be processed to obtain information (Hartmann et al., 2009). Data generated in NGS techniques are required to go through quality filtration followed by analysis.

Here, we have discussed various NGS techniques and their use in a systematic manner and developed a holistic understanding of different NGS platforms followed by how and when those can be used for having a clear knowledge about the plant–microbe interaction and practical application of that vision for disease control and enhancing crop productivity.

7.2 Sequencing Technology

Sequences of nucleic acids in the polynucleotide chain carry hereditary information and molecular identity. This information can be achieved through different biochemical methods that can be termed as sequencing. The scientific advancement of sequencing progresses through three generations. The very first generation is termed as shotgun sequencing, the second generation is high-throughput sequencing, and the third generation is single-molecule long-read sequencing (Loman et al., 2015). Heather and Chain (2016) described the history of sequencers in detail starting from the first generation to the third generation.

7.2.1 First-Generation Sequencing

The chain termination method developed by Sanger and Coulson in 1975 and the chain degradation method invented by Maxam and Gilbert during 1976 and 1977 are termed as the first generation of sequencing (Heather and Chain, 2016).

The principle behind the Sanger method is that due to the lack of hydroxyl in its 2′ and 3′ positions in ddNTP, it cannot form a phosphodiester bond during DNA synthesis (Sanger et al., 1975). This is used to interrupt the DNA synthesis reaction toward termination. Addition of a certain proportion of radiolabeled ddATP, ddCTP, ddGTP, and ddTTP, along with dNTP, into four DNA synthesis reaction systems, respectively, helps in chain termination. After that, gel electrophoresis and autoradiography are conducted to determine the sequence according to the position of the electrophoretic band. Later on, this method has been used by replacing the radiolabel and gel electrophoresis with a fluorescent dye and capillary electrophoresis, respectively. ABI PRISM developed from Leroy Hood's research, produced by Applied Biosystems, is the first-generation sequencer that helps in the production of Human Genome Project (Smith, 1986).

The Maxam and Gilbert technique relies on the degradation of DNA chain at a specific base. After gel electrophoresis, the sequence can be obtained by analyzing the sequence length on a gel (Maxam et al., 1977). This technique gives gold standard sequencing at a high cost of time and money. The efficiency of this sequencing is 99.99% with a maximum fragment length of 1,000 base pairs.

As the first-generation DNA sequencing produces nearly about 1,000 bp reads, it is very difficult to analyze the longer fragments. To overcome this shortcoming, shotgun sequencing protocol was implemented. In this method, DNA fragments that overlapped were cloned and sequenced separately. After that, bioinformatics tools were used to assemble them into one long contiguous sequence (or "contig") (Staden, 1979; Anderson, 1981). Later on, the development of techniques such as polymerase chain reaction (PCR) and other recombinant DNA technologies helped in the generation of a high amount of DNA sequence data, which is also time-consuming and costly (Saiki et al., 1985; Saiki, 1988; Jackson et al., 1972).

7.2.2 Second-Generation Sequencing

To overcome this problem of the first-generation sequencing method, a massive parallel sequencing method of second generation is developed. Three different platforms are developed by Roche 454, Illumina, and Life Technologies. Roche and Illumina used the method of sequencing by synthesis, whereas Life Technologies used the method of Sequencing by Oligonucleotide Ligation and Detection (SOLiD).

The SOLiD system was introduced by Applied Biosystems, which later on was named Life Technologies. In this method, DNA ligase is used for the ligation of oligonucleotide on DNA strand. After that, a fluorescent tag is used to detect the type of newly bound oligonucleotide per cycle. This was developed based on the principle of "polony" sequencing developed by George Church's group (Shendure, 2005).

Roche 454 uses the method of measurement of pyrophosphate concentration changes as oligonucleotide synthesis progresses. In this pyrosequencing method, once the pyrophosphate is generated during DNA synthesis, it is used to obtain the sequencing. Two enzymes play the key role of sequencing. ATP sulfurylase converts pyrophosphate into ATP, and this ATP is used as the substrate for luciferase to produce light proportional to the amount of pyrophosphate. This reaction is performed cyclically to obtain the sequence chronology through bead-based water-in-oil emulsion PCR. The first 454 high-throughput sequencing (HTS) machine was GS 20, which later on is known as 454 GS FLX.

The most widely used recent method is the Solexa method by Illumina. Here, adapter-linked DNA molecules are allowed to travel through a lawn of complementary oligonucleotides bound to a flow cell followed by solid-phase PCR-based bridge amplification. This process generates sequencing information in a cluster, based on the fluorophore signal detected by a laser detector. This produces a huge amount of short paired reads. Illumina introduced the first sequencer as Genome Analyzer (GA) that produces 35-bp-long sequences. Later on, HiSeq, MiSeq, and NovaSeq machines are introduced in the market to reduce the sequencing cost per base by producing a huge amount of sequencing per run.

The second-generation sequencing produces short reads within 500 bp with high accuracy, but it is not sufficient to understand the genomic structure. Here comes the Illumina introduced mate pair sequencing in rescue. In this method, long reads are inserted between adapter sequences followed by circularization of the DNA. Then the product is chopped off to get the adapter-associated sequence followed by sequencing. This produces an estimate of the distance between two sequences. But this process is somewhat error-prone, for which the third-generation sequencing platform is developed.

7.2.3 Third-Generation Sequencing

The two companies Pacific Biosciences (PacBio) and Oxford Nanopore Technologies (ONT) released two different platforms for the achievement of long-read generation. Both these platforms require high molecular weight DNA molecule to produce long DNA reads.

PacBio uses a single-molecule real-time (SMRT) protocol where DNA polymerization occurs in arrays called zero-mode waveguides (ZMWs), a microfabricated nanostructure. The ZMWs are tiny holes in a metallic film covering a chip. It measures the light passing through apertures during synthesis. The laser excitement at the bottom of the ZMWs was measured to determine the sequence. PacBio released sequencers called PacBio RS, RS-II, Sequel, and Sequel-II in the years 2010, 2013, 2015, and 2021 chronologically. All of these sequencers differ in the availability of the number of ZMWs on the chip.

ONT uses the basics of a lipid bilayer structure of membrane for sequencing purposes. In this method, α-hemolysin ion channel is used for sequencing purpose. As DNA proceeds through nanochannel during synthesis, it generates changes in electric potential as per the bases. This potential is measured to interpret the sequence. ONT released their platforms named GridION, PromethION, and MinION. The first two sequencers are called desktop sequencer, and the last one is a bit interesting having a mobile phone-like size and produces sequences with high quality.

The third-generation sequencer produces long reads up to an average of 15KB with high quality. Initially, these sequencers produced long error-prone reads. It was documented that PacBio has an error rate of 10%–15%, whereas ONT has up to 30%. But in a recent release, the error rate has improved dramatically. PacBio released a HiFi sequencing protocol that has a high sequencing quality with an accuracy of 99.8%. ONT also released the new R9 chemistry to achieve an error rate of sequencing of 99.995%.

7.3 Selection of NGS Technique

The very first step in microbiome research is to identify the objective and choose the right NGS method by understanding the advantages and limitations based on the objective. NGS methods are used at DNA or mRNA levels, and the optimum method should be selected based on the experimental objectives (Figure 7.1).

As DNA is one of the most stable molecules of a cellular organism, it is easily extractable and can be preserved for sequencing purposes. The commonly used method for microbial analysis is amplicon and metagenomic sequencing. To distinguish between rhizosphere and phyllosphere communities, amplicon sequencing is the most commonly used, which is based on microbiome analysis method. In this process, the 16S ribosomal DNA (rDNA) region of prokaryotes and 18S rDNA or internal transcribed spacer (ITS) region of eukaryotes are amplified through PCR and then sequenced. Depending on the available genomic information in different databases, the abundance of known microbes can be estimated and interpreted. After the amplification of specific marker region, sequencing is mostly performed on Illumina platforms such as MiSeq, HiSeq, and NovaSeq. Earlier, 100,000–2,000,000 single-end reads were used for the estimation of mitochondrial abundance, but nowadays, paired-end reads are used for abundance estimation. Amplicon sequencing can be advantageous as a low concentration of microbe sample is required, which can easily be amplified, but the major limitation of this process is that it gives only genus-level resolution and is also sensitive to primer specificity as well as the number of PCR cycles.

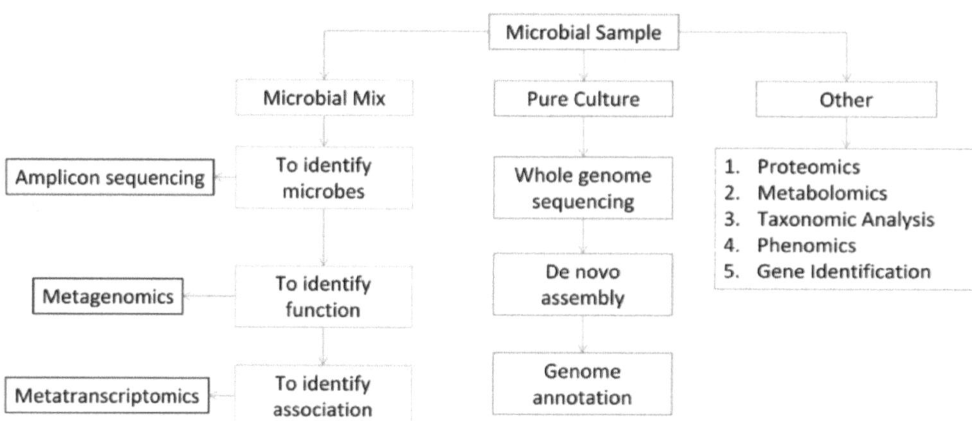

FIGURE 7.1 Application of sequencing in various fields of microbial analysis.

Metagenomic sequencing is more informative than amplicon sequencing, but expensive as well. In this case, the host genome may also be sequenced besides the metagenome. So overall, as more amount of sequences are available, it gives higher taxonomic resolution besides the functional information (Oulas et al., 2015). In this process, the total genomic content is isolated followed by sequencing on Illumina-based platform. The host genome is removed from the overall content and then subjected to abundance and functional annotation.

Metatranscriptomic sequencing is used for the estimation of mRNA profile. This process is carried out to estimate the gene expression and functional profile of the microbial community. Most commonly, the Illumina platform is used for sequencing purposes, but as nanopore sequencer is also able to sequence RNA molecule, recently, this has become a good alternative to metatranscriptomic analysis (Semmouri et al., 2019).

Whenever new beneficial microbial species is identified and isolated from the culture, it is subjected to de novo genome sequencing, followed by annotation. In this process, different sequencing platforms are used in combination. If genome size is more than 10 MB, the third-generation sequencing gives an advantage of long scaffolding, followed by Illumina-based polishing. In the case of shorter genome size, Illumina short reads help in the assembly process. After assembling, Illumina-based transcriptome sequencing promotes gene discovery.

Now the question arises: Depending on the situation, how a platform for gene sequencing can be selected? Based on the existing budget, single or multiple methods can be applied. Amplicon sequencing is applied for large-scale fundamental research, where abundance of different microbes is required to be estimated. Metagenomic sequencing comes into action for de novo identification of species from the sample. The process not only provides species-level resolution of the sample, but also gives insight into the functional profile. Metatranscriptomics helps in the identification of the process or method through which microbiome interacts with the host. Finally, de novo microbial genome sequencing helps in the identification of specific microbes' genome and its function. For choosing a suitable technique, it is required to know that amplicon works cost-effectively on samples with low biomass, whereas the other two methods may work on such samples if it is considered for high volume of sequencing. Once the sample is identified and isolated, then it will be subjected to *de novo* sequencing and annotation.

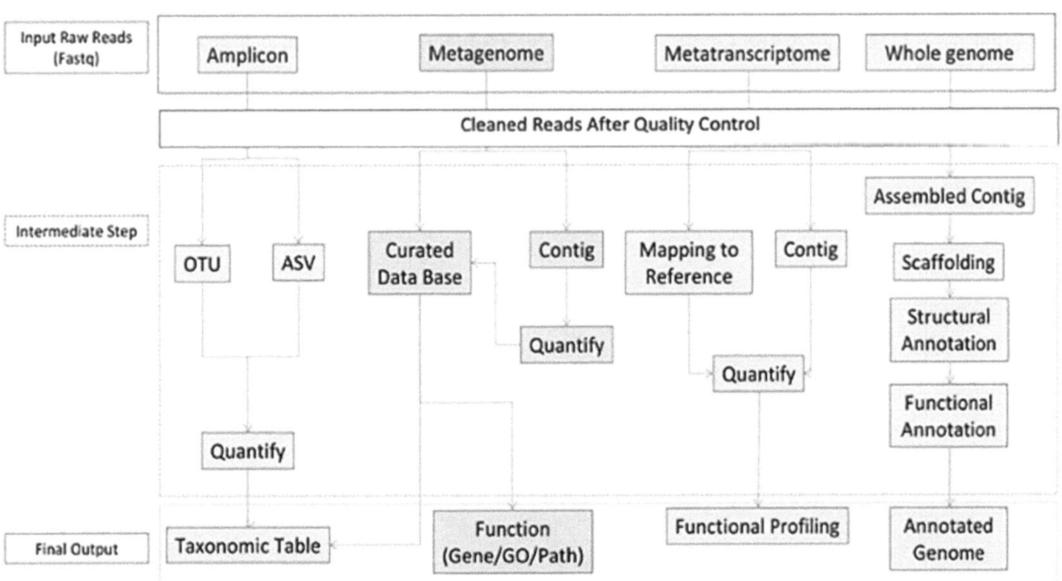

FIGURE 7.2 Workflow of commonly used methods applied for microbial study.

7.4 Analysis of Sequencing Results

Sequencing analysis goes through a step-by-step program or scripts that help to understand the information obtained from the sequencing result. The analysis pipeline can be divided into some groups such as amplicon sequencing, metagenome sequencing, metatranscriptome sequencing, and de novo genome sequencing (Figure 7.2). As the analysis pipeline requires basic script implementation, mostly in Linux environment, the implementation of such a pipeline requires some basic knowledge of the Linux operating system. Sometimes PERL, Python, or R language is also implemented for statistical test and analysis purposes. Nowadays, Anaconda/Bioconda environment is used for easy installation and implementation of such analysis pipelines (Table 7.1). The Conda environment can be implemented on Linux, MacOS, and Windows systems [https://docs.anaconda.com/anaconda/install/]. Some basic steps are starting from quality control to sample analysis, which are required to get the optimum information from the microbial sequence data.

The very first step is checking the sequencing quality for which FastQC, as well as NGS QC toolkit, is applied. Most of the time MultiQC is implemented upon multiple FastQC files to summarize the result of multiple sequencing data quality. This process helps to understand the sequencing quality. If all goes well, then these data are used for further analysis. Otherwise, it is used for quality trimming with the help of various tools such as seqtk, FastqCleaner, Trim Galore, and Trimmomatric etc.

7.4.1 Amplicon Analysis

After the first step of quality trimming of sequencing data, quantifying the representative sequences is required (Figure 7.3). There are two approaches for the selection of representative sequence quantification. The first is denoising to amplicon sequence variant (ASVs), and the second is clustering to operational taxonomic units (OTUs). The UPARSE algorithm is used for clustering the sequences to OTUs with 97% similarity (Edgar, 2013). But this method has some limitations such as the inability to differentiate among species and strains. To overcome this, a new method of denoising called DADA2 is implemented called DADA2. This denoising can be carried out through DADA2 (denoise-paired/denoise-single), QIIME2 (Deblur), and USEARCH (unoise3). Finally, the feature table is obtained for each of the samples. Moreover, taxonomic classification information such as kingdom, phylum, class, order, family, genus, and species may also be added to the feature table. There are some newly developed packages that help to add some functional information depending on databases. PICRUSt is one of such tools that take advantage of the Greengenes database (McDonald et al., 2011) and Kyoto Encyclopedia

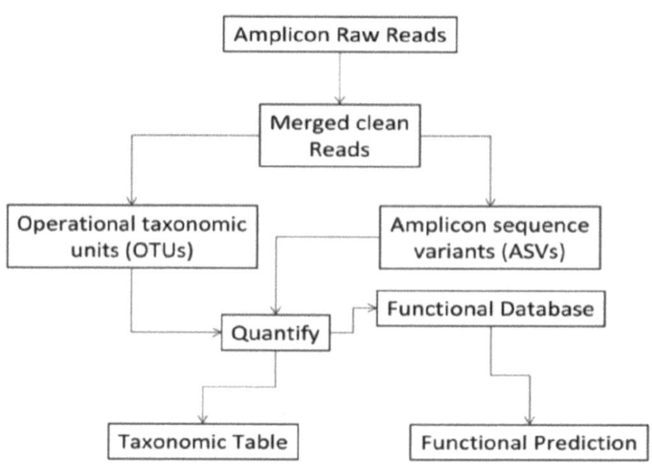

FIGURE 7.3 Workflow of commonly used methods applied for amplicon analysis.

TABLE 7.1
List of Tools Commonly Used for Microbiome Analysis

S. No.	Tool Name	Function	Conda Available	Reference	Web site
1	FastQC	QC	Yes		http://www.bioinformatics.babraham.ac.uk/projects/fastqc/
2	NGS QC toolkit	QC	No	Patel et al. (2012)	http://www.nipgr.ac.in/ngsqctoolkit.html
3	MultiQC	QC	Yes	Ewels et al. (2016)	http://multiqc.info
4	Trimmomatic	TR	No	Goldman et al. (2006)	http://www.usadellab.org/cms/?page=trimmomatic
5	Cutadapt	TR	Yes	Magoč et al. (2011)	https://cutadapt.readthedocs.io/en/stable/
6	seqtk	TR	Yes		https://github.com/lh3/seqtk
7	FastqCleaner	TR	Yes	Roser et al. (2019)	https://bioconductor.org/packages/3.13/bioc/html/FastqCleaner.html
8	Trim Galore	TR	Yes		http://www.bioinformatics.babraham.ac.uk/projects/trim_galore/
9	KneadData	TR	Yes		http://huttenhower.sph.harvard.edu/kneaddata
10	UPARSE	AD	No	Edgar et al. (2013)	http://drive5.com/uparse/
11	DADA2	AC	Yes	Ihrmark et al. (2012)	https://bioconductor.org/packages/3.13/bioc/html/dada2.html
12	QIIME2	AC	Yes	Estaki et al. (2020)	https://qiime2.org
13	USEARCH	AC	No	Edgar et al. (2011)	https://www.drive5.com/usearch/
14	PICRUSt	FA	Yes	Langille et al. (2013)	http://picrust.github.com
15	MetaPhlAn2	MTD, TP	Yes	Truong et al. (2015)	https://bitbucket.org/biobakery/metaphlan2
16	Kraken 2	TP	Yes	Wood et al. (2019)	https://ccb.jhu.edu/software/kraken2/
17	CLARK	TP	Yes	Ounit et al. (2015)	http://clark.cs.ucr.edu/
18	Bracken	TP	Yes		https://github.com/jenniferlu717/Bracken
19	MegaBLAST	TP	No		http://nebc.nox.ac.uk/bioinformatics/docs/megablast.html
20	MEGAN	MGM, FA	Yes	Huson et al. (2007)	http://ab.inf.uni-tuebingen.de/software/megan6/
21	ANASTASIA	MGM, FA	No	Koutsandreas et al. (2019)	https://galaxyproject.org/use/anastasia/
22	Nephele	MGM, FA	No		https://nephele.niaid.nih.gov/about/
23	Parallel-META 3	MGM, FA	No	Jing et al. (2017)	http://bioinfo.single-cell.cn/parallel-meta.html
24	MG-RAST	MGM, FA	Yes	Keegan et al. (2016)	https://www.mg-rast.org/
25	MEGAHIT	MGD	Yes	Liu et al. (2015)	https://github.com/voutcn/megahit
26	metaSPAdes	MGD	No	Nurk et al. (2017)	https://cab.spbu.ru/software/meta-spades/
27	Velvet	MGD	Yes	Zerbino et al. (2010)	https://www.ebi.ac.uk/~zerbino/velvet/
28	MetaVelvet	MGD	Yes	Hughes et al. (2001)	http://metavelvet.dna.bio.keio.ac.jp/
29	Ray Meta	MGD	No	Boisvert et al. (2012)	http://denovoassembler.sourceforge.net/manual.html
30	Edena	MGD	Yes	Hernandez et al. (2008)	http://www.genomic.ch/edena.php
31	MOCAT2	MGD, FA	No	Li et al. (2021a,b)	https://mocat.embl.de/

(Continued)

TABLE 7.1 (*Continued*)

List of Tools Commonly Used for Microbiome Analysis

S. No.	Tool Name	Function	Conda Available	Reference	Web site
32	ATLAS	MGD, FA	No	Kieser et al. (2020)	https://github.com/metagenome-atlas/atlas
33	SOAPdenovo	DG	Yes		https://github.com/aquaskyline/SOAPdenovo2
34	ABySS	DG	Yes	Jackman et al. (2017)	https://github.com/bcgsc/abyss
35	IDBA-UD	DG	No	Zhu et al. (2010)	http://www.cs.hku.hk/~alse/idba_ud
36	SPAdes	DG	Yes	Dvorkin et al. (2012)	https://github.com/ablab/spades
37	Trinity	DG	Yes	Grabherr et al. (2011)	https://github.com/trinityrnaseq/trinityrnaseq/
38	HGAP	DG	No	Jayakumar et al. (2019)	https://github.com/PacificBiosciences/Bioinformatics-Training/wiki/HGAP
39	FALCON	DG	Yes	Chin et al. (2016)	https://github.com/PacificBiosciences/falcon
40	Canu	DG	Yes	Heller et al. (2020)	https://canu.readthedocs.io/en/stable/
41	MaSuRCA	DG	Yes	Zimin et al. (2013)	http://masurca.blogspot.co.uk/
42	DBG2OLC	DG	Yes	Ye et al. (2016)	https://github.com/yechengxi/DBG2OLC
43	Flye	DG	Yes	Kolmogorov et al. (2020)	https://github.com/fenderglass/Flye
44	Unicycler	DG	Yes	Bidone et al. (2017)	https://github.com/rrwick/Unicycler
45	Trycycler	DG	Yes		https://github.com/rrwick/Trycycler
46	MetaGeneMark	FA	No	Zhu et al. (2010)	http://exon.gatech.edu/meta_gmhmmp.cgi
47	MetaGeneAnnotator	FA	No	Noguchi et al. (2008)	http://metagene.nig.ac.jp/
48	Prokka	FA	Yes	Seemann et al. (2014)	https://github.com/tseemann/prokka
49	Trinotate	FA	Yes	Bryant et al. (2017)	https://trinotate.github.io/
50	RAST	FA	No	Michaud et al. (2008)	https://rast.nmpdr.org/
51	DRAM	FA	Yes	Shaffer et al. (2020)	https://github.com/shafferm/DRAM/
52	PGAP	FA	No	Li et al. (2021a,b)	https://github.com/ncbi/pgap
53	MetaTrans	QC, MTM, FA	No	Martinez et al. (2016)	https://www.metatrans.org/
54	COMAN	QC, MTM, FA	No	Ni et al. (2016)	http://sbb.hku.hk/COMAN/
55	FMAP	QC, MTM, FA	No	Kim et al. (2016)	https://github.com/jiwoongbio/FMAP
56	SAMSA2	QC, MTM, FA	No	Westreich et al. (2018)	https://transcript.github.io/samsa2/
57	HUMAnN2	QC, MTD, MTM, FA	Yes		https://huttenhower.sph.harvard.edu/humann2/
58	SqueezeMeta	QC, MTD, FA	Yes	Tamames et al. (2019)	https://github.com/jtamames/SqueezeMeta

QC: quality control; TR: trimming; AD: amplicon denoising; AC: amplicon clustering; MGM: metagenome mapping; MGD: metagenome de novo assembly; MTM: metatranscriptome mapping; TP: taxonomic profiling; MTD: metatranscriptome de novo assembly; DG: de novo genome assembly; FA: functional annotation.

of Genes and Genomes (KEGG) database (Kanehisa and Goto, 2000) to assign functional annotation to the respective OTUs.

Phytohormone profiling was done with the help of amplicon sequencing to understand the structure of the fungal community in the root of tomatoes (Manzotti et al., 2020). The amplicon sequencing also helps in understanding the role of microbes in waste management and the interaction between microbes (Jiang et al., 2019). In both cases, OTU has been identified with the help of QIIME. Previously, RDP pyrosequencing pipeline (Cole et al., 2019) is used to understand the phylloepiphytic bacteria communities of spinach (Lopez-Velasco et al., 2011). In a larger study, the amplicon reanalysis was applied to draught-related rice microbiome with the help of QIIME2 and DADA2 to understand the rhizobium shift in rice due to draught condition over a period (Jang et al., 2020). USEARCH pipeline was used in combination with QIIME to reveal the plant–microbe interaction for natural herbivore management (Howard et al., 2020). Wang et al. showed the mechanism of microbial communication on seaweed fertilizer that helps in the improvement of the utility of natural fertilizer.

7.4.2 Metagenome Analysis

Shotgun metagenome data provide high-resolution taxonomic annotation and functional profiles. But the process of analysis requires more computational skills and time (Figure 7.4). Here, Conda aids in the preparation of analysis pipeline. The metagenome may be contaminated with the host genome as well as other genomes, so quality filtration is one of the most important steps in this process. Low-quality bases and adapter sequences are trimmed with Trimmomatic (Bolger et al., 2014), and the host genome is removed by mapping with the help of Bowtie 2 (Langmead et Al., 2012). This process can be done individually or with the help of KneadData pipeline (the Huttenhower Lab). It is required to provide an indexed host genome for Bowtie 2 for both of these pipelines. Taxonomic identification and functional annotation are the main steps for metagenome analysis. This can be done based on mapping of raw reads or assembled reads.

One of the common practices of doing taxonomic profiling and functional annotation from raw reads is performed by MetaPhlAn2 and HUMAnN2, respectively. The usage and the scripts for running both of these methods are available in Microbiome Helper (Comeau et al., 2017). There are other tools to perform taxonomic profiling, such as Kraken 2, CLARK, Bracken, and MegaBLAST. MEGAN (Huson et al., 2007, 2011) is graphics-based software that performs both taxonomic and functional analyses. ANASTASIA (Koutsandreas et al., 2019), Nephele (Weber et al., 2018), Parallel-META 3 (Jing et al., 2017), and MG-RAST (Lindgreen et al., 2016) are different tools that help in the analysis of metagenome.

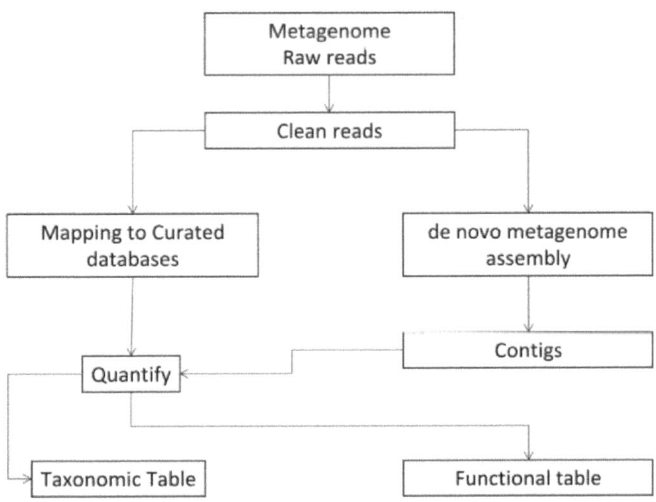

FIGURE 7.4 Workflow of commonly used methods applied for metagenome analysis.

The assembly-based method is performed in two steps: metagenome assembly followed by annotation. The metagenome assembly can be performed with the help of MEGAHIT, metaSPAdes, Velvet, MetaVelvet, Ray Meta, Edena, etc., followed by gene annotation with MetaGeneMark, Prokka, etc. The MOCAT2 and ATLAS are two pipelines that perform de novo metagenome assembly and annotation in one go.

Metagenomic shotgun technique in combination with amplicon sequencing helps in understanding the rhizobiome modulation in rice across different developmental stages (Imchen et al., 2019). Here, the MG-RAST server is used for the analysis of sequencing data. Fadiji et al. used MG-RAST to analyze shotgun metagenomic data for understanding the association of root microbiomes with maize plant. The combination of de novo metagenome assembly and functional annotation approach is utilized to understand viral interaction with woods (Bertazzon et al., 2020). A similar process has also been utilized to reveal the process of growth promotion of millet by bacteria (Kuramae et al., 2020). Metagenomics also helps in understanding the plant–pathogen interaction between banana and fusarium wilt (Kaushal et al., 2020). The combination of MEGAHIT and MetaGene was used to infer the soil microbiome during continuous sugarcane cropping (Pang et al., 2021).

7.4.3 De novo Microbial Genome Assembly

De novo genome assembly and annotation are used sequentially for the discovery of structure and function of the microbial genome. They can be applied when pure culture of a microbe is available. In this process, shotgun genome sequencing is performed on Illumina/ONT/PacBio platform or using a combination of different platforms. There are two approaches that can be applied for de novo assembly if data are generated simultaneously in different platforms (Figure 7.5). The first step is the assembly of short reads to contigs, followed by scaffolding with the help of long reads produced by ONT or PacBio. The second step is the assembly of long reads produced by ONT or PacBio, followed by polishing with Illumina-based short reads. There are short-read assemblers such as SOAPdenovo, ABySS, SPAdes, Velvet, and Trinity that work on Illumina data. HGAP and FALCON work on PacBio data, and Canu, MaSuRCA, and Flye can work on both ONT and PacBio data. A special type of de novo genome

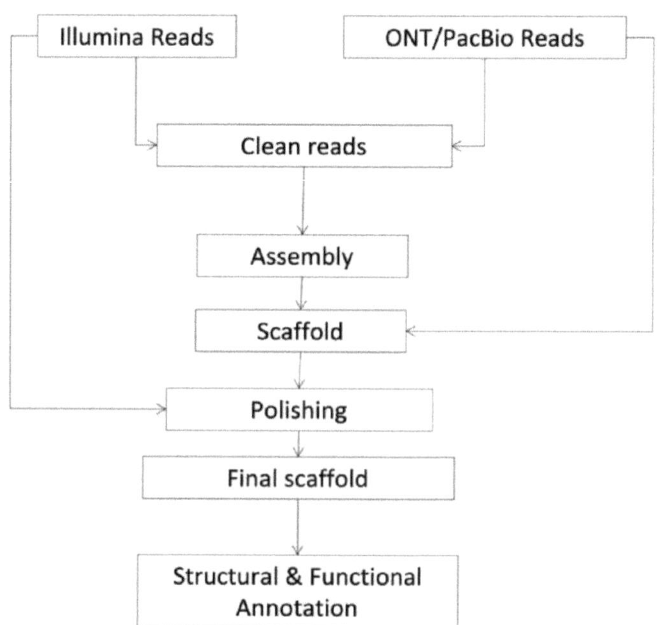

FIGURE 7.5 Workflow of commonly used methods applied for de novo assembly of microbial genome.

assembler called Unicycler is used for the assembly of circular genome of microbes with the help of short reads. Trycycler is the updated version of Unicycler that can work with long reads of ONT or PacBio data.

Annotation of the genome is the next important step after de novo genome sequencing. This can be done in two steps: structural annotation, to find out the structure of coding and noncoding regions, and functional annotation, with the help of different databases. There are different tools such as Prokka, MetaGeneMark, Trinotate, RAST, DRAM, and PGAP (Ruiz-Perez et al., 2021) that are used to annotate the genome. Additionally, InterProScan, eggNOG-mapper, and other custom database-specific annotation tools may help improve functional annotation.

Nowadays, sequencing small microbial genome is a common practice. This method is mostly used for the construction of annotated whole genome of pathogens. A combination of ONT and Illumina sequences is utilized to construct the genome of *Erwinia amylovora* Ea1189 with the help of Unicycler and Prokka (Singh et al., 2020). In a different study, high-depth Illumina sequencing along with PacBio sequencing helps in the preparation of brown rot fungal pathogen (Landi et al., 2020). Reference-quality genome of four barley spot fungus was prepared as a resource for disease association study with the help of de novo assembly of genome data produced on PacBio (Wyatt et al., 2021). To understand the mildew disease in cucurbits, genomes of the responsive pathogen were also constructed with the help of the MaSuRCA pipeline (Polonio et al., 2021).

7.4.4 Metatranscriptomic Analysis

The analysis of amplicon, metagenome, or de novo microbial genome gives knowledge regarding the character of the respective samples. But understanding the scenario of host–microbe or plant–microbe interaction is something different. It is like a story of how plants and microbes coexist and help each other. Metatranscriptomics gives a fair knowledge to understand this interaction story. In this process, RNA is isolated and converted to cDNA followed by sequencing on Illumina platform or ONT platform (Shukya et al. 2019). As the Illumina platform is mostly used nowadays, the analysis of such data will be discussed. After data generation, two different ways can be followed (Figure 7.6). The first one is the mapping of reads on existing databases and estimation of functions, and the second one is performing de novo assembly followed by mapping and annotation. There are different tools such as MetaTrans, COMAN, FMAP, SAMSA2, and HUMAnN2 that work on mapping-based approach for the analysis of metatranscripts. On the other hand, SqueezeMeta, IMP, and MOSCA work on the assembly followed

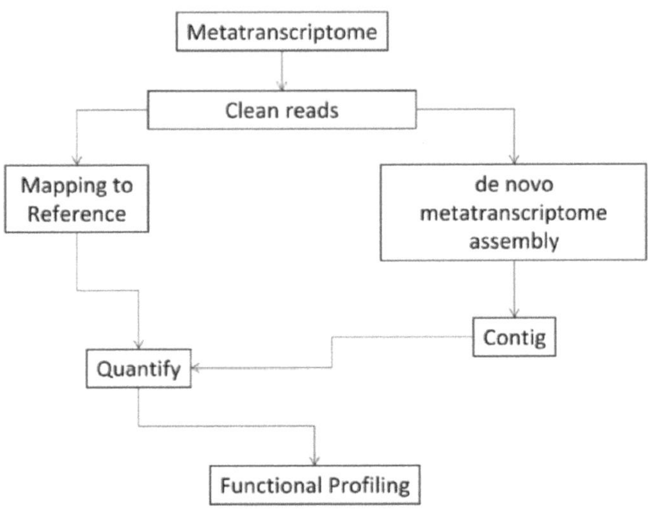

FIGURE 7.6 Workflow of commonly used methods applied for metatranscriptomic analysis.

by the analysis of metatranscripts. This gives an overall idea of how a microbe interacts with plants and establishes a beneficial connection.

Crump et al. (2018) used the metatranscriptomics sequencing method beside the amplicon analysis pipeline to understand the mutualisms in seagrass microbiomes. Martinez et al. (2016) reported a polymicrobial consortium using metatranscriptomic approach that infect olive tree through root damage instead of a single pathogen infection. A combination of metagenomic and metatranscriptomic approaches was utilized to investigate the post-harvest processing of cocoa in Costa Rica (Verce et al., 2021). Schneider et al. (2021) investigated the interaction between microbes and Norway spruce by utilizing combined metatranscriptomics and amplicon sequencing.

7.5 Conclusions

High-throughput screening-based methods help in the generation of high-quality massive data. This can be utilized as a tool for understanding the nature of microbial genome and transcriptome. This knowledge helps in understanding the phenomenon of host–microbe interaction. Plants are the major source of microbial development, and novel microbe can be identified using sequencing techniques in a short time. This sequencing technique can help in the identification of beneficial as well as pathogenic microbes. Besides identification, the mode of action and degree of responsiveness can also be estimated through metatranscriptomics and meta-metabolomics. This helps in the identification and discovery of different microbial environments and their effect on plant growth. Later on, this knowledge can be used for the improvement of plant health toward sustainable agriculture through the use of beneficial microbe or engineered microbe for the betterment of plant growth.

REFERENCES

Anderson S. Shotgun DNA sequencing using cloned DNase I-generated fragments. *Nucleic Acids Research.* 1981; 9:3015–3027.

Bai et al. N-3-Oxo-decanoyl-l-homoserine-lactone activates auxin-induced adventitious root formation via hydrogen peroxide- and nitric oxide-dependent cyclic gmp signaling in mung bean. *Plant Physiology* 2012; 158:725–736.

Bertazzon et al. Two new putative plant viruses from wood metagenomics analysis of an esca diseased vineyard. *Plants.* 2020; 9(7):835.

Bidone et al. Morphological transformation and force generation of active cytoskeletal networks. *PLoS Computational Biology.* 2017; 13(1):e1005277.

Boisvert et al. Ray Meta: scalable de novo metagenome assembly and profiling. *Genome Biology.* 2012; 13(12):1–3.

Bolger et al. Trimmomatic: a flexible trimmer for Illumina sequence data. *Bioinformatics.* 2014; 30(15): 2114–2120.

Bryant et al. A tissue-mapped axolotl de novo transcriptome enables identification of limb regeneration factors. *Cell Reports.* 2017; 18(3):762–776.

Chin et al. Phased diploid genome assembly with single-molecule real-time sequencing. *Nature Methods.* 2016; 13(12):1050–1054.

Cole et al. The Ribosomal Database Project: improved alignments and new tools for rRNA analysis. *Nucleic Acids Res.* 2009; 37(Database issue):D141–D145.

Comeau et al. Microbiome helper: a custom and streamlined workflow for microbiome research. *MSystems.* 2017; 2(1):e00127-16.

Crump et al. Metatranscriptomics and amplicon sequencing reveal mutualisms in seagrass microbiomes. *Frontiers in Microbiology.* 2018; 9:388.

Dvorkin M and Kulikov AS. SPAdes: a new genome assembly algorithm and its applications to single-cell sequencing. *Journal of Computational Biology.* 2012; 19:455477.

Edgar RC et al. UCHIME improves sensitivity and speed of chimera detection. *Bioinformatics.* 2011; 27(16): 2194–2200.

Edgar RC. UPARSE: highly accurate OTU sequences from microbial amplicon reads. *Nature Methods.* 2013; 10(10):996–998.

Estaki et al. QIIME 2 enables comprehensive end-to-end analysis of diverse microbiome data and comparative studies with publicly available data. *Current Protocols in Bioinformatics.* 2020; 70(1):e100.

Ewels et al. MultiQC: summarize analysis results for multiple tools and samples in a single report. *Bioinformatics.* 2016; 32(19):3047–3048.

Fadiji et al. Shotgun metagenomics reveals the functional diversity of root-associated endophytic microbiomes in maize plant. *Current Plant Biology.* 2021; 25:100195.

Goldman et al. Evolution of sensory complexity recorded in a myxobacterial genome. *Proceedings of the National Academy of Sciences.* 2006; 103(41):15200–15205.

Grabherr et al. Full-length transcriptome assembly from rNA-seq data without a reference genome. *Nature Biotechnology.* 2011; 29(7):644–652.

Hartmann et al. Plant-driven selection of microbes. *Plant Soil.* 2009; 321:235–257.

Heather JM and Chain B. The sequence of sequencers: The history of sequencing DNA. *Genomics.* 2016; 107(1):1–8.

Heller D and Vingron M. SVIM-asm: Structural variant detection from haploid and diploid genome assemblies. *bioRxiv.* 2020; 36(22–23):5519–5521.

Hernandez et al. De novo bacterial genome sequencing: millions of very short reads assembled on a desktop computer. *Genome Research.* 2008; 18(5):802–809.

Howard et al. Shifts in plant-microbe interactions over community succession and their effects on plant resistance to herbivores. *New Phytologist.* 2020; 226(4):1144–1157.

Hughes et al. Counting the uncountable: statistical approaches to estimating microbial diversity. *Applied and Environmental Microbiology.* 2001; 67(10):4399.

Huson et al. MEGAN analysis of metagenomic data. *Genome Research.* 2007; 17(3):377–386.

Huson et al. Integrative analysis of environmental sequences using MEGAN4. *Genome Research.* 2011; 21: 1552–1560.

Huttenhower Lab. KneadData. http://huttenhower.sph.harvard.edu/kneaddata. *Accessed Apr 20,* 2018.

Ihrmark et al. New primers to amplify the fungal ITS2 region–evaluation by 454-sequencing of artificial and natural communities. *FEMS Microbiology Ecology.* 2012; 82(3):666–677.

Imchen et al. 16S rRNA gene amplicon based metagenomic signatures of rhizobiome community in rice field during various growth stages. *Frontiers in Microbiology.* 2019; 10:2103.

Jackman et al. ABySS 2.0: resource-efficient assembly of large genomes using a Bloom filter. *Genome Research.* 2017; 27(5):768–777.

Jackson et al. Biochemical method for inserting new genetic information into DNA of simian virus 40: circular SV40 DNA molecules containing lambda phage genes and the galactose operon of Escherichia coli. *Proceedings of the National Academy of Sciences of the United States of America.* 1972; 69:2904–2909.

Jang et al. Re-analysis of 16S amplicon sequencing data reveals soil microbial population shifts in rice fields under drought condition. *Rice.* 2020; 13: 44.

Jayakumar V and Sakakibara Y. Comprehensive evaluation of non-hybrid genome assembly tools for third-generation PacBio long-read sequence data. *Briefings in Bioinformatics.* 2019; 20(3):866–76.

Jing et al. Parallel-META 3: Comprehensive taxonomical and functional analysis platform for efficient comparison of microbial communities. *Science Reports.* 2017; 7: 40371.

Jiang et al. Exploring the roles of and interactions among microbes in dry co-digestion of food waste and pig manure using high-throughput 16S rRNA gene amplicon sequencing. *Biotechnol Biofuels.* 2019; 12:5.

Kanehisa M and Goto S. KEGG: Kyoto encyclopedia of genes and genomes. *Nucleic Acids Research.* 2000; 28(1):27–30.

Kaushal et al. Metagenomic insights of the root colonizing microbiome associated with symptomatic and non-symptomatic bananas in fusarium wilt infected fields. *Plants (Basel).* 2020; 9(2):263.

Keegan et al. MG-RAST, a metagenomics service for analysis of microbial community structure and function. In: Microbial Environmental Genomics (MEG), 2016, pp. 207–233. Humana Press, New York.

Kieser et al. ATLAS: a Snakemake workflow for assembly, annotation, and genomic binning of metagenome sequence data. *BMC Bioinformatics.* 2020; 21(1):1–8.

Kim et al. FMAP: functional mapping and analysis pipeline for metagenomics and metatranscriptomics studies. *BMC Bioinformatics.* 2016; 17(1):1–8.

Kolmogorov et al. metaFlye: scalable long-read metagenome assembly using repeat graphs. *Nature Methods.* 2020; 17(11):1103–1110.

Koutsandreas et al. ANASTASIA: an automated metagenomic analysis pipeline for novel enzyme discovery exploiting next generation sequencing data. *Frontiers in Genetics.* 2019; 10:469.

Kuramae et al. Sorghum growth promotion by *Paraburkholderia tropica* and *Herbaspirillum frisingense*: Putative Mechanisms Revealed by Genomics and Metagenomics. *Microorganisms.* 2020; 8(5):725.

Landi et al. Draft genomic resources for the brown rot fungal pathogen Monilinia laxa. *Molecular Plant-Microbe Interactions.* 2020; 33(2):145–148.

Langille et al. Predictive functional profiling of microbial communities using 16S rRNA marker gene sequences. *Nature Biotechnology.* 2013; 31(9):814–821.

Langmead B and Salzberg SL. Fast gapped-read alignment with Bowtie 2. *Nature Methods.* 2012; 9(4):357–359.

Li et al. Comparison of gut microbiota structure and Actinobacteria abundances in healthy young adults and elderly subjects: a pilot study. *BMC Microbiology.* 2021a; 21(1):1.

Li et al. RefSeq: expanding the prokaryotic genome annotation pipeline reach with protein family model curation. *Nucleic Acids Research.* 2021b; 49(D1):D1020–D1028.

Lindgreen et al. An evaluation of the accuracy and speed of metagenome analysis tools. *Science Reports.* 2016; 6:19233.

Liu et al. MEGAHIT: an ultra-fast single-node solution for large and complex metagenomics assembly via succinct de Bruijn graph. *Bioinformatics.* 2015; 31(10):1674–1676.

Loman NJ and Pallen MJ. Twenty years of bacterial genome sequencing. *Nature Reviews Microbiology.* 2015; 13(12):787–794.

Lopez-Velasco et al. Changes in spinach phylloepiphytic bacteria communities following minimal processing and refrigerated storage described using pyrosequencing of 16S rRNA amplicons. *Journal of Applied Microbiology.* 2011; 110(5):1203–1214.

Magoč T and Salzberg SL. FLASH: fast length adjustment of short reads to improve genome assemblies. *Bioinformatics.* 2011; 27(21):2957–2963.

Manzotti et al. Insights into the community structure and lifestyle of the fungal root endophytes of tomato by combining amplicon sequencing and isolation approaches with phytohormone profiling. *FEMS Microbiology Ecology.* 2020; 96(5): fiaa052, https://doi.org/10.1093/femsec/fiaa052

Martí et al. Metatranscriptomic dynamics after Verticillium dahliae infection and root damage in Olea europaea. *BMC Plant Biology.* 2020; 20(1):1–5.

Martinez et al. MetaTrans: an open-source pipeline for metatranscriptomics. *Scientific Reports.* 2016; 6(1):1–2.

Maxam et al. A new method for sequencing DNA. *Proceedings of the National Academy of Sciences of the United States of America.* 1977; 74(2):560–564.

McDonald et al. Clusters of nucleotide substitutions and insertion/deletion mutations are associated with repeat sequences. *PLoS Biology.* 2011; 9(6):e1000622.

Michaud et al. Integrative analysis of RUNX1 downstream pathways and target genes. *BMC Genomics.* 2008; 9(1):1–7.

Ni et al. COMAN: a web server for comprehensive metatranscriptomics analysis. *BMC Genomics.* 2016; 17(1):1–7.

Noguchi et al. MetaGeneAnnotator: detecting species-specific patterns of ribosomal binding site for precise gene prediction in anonymous prokaryotic and phage genomes. *DNA Research.* 2008; 15(6):387–396.

Nurk et al. metaSPAdes: a new versatile metagenomic assembler. *Genome Research.* 2017; 27(5):824–834.

Ortíz-Castro et al. N-acyl-L-homoserine lactones: a class of bacterial quorum-sensing signals alter post-embryonic root development in Arabidopsis thaliana. *Plant, Cell & Environment.* 2008; 31(10):1497–1509.

Oulas et al. Metagenomics: tools and insights for analyzing next-generation sequencing data derived from biodiversity studies. *Bioinformatics and Biology Insights.* 2015; 9:75–88.

Ounit et al. CLARK: fast and accurate classification of metagenomic and genomic sequences using discriminative k-mers. *BMC Genomics.* 2015; 16(1):1–3.

Pang et al. Soil metagenomics reveals effects of continuous sugarcane cropping on the structure and functional pathway of rhizospheric microbial community. *Frontiers in Microbiology.* 2021; 12:627569.

Patel RK and Jain M. NGS QC Toolkit: a toolkit for quality control of next generation sequencing data. *PloS One.* 2012; 7(2):e30619.

Polonio et al. A hybrid genome assembly resource for Podosphaera xanthii, the main causal agent of powdery mildew disease in cucurbits. *Molecular Plant-Microbe Interactions.* 2021; 34(3):319–324.

Roser et al. FastqCleaner: an interactive Bioconductor application for quality-control, filtering and trimming of FASTQ files. *BMC Bioinformatics.* 2019; 20(1):1–7.
Ruiz-Perez et al. MicrobeAnnotator: a user-friendly, comprehensive functional annotation pipeline for microbial genomes. *BMC Bioinformatics.* 2021; 22:11.
Saiki RK. Primer-directed enzymatic amplification of DNA with a thermostable DNA polymerase. *Science.* 1988; 239:487–491.
Saiki et al. Enzymatic amplification of β-globin genomic sequences and restriction site analysis for diagnosis of sickle cell anemia. *Science.* 1985; 230(4732):1350–1354.
Sanger F and Coulson AR. A rapid method for determining sequences in DNA by primed synthesis with DNA polymerase. *Journal of Molecular Biology.* 1975; 94(3):441–448.
Schneider et al. Comparative fungal community analyses using metatranscriptomics and ITS-amplicon sequencing from Norway spruce. *mSystems.* 2021; 6(1):e00884-20.
Seemann T. Prokka: rapid prokaryotic genome annotation. *Bioinformatics.* 2014; 30(14):2068–2069.
Semmouri et al. Evaluating the potential of direct RNA nanopore sequencing: Metatranscriptomics highlights possible seasonal differences in a marine pelagic crustacean zooplankton community. *Marine Environmental Research.* 2019; 104836. doi: 10.1016/j.marenvres.2019.104836.
Shaffer et al. DRAM for distilling microbial metabolism to automate the curation of microbiome function. *Nucleic Acids Research.* 2020; 48(16):8883–8900.
Shakya et al. Advances and Challenges in Metatranscriptomic Analysis. *Frontiers in Genetics.* 2019; 10:904.
Shendure J. Accurate multiplex polony sequencing of an evolved bacterial genome. *Science.* 2005; 309: 1728–1732.
Smith LM. Fluorescence detection in automated DNA sequence analysis. *Nature.* 1986; 321:674–679.
Staden R. A strategy of DNA sequencing employing computer programs. *Nucleic Acids Research.* 1979;6: 2601–2610.
Tamames J and Puente-Sánchez F. SqueezeMeta, a highly portable, fully automatic metagenomic analysis pipeline. *Frontiers in Microbiology.* 2019; 9:3349.
Tran et al. The bacterial quorum sensing signal DSF hijacks Arabidopsis thaliana sterol biosynthesis to suppress plant innate immunity. *Life Science Alliance.* 2020; 3(10):e202000720.
Truong et al. MetaPhlAn2 for enhanced metagenomic taxonomic profiling. *Nature Methods.* 2015; 12(10):902–903.
Verce et al. Combined metagenomics and metatranscriptomics approach to unravel costa rican cocoa box fermentation processes reveals yet unreported microbial species and functionalities. *Frontiers in Microbiology.* 2021; 12:245.
Wang et al. Responses of soil microbial communities to a short-term application of seaweed fertilizer revealed by deep amplicon sequencing. *Applied Soil Ecology.*2018; 125:288–296.
Weber et al. Nephele: a cloud platform for simplified, standardized and reproducible microbiome data analysis. *Bioinformatics.* 2018; 34(8):1411–1413.
Westreich et al. SAMSA2: a standalone metatranscriptome analysis pipeline. *BMC Bioinformatics.* 2018; 19(1):1.
Wood et al. Improved metagenomic analysis with Kraken 2. *Genome Biology.* 2019; 20(1):1–3.
Wyatt NA and Friesen TL. Four reference quality genome assemblies of Pyrenophora teres f. maculata: a resource for studying the barley spot form net blotch interaction. *Molecular Plant-Microbe Interactions.* 2021; 34(1):135–139.
Ye et al. DBG2OLC: efficient assembly of large genomes using long erroneous reads of the third generation sequencing technologies. *Scientific Reports.* 2016; 6(1):1–9.
Yu et al. Complete genome sequence of the fire blight pathogen strain Erwinia amylovora Ea1189. *Molecular Plant-Microbe Interactions.* 2020; 33(11):1277–1279.
Zerbino DR. Using the Velvet de novo assembler for short-read sequencing technologies. *Current Protocols in Bioinformatics.* 2010; 31(1):11–15.
Zhu et al. Ab initio gene identification in metagenomic sequences. *Nucleic Acids Research.* 2010; 38(12):e132.
Zimin et al. The MaSuRCA genome assembler. *Bioinformatics.* 2013; 29(21):2669–2677.

8

Metaomics Technologies in Understanding Ethnomedicinal Plants and Endophyte Microbiome

Bishal Pun, Fenella M.W. Nongkhlaw, and S.R. Joshi*
North-Eastern Hill University

CONTENTS

8.1 Introduction

With an alarming worldwide resistance of infectious microorganisms to existing compounds, there is an urgent need for new antibiotics, agrochemicals, and novel compounds that are highly effective without toxicity (Tenguria and Khan, 2011). Endophytes are microorganisms that are found residing within the interior plant tissues, particularly in the leaves, branches, roots, and stems, without causing visible damage to their hosts.

The discovery of novel metabolites from endophytes is an important fallback to increasing level of drug-resistant pathogens inadequate number of effective antibiotics against diverse bacterial and fungal pathogens and biotic and abiotic stresses. However, identifying endophytes and defining its functions is a challenge. Techniques, such as 16S amplicon sequencing, whole-genome shotgun (WGS) sequencing, and metaomics approaches, have been used to identify endophytes and evaluate their diversity and function.

Modern genomics reveals both culturable and unculturable endophytes and subsequently characterization of endophytes from the studied environment. It is also possible to further evaluate the evolutionary trend of the associated microbes in relation to one another. As a result of long-held association between endophytes and their hosts, it is possible that some of these microbes devised genetic systems allowing for the transfer of information between themselves and higher plants (Stierle et al. 1993). Advancements in "omics" technologies have allowed the quantitative surveillance of the diverse biomolecules in biological system in a high-throughput system. Considering the complexities of host microbiomes, different

* Corresponding author's email: srjoshi2006@gmail.com

DOI: 10.1201/9781003171416-8

"omics" approaches have been introduced to get insights into the host microbiome systems biology (Zhang et al. 2010). Combination of individual sets of data acquired from several omics platforms is used to understand the dynamic, functional host–microbiome interaction (Tlaskalova-Hogenova et al. 2011).

Endophytes show host-specificity, in which microorganisms are restricted to a single host or a group of related species, and this indicates the occurrence of complex biochemical interactions between host and its associated microorganisms (Strobel, 2000; Strobel et al. 2004), raising enormous variability between endophytes, through mutation and genetic crossing, and thereby allowing transfer of information between themselves and host plants (Tan and Zou, 2001; Firáková et al. 2007). Metaomics studies give a better understanding of plant-associated microbial communities and their functional diversity. The central interest is the predictability of the composition of the endophytes from host plant identity. The composition of the endophytic communities in *Rhododendron* was similar among host species and clearly distinguishable from other plants (Raizen, 2013). However, endophytic communities may vary among genotypes of a certain host species and therefore not common among host taxa (Lamit et al. 2014).

8.2 Metaomics

The etymology of "omics" dates back to 1920 when Hans Winkler merged the term "gene" with "-ome" and coined the term "genome" to express the totality of genes in an organism (Winkler 1920). With the knowledge of the presence of noncoding DNA elements, the present meaning of the term "genome" represents the "complete genetic makeup of an organism" (Yadav 2007). In the following years, "genomics" was introduced as title for a journal in 1987 (Kuska 1998), and subsequently, the suffix "omics" was also adopted by various subjects such as transcriptomics, proteomics, and metabolomics (Abbot 1999; Joye and Palsson 2006). The term "metagenome" was first defined in 1998 by Handelsman et al. (1998) to describe the collective genome of soil microflora. Metagenomics was later used to specify culture-independent genomic analysis of microbial DNA from environmental communities (Schloss and Handelsman 2003). This was also taken over by other disciplines, which led to the emergence of the terms such as "metatranscriptomics," "metaproteomics," and "metametabolomics" (Schneider and Riedel 2010, Zengler and Palsson 2012, Jones et al. 2014). The term "metaomics" has recently been introduced as an umbrella term for all these techniques. It identifies the study of collection of organisms in the sense of Handelsman's metagenomics by any "omics" discipline (Figure 8.1).

8.3 Endophytic Metagenomics

The fact that less than 1% of the microorganisms are culturable in laboratory calls for the exploration of the remaining majorities of microbial population that are only viable in environment and are thus not culturable in laboratory conditions (Ward et al. 1990). Metagenomics is a contemporary genomic technique of culture-independent genomic analysis to acquire the knowledge of both the culturable and unculturable microorganisms. This method provides information to explore the untouched genetic reservoir of natural environment, circumventing the requirement for laboratory isolation and cultivation of individual species (Chen and Pachter 2005). It links to the environment genomics to prospect uncultivated microorganisms of the ecosystem (Cowan et al. 2015). It is vital to unmask the metabolic potential and important characteristics of the unculturable endophytes in order to realize the host–endophyte interaction. The application of metagenomics involves the determination of unculturable microbial communities and their putative functional characteristics (Sessitsch et al. 2012). It circumvents the requirement for isolation and cultivation of each species. Here, whole population DNA is extracted for the analysis of sequence information for microbial identification. Isolation of DNA of endophytic microbial community demands technical skills. It is essential to isolate good quality and quantity of total genomic DNA. It is challenging to isolate and sequence DNA of only bacterial community from the tissue of the target host plant with large number of plant cells. To prevent hindrance of host plant DNA, enrichment of endophytes host DNA is important before DNA amplification (Govindasamy et al. 2014). Metagenomic DNA can be analyzed for functional screening by libraries construction followed by the expression in suitable vector and screening of novel traits (Rondon et al. 2000).

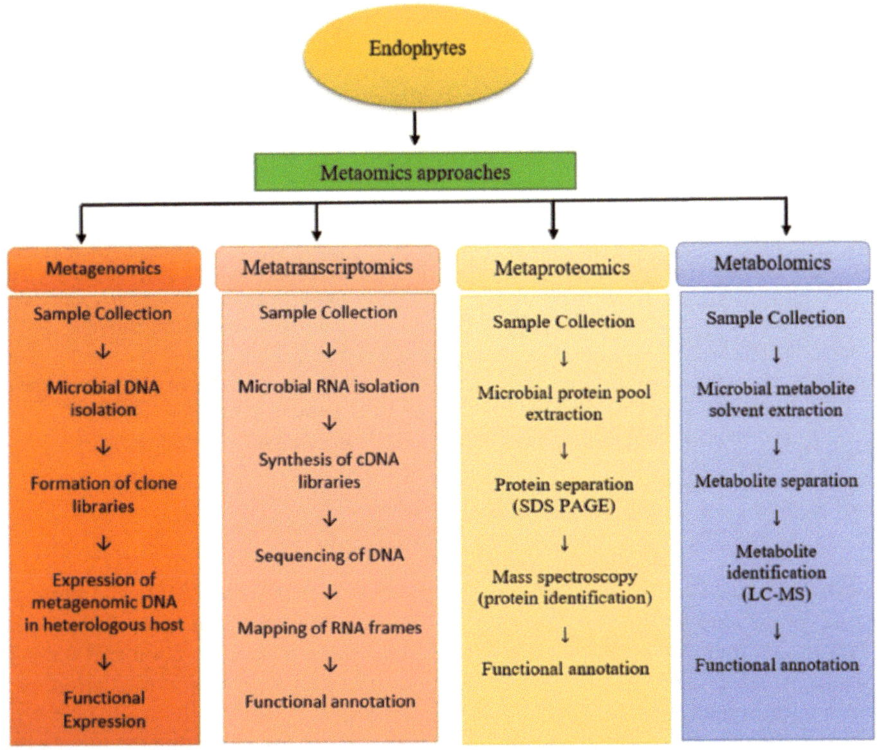

FIGURE 8.1 Metaomics approaches to understand endophytes associated with host plant.

8.4 Approaches to Studying Metagenome

Initially, metagenomics was applied for taxonomic and functional analysis of genome of microbial community from environmental sample. This is called shotgun metagenomics (Handelsman 2004). Shotgun sequencing is useful in the identification of diversely present microorganisms in environmental samples. It has the advantage of connecting microbial diversity with functional and taxonomical analysis to understand the microbial community in an ecological niche (Brenig et al. 2010).

With time, metagenomics is broadly used to analyze targeted PCR-amplified genes of interest. This is referred to as targeted metagenomics. It is a low-cost and faster approach to studying microbial community profile using PCR-amplified conserved marker genes such as bacterial 16S rRNA gene and fungal internal transcribed spacer (ITS) (Shah et al. 2011) (Table 8.1).

The functions performed by endophytes as a result of their interaction with the host plant and other members of the host microbiome could be traced using modern genomic technologies. Studies in bioinformatics and next-generation sequencing (NGS) technologies have revealed the genes present in endophyte genomes that may be due to the effect of mutualistic association with their host; such genes are nitrogen fixation, phytohormone production (auxin, abscisins, ethylene, gibberellins, and indole acetic acid), mineral acquisition (Fe, P, etc.), acid phosphatase, 1-aminocyclopropane-1-carboxylic acid (ACC) deaminase, stress tolerance, adhesion, and other colonization genes (Fouts et al. 2008; Firrincieli et al. 2015; Martinez-Garcia et al. 2015). High-throughput sequencing has prompted metagenomic studies comparatively easier and rapid for the characterization of endophytes (Akinsanya et al. 2015). It has enabled the researchers to perform quick and affordable studies of DNA sequences from various ecological environments (Jones 2010). Endophytes produce enzymes necessary for the colonization of plant tissues, such as xylanase and pectinase, and produce non-specific peroxidases, laccases, chitinase,

TABLE 8.1

Comparison of Shotgun and Targeted Sequencing

Parameters	Shotgun Sequencing	Targeted Sequencing
Data generated	Functional categorization of microbial niche and reform of draft genome sequences for each member of microbial community	Phylogenetic structure of a microbial neighborhood indicated as OTUs.
Sensitivity	Sensitivity increased only through deep and intense sequencing	Very sensitive.
Biasedness	Sequence content bias	Prone to probe and PCR biasedness. The amplicon may not strictly depict the whole genome due to mutation or horizontal transfer.
Gene composition	Useful in new genes and biochemical pathways discoveries and produce extensive gene inventories and draft genomes.	The gene content and functional data of many species of microorganisms are unexplored with variability among strains.
Applied in	Detection of new species members, genes, and compound taxonomies	Population scrutinization.

glucanase, extracellular cellulase, and hemicellulases (Promputtha et al. 2011). The presence of these genes explains the ecology and evolution of endophytes and their role in nutrient cycling and ability to colonize plant tissue. Metagenomics reveals the mechanisms of host association and the role of endophytes in plant growth enhancement, biotic and abiotic stress tolerance, bioremediation, or protection from pathogens (Rodriguez et al. 2006; Choudhary et al. 2009).

Metagenomics allows researchers to use microbial communities to determine the metabolic processes of endophytes. Genome analysis has revealed the genes involved in the mechanism of production of antimicrobial and antioxidant agents, *N*-acyl homoserine lactone synthases and hydrolases, siderophores, and plant hormones, which shows the biotechnological potential of endophytes. Obtaining antimicrobial compounds among plant-associated microbes is a promising way to overcome the increasing threat of drug-resistant microbes against human and plant pathogens (Tan and Zou 2001; Yu et al. 2010). Antimicrobial compounds are low molecular weight organic compounds made by microorganisms. One such example is cryptocandin, which is an antifungal metabolite obtained from endophytic fungus *Cryptosporiopsis* (Selim et al. 2012). Cryptocandin showed antifungal activity against human fungal pathogens, including *Candida albicans* and *Trichophyton* spp., and against a number of plant pathogens, including *Sclerotinia sclerotiorum* and *Botrytis cinerea*. Cryptocandin and its related compounds are currently used against a number of pathogens causing diseases of the skin and nails (Strobel et al. 2004).

Lu et al. (2000) reported three compounds with antimicrobial activity against both human and plant pathogens from endophytic fungus *Colletotrichum* sp., isolated from medicinal plant *Artemisia annua*. *Muscodor* is a novel endophytic fungal genus that produces bioactive volatile organic compounds (VOCs). The VOCs produced by this fungus has extensive potential in agriculture, industry, and medicine (Selim et al. 2012). Endophytic *Muscodor albus* and *Muscodor crispans* produce a mixture of VOCs that act synergistically against a wide variety of plants and human pathogenic fungi and bacteria. Phomoenamide was isolated and showed *in vitro* antimycobacterial activity against *Mycobacterium tuberculosis*, the causative organism of tuberculosis (Rukachaisirikul et al. 2008). It is evident that plant-associated microbes could be utilized as a source for the production of reliable and novel antimicrobial agents, to meet the emergent demands of effective, less toxic, and environmentally friendly antibiotics.

Researchers use metagenomics to determine antioxidant and microbial characteristics in an attempt to develop superfoods from sprouted soybeans (Oyedeji 2018). Antioxidants have been considered a promising therapy for the prevention and treatment of reactive oxygen species (ROS)-linked diseases such as cancer, cardiovascular disease, hypertension, ischemia/reperfusion injury, diabetes mellitus, neurodegenerative diseases, rheumatoid arthritis, and aging (Valko et al. 2007). Many antioxidant compounds possess anti-inflammatory, antitumor, antimutagenic, anticarcinogenic, antibacterial, or antiviral activities in higher or lower level (Owen and Hundley 2004). Recently, the research on the potential application of natural antioxidants in stabilizing food stuffs against oxidation has received due attention (Gu and Weng 2001). The capacity of endophytic microorganisms to produce polysaccharides with antioxidant activity was observed in the bacterium endophyte *Paenibacillus polymyxa* isolated from the root tissue of *Stemona japonica*, a traditional Chinese medicine, producing exopolysaccharides (EPS) that demonstrate strong scavenging activities on superoxide and hydroxyl radicals (Liu et al. 2009). Pestacin and isopestacin isolated from endophytic fungus *Pestalotiopsis microspore* associated with the plant *Terminalia morobensis* display potent antioxidant activity. It was able to scavenge superoxide and hydroxyl free radicals in solution. Besides antioxidant activity, pestacin and isopestacin also showed antimycotic and antifungal activities (Strobel et al. 2002; Harper et al. 2003). Phomol and mevinic acid are anti-inflammatory compounds isolated from endophytic fungus *Phomopsis* (Weber et al. 2004). Phenylpropanoid compounds of endophytes have received much interest for medicinal applications due to their anticancer, antioxidant, antimicrobial, anti-inflammatory, and immunosuppressive properties (Korkina, 2007).

Endophytic microorganisms have high ability to produce various novel enzymes that could be used in various biotechnological applications (Firáková et al. 2007). Polyketides (PKs) and non-ribosomal peptides (NRPs) represent a large class of microbial natural products with important clinical and ecological impacts (Gulder and Moore 2009). Non-ribosomal peptide synthetase (NRPS) gene cluster is involved in the biosynthesis of antibiotics (penicillin and cephalosporin), and anti-inflammatory and immunosuppressive compounds (cyclosporine A) (Mankelow and Neilan 2000; Miller et al. 2012). Polyketide synthases (PKSs) and NRPSs are large modular enzymes, each of which contains the catalytic units required for recognition, activation, and peptide bond formation of the growing peptide chain by adenylation (A), thiolation (T), and condensation (C) (Mootz et al. 2002; Miller et al. 2012). Metagenomics and genome mining techniques have helped in the discovery of novel natural product biosynthetic gene clusters from diverse environments (Amoutzias et al. 2016; Wei et al. 2018). Studies reveal that PKs and NRPSs possess remarkable similarities in structure and catalytic activity (Du et al. 2001). Hybrid NRPS–PKS compounds have been isolated from numerous marine bacteria.

Phylogenetic analysis of the NRPS gene obtained from endophytes associated with medicinal plants of Meghalaya and the 16S rRNA encoding gene was used to predict the horizontal gene transfer (HGT) during gene evolution (Nongkhlaw and Joshi 2015a). The study involved the use of oligonucleotide primers that are suitable for the genetic screening of biologically active natural compounds that are pharmacologically important and utilized phylogenetic analysis of the NRPS and 16S rRNA encoding genes to predict the horizontal gene transfer (HGT) of NRPS among endophytic bacteria associated with ethnomedicinal plants (Nongkhlaw and Joshi 2015a) (Figure 8.2).

8.5 Comparative Metagenomics

Endophytes occupying internal plant tissue were visualized as single cells, doublets, or forming microcolonies (Hansen et al. 1997; Benizri et al. 2001). Bacteria sense the presence of neighboring cells by detecting an increase in the concentration of extracellular molecules, *N*-acyl homoserine lactones (AHLs) (Greenberg 1997). Plant recognition by microbes is considered to be the key initial event that occurs through adhesins, fimbriae, flagella, and type III and IV secretion systems (Lugtenberg et al. 2002). The commonly used methods in comparative metagenomics are 16S amplicon sequencing and WGS sequencing, which have provided insight into DNA sequencing analysis to identify endophytes and evaluate their diversity and abundance in various ecological communities (Huse et al. 2008). 16S rRNA amplicon consists of the conserved region interspersed by variable regions that promote sequencing and phylogenetic analysis (Yergeau et al. 2014). Here, the task of isolation and cultivation of individual species is excluded. Metagenomics approach is helpful in revealing the potential of uncultured endophytic communities by providing insight into groups of endophytes that are otherwise entirely unknown (Dinsdale et al. 2008). The first genome reconstruction of an uncultured organism is that of *Buchnera aphidicola*, which is an endosymbiont of an insect, aphid (Handelsman 2004). Genomic analysis has revealed that each partner in this relationship between the bacterium and insect functions dependently of the other.

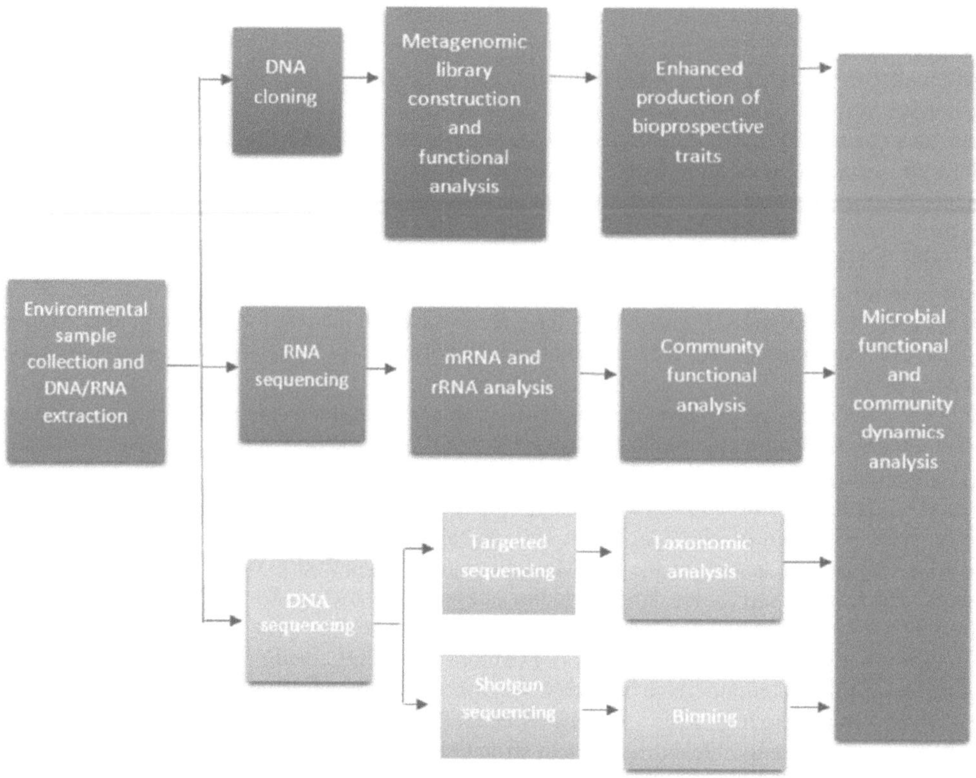

FIGURE 8.2 Standard metagenomics pipeline for microbiomes research.

Comparative metagenomics reveals the functional diversity among endophytes associated with the same or different host plants (Dinsdale et al. 2008). High-throughput sequencing called next-generation sequencing (NGS) has been a useful tool for rapid and economical metagenomic studies. Co-habitation of root-associated mycorrhizal and endophytic fungi of different plant species was displayed with the help of 454 pyrosequencing technique (Toju et al. 2013). Functional analysis requires successful transcription and translation of the gene of interest and secretion of the gene product. Functional analysis has identified novel antibiotics (Gillespie et al. 2002; Courtois et al. 2003; Venter et al. 2004), antibiotic resistance genes (Riesenfeld et al. 2004), and degradative enzymes (Henne et al. 2000). Comparative studies of differential expression profiles of endophytes within host plant can be helpful to identify interaction factors involved in maintaining the relationship. Researchers have revealed a potent antitumor compound, salinosporamide K, which was discovered with the help of comparative genomics between *Salinispora pacifica* genome and *Salinispora tropica* (Amoutzias et al. 2016). Reports on metagenomics and metaproteomics approach unraveled that alphaproteobacteria were found more frequently in rice phyllosphere than in rice rhizosphere (Knief et al. 2012), with the most abundant genera being *Rhizobium* and *Methylobacterium*, indicating that these genera prefer the host plant environment.

With the increasing amount of metagenomics data, the need for standardized procedures that allow projects to be analyzed and compared is becoming increasingly important in the field of metagenomics. Apart from NCBI, there are other fundamental tools and databases for analyzing microbial genomic/metagenomic data (Johannes et al. 2018). One of the most prominent is the Joint Genome Institute (JGI), which hosts the Genomes Online Database (GOLD), with information about the sequenced genomes (Mende et al. 2016). The Integrated Microbial Genomes (IMG) computational resources provide a framework for analyzing and reviewing the structural and functional annotations of genomes and metagenomes in a comparative way (Herlemann et al. 2011; Becraft et al. 2015). Other metagenomic computational resources include MG-RAST, which is an open-source Web application server that performs automatic phylogenetic and functional analysis of metagenomes (Andersson et al. 2010). The National Ecological Observatory Network (NEON), established by the National Science Foundation (NSF), is useful for characterizing and quantifying complex ecological processes associated with changing ecosystems.

8.6 Techniques Coupled with Metagenomics

Molecular techniques are effectively coupled with metagenomic studies for the characterization of endophytic microorganisms. These include polymerase chain reaction (PCR), denaturing gradient gel electrophoresis (DGGE), cloning and sequencing of ribosomal genes, fluorescence *in situ* hybridization (FISH), and restriction fragment length polymorphism (RFLP) (Sugiyama et al. 2014; Tschaplinski et al. 2014; Rivero et al. 2015).

Several advanced techniques such as gas chromatography–mass spectrometry (GC–MS), liquid chromatography–mass spectrometry (LC–MS), and capillary electrophoresis–mass spectrometry (CE–MS) are used for metabolomic profiling of endophytes associated with medicinal plants (Zhang et al. 2012).

8.7 Software for Metagenomic Analysis

The advancement in the NGS technologies has resulted in the widespread application of the techniques in different areas of research (Parmar et al. 2017). These techniques produce a huge amount of data with great complexity that demands proper data analysis and explanation. Vast nucleotide sequence data are generated from metagenome sequencing, which need to be analyzed by bioinformatics software to get meaningful results. Therefore, an authentic and dependable bioinformatics tool is essential to translate the raw data into a biologically meaningful manner (Goodwin et al. 2016). Software packages such as mothur (https://www.mothur.org), Quantitative Insights Into Microbial Ecology (QIIME), MEGAN, and CARMA are available for amplicon analysis. In silico predictive tools such as PICRUSt (Phylogenetic Investigation of Communities by Reconstruction of Unobserved States) software link taxonomic classification with metabolic information from meta-profiling data (Langille et al. 2013). Community enabling platforms to address analysis of different types of metagenomic data is available, which includes IMG/M (Integrated Microbial Genome and Microbiomes), CAMERA (Cyberinfrastructure for Advanced Marine Microbial Ecology Research and Analysis) and WebCARMA (Gerlach et al. 2009). The availability of new and reliable bioinformatics tools has resulted in easier bioinformatics analysis of metaomics (Figure 8.3).

8.8 Endophytic Metatranscriptomics

While metagenomic studies unveil the presence or absence of particular genes, transcriptomics helps to study the expression of the genes. Transcriptomic approach is useful for the analysis of differently expressed genes of microbial communities in the host plant and for understanding their interaction in different microenvironments (Molina et al. 2012). Metatranscriptomic approach is important to realize the true significance of endophytism. Metatranscriptomics of several important plants has let out the existence of several RNA sequences of various associative and symbiotic microbes unrelated to the host genome (Camilious-Neto et al. 2014). This technique unveils information of the expressed genes and active metabolic pathways of the endophytes in a given condition in a host plant (Simon and Daniel 2011). Comparative analysis of differential expression of the host plant genes in endophyte-free and with endophytes as well as expression profile of endophytes within the host plant is a useful approach to understanding the symbiotic interaction (Johnson et al. 2004) (Figure 8.4, Table 8.2).

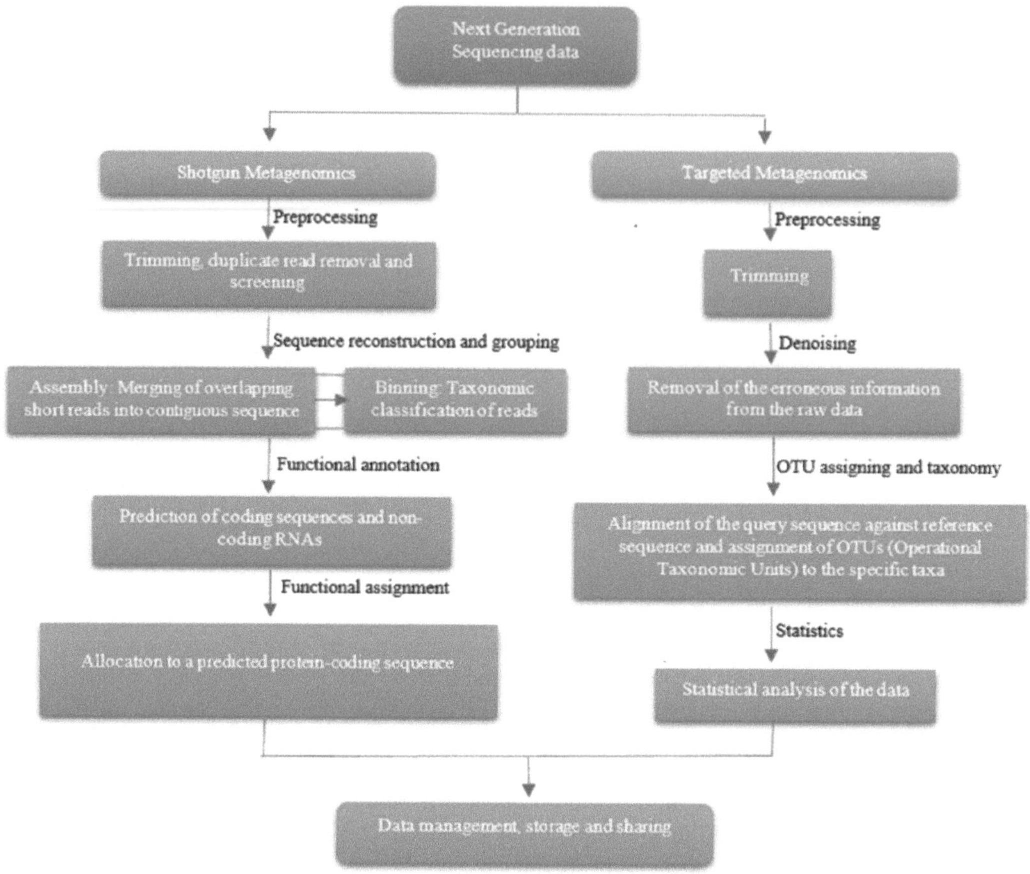

FIGURE 8.3 Flowchart of basic steps in metagenomic data analysis.

8.9 Endophytic Metaproteomics

Proteomics is the study of the entire protein complement expressed by an organism (Wilkins et al. 1995). Metaproteomics or whole genome proteomics is the large-scale identification and functional expression of metagenome from microbial communities to understand their metabolic activities (Maron et al. 2007). Metaproteomics employs high-performance mass spectroscopy as a dominant technological platform for proteomic measurement from microbial community. Proteomics studies also require genomic information to provide data connecting genetic and functional diversity of microbial communities. Metaproteogenomics links the proteome and the genome of the sample by combinatorial analysis of metagenome and metaproteome of the same environmental sample. This approach enables the spotting of a greater number of proteins to understand the functional diversity than proteomics alone (Delmotte et al. 2009). The technique overcomes the obstacles of protein identification as in metaproteomic studies because of the unavailability of closely related reference genomes.

Metaproteomics provides a suitable approach to studying the microbial communities associated with different medicinal plants. While genome and metagenome studies reveal the presence or absence of specific genes, the expression of specific genes in an endophytic community of an environmental sample is essential to understand the metabolic activities occurring within a community and elucidate endophytic function. A study on metaproteome data sets shows the information about microbial community

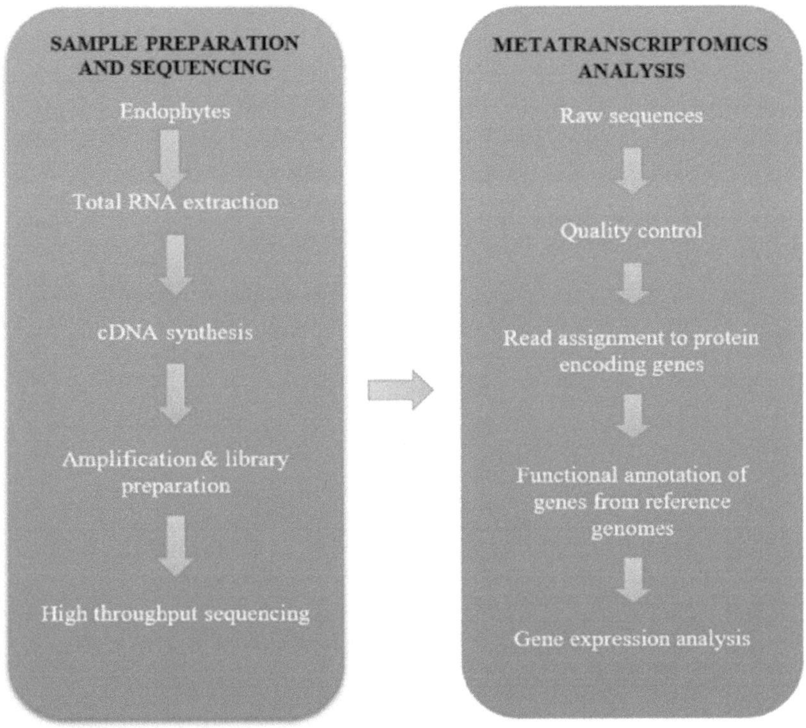

FIGURE 8.4 Flowchart of major steps in metatranscriptomics analysis.

structure, dynamics, and functional activities that provide knowledge on various community aspects, such as microbial cooperation and competition for nutrients, and energy transfer across the community (Hettich et al. 2013). Metagenomics data in combination with metaproteomics data could produce substantial evidence on the activities of endophytes (Figure 8.5).

Metaproteomics exploits the power of high-performance mass spectrometry for extensive study of proteins expressed by endophytes in an environmental sample. However, to obtain desirable metaproteome data sets, efficient protein extraction from environmental samples is required; peptide/protein separation/fractionation is required prior to detection and followed by high-throughput peptide/protein identification (Hettich et al. 2013). Proteins involved in plant–endophyte interactions that could not be studied by cultivated methods are now targets for metaproteomic studies. Proteomics has revealed that despite the presence of *nifH* genes in both the phyllosphere and rhizosphere microenvironments of rice, however, the expression of *nifH* gene was only found in the rhizosphere (Knief et al. 2012).

Endophytic protein secretion system can be used to ascertain plant–bacterial interactions (Downie 2010). Reports on metaproteomics on the proteome differences between laboratory-cultured microbes and plant symbiont microbes have revealed such interactions (Knief et al. 2011). Other studies include the effect of proteins on crop rhizosphere communities (Wang et al. 2011; Wu et al. 2011).

Additionally, the application of metaproteomics in aquatic environmental microbiology has successfully revealed the importance of bacteriochlorophyll in the adaptation to low light intensity (Ng et al. 2010), the metabolic activity that allows life in the cold oligotrophic environments (Lauro et al. 2011), and the significance of high-affinity transporters for substrate acquisition in marine bacteria (Sowell et al. 2011). Metaproteomics study has revealed ATP-binding cassette (ABC) transporter proteins in alpha- and betaproteobacteria to be more abundant in oligotrophic than in eutrophic conditions in Atlantic Ocean, where the environment was rich in algal and cyanobacterial exudates (Russo et al. 2016) (Table 8.3).

TABLE 8.2

List of Tools Used for Metagenomics

Tool	Validated on	Links
OTU Clustering		
UPARSE	16S, ITS	drive5.com/uparse
GramCluster	16S	bioinfo.unl.edu/gramcluster.php
DNACLUST	16S	dnaclust.sourceforge.net
DOTUR	16S, rpoB	github.com/mothur/DOTUR
Swarm	16S	github.com/torognes/swarm
BACDNAS	16S	www.helsinki.fi/bsg/software/BACDNAS/
mothur	16S	github.com/mothur/Mothur
CD-HIT	16S	cd-hit.org
OTUCLUST	16S, ITS	github.com/compmetagen/micca/wiki
Binning		
MaxBin	Bacterial genomes	sourceforge.net/projects/maxbin/
MetaCluster	Bacterial, whole metagenome (viral included)	i.cs.hku.hk/~alse/MetaCluster
TETRA	Only 16s	www.megx.net/tetra
DNACLUST	AMD data set	dnaclust.sourceforge.net
LSA	Bacterial genomes, some remarks on phage analysis	github.com/brian-cleary/LatentStrainAnalysis
TOSS	Bacterial genomes	www.cs.ucr.edu/~tanaseio/toss.htm
CompostBin	Bacterial genomes	bobcat.genomecenter.ucdavis.edu/souravc/compostbin/
MetaBAT	Bacterial genomes	bitbucket.org/berkeleylab/metabat
SCIMM	Only 16s	www.cbcb.umd.edu/software/scimm
LikelyBin	Bacterial genomes	ecotheory.biology.gatech.edu/downloads/likelybin
Diversity Profiling and Taxonomic Assignment		
TAC-ELM	Whole metagenome	cs.gmu.edu/~mlbio/TAC-ELM
RDP classifier	16S, LSU	rdp.cme.msu.edu
PhyloSift	16S	github.com/gjospin/PhyloSift
CSSS	Viral, bacterial	collaborators.oicr.on.ca/vferretti/borozan_csss/csss.html
NBC	16S, ITS, LSU, viral, fungal	nbc.ece.drexel.edu
FCP	Bacterial genomes	kiwi.cs.dal.ca/Software/FCP
Taxy	Whole metagenome	gobics.de/peter/taxy
16S Classifier	16S	metabiosys.iiserb.ac.in/16Sclassifier/application.php
Phymm	Whole metagenome	www.cbcb.umd.edu/software/phymm/
WGSQuikr	Whole metagenome	sourceforge.net/projects/wgsquikr
Gene Detection		
MetaGene	Bacterial	omics.informatics.indiana.edu/FragGeneScan
MetaGeneAnnotator	Bacteria, archaea, prophage	metagene.cb.k.u-tokyo.ac.jp
MetaGeneMark	Bacteria, archaea	Web access
MetaProdigal	Bacterial	github.com/hyattpd/Prodigal
Orphelia	Bacterial	orphelia.gobics.de
Glimmer-MG	Whole metagenome	www.cbcb.umd.edu/software/glimmer-mg
FragGeneScan	Bacteria, whole metagenome	omics.informatics.indiana.edu/FragGeneScan
Comparative Metagenomics		
R-SVM	Plants, bacteria	None
MetaPhyl	16S	www.cs.ucr.edu/~tanaseio/ metaphyl.htm
DectICO	Whole bacterial metagenome	github.com/dingxiao8715/DectICO
MetaDistance	16S	metadistance.igs.umaryland.edu/Welcome.html

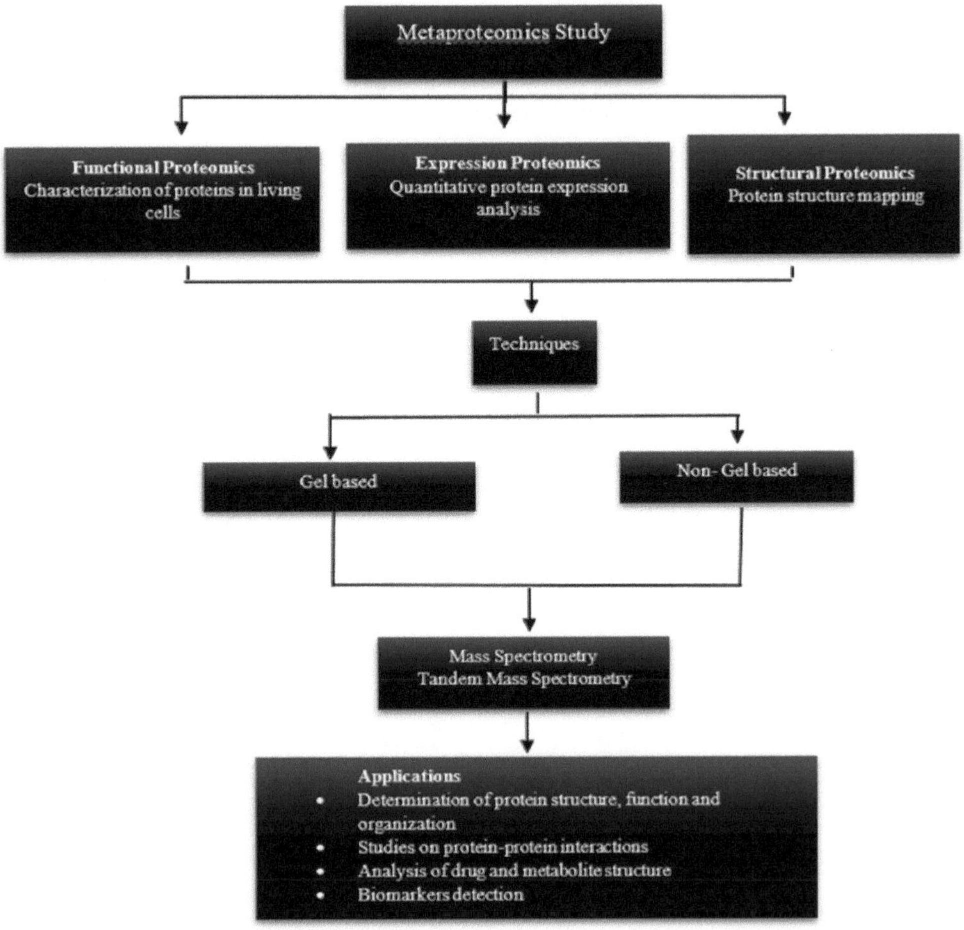

FIGURE 8.5 Techniques and applications of metaproteomics.

8.10 Endophytic Metabolomics

Metabolomics presents a useful approach to studying endophyte–plant interactions. Endophytes coexisting with medicinal plants produce numerous bioactive compounds, which regulate the plant species in terms of growth, metabolic pathways, and defense systems. Metabolomics is generally coupled with other omics techniques to validate the study by connecting the genotype with the phenotype.

According to Goodacre et al. (2004), metabolomics studies of living systems can be classified as follows:

1. Metabolite target analysis: study of metabolites confined to a single-enzyme system.
2. Metabolomics: whole metabolome analysis under a given condition.
3. Metabolite profiling: study of a set of metabolites from a selected pathway.
4. Metabolic profiling: majorly applied in the clinical sense to analyze and follow the fate of a drug or metabolite.
5. Metabolic fingerprinting: sanalysis of chemicals or metabolites based on their biological origin.
6. Metabonomics: study of biochemical profile of living systems to pathophysiological stimuli in a disease condition or upon the effect of a drug or toxin.

TABLE 8.3

Endophytes of Medicinal Plants and Their Potential Applications

Endophyte	Type of Endophyte	Medicinal Host Plant	Biological Activity of Endophyte	References
Actinoallomurus caesius	Bacteria	*Acacia auriculiformis*	Antibacterial activity	Bunyoo et al. (2009)
Actinopolyspora spp.	Bacteria	*Ocimum sanctum* and *Mentha arvensis*	Antagonistic activity against phytopathogenic fungi	Gangwar et al. (2014)
Alternaria sp.	Fungi	*Ricinus communis*	Insecticidal activity	Parthasarathy and Sathiyabama (2014)
Arthrobacter sp.	Bacteria	*Leucas ciliata*	Antioxidant activity	Akshatha et al. (2016)
Aspergillus flavipes	Fungi	*Acanthus ilicifolius*	Cytotoxicity	Taechowisan et al. (2006)
Bacillus subtilis	Bacteria	*Centella asiatica*	Antifungal	Zhou et al. (2013)
Botryosphaeria sp.	Fungi	*Garcinia atroviridis*	Antimicrobial	Gangwar et al. (2017)
Colletotrichum gloeosporioides	Fungi	*Justicia gendarussa*	Anticancer	Gangadevi and Muthumary (2008)
Fusarium decemcellulare	Fungi	*Flacourtia inermis*	Anticoagulant	Qader et al. (2018)
Fusarium oxysporum	Fungi	*Dendrobium lindleyi*	Antimutagenic	Bungtongdee et al. (2019)
Fusarium solani	Fungi	*Phaius tankervilleae*	Antimalarial	Yan et al. (2010)
Microbispora sp.	Bacteria	Mandarin	Plant growth-promoting activity	Shutsrirung et al. (2013)
Nocardiopsis alba	Bacteria	Mandarin	Plant growth-promoting activity	Shutsrirung et al. (2013)
Nocardiopsis sp.	Bacteria	*Triticum aestivum*	Improve plant growth	Jog et al. (2014)
Nocardiopsis umidischolae	Bacteria	*Aloe arborescens*	Antibiotic activity against gram-positive bacteria	Machavariani et al. (2014)
Penicillium oxalicum	Fungi	*Gymnema sylvestre*	Antidiabetic agent	Deshmukh Sunil et al. (2015)
Streptomyces albosporus	Bacteria	*Aloe vera*, *Mentha arvensis*, and *Ocimum sanctum*	Antagonistic activity against one or more phytopathogenic fungi	Gangwar et al. (2014)
Streptomyces albidoflavus	Bacteria	*Bruguiera gymnorrhiza*	Antifungal activity	Igarashi (2004)
Streptomyces globosus	Bacteria	*Rauwolfia densiflora*	Antioxidant activity	Akshatha et al. (2016)
Streptomyces parvulus	Bacteria	*Abutilon indicum*	Antibiotic	Chandrakar and Gupta (2019)
Streptomyces sp.	Bacteria	Maize plant	Biocontrol of phytopathogenic fungi	Costa et al. (2013)

Metabolomics studies require the identification of chemical compounds for metabolite profiling. Mass spectroscopy coupled with gas chromatography (GC–MS) and liquid chromatography (LC–MS) is central to metabolite profiling. Table 8.4 lists out links of some popular tools helpful in metabolomics analysis. This analysis consists of preprocessing, statistical analysis, and machine learning techniques for pattern similarity with available databases (Aguiar-Polido et al. 2016).

Phenolics, flavonoids, and alkaloids are secondary plant metabolites that are ubiquitously present in plants; however, they are also observed among endophytic bacteria associated with medicinal plants (Nongkhlaw and Joshi 2015b). The presence of such metabolites enhances the level of antioxidant (Rice-Evans et al. 1996) and antagonistic activities (Pereira et al. 2007) of endophytes. Endophytes are known to produce a variety of novel biologically active secondary metabolites that are of potential use in agricultural and pharmaceutical industries (Tan and Zou 2001; Zhang et al. 2006; Lin et al. 2007). This has attracted global attention, making endophytes good candidates for the isolation and characterization of

TABLE 8.4

List of Software Tools for Metabolomics Analysis

Software Tools	Source Code	References
Preprocessing		
MetaboliteDetector	http://md.tu-bs.de	Hiller et al. (2009)
MetAlign	http://www.wageningenur.nl/en/show/MetAlign-1.htm	Lommen and Kools (2012)
MSeasy	https://cran.r-project.org/web/packages/MSeasy/ index.	Nicolè et al. (2012)
XCMS	http://bioconductor.org/packages/release/bioc/html/xcms.html	Smith et al. (2006)
Metabolite Annotation		
batman	http://batman.r-forge.r-project.org	Hao et al. (2012)
FingerID	https://github.com/icdishb/fingerid	Heinonen et al. (2012)
MAGMa	http://www.emetabolomics.org/magma	Ridder et al. (2013)
MetaboMiner	http://wishart.biology.ualberta.ca/metabominer	Xia et al. (2008)
SIRIUS	https://bio.informatik.uni-jena.de/software/sirius	Kim et al. (2016)
Post-Processing		
batchCorr	https://gitlab.com/CarlBrunius/batchCorr	Brunius et al. (2016)
EigenMS	https://sourceforge.net/projects/eigenms	Karpievitch et al. (2014)
Metabolomics	https://cran.r-project.org/web/packages/ metabolomics/	De Livera et al. (2012)
metabnorm	https://sourceforge.net/projects/metabnorm	Jauhiainen et al. (2014)
MetabR	http://metabr.r-forge.r-project.org/	Ernest et al. (2012)
MSPrep	https://sourceforge.net/projects/msprep/	Hughes et al. (2014)
Statistical Analysis		
Ionwinze	https://sourceforge.net/projects/ionwinze	Kokubun and D'Costa (2013)
MetabolAnalyze	https://cran.r-project.org/web/packages/MetabolAnalyze	Nyamundanda et al. (2010)
muma	https://cran.r-project.org/web/packages/muma/	Gaude et al. (2013)

bioactive compounds. Some of the metabolites reported among endophytes are L-aspartic acid, L-lysine, L-threonine, 4-methyl-2-oxopentanoate, pantothenic acid, 3-pyrrolidin-2-yl-propionic acid, pyrrolo[1,2-a]pyrazine-1,4-dione, hexahydro-3-(2-methylpropyl), etc. (Prasanna et al. 2012; Wei et al. 2020).

Novel products from microorganisms are produced differently from artificial chemical synthesis such that endophytes can synthesize a variety of secondary metabolites commercially at a large scale *via* fermentation (Strobel 2000). The importance of endophytic metabolites is for their usage not only as drugs by humankind, but also as food preservatives in the control of food spoilage and food-borne diseases, which are a serious concern in the food chain (Liu et al. 2008). Many studies have proved that the metabolites produced by endophytes enhance the defense response to both biotic and abiotic stresses and effectively decrease the survival rate of tumor cells (Zou et al. 2001; Strobel et al. 2004). Endophytes are a source of L-asparaginase and natural antioxidants that could have great relevance as therapeutic agents (Nongkhlaw and Joshi 2015b). L-Asparaginase is an important chemotherapeutic agent used for the treatment of a variety of lymphoproliferative disorders, such as lymphosarcoma and acute lymphoblastic leukemia (Gallagher et al. 1999). It is preferred over the chemical drugs for its biodegradability and nontoxicity and could be administered at local site quite easily (Stecher et al. 1999; Ferrare et al. 2004).

L-Asparaginase is also being used in food industry, as it is known to reduce the formation of carcinogenic acryl amides in deep-fried potato recipes (Kornbrust et al. 2010). Pre-treatment of unprocessed food materials with the enzyme L-asparaginase leads to the reduction in free asparagine to a significant level, thus reducing the imminent risk of the synthesis of acrylamide (Lindsay and Jang 2005).

8.11 Methodological Challenges

The frequent problem for most metaomics approaches is extracting target molecules with good quality and reproducibility from the sample. In the past few years, significant advancements have been made in the purification of target molecules (Tan and Yiap 2009). The suitability of the many protocols available for the nucleic acid (Fredricks et al. 2005; Kennedy et al. 2014) and protein (Wu et al. 2014; Zheng et al. 2007) extraction varies among different sample types. A universal protocol for the extraction of proteins yields only satisfactory recovery rates compared to nucleic acids (Taylor and Williams 2010). Thus, protocols yielding reproducible results especially among varying matrices are an urgent requisite for metaproteomics (VerBerkmoes et al. 2009). To circumvent post-sampling alteration, sample preparation for metabolite extraction demands instant freezing and lyophilization. Each extraction protocol has to be generated for individual sample type and metabolite composition (Lankadurai et al. 2013).

A major problem for every sequence-based metaomics studies is connecting biological information to sequence data through their taxonomic assignment. The reliability of publicly available reference databases and software pipelines for biological and ecological assignment must be determined from scores on sequence similarities and alignment coverage (Peršoh et al. 2010). Genome and transcriptome sequencing has become more practical, attributable to cost reduction and expanding capacities of NGS technologies (Segata et al. 2013). The assembly of several sequence reads from complex communities is a prime challenge (Wooley et al. 2010). This step is error-prone and can be overcome by directly mapping the raw read sequences against a reference database (Davenport and Tummler 2013).

Metatranscriptomics studies need significant care during sample preparation and processing due to the short half-life of mRNA. Following sampling, the mRNA has to be stabilized to circumvent the problem of quantitative biases (Persoh et al. 2012). Metaproteomics studies to quantify proteins or peptides by applying normalized spectral abundance factor (NSAF), in some cases, did not correlate to the actual independently estimated enzyme activities (Schneider et al. 2012).

The next major challenge is the quantification of the respective molecules in the extracts. The actual abundance of a molecule is usually not predictable simply from the presence of a certain gene and its transcript because of transcriptional, post-transcriptional, and post-translational modifications. Inference of biomolecule abundance from the abundance of amplicons also demands strenuous experiments (Raidl et al. 2005). Also PCR amplification rates are prone to biases caused by several factors such as primer selectivity; binding kinetics; and length, composition, and nucleotide sequence of the amplicons (U'Ren et al. 2014).

8.12 Conclusions

Metaomics has provided abundance of knowledge attained from cultured and uncultured microbial life. Metaomics analysis of endophytic microorganisms has a significant influence on our knowledge of understanding the benefits of endophytes associated with medicinal plants. It has brought new insights with more resolution in the structure and function of the microbial community. Many novel compounds and secondary metabolites produced by endophytes possess therapeutic properties that can be utilized for human welfare. However, improvement in metadata collection and computational tools is required to make the experimental protocols more efficient and useful in biotechnology. However, with enormous research and findings on medicinal plant-associated endophytes with unprecedented bioactivity, the commercial production of such compounds is still at its early stages. Thus, the vast diversity of microorganisms present in metagenome can be made available to its full potential by the application of metaomics approaches.

REFERENCES

Abbot, A. (1999). Proteomics, transcriptomics: what's in a name? *Nature*. 402: 715.

Aguiar-Pulido, V., Huang, W., Suarez-Ulloa, V., Cickovski, T., Mathee, K., and Narasimhan, G. (2016). Metagenomics, metatranscriptomics, and metabolomics approaches for microbiome analysis. *Evol. Bioinformat. Onli.* 12(1): 5.

Akinsanya, M. A., Goh, J. K., Lim, S. P., and Ting, A. S. Y.. (2015). Metagenomics study of endophytic bacteria in *Aloe vera* using next-generation technology. *Genom. Data.* 6: 159–163.

Akshatha, J. V., Prakash, H. S., and Nalini, M. S.. (2016). Actinomycete endophytes from the ethno-medicinal plants of southern India: antioxidant activity and characterization studies. *J. Biol. Activ. Prod. Nat.* 6(2): 166–172.

Andersson, A. F. and Riemann, L. (2010). Bertilsson S. Pyrosequencing reveals contrasting seasonal dynamics of taxa within Baltic Sea bacterioplankton communities. *ISME J.* 4: 171–181.

Andreote, F. D., Carneiro, R. T., Salles, J. F., Marcon, J., Labate, C. A., Azevedo, J. L., and Araújo, W. L.. (2009). Culture independent assessment of rhizobiales-related alpha proteobacteria and the diversity of *Methylobacterium*in the rhizosphere and rhizoplane of transgenic eucalyptus. *Microbi. Ecol.* 57 (1): 82–93.

Araújo, W. L., Marcon, J., Maccheroni Jr, W., Van Elsas, J. D., Van Vuurde, J. W., and Azevedo, J. L.. (2002). Diversity of endophytic bacterial populations and their interaction with *Xylella fastidiosa* in citrus plants. *Appl. Environ. Microbiol.* 10: 4906–4914.

Becraft, E.D., Dodsworth, J.A., Murugapiran, S.K., Ohlsson, J.I., Briggs, B.R., Kanbar, J., De Vlaminck, I., Quake, S.R., Dong, H., Hedlund, B.P., and Swingley, W.D. (2015). Single-cell-genomics-facilitated read binning of candidate phylum EM19 genomes from geothermal spring metagenomes. *Appl. Environ. Microbiol.* 82: 992–1003.

Brenig, B., Beck, J., and Schütz, E. (2010). Shotgun metagenomics of biological stains using ultra-deep DNA sequencing. *Forens. Sci. Int. Genet.* 4(4): 228–231.

Brunius, C., Shi, L., and Landberg, R. (2016). Large-scale untargeted LC-MS metabolomics data correction using between-batch feature alignment and cluster-based within-batch signal intensity drift correction. *Metabology* 12(11): 173.

Bungtongdee, N., Sopalun, K., Laosripaiboon, W., and Iamtham, S. (2019). The chemical composition, antifungal, antioxidant and antimutagenicity properties of bioactive compounds from fungal endophytes associated with Thai orchids. *J Phytopathol.* 167:56–64.

Bunyoo, C., Duangmal, K., Nuntagij, A., and Thamchaipenet, A. (2009). Characterisation of endophytic actinomycetes isolated from wattle trees (*Acacia auriculiformis*A. Cunn. ex Benth.) in Thailand. *Thai. J. Genet.* 2: 155–163.

Camilios-Neto, D., Bonato, P., Wassem, R., Tadra-Sfeir, M.Z., Brusamarello-Santos, L.C., Valdameri, G., Donatti, L., Faoro, H., Weiss, V.A., Chubatsu, L.S., and Pedrosa, F.O. (2014). Dual RNA-seq transcriptional analysis of wheat roots colonized by *Azospirillum brasilense* reveals up-regulation of nutrient acquisition and cell cycle genes. *BMC Genom.* 15: 378.

Chandrakar, S. and Gupta, A. K. (2019). Studies on the production of broad spectrum antimicrobial compound polypeptide (actinomycins) and lipopeptide (fengycin) from *Streptomyces* sp. K-R1 associated with root of *Abutilon indicum* against multidrug resistant human pathogens. *Int. J. Pept. Res. Ther.* 25:779–798.

Chen, K. and Pachter, L. (2005) Bioinformatics for whole-genome shotgun sequencing of microbial communities. *PLoS Comp. Biol.* 1(2): 24.

Choudhary, D. K. and Johri. B. N. (2009). Interactions of *Bacillus* spp. and plants-with special reference to induced systemic resistance (ISR). *Microbiol. Res.* 164: 493–513.

Costa, F.G., Zucchi, T.D., and Melo, I.S.D. (2013). Biological control of phytopathogenic fungi by endophytic actinomycetes isolated from maize (*Zea mays* L.). *Braz. Arch. Biol. Technol.* 56(6): 948–955.

Courtois, S., Cappellano, C.M., Ball, M., Francou, F.X., Normand, P., Helynck, G., Martinez, A., Kolvek, S.J., Hopke, J., Osburne, M.S., and August, P.R. (2003). Recombinant environmental libraries provide access to microbial diversity for drug discovery from natural products. *Appl. Environ. Microbiol.* 69: 49–55.

Cowan, D.A., Ramond, J.B., Makhalanyane, T.P., and De Maayer, P. (2015). Metagenomics of extreme environments. *Curr. Opin. Microbiol.* 25: 97–102.

Davenport, C. F. and Tümmler, B. (2013). Advances in computational analysis of metagenome sequences. *Environ. Microbiol.* 15: 1–5.

De Livera, A.M., Dias, D.A., De Souza, D., Rupasinghe, T., Pyke, J., Tull, D., Roessner, U., McConville, M., and Speed, T.P. (2012). Normalizing and integrating metabolomics data. *Analyt. Chemist.* 84(24): 10768–10776.

Delmotte, N., Knief, C., Chaffron, S., Innerebner, G., Roschitzki, B., Schlapbach, R., von Mering, C., and Vorholt, J.A. (2009). Community proteogenomics reveals insights into the physiology of phyllosphere bacteria. *Proc. Natl. Acad. Sci.* 106: 16428–16433.

Deshmukh, S.K., Verekar, S.A., and Bhave, S.V. (2015). Endophytic fungi: a reservoir of antibacterials. *Front. Microbiol.* 5: 1–43.

Dinsdale, E.A., Edwards, R.A., Hall, D., Angly, F., Breitbart, M., Brulc, J.M., Furlan, M., Desnues, C., Haynes, M., Li, L., and McDaniel, L. (2008). Functional metagenomic profiling of nine biomes. *Nature* 452: 629–632.

Downie, J. A. (2010). The roles of extracellular proteins, polysaccharides and signals in the interactions of rhizobia with legume roots. *FEMS Microbiol. Rev.* 34: 150–170.

Ernest, B., Gooding, J.R., Campagna, S.R., Saxton, A.M., and Voy, B.H. (2012). MetabR: an R script for linear model analysis of quantitative metabolomic data. *BMC Res. Not.,* 5(1): 596.

Firáková, S., Šturdíková, M., and Múčková, M. (2007). Bioactive secondary metabolites produced by microorganisms associated with plants. *Biologia* 62: 251–257.

Firrincieli, A., Otillar, R., Salamov, A., Schmutz, J., Khan, Z., Redman, R.S., Fleck, N.D., Lindquist, E., Grigoriev, I.V., and Doty, S.L. (2015). Genome sequence of the plant growth promoting endophytic yeast *Rhodotorula graminis* WP1. *Front. Microbiol.* 6: 978.

Fouts, D.E., Tyler, H.L., DeBoy, R.T., Daugherty, S., Ren, Q., Badger, J.H., Durkin, A.S., Huot, H., Shrivastava, S., Kothari, S., and Dodson, R.J. (2008). Complete genome sequence of the N_2-fixing broad host range endophyte *Klebsiella pneumoniae* 342 and virulence predictions verified in mice. *PLoS Genet.* 4: 1000141.

Fredricks, D.N., Smith, C., and Meier, A. (2005). Comparison of six DNA extraction methods for recovery of fungal DNA as assessed by quantitative PCR. *J. Clin. Microbiol.* 43: 5122–5128.

Gallagher, M. P., Marshall, R. D., and Wilson, R. (1999). Asparaginase as a drug for treatment of acute lymphoblastic leukaemia. *Essa. Biochem.* 24: 1–40.

Gangadevi, V. and Muthumary, J. (2008). Isolation of *Colletotrichum gloeosporioides*, a novel endophytic taxol-producing fungus from the leaves of a medicinal plant. *Justicia Gendarussa. Mycol. Balc.* 5:1–4.

Gangwar, M., Dogra, S., and Sharma, N. (2011). Antagonistic bioactivity of endophytic actinomycetes isolated from medicinal plants. *J. Adv. Lab. Res. Biol.* 2(4): 154–157.

Gangwar, M., Dogra, S., Gupta, U.P., and Kharwar, R.N.. (2014) Diversity and biopotential of endophytic actinomycetes from three medicinal plants in India. *Afr. J. Microbiol. Res.* 8(2): 184–191.

Madhurama, G., Priyanka, K., Preeti, S., and Anu, K. (2017). Assessment of bioactivity of endophytic actinomycetes from some medicinal plants. *Agric. Res. J.* 54:58.

Gaude, E., Chignola, F., Spiliotopoulos, D., Spitaleri, A., Ghitti, M., Garcìa-Manteiga, J.M., Mari, S., and Musco, G. (2013). muma, An R package for metabolomics univariate and multivariate statistical analysis. *Curr. Metabol.,* 1(2): 180–189.

Gerlach, W., Jünemann, S., Tille, F., Goesmann, A., and Stoye, J. (2009). WebCARMA: a web application for the functional and taxonomic classification of unassembled metagenomic reads. *BMC Bioinform.* 10: 430.

Gillespie, D.E., Brady, S.F., Bettermann, A.D., Cianciotto, N.P., Liles, M.R., Rondon, M.R., Clardy, J., Goodman, R.M., and Handelsman, J. (2002). Isolation of antibiotics turbomycin A and B from a metagenomic library of soil microbial DNA. *Appl. Environ. Microbiol.* 68: 4301–4306.

Goodacre, R., Vaidyanathan, S., Dunn, W.B., Harrigan, G.G., and Kell, D.B. (2004). Metabolomics by numbers: acquiring and understanding global metabolite data. *Trend. Biotechnol.* 22(5): 245–252.

Goodwin, S., McPherson, J.D., and McCombie, W.R. (2016). Coming of age: ten years of next-generation sequencing technologies. *Nat. Rev. Genet.* 17: 333–351.

Govindasamy, V., Franco, C.M., and Gupta, V.V. (2014). Endophytic actinobacteria: diversity and ecology. *Adv. Endophyt. Res.* 27–59.

Handelsman, J. (2004). Metagenomics: application of genomics to uncultured microorganisms. *Microbiol. Mol. Biol. Rev.* 68(4): 669–685.

Hao, J., Astle, W., De Iorio, M., and Ebbels, T.M. (2012). Batman: an R package for the automated quantification of metabolites from nuclear magnetic resonance spectra using a bayesian model. *Bioinformatics* 28(15): 2088–2090.

Heinonen, M., Shen, H., Zamboni, N., and Rousu, J. (2012). Metabolite identification and molecular fingerprint prediction through machine learning. *Bioinformatics* 28(18): 2333–2341.

Henne, A., Schmitz, R.A., Bömeke, M., Gottschalk, G., and Daniel, R. (2000). Screening of environmental DNA libraries for the presence of genes conferring lipolytic activity on *Escherichia coli. Appl. Environ. Microbiol.* 66: 3113–3116.

Herlemann, D.P., Labrenz, M., Jürgens, K., Bertilsson, S., Waniek, J.J., and Andersson, A.F. (2011). Transitions in bacterial communities along the 2000 km salinity gradient of the Baltic Sea. *ISME J.* 5: 1571–1579.

Hettich, R.L., Pan, C., Chourey, K., and Giannone, R.J.. (2013). Metaproteomics: harnessing the power of high performance mass spectrometry to identify the suite of proteins that control metabolic activities in microbial communities. *Anal. Chem.* 85: 4203–4214.

Hiller, K., Hangebrauk, J., Jäger, C., Spura, J., Schreiber, K., and Schomburg, D. (2009). Metabolite detector: Comprehensive analysis tool for targeted and nontargeted GC/MS based metabolome analysis. *Analyt. Chemist.* 81(9): 3429–3439.

Hughes, G., Cruickshank-Quinn, C., Reisdorph, R., Lutz, S., Petrache, I., Reisdorph, N., Bowler, R., and Kechris, K. (2014). MSPrep–summarization, normalization and diagnostics for processing of mass spectrometry-based metabolomic data. *Bioinformatics* 30(1): 133–134.

Huse, S.M., Dethlefsen, L., Huber, J.A., Welch, D.M., Relman, D.A., and Sogin, M.L. (2008). Exploring microbial diversity and taxonomy using SSU rRNA hypervariable tag sequencing. *PLoS Genet.* 4: 1000255.

Huttenhower, C. (2013). Predictive functional profiling of microbial communities using 16S rRNA marker gene sequences. *Nat. Biotechnol.* 31: 814–821.

Igarashi, Y. (2004). Screening of novel bioactive compounds from plant-associated actinomycetes. *Actinomycetologica* 18: 63–66.

Jauhiainen, A., Madhu, B., Narita, M., Narita, M., Griffiths, J., and Tavare, S. (2014). Normalization of metabolomics data with applications to correlation maps. *Bioinformatics* 30(15): 2155–2161.

Jog, R., Pandya, M., Nareshkumar, G., and Rajkumar, S. (2014). Mechanism of phosphate solubilization and antifungal activity of *Streptomyces* spp. isolated from wheat roots and rhizosphere and their application in improving plant growth. *Microbiology* 160(4): 778–788.

Jones, W. J. (2010). High through-put sequencing and metagenomics. *Estuaries Coast.* 33: 944–952.

Jones, O.A., Sdepanian, S., Lofts, S., Svendsen, C., Spurgeon, D.J., Maguire, M.L., and Griffin, J.L. (2014). Metabolomic analysis of soil communities can be used for pollution assessment. *Environ. Toxicol. Chem.* 33: 61–64.

Joyce, A. R. and Palsson, B.O. (2006). The model organism as a system: integrating 'omics' data sets. *Nat. Rev. Mol. Cell. Biol.* 7: 198–210.

Karpievitch, Y.V., Nikolic, S.B., Wilson, R., Sharman, J.E., and Edwards, L.M. (2014). Metabolomics data normalization with EigenMS. *PLoS One* 9(12): 1–10.

Kennedy, N.A., Walker, A.W., Berry, S.H., Duncan, S.H., Farquarson, F.M., Louis, P., Thomson, J.M., Satsangi, J., Flint, H.J., and Parkhill, J. (2014). The impact of different DNA extraction kits and laboratories upon the assessment of human gut microbiota composition by 16S rRNA gene sequencing. *PLoS ONE.* 9: 88982.

Kim, S., Thiessen, P.A., Bolton, E.E., Chen, J., Fu, G., Gindulyte, A., Han, L., He, J., He, S., Shoemaker, B.A., and Wang, J. (2016). PubChem substance and compound databases. *Nucl. Aci. Res.* 44(D1): D1202–D1213.

Knief, C., Delmotte, N., and Vorholt, J.A. (2011). Bacterial adaptation to life in association with plants - A proteomic perspective from culture to in situ conditions. *Proteomics.* 1(15): 3086–3105.

Knief, C., Delmotte, N., Chaffron, S., Stark, M., Innerebner, G., Wassmann, R., Von Mering, C., and Vorholt, J.A. (2012). Metaproteogenomic analysis of microbial communities in the phyllosphere and rhizosphere of rice. *ISMEJ.* 6: 1378–1390.

Kokubun, T. and D'Costa, L. (2013). Direct and unbiased information recovery from liquid chromatography-mass spectrometry raw data for phenotype-differentiating metabolites based on screening window coefficient of ion currents. *Analyt. Chemist.* 85(18): 8684–8691.

Kornbrust, B.A., Stringer, M.A., Lange, N.E.K., Hendriksen, H.V., Whitehurst, R., and Oort, M.V. (2010). Asparaginase–an enzyme for acrylamide reduction in food products. In: Whitehurst RJ and Maarten VO (eds.) *Enzymes in Food Technology.* 2nd edn, Wiley-Blackwell, Hoboken, NJ. 59–87.

Kuska, B. (1998). Beer, Bethesda, and biology: how Bgenomics came into being [news]. *J. Natl. Cancer Inst.* 90: 93.

Lamit, L.J., Lau, M.K., Sthultz, C.M., Wooley, S.C., Whitham, T.G., and Gehring, C.A. (2014). Tree genotype and genetically based growth traits structure twig endophyte communities. *Am. J. Bot.* 101: 467–478.

Lankadurai, B.P., Nagato, E.G., and Simpson, M.J. (2013). Environmental metabolomics: an emerging approach to study organism responses to environmental stressors. *Environ. Rev.* 21: 180–205.

Lauro, F.M., DeMaere, M.Z., Yau, S., Brown, M.V., Ng, C., Wilkins, D., Raftery, M.J., Gibson, J.A., Andrews-Pfannkoch, C., Lewis, M., and Hoffman, J.M. (2011). An integrative study of a meromictic lake ecosystem in Antarctica. *ISME J.* 5: 879–895.

Li, L., Josef, B.A., Liu, B., Zheng, S., Huang, L., and Chen, S. (2017). Three-dimensional evaluation on ecotypic diversity of traditional Chinese medicine: a case study of *Artemisia annua* L. *Front. Plant. Sci.* 8: 1225.

Lin, X., Lu, C., Huang, Y., Zheng, Z., Su, W., and Shen, Y. (2007). Endophytic fungi from a pharmaceutical plant, *Camptotheca acuminata*: Isolation, identification and bioactivity. *Wor. J. Microbiol. Biotechnol.* 23: 1037–1040.

Lindsay, R. C. and Jang, S. (2005). Chemical intervention strategies for substantial suppression of acrylamide formation in fried potato products. *Adv. Exp. Med. Biol.* 561: 393–404.

Liu, X., Dong, M., Chen, X., Jiang, M., Lv, X., and Zhou, J. (2008). Antimicrobial activity of an endophytic *Xylaria* sp. YX-28 and identification of its antimicrobial compound 7-amino-4-methylcoumarin. *Appl. Microbiol. Biotechnol.* 78: 241–247.

Lommen, A. and Kools, H. J. (2012). MetAlign 3.0: Performance enhancement by efficient use of advances in computer hardware. *Metabolism* 8(4): 719–726.

Lu, H., Zou, W.X., Meng, J.C., Hu, J., and Tan, R.X. (2000). New bioactive metabolites produced by *Colletotrichum* sp., an endophytic fungus in *Artemisia annua*. *Plan. Sci.* 151: 67–73.

Machavariani, N.G., Ivankova, T., Sineva, O.N., Terekhova, L.P., and Federation., R. (2014). Isolation of endophytic actinomycetes from medicinal plants of the Moscow region, Russia. *Wor. Appl Sci J.* 30(11): 1599–1604.

Maron, P.A., Ranjard, L., Mougel, C., and Lemanceau, P. (2007). Metaproteomics: a new approach for studying functional microbial ecology. *Microb. Ecol.* 53: 486–493.

Martínez-García, P.M., Ruano-Rosa, D., Schilirò, E., Prieto, P., Ramos, C., Rodríguez-Palenzuela, P., and Mercado-Blanco, J. (2015). Complete genome sequence of *Pseudomonas fluorescens* strain PICF7, an indigenous root endophyte from olive (*Oleaeuropaea*L.) and effective biocontrol agent against *Verticillium dahlia*. *Stand. Genomic. Sci.* 10: 10.

Mende, D.R., Aylward, F.O., Eppley, J.M., Nielsen, T.N., and DeLong, E.F. (2016). Improved environmental genomes via integration of metagenomic and single-cell assemblies. *Fron. Microbiol.* 7: 1–9.

Molina, L.G., Cordenonsi da Fonseca, G., Morais, G.L.D., de Oliveira, L.F.V., Carvalho, J.B.D., Kulcheski, F.R., and Margis, R. (2012). Metatranscriptomic analysis of small RNAs present in soybean deep sequencing libraries. *Genet. Mol. Biol.* 35: 292–303.

Ng, C., DeMaere, M.Z., Williams, T.J., Lauro, F.M., Raftery, M., Gibson, J.A., Andrews-Pfannkoch, C., Lewis, M., Hoffman, J.M., Thomas, T., and Cavicchioli, R. (2010). Metaproteogenomic analysis of a dominant green sulfur bacterium from Ace Lake, Antarctica. *ISME J.* 4: 1002–1019.

Nicole, F., Guitton, Y., Courtois, E.A., Moja, S., Legendre, L., and Hossaert-McKey, M. (2012). MSeasy: unsupervised and untargeted GC-MS data processing. *Bioinformatics* 28(17): 2278–2280.

Nongkhlaw, F. M. W. and Joshi, S.R. (2015a). Horizontal gene transfer of the non-ribosomal peptide synthetase gene among endophytic and epiphytic bacteria associated with ethnomedicinal plants. *Curr. Microbiol.* 72: 1–11.

Nongkhlaw, F. M. W. and Joshi, S. R. (2015b). L-asparaginase and antioxidant activity of endophytic bacteria associated with ethnomedicinal plants. *Ind. J. Biotechnol.* 15: 59–64.

Nyamundanda, G., Brennan, L., and Gormley, I.C. (2010). Probabilistic principal component analysis for metabolomic data. *BMC Bioinform.* 11(1): 571.

Oulas, A., Pavloudi, C., Polymenakou, P., Pavlopoulos, G.A., Papanikolaou, N., Kotoulas, G., Arvanitidis, C., and Iliopoulos, L. (2015) Metagenomics: tools and insights for analyzing next-generation sequencing data derived from biodiversity studies. *Bioinform. Biol. Insight.* 9: 75–88.

Parthasarathy, R. and Sathiyabama, M. (2014). Gymnemagenin-producing endophytic fungus isolated from a medicinal plant *Gymnema sylvestre* R. Br. *Appl. Biochem. Biotechnol.* 172: 3141–3152.

Pereira, J.A., Oliveira, I., Sousa, A., Valentão, P., Andrade, P.B., Ferreira, I.C., Ferreres, F., Bento, A., Seabra, R., and Estevinho, L. (2007). Walnut (*Juglans regia L.*) leaves: phenolic compounds, antimicrobial activity and antioxidant potential of different cultivars. *Food Chem. Toxicol.* 45: 2287–2295.

Peršoh, D. and Rambold, G. (2012). Lichen-associated fungi of the *Letharietum vulpinae*. *MycolProgre.* 11: 753–760.

Peršoh, D., Melcher, M., Flessa, F., and Rambold, G. (2010). First fungal community analyses of endophytic ascomycetes associated with *Viscum album* ssp. *austriacum* and its host *Pinus sylvestris*. *Fung. Biol.* 114: 585–596.

Promputtha, I., Jeewon, R., Lumyong, S., McKenzie, E.H., and Hyde, K.D. (2005). Ribosomal DNA fingerprinting in the identification of non-sporulating endophytes from *Magnolia liliifera* (Magnoliaceae). *Funga. Divers* 20: 167–186.

Qader, M.M., Kumar, N.S., Jayasinghe, L., and Fujimoto, Y. (2018). Shikimic acid production by *Fusarium decemcellulare*, an endophytic fungus isolated from *Flacourtia inermis* fruits. *J. Biol. Active. Prod. Nat.* 8: 43–50.

Raidl, S., Bonfigli, R., and Agerer, R. (2005). Calibration of quantitative real-time TaqMan PCR by correlation with hyphal biomass and ITS copies in mycelia of *Pilodermacroceum. Plan. Biol.* 7: 713–717.

Raizen, N. (2013). Fungal endophyte diversity in foliage of native and cultivated *Rhododendron* species determined by culturing, ITS sequencing, and pyrosequencing. *Masters Thesis*, Oregon State University, Corvallis.

Rice-Evans, C.A., Miller, N.J., and Paganga, G. (1996). Structure-antioxidant activity relationships of flavonoids and phenolic acids. *Fre. Radic. Biol. Med.* 20: 933–956.

Ridder, L., van der Hooft, J.J., Verhoeven, S., de Vos, R.C., Bino, R.J., and Vervoort, J. (2013). Automatic chemical structure annotation of an LC–MSn based metabolic profile from green tea. *Analyt. Chemist.* 85(12): 6033–6040.

Riesenfeld, C.S., Goodman, R.M., and Handelsman, J. (2004). Uncultured soil bacteria are a reservoir of new antibiotic resistance genes. *Environ. Microbiol.* 6: 981–989.

Rivero, J., Gamir, J., Aroca, R., Pozo, M.J., and Flors, V. (2015). Metabolic transition in mycorrhizal tomato roots. *Front. Microbiol.* 6: 598.

Rodríguez, H., Fraga, R., Gonzalez, T., and Bashan, Y. (2006). Genetics of phosphate solubilization and its potential applications for improving plant growth promoting bacteria. *Plan. Soi.* 287: 15–21.

Rondon, M.R., August, P.R., Bettermann, A.D., Brady, S.F., Grossman, T.H., Liles, M.R., Loiacono, K.A., Lynch, B.A., MacNeil, I.A., Minor, C., and Tiong, C.L. (2000). Cloning the soil metagenome: a strategy for accessing the genetic and functional diversity of uncultured microorganisms. *Appl. Environ. Microbiol.* 66(6): 2541–2547.

Russo, D.A., Couto, N., Beckerman, A.P., and Pandhal, J. (2016). Metaproteomic analysis of the response of a freshwater microbial community under nutrient enrichment. *Front. Microbiol.* 7: 1172.

Schloss, P. D. and Handelsman, J. (2003). Biotechnological prospects from metagenomics. *Curr. Opin. Biotechnol.* 14: 303–310.

Schneider, T. and Riedel, K. (2010). Environmental proteomics: analysis of structure and function of microbial communities. *Proteomics* 10: 785–798.

Schneider, T., Keiblinger, K.M., Schmid, E., Sterflinger-Gleixner, K., Ellersdorfer, G., Roschitzki, B., Richter, A., Eberl, L., Zechmeister-Boltenstern, S., and Riedel, K. (2012). Who is who in litter decomposition? Metaproteomics reveals major microbial players and their biogeochemical functions. *ISME J* 6: 1749–1762.

Segata, N., Boernigen, D., Tickle, T.L., Morgan, X.C., Garrett, W.S., and Huttenhower, C. (2013). Computational meta'omics for microbial community studies. *Mol. Syst. Biol.* 9:6 66.

Selim, K.A., El-Beih, A.A., AbdEl-Rahman, T.M., and El-Diwany, A.I. (2012). Biology of Endophytic Fungi. *Curr. Res. Environ. Appl. Mycol.* 2: 31–82.

Sessitsch, A.N.G.E.L.A., Hardoim, P.A.B.L.O., Döring, J., Weilharter, A., Krause, A.N.D.R.E.A.S., Woyke, T.A.N.J.A., Mitter, B., Hauberg-Lotte, L., Friedrich, F., Rahalkar, M., and Hurek, T. (2012). Functional characteristics of an endophyte community colonizing rice roots as revealed by metagenomic analysis. *Mol. Plan. Micro. Interact.* 25: 28–36.

Shah, N., Tang, H., Doak, T.G., and Ye, Y. (2011). Comparing bacterial communities inferred from 16S rRNA gene sequencing and shotgun metagenomics. *Pac. Symp. Biocomput.* 165–176.

Shutsrirung, A., Chromkaew, Y., Pathom-Aree, W., Choonluchanon, S., and Boonkerd, N. (2013). Diversity of endophytic actinomycetes in mandarin grown in northern Thailand, their phytohormone production potential and plant growth promoting activity. *Soil. Sci. Plant. Nutr.* 59(3): 322–330.

Simon, C. and Daniel, R. (2011). Metagenomic analyses: past and future trends. *Appl. Environ. Microbiol.* 77(4): 1153–1161.

Smith, C.A., Want, E.J., O'Maille, G., Abagyan, R., and Siuzdak, G. (2006). XCMS: processing mass spectrometry data for metabolite profiling using nonlinear peak alignment, matching, and identification. *ACS Publications* 78(3): 779–787.

Sowell, S.M., Abraham, P.E., Shah, M., Verberkmoes, N.C., Smith, D.P., Barofsky, D.F., and Giovannoni, S.J. (2011). Environmental proteomics of microbial plankton in a highly productive coastal upwelling system. *ISME J.* 5: 856–865.

Stecher, A.L., De Deus, P.M. (1999). Stability of L-asparaginase-an enzyme used in Leukemia treatment. *Pharmaceut. Acta. Helv.* 74: 1–9.

Stierle, A., Strobel, G., and Stierle, D. (1993). Taxol and taxane production by *Taxomyces andreanae*, an endophytic fungus of pacific yew (*Taxusbrevifolia*). *Science* 260: 214–216.

Strobel, G. A. (2000). Microbial gifts from rain forests. *Can. J. Plan. Pathol.* 24: 14–20.

Strobel, G., Daisy, B., Castillo, U., and Harper, J. (2004). Natural products from endophytic microorganisms. *J. Nat. Prod.* 67: 257–268.

Sugiyama, A., Ueda, Y., Zushi, T., Takase, H., and Yazaki, K. (2014). Changes in the bacterial community of soybean rhizospheres during growth in the field. *PLoS One* 9: 100709.

Taechowisan, T., Wanbanjob, A., Tuntiwachwuttikul, P., and Taylor, W.C. (2006). Identification of *Streptomyces* Sp. Tc022, an endophyte in *Alpinia galanga*, and the isolation of Actinomycin D. *Ann. Microbiol.* 56: 113–117.

Tan, R. X. and Zou, W. X. (2001). Endophytes: A rich source of functional metabolites. *Nat. Prod. Rep.* 18: 448–459.

Tan, S. C. and Yiap, B. C. (2009). DNA, RNA, and protein extraction: the past and the present. *J. Biomed. Biotechnol.* 574398.

Taylor, E. B. and Williams, M. A. (2010). Microbial protein in soil: influence of extraction method and amendment on extraction and recovery. *Microb. Ecol.* 59: 390–399.

Tenguria, R. K. and Khan F. N. (2011). Distribution of Endophytic Fungi in Leaves of *Azadirachta indica* JUSS. (Neem) of Panchmarhi Biosphere Reserve. *J. Phyto. Curr. Bot.* 2: 27–29.

Tlaskalová-Hogenová, H., Štěpánková, R., Kozáková, H., Hudcovic, T., Vannucci, L., Tučková, L., Rossmann, P., Hrnčíř, T., Kverka, M., Zákostelská, Z., and Klimešová, K. (2011). The role of gut microbiota (commensal bacteria) and the mucosal barrier in the pathogenesis of inflammatory and autoimmune diseases and cancer: contribution of germ-free and gnotobiotic animal models of human diseases. *Cel. Mol. Immunol.* 8: 110–120.

Toju, H., Yamamoto, S., Sato, H., Tanabe, A.S., Gilbert, G.S., and Kadowaki, K. (2013). Community composition of root-associated fungi in a *Quercus*-dominated temperate forest: "codominance" of mycorrhizal and root-endophytic fungi. *Ecol. Evol.* 3: 1281–1293.

Tschaplinski, T.J., Plett, J.M., Engle, N.L., Deveau, A., Cushman, K.C., Martin, M.Z., Doktycz, M.J., Tuskan, G.A., Brun, A., Kohler, A., and Martin, F. (2014). *Populu strichocarpa* and *Populus deltoids* exhibit different metabolomic responses to colonization by the symbiotic fungus *Laccaria bicolor*. *Mol. Plant-Micro. Interact.* 27: 546–556.

U'Ren, J.M., Riddle, J.M., Monacell, J.T., Carbone, I., Miadlikowska, J., and Arnold, A.E. (2014). Tissue storage and primer selection influence pyrosequencing-based inferences of diversity and community composition of endolichenic and endophytic fungi. *Mol. Ecol. Resour.* 14: 1032–1048.

Venter, J.C., Remington, K., Heidelberg, J.F., Halpern, A.L., Rusch, D., Eisen, J.A., Wu, D., Paulsen, I., Nelson, K.E., Nelson, W., and Fouts, D.E. (2004). Environmental genome shotgun sequencing of the SargassoSea. *Science* 304: 66–74.

VerBerkmoes, N.C., Denef, V.J., Hettich, R.L., and Banfield, J.F. (2009). Systems biology: Functional analysis of natural microbial consortia using community proteomics. *Nat. Rev. Microbiol.* 7: 196–205.

Wang, H.B., Zhang, Z.X., Li, H., He, H.B., Fang, C.X., Zhang, A.J., Li, Q.S., Chen, R.S., Guo, X.K., Lin, H.F., and Wu, L.K., 2011. Characterization of metaproteomics in crop rhizospheric soil. *J. Proteom. Res.* 10: 932–940.

Ward, D.M., Weller, R., and Bateson, M.M. (1990). 16S rRNA sequences reveal numerous uncultured microorganisms in a natural community. *Nature* 345(6270): 63.

Wei, J., Chen, F., Liu, Y., Abudoukerimu, A., Zheng, Q., Zhang, X., Sun, Y., and Yimiti, D. (2020). Comparative metabolomics revealed the potential antitumor characteristics of four Endophytic Fungi of *Brassica rapa* L. *ACS Omega.* 5(11): 5939–5950.

Wilkins, M.R., Sanchez, J.C., Gooley, A.A., Appel, R.D., Humphery-Smith, I., Hochstrasser, D.F., and Williams, K.L. (1995). Progress with proteome projects: why all proteins expressed by a genome should be identified and how to do it. *Biotechnol. Genet. Eng. Rev.* 13: 19–50.

Winkler, H. (1920). *Verbreitung und Ursache der Parthenogenesis imPflanzen- und Tierreiche.* Fischer, Jena.

Wooley, J.C., Godzik, A., and Friedberg, I. (2010). A primer on metagenomics. *PLoS Comput. Biol.* 6: 1000667.

Wu, X., Gong, F., and Wang, W. (2014). Protein extraction from plant tissues for 2DE and its application in proteomic analysis. *Proteome* 14: 645–658.

Xia, J., Bjorndahl, T.C., Tang, P., and Wishart, D.S. (2008). MetaboMiner-semi-automated identification of metabolites from 2D NMR spectra of complex biofluids. *BMC Bioinform.* 9: 507.

Yadav, S. P. (2007). The Wholeness in Suffix -omics, –omes, and the Word Om. *J. Biomol. Tech.* 18: 277.

Yan, L.L., Han, N.N., Zhang, Y.Q., Yu, L.Y., Chen, J., Wei, Y.Z., Li, Q.P., Tao, L., Zheng, G.H., Yang, S.E., and Jiang, C.X. (2010). Antimycin A 18 produced by an endophytic *Streptomyces albidoflavus* isolated from a mangrove plant. *J. Antibiot.* 63: 59–261.

Yergeau, E., Sanschagrin, S., Maynard, C., St-Arnaud, M., and Greer, C.W. (2014). Microbial expression profiles in the rhizosphere of willows depend on soil contamination. *ISME J.* 8: 344–358.

Yu, H., Zhang, L., Li, L., Zheng, C., Guo, L., Li, W., Sun, P., and Qin, L. (2010). Recent developments and future prospects of antimicrobial metabolites produced by endophytes. *Microbiol. Res.* 165: 437–449.

Zengler, K. and Palsson, B. O. (2012). A road map for the development of community systems (CoSy) biology. *Nat. Rev. Microbiol.* 10: 366–372.

Zhang, H.W., Song, Y.C., and Tan, R.X. (2006). Biology and chemistry of endophytes. *Nat. Prod. Rep.* 23: 753.

Zhang, W., Li, F., and Nie, L. (2010). Integrating multiple 'omics' analysis for microbial biology: application and methodologies. *Microbiology* 156: 287–301.

Zhang, A., Sun, H., Wang, P., Han, Y., and Wang, X. (2012). Modern analytical techniques in metabolomics analysis. *Analyst* 137: 293–300.

Zheng, Q., Song, J., Doncaster, K., Rowland, E., and Byers, D.M. (2007). Qualitative and quantitative evaluation of protein extraction protocols for apple and strawberry fruit suitable for two-dimensional electrophoresis and mass spectrometry analysis. *J. Agric. Food Chem.* 55: 1663–1673.

Zhou, H., Yang, Y., Peng, T., Li, W., Zhao, L., Xu, L., and Ding, Z. (2013). Metabolites of *Streptomyces* sp., an endophytic actinomycete from *Alpinia oxyphylla. Nat. Prod. Res.* 28: 265–267.

Zou, W. X. and Tan, R. X. (2001). Recent advances on endophyte research. *Acta. Bot. Sin.* 43: 881–892.

9

Plant–Rhizomicrobiome Interactive Ecology through the Lenses of Multi-Omics and Relevant Bioinformatics Approaches

Siddhartha Pal*
National Centre for Cell Science

Kriti Sengupta
Agharkar Research Institute

CONTENTS

9.1 Introduction

Recent advancements in integrated omics technologies have led to multi-omics revolution to understand microbial ecology and their functional interplay in several agroecosystems. Next-generation sequencing (NGS) approaches accelerated the study of microbial diversity, metabolic diversity, genome-scale metabolic reconstruction, and transcriptional activities of a single microbe or a complex community-dweller in a particular environment (Gutleben et al. 2018). However, the limited availability of databases and bioinformatics tools is a significant constraint in the gap between culture-dependent techniques and multi-omics analyses. Analyzing laboratory-generated high-throughput data on microbial metabolites is very challenging. It is presently not achievable to construct such networks based on a large set of multi-omics data. Therefore, *in silico* study of metabolic pathway networking in contrast to other biological levels came into the picture, curated from gene annotations available in public databases (Kint et al. 2010). DNA sequence data from genome and metagenome and RNA sequence data from metatranscriptome are available in public databases. The forthcoming decades should be utilized for the advancement of bioinformatics tools for explanatory purposes and accurate drawing of testing hypotheses related to microbial interactions in agroecosystems. Moreover, plant-associated microbiome, especially plant growth-promoting rhizobacteria (PGPR) and mycorrhizal fungi, has been well studied, but

* Corresponding author's email: palsiddhartha09@gmail.com

DOI: 10.1201/9781003171416-9

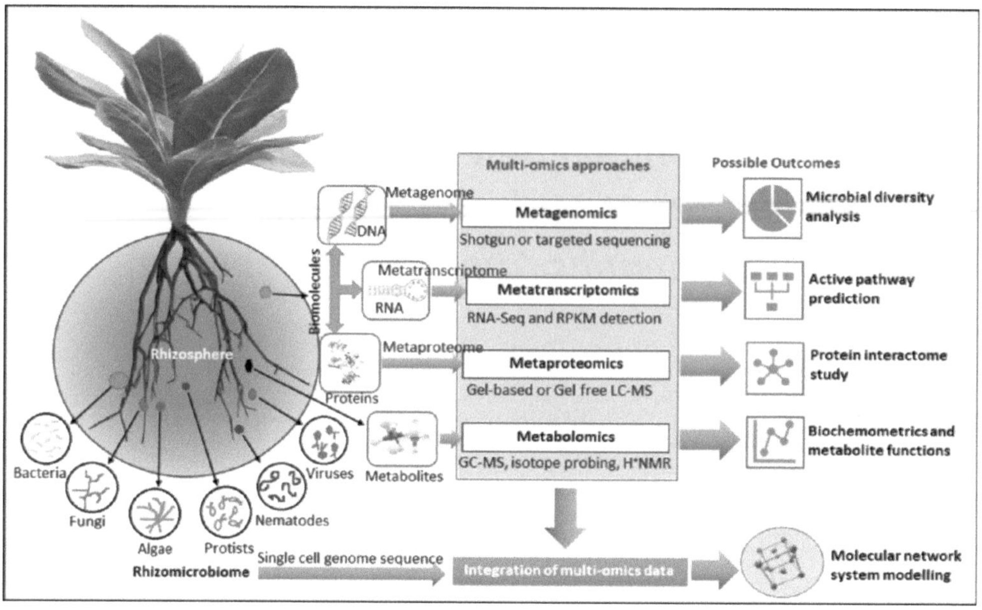

FIGURE 9.1 A pictorial representation of rhizospheric components and multi-omics approaches applied in rhizomicrobiome studies.

without proper understanding of functional interplay among diverse rhizomicrobiomes. In this regard, the metagenomic analysis only reveals the microbial diversity, but when integrated with metatranscriptomics or metaproteomics studies, it unravels the inducible pathways and metabolic interactions among plant microbiome. Some other breakthroughs of applying multi-omics technologies are resolving the complexity of plant-microbe-associated pathogenicity, plant stress response, and defense mechanisms (Sarrocco et al. 2020). Sharma et al. (2020) have recently demonstrated several multi-omics techniques used in exploring some beneficial facets of plant–microbe interactions, such as nitrogen fixation, stress response, and plant growth promotions.

Additionally, Ichihashi et al. (2020) used integrated multi-omics methods to explain the heterogenicity in plant traits due to complex interaction with soil metabolites and microbes. These recent findings suggested the effectiveness of integrating bioinformatics with several omics tools such as genomics, transcriptomics, proteomics, meta-omics, and metabolomics, which would be helpful to get a deeper insight into plant–microbe interacting networks at the molecular level. In this chapter, several multi-omics techniques used in the study of rhizomicrobiome interaction have been discussed while mainly focusing on the importance of omics data integration applied in understanding rhizospheric plant–microbe interactions. A pictorial representation (Figure 9.1) of integrated multi-omics approaches related to plant–rhizomicrobiome study would simplify the projection drawn in this chapter.

9.2 Genomics and Metagenomics

Microorganisms colonizing different plant compartments constitute the second genome of the plant. These microorganisms have been considered critical players in shaping plant health, community composition, productivity, and ecosystem functioning (Berg et al. 2014). There is a clear potential for metagenomics to contribute to the study of microbial communities of the rhizosphere, particularly PGPR, as comparing plant-associated communities may lead to phylogenetic and functional insights. Possible contributions include (i) the discovery of novel plant growth-promoting genes and gene products and (ii) the characterization of (not-yet-) culturable PGPR (Leveau 2007). Unno and Shinano (2013) elucidated the mechanisms by which plants could interact with the microbial community within the rhizosphere that could facilitate the conversion of phytic acid into the plant-available form of phosphorus. To better assess this complex and multifactorial microbial contribution, shotgun metagenomics was conducted using two molecular ecological tools. First was the molecular fingerprint approach based on comparing the rRNA gene diversity, permitting the analysis of phylogenetic diversity, and the second, the high-throughput shotgun sequencing for determining the functional gene diversity in soil microbiome. Metagenomic analyses of plant-associated bacteria may help not only to explain the high productivity of sugar beet, but also to overcome the difficulties associated with surveying practical plant growth-promoting bacteria (PGPB) (Tsurumaru et al. 2015). The metagenomic study not only sheds light on the microbial diversity present within the rhizosphere, but also helps in investigating the genes involved in plant growth-promoting traits (PGPTs) (Tsurumaru et al. 2015). The study performed by Tsurumaru et al. (2015) showed the presence of functional genes for 1-aminocyclopropane-1-carboxylic acid (ACC) deaminase, phosphate solubilization, pyrroloquinoline quinone, siderophores, and plant disease suppression within the taproot-associated microbiome of sugar beet. These functional genes were mainly associated with highly abundant *Mesorhizobium, Bradyrhizobium*, and *Streptomyces,* suggesting their ecologically essential roles within the taproot of sugar beet. Whole-genome analysis of three PGPR (belonging to *Enterobacteriaceae* and *Pseudomonadaceae* families) from coconut, cocoa, and areca nut exhibited the presence of genes encoding for mineral phosphate solubilization, siderophores, butanediol, acetoin, ACC deaminase, etc. (Gupta et al. 2014). Most importantly, the genomes encoded genes for quorum sensing molecules and those required for hydrogen sulfide synthesis, proposed recently to enhance plant growth. Pramanik et al. (2020), through metagenomic investigation of the rhizospheric microbiome, showed the presence of cultivar-specific soil microorganisms that could make a difference in the adaptability of lentil cultivars under rice-fallow ecology. Although functional prediction profiles of both the cultivars (namely, "Farmer-2" and "Moitree") mostly revealed the presence of housekeeping genes and genes for general metabolism, the microbiome composition within the rhizosphere of both the cultivars was quite variable. Due to this variability, the strategies for nutrient acquisition were different for the two cultivars, along with differences in their genetic makeup. The cultivar–microbe interaction has thus proved to enhance the nitrogen fixation and phosphate solubilization within the rhizospheric environment (Pramanik et al. 2020). Mineral phosphate solubilization (MPS), an important phenomenon carried out by rhizospheric soilborne microorganisms, provides orthophosphate anions required for plant growth. Functional metagenomics analyzed through 454 Roche sequencing, performed by Chhabra et al. (2013) on the barley rhizosphere soil microbiome, revealed the presence of functional and regulatory genes related to phosphorous uptake and solubilization, mostly from the uncultivated microbiome. Due to the advent of NGS technology, it has been profusely used in recent times to sequence the whole genome of PGPR to gain insight into its beneficial attributes (Gupta et al. 2014). Genome sequence of rhizosphere plant growth-promoting *Pseudomonas, Bacillus, Paenibacillus*, etc., revealed the presence of genes involved in IAA biosynthetic pathway, solubilizing inorganic phosphate, antimicrobial activity, and siderophore production (Matilla et al. 2011; Wang et al. 2012; Eastman et al. 2014).

9.3 Transcriptomics and Metatranscriptomics

Rhizomicrobiome is a complex microbial community that responds to the root exudate, which is the critical step in establishing the plant–microbe association. In response to the particular root exudates, a specific group of microorganisms showed differential gene expression, which could be examined by transcriptome analysis (Yi et al. 2018). Transcriptomics deals with RNA sequencing (RNA-Seq), a persuasive high-throughput technique used for transcript profiling in diverse organisms under various environments and treatments. The basic steps in transcriptomics protocol include isolation of total RNA by TRIzol method, cDNA synthesis, and RNA-Seq followed by data acquisition in RPKM value (reads per kilobase of transcript per million mapped reads), which signifies the number of cDNAs obtained per sample (Sengupta et al. 2019). Initially, microarray techniques were available to measure specific gene expression levels, which had just faded away with the emergence of NGS technologies in the last decade (Tan et al. 2009). Yi et al. (2018) showed that several multidrug efflux proteins, sporulation protein, acetolactate synthase, pullulanase, cardiolipin synthase, and many transcription regulatory proteins were overexpressed in rhizobacteria *Bacillus mycoides* in the presence of potato exudate. Zhang et al. (2020) have recently identified several genes involved in salt tolerance in a PGPR strain of *Halomonas* species. Other gene expression profiling using transcriptomics analysis to study plant–microbe interaction has been done in the past few years (Hennessy et al. 2017; Nobori et al. 2018; Pombo et al. 2019). On the other hand, plant transcriptomics has also been popularized as this technique enables exploring the plant response in various conditions such as pathogen, stress, or beneficial rhizome microbiota (Imadi et al. 2015). When integrated with genomics, transcriptomics mainly improved the understanding of the genetic and molecular response of cash crops such as sugarcane, rice, barley, wheat, and maize. Hence, this method was proved efficient to detect the abundance of essential genes and regulons involved in various plant–microbe interactions and their role in withstanding multiple environmental conditions. However, stand-alone transcriptomics is insufficient to provide the holistic picture of functional gene expression of a complex rhizomicrobiome. In such cases, metatranscriptomics, which includes studying total environmental RNA profile, is helpful to get a deeper insight into gene expression at the microbial community level (Gutleben et al. 2018). The implementation of metatranscriptomics-based evidence offers additional information regarding active or passive community group members and their highly expressed metabolic pathways, which help establish robust culture conditions (Frias-Lopez et al. 2008). This technique could detect significant and minor gene categories involved in plant–microbe interaction, which is impossible to figure out solely with genomics or metagenomics approaches. Moreover, a comparative study of metatranscriptomes obtained from various samples exposed to different environmental conditions could provide relationships among gene expression and several environmental factors (Bomar et al. 2011). Thus, metatranscriptomics can benefit the research in understanding the novel gene expression and function of the significant genes and their corresponding proteins involved in constructing the rhizomicrobiome. Furthermore, metatranscriptomics refers to the active transcriptional stage of several microorganisms present in a rhizosphere. For example, metatranscriptome profiling was used to detect the rhizobacterial genes actively expressed during the development of *Arabidopsis* (Chaparro et al. 2014). Another study with willow plant where rhizospheric microbial community influenced the phytoremediation process became evident from transcriptomics analysis (Yergeau et al. 2014). Moreover, metatranscriptomics is also an effective tool to capture the function of uncultivable microbes. The practice of transcriptomics and metatranscriptomics approaches has been improved to gather information regarding the gene expression of rhizobacteria. Transcriptomics has proved to be an advantageous method over whole-genome analysis. However, the availability of reference genome sequence quality in public databases limits the extensive use of metatranscriptomics analysis in this field.

9.4 Proteomics and Metaproteomics

Whole-cell and communal protein profiling, known as proteomics and metaproteomics, respectively, have emerged as indispensable techniques to study plant–microbe interactions even in the post-genomic era. Proteome-level studies were performed to investigate the cellular or extracellular enzymatic factors induced in symbiotic partners during host-specific associations (Kav et al. 2007). Khatabi et al. (2019) have recently reviewed the multi-purpose proteomics techniques used in plant–microbe studies, such as the role of a specific set of symbiotic proteomes in *Rhizobium* with legume and non-legume plants. Traditionally, proteomics analysis has been performed by the separation of proteins by 2-D gel electrophoresis or via liquid chromatography (LC) followed by mass spectroscopy (MS) after peptide digestion (Cooper et al. 2006; Lee et al. 2009). Proteomics has been proved to catalog the proteins involved in pathogenicity, symbiotic relationship, and secretory pathways present in the rhizomicrobiome. One of the most steadfast approaches is quantitative proteomic analysis integrated with metabolite labeling of cells. The procedure is based on a few crucial steps such as standardized incorporation of isotopes during exponential cell growth, protein extraction, separation and consecutive digestion of proteins, and peptide separation by LC and MS/MS analysis (Beynon and Pratt 2005; Becker 2008). The two standard proteomic methods are 2-D gel electrophoresis (combined isoelectric focusing and SDS-PAGE) and LC–MS/MS (Ibort et al. 2018). However, the 2-D gel-based method is associated with few problems, such as poor resolution ability for hydrophobic, alkaline, and minimum abundant proteins. Nonetheless, shotgun proteome analysis could be helpful to accomplish an intertwined analysis of extracted cellular, subcellular, and membrane proteins from plant cells (Fournier et al. 2007; Takahashi et al. 2014). Most effectively, integrating proteomics with genomics, sometimes collectively known as proteogenomics, has been popularized to understand the complex signaling symbiotic networks. Furthermore, practically single-cell proteome, genome, or proteogenomic is not enough to explore beyond one-on-one symbiotic relationships. To study the functional role of a whole rhizospheric microbial community, metaproteomics, also known as environmental proteomics, is a powerful technique. Rhizospheric soil metaproteomics could be useful to operate enzymatic or proteinaceous components of the root-associated microflora. Lin et al. (2013) performed a comparative metaproteomics study of microflora associated with sugarcane yield in ratoon soil. More than 30 proteins were characterized as differentially expressed in ratoon soil in comparison with control soil. Among these proteins, upregulated plant proteomes include stress response, carbohydrate, and amino acid metabolism, whereas microbial proteomes were related to membrane transport and regulatory proteins. There are few comparative metaproteomics studies that have been executed to address the essential aspects of the plant–microbe association. One of these studies showed that microbial dinitrogenase was exclusively found in the rice rhizosphere, whereas dinitrogenase reductase was identified in phyllosphere bacteria (Knief et al. 2011). Mendes et al. (2013) studied the metaproteomics profile of rhizomicrobiome under several environmental conditions. Such analysis of the rhizosphere soil of different crop varieties help recognize secretome and microbial cell signaling pathways. Some reports also suggested that the metaproteome approaches were benefited to identify several factors such as cellular biomolecules, soil micronutrient heavy metal contents, and the high diversity of rhizospheric bacteria (Mattarozzi et al. 2017; Bona et al. 2019).

9.5 Metabolomics

Plant–microbe metabolomics often deals with the extraction and identification of induced metabolites either in plants or released from microorganisms due to plant–microbe association. As a result of an emerging interest in metabolomics, which provides a chemical fingerprint of plant–microbe ecology and bioactivity, it provided a lead approach known as "biochemometrics" when integrated with statistical data analysis (Kellogg and Kang 2020). A comparative metabolomic analysis between healthy and diseased plants carries the potential to unravel signaling pathways involved in the resistance mechanism of plants from their pathogen interaction (Castro-moretti et al. 2020). Recent studies of real-time changes of metabolites during plant–microbe associations have suggested that several metabolites commonly known as photostimulation compounds of microbial origin directly improve plant health (Adeniji et al. 2020; Chen et al. 2021). Derrien et al. (2003) studied the rhizospheric positive effect on plant sugar dynamics by using gas chromatography/isotope-ratio mass spectrometry. The targeted plant–microbe metabolomics using isotope labeling-based metabolite profiling reduces the risk of coexisting contaminant metabolite. Several reports are available demonstrating the effectiveness of stable isotope labeling of reference metabolites for accurate identification and quantification in a different group of organisms (Creek et al. 2012; Batista Silva et al. 2016). Another metabolite profiling strategy includes solvent extraction followed by liquid chromatography–high-resolution mass spectrometry (LC–HRMS) and headspace solid-phase microextraction–gas chromatography–mass spectrometry (HS-SPME–GC–MS), which have been used to examine the dynamics of volatile and non-volatile metabolites induced during plant–fungal association (Azzollini et al. 2018). Also, proton nuclear magnetic resonance-based metabolite identification has been used for examining a positive correlation of metabolites such as glucose, glutamate, methionine, serine, and citrate on a carbendazim-resistant strain of plant fungal pathogen (Sevastos et al. 2018). Two subsets of plant metabolomics study include volatilomics. The volatilomics deals with the analysis of volatile substances such as plant hormones, terpenoids and other gaseous induced compounds (Crandall et al. 2020). Although metabolomics is a productive technique for identifying and quantifying small molecular weight metabolites and their dynamics during plant–microbe interaction, this method has got some limitations. The limitations such as the cost of instruments, unavailability of technical experts, and reference databases are challenging to avail this approach (Sharma et al. 2020). It has been studied that nearly 2,500 unique metabolites are present in bacteria, while approximately 20,000 unique metabolites are found in the plant kingdom, whose library preparation itself is a time-consuming task (Crandall et al. 2020). Nevertheless, metabolites are the ultimate products and are unable to relate the outcome directly from gene expression data; hence, linking metabolomics study with other omics such as proteomics or transcriptomics would be more helpful. However, building online platforms for omics integration with a large set of diverse metabolite profiles is a tedious job, which demands high bioinformatics expertise and machine learning skills.

9.6 Genome Editing Tools

Clustered regularly interspaced short palindromic repeats (CRISPRs) or CRISPR/Cas-mediated genome editing (GE) has been used in recent years to enhance the agroeconomic traits of crop plants (Shelake et al. 2019). This GE tool has been considered as the modern tool to understand the plant-specific rhizomicrobiome playing roles in the expression of suitable phenotype (Muñoz et al. 2019). In previous years, genetic modification was promising with the fastest outcomes, but was carried out by incorporation of foreign genes. On the other hand, this GE tool edits the genome in a precise manner without the incorporation of a foreign gene and thus acquires greater interest among researchers. In GE techniques, an engineered endonuclease is implicated in making a double-stranded cleavage, which undergoes DNA repair and forms various mutations. The joining of double-strand break would be possible either by homologous recombination or by non-homologous end-to-end joining. Several nucleases are being used in targeted genome modification technologies, such as CRISPR/Cas, transcription activator-like effector nucleases (TALENs), and zinc finger nucleases (ZFNs). The most specific, versatile, cost-effective, and straightforward is CRISPR/Cas and well adapted to eukaryotic GE methods (Wright et al. 2016). The understanding of the molecular mechanism of root–rhizobial interaction and phytopathogen association has been deepened by GE technology. For example, CRISPR/Cas-based modification has been engineered in rhizobacterium *Pseudomonas putida* KT2440 for gene deletion, insertion, and transcription repression (Sun et al. 2018). Additionally, CRISPR/Cas system has been applied for targeted GE in some biocontrol agent fungi such as *Beauveria bassiana* and *Purpureocillium lilacinum*, controlling insect and plant nematodes, respectively (Chen et al. 2017; Jiao et al. 2019). The CRISPR/Cas skill has a massive potential to benefit researchers to recognize the fundamentals of plant–microbe interactions and to develop model plant/microbes applicable for agronomy. Subsequently, studying a greater number of plant species, much more NGS data, and meta-omics data and analyses of the plant rhizomicrobiome are added to comprehend the community-level molecular interplay mechanisms under field environments. Documentation of specific plant or microbial genes determining agronomic traits will enable CRISPR-based applications in future sustainable agricultural practices.

9.7 Bioinformatics Tools and Online Databases

Bioinformatics and machine learning tools have significantly contributed to the in-depth analysis of high-throughput sequence data obtained from any biological system (Lucaciu et al. 2019). For example, some complicated questions raised in the microbial genome, such as horizontal gene transfer, mutation rate, or an evolutionary genetic modification, could be solved nowadays using some advanced bioinformatics tools (Koutsovoulos et al., 2016). Moreover, computational methods are required for biologically meaningful interpretation of complex NGS data. For instance, multiple software packages for reads trimming, assembly, binning, and annotation are available to process the raw reads obtained from nanopore, Ion Torrent, Illumina HiSeq/MiSeq platform, or PacBio. The NGS data interpretation for metagenomics analysis, such as determination of OTUs (operational taxonomic units) and ASVs (amplicon sequence variants), is widely performed by software packages such as mothur, QIIME, and DADA2 pipelines (Caporaso et al. 2010; Callahan et al. 2016). These tools could be implemented with reference databases such as SILVA, Greengenes, and RDP for 16S rRNA gene classification (Wang et al. 2007); UNITE and WARCUP for fungal ITS sequence classification (Abarenkov et al. 2010; Deshpande et al. 2016). The RNA-Seq investigation entails data preprocessing such as separation of rRNA, long poly-A tail removal, and trimming of low-quality bases. In plant rhizomicrobiome study, the total RNA from the plant host

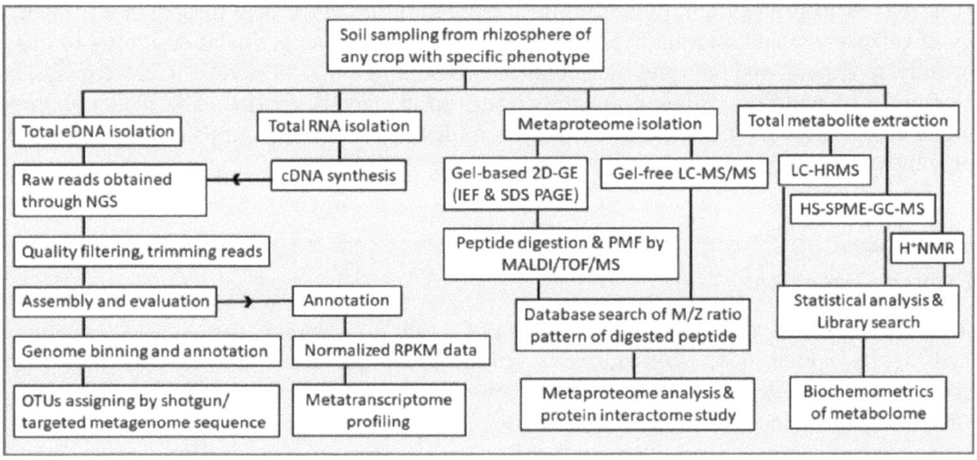

FIGURE 9.2 A schematic workflow chart of meta-omics methodology used to link specific plant phenotype with rhizospheric metagenome, metatranscriptome, metaproteome, and metabolome profiles.

could be distinguished by mapping the reads to the closest reference genome or transcriptome available in databases. As per the read-based method, rRNA and non-rRNA reads are identified and separated by aligning to a reference database. Reference-free assembly approaches executed by *de novo* transcriptome assemblers such as Trinity can produce the gene isoforms (Grabherr et al. 2011). Transcriptome-based taxonomic classification is generally performed by MEGAN, which is based on the lowest common ancestor (LCA) algorithm (Huson et al. 2007). Comparative transcriptome analysis among numerous samples often requires computation-based data normalization. In order to analyze differential gene expression level as an aim of transcriptome study, RPKM (reads per kilobase of transcript per million mapped reads), which is directly proportional to the transcript level in a sample, could be obtained by using edgeR tool (Robinson et al. 2009). Metaproteome also requires computational processing of peptide identification by matching the mass-by-charge (M/Z) ratio to a spectrum searched in the library, e.g., Mascot Search or SpectraST (Lam et al. 2007). UniProtKB, RefSeq, or Ensemble are the protein databases that are used for downstream validation purposes. Several online tools for statistical analysis and visualization of metabolome data are available, such as MetaboAnalyst and XCMS (Gowda et al. 2014; Xia et al. 2015). Plant metabolic network and plant genome annotation-based metabolic reconstruction tools are KEGG, PlantSEED, Gramene/Plant Reactome, and PlantCyc database (Kanehisa et al. 2014; Seaver et al. 2014; Naithani et al. 2017; Schläpfer et al. 2017; Tello-Ruiz et al. 2018). A schematic workflow of multiple meta-omics approaches applied to collect samples from the rhizosphere related to specific plant phenotype is provided in Figure 9.2. Hence, bioinformatics and online database development in the future would enrich the quality analysis of multi-omics data and would be a guide to predict the system ecology of the rhizosphere when integrated with other physiochemical parameters of soils.

9.8 Conclusive Remarks

This chapter outlines the integrative designing of multi-omics approaches used in plant-associative rhizomicrobiome. In this chapter, multiple meta-omics through NGS technologies and their bioinformatics analysis using various tools and databases to explore the rhizosphere diversity, active genes, and pathway networks involved in plant–microbe relations have been discussed systematically. Briefing all relevant meta-omics technologies and complementary bioinformatics approaches would provide a visionary effect to readers regarding the powerful impact of multi-omics methods in several plant–microbe studies. There are several promises of genomics, metagenomics, metatranscriptomics, metaproteomics, metabolomics, and a combination of more than one omics technique to solve the problem in plant pathology ecology, plant stress response, and understanding plants' communication with beneficial rhizomicrobiome. The current development in integrating multi-omics large dataset provided new insight into the microbial ecology of rhizosphere and relevant hypothesis for *in silico* guided experiment designing in this field. Conclusively, to explain and interpret molecular phenomena in a plant–microbe interactive system, an extensive range of multi-omics-based methods has been eloquently applied. The fusion approach of merging bioinformatics and multidimensional meta-omics data would permit us to discover unmapped pathways and networks of plant–microbe interactions.

9.9 Future Directions

The plant rhizosphere is a treasure box consisting of a rich microbiome, metabolites, and other biochemical factors, which directly or indirectly help improve plant health. Although there has been significant progress in exploring plant rhizomicrobiome, there is a requirement of process development that manipulates the active functional microbial community in order to bring out of best for crop yields. Subsequently, perusing a higher number of plant varieties, further in-depth sequence analyses of the plant rhizomicrobiome metagenome and metatranscriptome data are added advantages to recognize the molecular mechanisms at community level under field environments. However, the hybrid multi-omics approaches, including the integration of a large set of sequence data with high metabolite profiles obtained from rhizospheric soil, are a challenging task with currently available databases and software or online tools. In the upcoming years, there is a need to develop machine learning and database creation tools related to the systems biology of the plant rhizosphere. In the genome editing era, few important questions should be addressed regarding the genes involved in shaping rhizomicrobiome, physiochemical and biological factors affecting the host–microbe association, communicative pathways, and network between plant root exudates and associated microbiota. This information will directly establish a connection among various phenotypes and genotypes of the plant or microbe and thus facilitate the bioengineering of crucial microbial communities for enhanced crop production.

REFERENCES

Abarenkov et al. (2010) The UNITE database for molecular identification of fungi - recent updates and future perspectives. *New Phytol.* 186:281–285.

Adeniji et al. (2020) Metabolomic applications for understanding complex tripartite plant-microbes interactions: Strategies and perspectives. *Biotechnol Reports* 25:e00425.

Azzollini et al. (2018) Dynamics of metabolite induction in fungal co-cultures by metabolomics at both volatile and non-volatile levels. *Front Microbiol* 9:72.

Batista Silva et al. (2016) Can stable isotope mass spectrometry replace radiolabelled approaches in metabolic studies? *Plant Sci.* 249:59–69.

Becker GW (2008) Stable isotopic labeling of proteins for quantitative proteomic applications. *Briefings Funct Genomics Proteomics* 7:371–382.

Berg et al. (2014) Unraveling the plant microbiome: Looking back and future perspectives. *Front Microbiol* 5:1–7.

Beynon RJ and Pratt JM (2005) Metabolic labeling of proteins for proteomics. *Mol Cell Proteomics* 4:857–872.

Bomar et al. (2011) Directed culturing of microorganisms using metatranscriptomics. *MBio* 2(2):e00012-11.

Bona et al. (2019) Metaproteomic characterization of the Vitis vinifera rhizosphere. *FEMS Microbiol Ecol* 95. https://doi.org/10..1093/femsec/fiy204.

Callahan et al. (2016) DADA2: High-resolution sample inference from Illumina amplicon data. *Nat Methods* 13:581–583.

Caporaso et al. (2010) QIIME allows analysis of high-throughput community sequencing data. *Nat Methods* 7:335–336.

Castro-moretti et al. (2020) Metabolomics as an emerging tool for the study of plant–pathogen interactions. *Metabolites* 10:1–23.

Chaparro et al. (2014) Rhizosphere microbiome assemblage is affected by plant development. *ISME J* 8:790–803.

Chen et al. (2017) CRISPR/Cas9-mediated efficient genome editing via blastospore-based transformation in entomopathogenic fungus *Beauveria bassiana*. *Sci Rep* 8. https://doi.org/10.1038/srep45763.

Chen et al. (2021) *Metabolomics in Plant-Microbe Interactions in the Roots*, 1st edn. Elsevier Ltd. https://doi.org/10.1016/bs.abr.2020.09.018

Chhabra et al. (2013) Characterization of mineral phosphate solubilization traits from a barley rhizosphere soil functional metagenome. *Microbiologyopen* 2:717–724.

Cooper et al. (2006) Shotgun identification of proteins from uredospores of the bean rust Uromyces appendiculatus. *Proteomics* 6:2477–2484.

Crandall et al. (2020) A multi-omics approach to solving problems in plant disease ecology. *PLoS One* 15:1–23.

Creek et al. (2012) Stable isotope-assisted metabolomics for network-wide metabolic pathway elucidation. *Anal Chem* 84:8442–8447.

Derrien et al. (2003) Measurement of the 13C/12C ratio of soil-plant individual sugars by gas chromatography/combustion/isotope-ratio mass spectrometry of silylated derivatives. Rapid Communications in Mass Spectrometry. 17(23):2626–2631. doi: 10.1002/rcm.1269

Deshpande et al. (2016) Fungal identification using a Bayesian classifier and the Warcup training set of internal transcribed spacer sequences. *Mycologia* 108:1–5.

Eastman et al. (2014) Comparative and genetic analysis of the four sequenced *Paenibacillus polymyxa* genomes reveals a diverse metabolism and conservation of genes relevant to plant-growth promotion and competitiveness. *BMC Genomics* 15:1–22.

Fournier et al. (2007) Multidimensional separations-based shotgun proteomics. *Chem Rev* 107:3654–3686

Frias-Lopez et al. (2008) Microbial community gene expression in ocean surface waters. *Proc Natl Acad Sci U S A* 105:3805–3810.

Gowda et al. (2014) Interactive XCMS online: simplifying advanced metabolomic data processing and subsequent statistical analyses. *Anal Chem* 86:6931–6939.

Grabherr et al. (2011) Full-length transcriptome assembly from RNA-Seq data without a reference genome. *Nat Biotechnol* 29:644–652.

Gupta et al. (2014) Whole genome sequencing and analysis of plant growth promoting bacteria isolated from the rhizosphere of plantation crops coconut, cocoa and areca nut. *PLoS One* 9. https://doi.org/10.1371/journal.pone.0104259.

Gutleben et al. (2018) The multi-omics promise in context: from sequence to microbial isolate. *Crit Rev Microbiol* 44:212–229.

Hennessy et al. (2017) Transcriptomic profiling of microbe-microbe interactions reveals the specific response of the biocontrol strain P. fluorescens In5 to the phytopathogen *Rhizoctonia solani*. *BMC Res Notes* 10:1–9.

Huson et al. (2007) MEGAN analysis of metagenomic data. *Genome Res* 17:377–386.

Ibort et al. (2018) Postprint 2018. *J Plant Physiol* 220:43–59.

Ichihashi et al. (2020) Multi-omics analysis on an agroecosystem reveals the significant role of organic nitrogen to increase agricultural crop yield. *Proc Natl Acad Sci U S A* 117:14552–14560.

Imadi et al. (2015) Plant transcriptomics and responses to environmental stress: an overview. *J Genet* 94:525–537.

Jiao et al. (2019) Functional genetic analysis of the leucinostatin biosynthesis transcription regulator lcsL in Purpureocillium lilacinum using CRISPR-Cas9 technology. *Appl Microbiol Biotechnol* 103:6187–6194.

Kanehisa et al. (2014) Data, information, knowledge and principle: back to metabolism in KEGG. *Nucleic Acids Res* 42. https://doi.org/10..1093/nar/gkt1076.

Kellogg J *and* Kang S (2020) Metabolomics, an essential tool in exploring and harnessing microbial chemical ecology. *Phytobiomes J* 4:195–210.

Khatabi et al. (2019) Plant–microbe symbiosis: what has proteomics taught us? *Proteomics* 19:1–20.

Kint et al. (2010) Integration of 'omics data: Does it lead to new insights into host-microbe interactions? *Future Microbiol* 5:313–328.

Knief et al. (2011) Bacterial adaptation to life in association with plants - A proteomic perspective from culture to in situ conditions. *Proteomics* 11:3086–3105.

Lam et al. (2007) Development and validation of a spectral library searching method for peptide identification from MS/MS. *Proteomics* 7:655–667.

Lee et al. (2009) Quantitative proteomic analysis of bean plants infected by a virulent and avirulent obligate rust fungus. *Mol Cell Proteomics* 8:19–31.

Leveau JHJ (2007) The magic and menace of metagenomics: prospects for the study of plant growth-promoting rhizobacteria. *Eur J Plant Pathol* 119:279–300.

Lin et al. (2013) Metaproteomic analysis of ratoon sugarcane rhizospheric soil. *BMC Microbiol* 13:135.

Lucaciu et al. (2019) A bioinformatics guide to plant microbiome analysis. *Front Plant Sci* 10:1–18.

Matilla et al. (2011) Complete genome of the plant growth-promoting rhizobacterium *Pseudomonas putida* BIRD-1. *J Bacteriol* 193:1290.

Mattarozzi et al. (2017) A metaproteomic approach dissecting major bacterial functions in the rhizosphere of plants living in serpentine soil. *Anal Bioanal Chem* 409:2327–2339.

Mendes et al. (2013) The rhizosphere microbiome: significance of plant beneficial, plant pathogenic, and human pathogenic microorganisms. *FEMS Microbiol Rev* 37:634–663.

Muñoz et al. (2019) CRISPR-CAS for fungal genome editing: a new tool for the management of plant diseases. *Front Plant Sci* 10:135.

Naithani et al. (2017) Plant reactome: a resource for plant pathways and comparative analysis. *Nucleic Acids Res* 45:D1029–D1039.

Nat et al. (2007) Application of proteomics to investigate plant-microbe interactions. *Curr Proteomics* 4:28–43.

Nobori et al. (2018) Transcriptome landscape of a bacterial pathogen under plant immunity. *Proc Natl Acad Sci U S A* 115:E3055–E3064.

Pombo et al. (2019) Transcriptome-based identification and validation of reference genes for plant-bacteria interaction studies using Nicotiana benthamiana. *Sci Rep* 9:1–10.

Pramanik et al. (2020) Metagenomic insights into rhizospheric microbiome profiling in lentil cultivars unveils differential microbial nitrogen and phosphorus metabolism under rice-fallow ecology. *Int J Mol Sci* 21:1–22.

Robinson et al. (2009) edgeR: A Bioconductor package for differential expression analysis of digital gene expression data. *Bioinformatics* 26:139–140.

Sarrocco et al. (2020) Editorial: plant disease management in the post-genomic era: from functional genomics to genome editing. *Front. Microbiol.* 11:107. doi: 10.3389/fmicb.2020.00107

Schläpfer et al. (2017) Genome-wide prediction of metabolic enzymes, pathways, and gene clusters in plants. *Plant Physiol* 173:2041–2059.

Seaver et al. (2014) High-throughput comparison, functional annotation, and metabolic modeling of plant genomes using the PlantSEED resource. *Proc Natl Acad Sci U S A* 111:9645–9650.

Sengupta et al. (2019) Genome sequencing and comparative transcriptomics provide a holistic view of 4-nitrophenol degradation and concurrent fatty acid catabolism by *Rhodococcus* sp. strain BUPNP1. *Front Microbiol* 10. https://doi.org/10.3389/fmicb.2018.03209.

Sevastos et al. (2018) Implication of Fusarium graminearum primary metabolism in its resistance to benzimidazole fungicides as revealed by 1H NMR metabolomics. *Pestic Biochem Physiol* 148:50–61.

Sharma et al. (2020) Deciphering the omics of plant-microbe interaction: perspectives and new insights. *Curr Genomics* 21:343–362.

Shelake et al. (2019) Exploration of plant-microbe interactions for sustainable agriculture in CRISPR era. *Microorganisms* 7:1–32.

Sun et al. (2018) Genome editing and transcriptional repression in Pseudomonas putida KT2440 via the type II CRISPR system. *Microb Cell Fact* 17. https://doi.org/10..1186/s12934-018-0887-x.

Takahashi et al. (2014) Shotgun proteomics of plant plasma membrane and microdomain proteins using Nano-LC-MS/MS. *Methods Mol Biol* 1072:481–498.

Tan et al. (2009) Assessing the impact of transcriptomics, proteomics and metabolomics on fungal phytopathology. *Mol Plant Pathol* 10:703–715.

Tello-Ruiz et al. (2018) Gramene 2018: Unifying comparative genomics and pathway resources for plant research. *Nucleic Acids Res* 46:D1181–D1189.

Tsurumaru et al. (2015) Metagenomic analysis of the bacterial community associated with the taproot of sugar beet. *Microbes Environ* 30:63–69.

Unno Y and Shinano T (2013) Metagenomic analysis of the rhizosphere soil microbiome with respect to phytic acid utilization. *Microbes Environ* 28:120–127.

Wang et al. (2007) Naïve Bayesian classifier for rapid assignment of rRNA sequences into the new bacterial taxonomy. *Appl Environ Microbiol* 73:5261–5267.

Wang et al. (2012) Genome sequence of the plant growth-promoting rhizobacterium *Bacillus* sp. strain 916. *J Bacteriol* 194:5467–5468.

Wright et al. (2016) Biology and applications of CRISPR systems: harnessing nature's toolbox for genome engineering. *Cell* 164:29–44.

Xia et al. (2015) MetaboAnalyst 3.0-making metabolomics more meaningful. *Nucleic Acids Res* 43:W251–W257.

Yergeau et al. (2014) Microbial expression profiles in the rhizosphere of willows depend on soil contamination. *ISME J* 8:344–358.

Yi et al. (2018) Plant-microbe interaction: Transcriptional response of bacillus mycoides to potato root exudates. *J Vis Exp* 2018:1–7.

Zhang et al. (2020) Transcriptome analysis of ice plant growth-promoting endophytic bacterium *Halomonas* sp. Strain MC1 to identify the genes involved in salt tolerance. *Microorganisms* 8:1–18.

10

Future Prospects of Next-Generation Sequencing

Abdul Rehman*
Zhengzhou University
Institute of Cotton Research Chinese Academy of Agricultural Science

Hafiza Iqra Almas
University of Agriculture

Abdul Qayyum
Bahauddin Zakariya University

Hongge Li and Zhen Peng
Institute of Cotton Research Chinese Academy of Agricultural Science

Guangyong Qin
Zhengzhou University

Yinhua Jia and Zhaoe Pan
Institute of Cotton Research Chinese Academy of Agricultural Science

Shoupu He and Xiongming Du*
Zhengzhou University
Institute of Cotton Research Chinese Academy of Agricultural Science

CONTENTS

* Corresponding author's email: abdulpbg@gmail.com & dujeffrey8848@hotmail.com

DOI: 10.1201/9781003171416-10

10.1 Introduction

The nucleic acid techniques are used for the identification of the exact position of nucleotide sequences in genomic RNA and DNA. In the past few years, different techniques of sequencing have actively been used in many kinds of research related to molecular biology because of their maximum outputs, practical developments, and cost-effectiveness. Next-generation sequencing (NGS) techniques play a crucial role in knowing the chemistry of plants in response to environmental stresses while their major function in transcriptomics, genomics, and epigenomics reports (Egan et al. 2012). Nanofluidics, signal transduction systems and assembly of major computational data are the basic sources for the achievement of these NGS techniques (Mardis 2013).

These techniques of NGS are frequently categorized in the sequencing of second and third generations (Liu et al. 2012). However, a reliable way for the categorization of these techniques is not present still (Schadt et al. 2010). Commonly and specifically in second-generation sequencing method, PCR for the intensification of signal is the first step after sequencing, while in third-generation sequencing method, the sequencing of only a single molecule can be done. The initiation of these technologies of NGS takes advantage of numerous unique techniques for genomic studies in plants, namely ChIP-seq, WGB-seq, sequencing of transcriptome, and sequencing of exome; all of these are utilized to study the binding protein domain of DNA and protein, alteration of epigenetics, profiling of gene expression, and alteration in alleles, respectively (Van Dijk et al. 2014).

10.2 First-Generation Sequencing Techniques (FGSTs)

Different sequencing techniques are present in this generation, such as termination of Sanger chain and chemical-based technique, i.e., Maxam–Gilbert sequencing. Sanger and classical DNA sequencing techniques were widely used because of their less toxic chemicals and numerous other developments (Hutchison III 2007).

10.3 Second-Generation Sequencing Techniques (SGSTs)

SGSTs are announced as technologies of NGS with cost-effectiveness, accuracy, and maximum output compared to FGSTs. Different SGST techniques were introduced in different years; 454 pyrosequencing, Solexa or Illumina (synthetic sequencing), SOLiD (ligation sequencing), and Ion Torrent was introduced in 2005, 2006, 2007, and 2008, respectively. FGSTs and all of the above techniques are diverse from each other and need simpler steps of amplification based on PCR for assembling the libraries of sequencing. Commonly, these PCRs are prepared in minute quantities with template DNA attached to the surface of glass plate and in microbeads (Head et al. 2014). Because of changes in their sequencing cost, precision, chemistry, read length and speed are changed in various SGSTs. Overnight, SGSTs can generate billions of (25–800) nucleotides in length of reads with a lower cost than Sanger sequencing technique. But, the precision of SGSTs differs because of their dependency on many proliferation steps in preparation of libraries. Each alteration of natural DNA results in a variety of artifacts in DNA assays. Moreover, short reads formed by these methods are not appropriate for the preparation of genome *de novo*. Hence, unique techniques are being intended so that there is no or low manipulation of natural molecules of DNA (Ozsolak 2012).

Meanwhile, the complete data on several SGSTs are comprehensively studied (Mardis 2013; Yadav et al. 2014), so third-generation sequencing techniques (TGSTs) are predominantly discussed herewith.

10.4 Third-Generation Sequencing Techniques (TGSTs)

The current advancements in imaging of programmed signal molecules and nanofluidics technologies have covered the progress of TGSTs that help in observing the natural RNA and DNA without any alteration in maximum output method (Ozsolak 2012). Direct sequencing inhibits the wrongness in natural DNA with low or no alterations produced through SGSTs and causes extended reads. The third-generation sequencing technologies have normal length of reads almost greater than 10 kb, while it can make length of reads longer than 150 kb (Yadav et al. 2014).

10.5 Single-Molecule DNA Sequencing (SMDS)

The SMDS technology is introduced by laboratories of Stephen R Quake and is founded on SBS (Braslavsky et al. 2003). It is the first marketable product of TGSTs developed by Helicos Biosciences and can run SMDS and DRS (direct RNA sequencing) (Ozsolak et al. 2009).

Ligated and fractionated DNA with poly-A tail is extracted and isolated. The stage of sequencing has a flow cell treated by oligonucleotide, poly-dT; at that phase, poly-A tail is fixed with ligated DNA. Subsequently, template trapped the flow cells; the sequencing-by-synthesis is initiated by incubating the molecule of DNA template with terminator nucleotide that is marked with cleavable chemical inhibitor and fluorescent dye and, in cyclic method, the DNA polymerase that manipulates the nucleotide in progressive cycles (Bowers et al. 2009). After every cycle, the moiety-terminating dye is removed, while the cycles are continuing until the length of read is achieved (Ozsolak 2012).

The nature of SMDS prevents more alteration steps of DNA template causing no bias of GC and better analysis than SGSTs. Although the minimum length of read is 55 nucleotides, it causes a certain difficulty for several applications, including alternative splicing identification, microbiome sequencing, and sequencing of genome de novo.

10.6 Single-Molecule Real-Time Sequencing (SMRT)

In the mid-1990s at Cornell University, the SMRT sequencing technique was introduced by Jonas Korlach and Steve Turner. The central idea was to detect the combination of real-time marked nucleotides with fluorescence in growing strand via DNA polymerase. The progress in zero-mode waveguide (ZMW) technique, nanofluidics, biochemistry, incorporation in processing of semiconductors and photonics make it conceivable to study the fast integration polymerase kinetics to a scale of single molecule (Levene et al. 2003). ZMW is microwells placed on the surface of silica in a metal. Because of the unique behaviour of light in tiny wells, lighting of the bottom of a ZMW allows signal identification without severe interference from the top wells. At the lower part of the wells, only the DNA polymerase molecule is retained; the sequence of DNA is encumbered in the wells. At this moment, the differently labeled fluorescent nucleotides are loaded in growing thread via polymerase of DNA and the analysis of RT ends by observing the method of ZMW. The nucleotides used in SMRT are marked with fluorescent dye on their moiety phosphate; thus, it comes to be released by the PPi in integration of nucleotide by growing thread, and by dispersion, it comes separate from the well and leaves the cavity for novel nucleotide (Eid et al. 2009). With the sequence of DNA, the SMRT emission spectra of sequencing technology provide data around the variation in epigenetics additionally, because of minimal differences between normal and methylated loci of normal emission spectra (Flusberg et al. 2010).

The breakdown of polymerase by using maximum potential laser for imaging is the main disadvantage of SMRT sequencing technique. In addition to the drawbacks of existing optical imaging technologies that cannot detect the nucleotide integration through DNA polymerase at the normal speed of approximately 1,000 nt/s, it is an alternative restriction in this technology (Ozsolak 2012). Currently, the error percentage rate of this sequencing technology is immense, above 20%, but it is used in many applications, including the discovery of SNPs and studies of phylogenetics (Hackl et al. 2014).

10.7 Nanopore Sequencing

Nanopore sequencing is the process of electrophoretically passing a single DNA molecule through nanoscale pores to sequence it. Similar to TGSTs, this technology does not need a step for amplification; this step requires assembling sequencing libraries. In this sequencing, a bias voltage is passed through the nanopores, which creates an electric field in them, and in this processing, single-stranded DNA (ssDNA) is obtained. Changes in ionic current and duration across the nanopore are used to detect the passage of ssDNA (Healy 2007). This technology has massive potential as it can use minimal amount of DNA and it can directly sequence this sample with very low or absence of alternation to get highly extended reads. In this technology, the natural RNA/DNA can be identified directly although in other sequencing methods such as sequencing-by-synthesis technology, normally the template copy is formed in the reaction and is identified. Oxford Nanopore Technologies introduced the first commercially accessible platform of nanopore sequencing; it is called MinION and has a small apparatus like mobile phone, which can easily be attached to a laptop via the USB (Bleidorn 2016). It can generate longer reads of about 98 kb enfolding several genes in individual reads. Another highly efficient nanopore sequencer whose development is in progress is GridION (Heather and Chain 2016). Because of the length of tunnel and sequencing speed, numerous nucleotides are organized in the tunnel due to this maximum percentage of error of approximately 20% (Ip et al. 2015). But, in some modern technologies such as intramolecular-ligated nanopore consensus sequencing (INC-Seq) and Illumina hybrid nanopore sequencing, the precision of sequencing can be improved in a way that it can be used to study many biological issues (Li et al. 2016; Madoui et al. 2015).

10.8 Potential TGSTs

Currently, several private companies and educational institutes use some TGSTs, yet they are unable to launch this technique commercially. One method is to quantify the movement of RNA polymerase through a template of DNA, upgraded at Stanford by the Block group (Greenleaf and Block 2006). There are several other methods of sequencing DNA through direct imaging systems using transmission electron microscopy and tunnel scanner, which are about to be launched by several companies such as Reveo and Genetics ZS (Ozsolak 2012).

10.9 NGS Data Analysis

Due to many reasons such as cost-effectiveness, short processing time, use in many technologies, and high efficiency, the NGS techniques are acquiring advancements. That way, recently the sequence of data assembly has become easier. However, the study of such information still causes numerous significant challenges that control the progress of many techniques for managing and studying information in a particular method (Rusk 2009). Nearly every technology of NGS generates a massive amount of data sequence per cycle efficiently. In NGS analysis of data, the common phases are (i) measurement of quality, (ii) alignment, (iii) identification of variant, (iv) annotation of variant, and (v) visualization of data (Pabinger et al. 2014; Rusk 2009).

10.10 Use of NGS Techniques in Molecular Plant Biology and Crop Improvement

Due to the maximum efficiency and cost-effectiveness, NGS technologies are made simpler and available for small research centers. This technique is used in molecular markers, the sequence of DNA segment that does not translate automatically any traits while transferred in Mendelian pattern and is not exaggerated through environment. Molecular markers depicted various kinds of basic and applied roles in crop biology such as fingerprinting of genotype, development of map linkage, analysis of parentage, mapping of QTL, studies of evolutionary and genetic diversity, MAS (marker-assisted selection), and gene flow. Microsatellites are also called SSRs (simple sequence repeats), repetition of 1–6 nucleotides in DNA sequences that are set up in tandem arrays of entire genome and widely utilized in the development of different crops. SSRs were previously created by building genomic libraries, then isolating and sequencing SSR-containing clones, which is a time-consuming and tedious method. At present, using NGS technologies, anyone can directly sequence the genomic DNA and transcriptomic RNA and analyze the SSRs by using many analytical methods (Zalapa et al. 2012). With the help of many techniques of NGS, a massive amount of SSR markers have been produced in various crops, including *Larix*, *Jatropha*, and *Betula* (Chen et al. 2015; Hao et al. 2015; Kumari et al. 2013). Currently, Singh et al. (2016) classified 2863 genic SSR markers in B. juncea by Illumina paired-end sequencing technique. From these, 460 EST-SSRs and EST-SSR loci were confirmed in different genotypes of Indian mustard; among these, 339 loci amplified significantly, while 134 EST-SSRs presented polymorphisms.

The variation in single nucleotide at analogous loci of other linked molecules of DNA is called SNPs (single nucleotide polymorphisms). SNPs have many important roles in the identification of variation in genomes and assembly of genetic maps of linkage. They are linked with the traits of phenotype and also application in plant breeding (Mammadov et al. 2012). SNPs were previously discovered by mining Sanger sequencing amplicons or Sanger expressed sequence tags (ESTs), but presently an emerging SNP discovery method involves mining NGS sequencing data of DNA/RNA molecules. The reads are acquired by NGS sequencing; these germplasm sequences are assembled for the standard plants genome to analyze the SNPs present in gene pool. Identification of SNPs is very difficult based upon NGS sequencing because the species of plants have no standard genome as a reference; moreover, short read sequences are present having weak alignment for different genotypes and individuals (Kumar et al. 2012). But, Azam et al. (2012) studied the consequence of four commonly used short read alignment techniques, viz. Maq, Bowtie, SOAP2, and Novoalign with the help of CbCC method for the discovery of SNPs in chickpea, lacking standard genome at that moment. In conclusion, they described that Maq is the most sensitive and precise tool at minimal depth of read. Mostly, the detection accuracy of SNPs is based on the depth of reads; the accuracy increases with the increase in depth of reads. Table 10.1 summarizes the identification of SNPs by bioinformatics software and the technique of NGS in several plant species.

Diversity array technology (DArT) relies on restriction enzymes to reduce genomic complexity, then hybridization on microarrays to simultaneously screen hundreds of markers throughout the genome (Jaccoud et al. 2001). Moreover, this technology is benefits from the progress of NGS techniques, and recently, NGS-DArT marker has replaced the microarray markers of DArT. Sansaloni et al. (2011) explained that the combined use of NGS-DArT markers provide three times more markers as compared to traditional markers, including arrays-based DArT markers with an extra unit of SNP markers. Denser genetic maps are obtained by the combination of DArT markers with other molecular markers in several crops such as sorghum, wheat, and barley (Crossa et al. 2007; Mace et al. 2009; Peleg et al. 2008).

TABLE 10.1
List of iIdentified SNPs in vVarious cCrops by uUsing NGS tTechniques

Sr. No.#	Species	Sequencer	No. of SNPs	References
1	*Arachis* spp.	Roche 454	8,486	(Khera et al. (2013)
2	*Arachis* spp.	Affymetrix GeneTitan	58,233	(Pandey et al. (2017)
3	*Brassica napus*	Illumina	892,536	(Huang et al. (2013a)
4	*Camellia sinensis*	Illumina HiSeq 2000	6,042	(Ma et al. (2015)
5	*Elaeis guineensis*	Ion Torrent Ion	21,471	(Pootakham et al. (2015)
6	*Ginkgo biloba*	Illumina	139,854	(Wu et al. (2019)
7	*Gossypium hirsutum*	Illumina HiSeq 2000	40,503	(Zhu et al. (2014)
8	*Gossypium hirsutum*	Illumina GAII and Illumina HiSeq 2000	1,873	(Logan-Young et al. (2015)
9	*Helianthus annuus*	Illumina GAIIx	16,467	(Pegadaraju et al. (2013)
10	*Manihot esculenta*	Illumina HiSeq	73,458	(Hamblin and Rabbi (2014)
11	*Medicago sativa*	Illumina	872,384	(Li et al. (2012)
12	*Oryza sativa*	Illumina	3,202,922	(Lim et al. (2014)
13	*Oryza sativa*	Illumina HiSeq 2000	456,777	(Yang et al. (2013)
14	*Punica granatum*	Roche 454	6,500	(Ophir et al. (2014)
15	*Raphanus sativus*	Illumina	1,953	(Zou et al. (2013)
16	*Solanum tuberosum*	Illumina HiSeq 2000	42,625	(Uitdewilligen et al. (2013)
17	*Sorghum bicolor*	Illumina	1,243,151	(Zou et al. (2012)
18	*Triticum aestivum*	Illumina	10,251	(Allen et al. (2013)
19	*Triticum aestivum*	Affymetrix GeneTitan	35,143	(Allen et al. (2017)
20	*Triticum aestivum*	Illumina HiSeq 2000	3,977	(Cubizolles et al. (2016)

10.11 Transcriptome Investigation

The isolated mRNA sequencing of various plant tissues during different periods is called sequencing of transcriptome, which emphasizes the studies of transcribed genome part (Wolf 2013). Sequencing of transcriptome has many uses to understand the concept of molecular mechanisms in plants, such as profiling expression of gene, annotation of genome, splicing of intron, rearrangement detection of genome, and quantification and identification of noncoding RNA. The flexibility of transcribed data allows synchronized analysis of several biological mechanisms. Table 10.2 concludes the plant species that sequence is transcribed. Qu et al. carried out research on transcriptome (Qu et al. 2015) in herbaceous flower (*Phlox subulata*) that can tolerate –30 temperature by Illumina HiSeq 2000 and reported the major cold-responsive genes such as signaling genes of ABA and Ca^{2+}, protein genes in osmoregulation, antioxidant genes, and transcription factors. In non-modal species of plants, the studies of advancement of transcriptome sequencing can also be conducted.

10.12 Gene Expression Profiling

One of the most popular ways of identifying the gene(s) responsible for a given feature of interest in plants is to identify and quantify mRNA species under various circumstances or in distinct cell types. The microarray-based technique and the sequencing of cDNA fragments method are two alternative methodologies for high throughput gene expression profiling. The cDNA is hybridized to an array of complementary oligonucleotide probes corresponding to a gene of interest in microarrays. A fundamentally distinct method involves sequencing cDNA fragments and then counting the number of times each fragment is found in the sample. Different other techniques are based on transcriptomic analysis, such as MPSS and SAGE. In SAGE, the cDNA is cut into 14–17 base pairs of short DNA segments by using

TABLE 10.2

Transcriptome sSequencing of sSome iImportant cCrop sSpecies with NGS

Sr. No.#	Species	Platform	References
1	*Arabidopsis thaliana*	Illumina	(Bernal et al. (2012)
2	*Avena barbata*	454- pPyrosequencing	(Swarbreck et al. (2011)
3	*Brassica juncea*	Illumina	(Liu et al. (2013)
4	*Brassica juncea*	Illumina	(Sharma et al. (2015)
5	*Brassica napus*	Illumina GAIIx	(Lu et al. (2014)
6	*Brassica napus*	Illumina	(Trick et al. (2009)
7	*Caragana korshinskii*	Illumina HiSeq 2000	(Lin et al. (2016)
8	*Cicer arientinumarietinum*	Illumina, 454, and Sanger sequencing	(Kudapa et al. (2014)
9	*Cicer arietinum*	Illumina	(Kumar et al. (2019)
10	*Clerodendrum trichotomum*	Illumina 4000	(Chen et al. (2019)
11	*Gossypium hirsutum*	Illumina	(Yang et al. (2014)]
12	*Crucihimalaya himalaica*	Illumina HiSeq 2500	(Qiao et al. (2016)
13	*Cucumis sativus*	454	(Guo et al. (2010)
14	*Cucumis sativus*	Illumina HiSeq 2000	(Zhao et al. (2015)
15	*Fagopyrum esculentum*	Illumina HiSeq 2500	(Zhu et al. (2015)
16	*Halogeton glomeratus*	PacBio	(Yao et al. (2018)
17	*Ipomoea batatas*	Illumina	(Arisha et al. (2020)
18	*Zea mays*	SOLiD	(Li et al. (2014b)
19	*Medicago falcata*	Illumina	(Cui et al. (2019)
20	*Medicago ruthenica*	Illumina	(Shu et al. (2018)
21	*Medicago truncatula*	Illumina	(Cabeza et al. (2014)
22	*Medicago truncatula*	Illumina HiSeq 2000	(Cabeza et al. (2014)
23	*Nepenthes*	Illumina HiSeq 2500	(Zakaria et al. (2016)
24	*Olea europaea*	Illumina	(Iaria et al. (2016)
25	*Pyrus colleryanacalleryana*	Illumina HiSeq 2000	(Xu et al. (2015)
26	*Pisum sativum*	454	(Franssen et al. (2011)
27	*Raphanus sativus*	Illumina HiSeq 2500	(Gao et al. (2019a)
28	*Oryza sativa*	Illumina	(Wakasa et al. (2014)
29	*Solanum lycopersicum*	Illumina GAIIx	(Sundaresan et al. (2016)
30	*Solanum tuberosum*	Illumina	(Moon et al. (2018)
31	*Solanum lycopersicum*	Illumina	(Chen et al. (2013)

Type IIs restriction enzymes followed by concatenation of these fragments. Expression profiling of resultant mRNAs are observed by sequencing of these concatemers. Similarly, the MPSS approach generates small segments of every mRNA species after sequencing in a very highly efficient process. In the past, because of the low efficiency and high cost-effectiveness, the technique of ESTs used for sequencing of these tags was incompetent to characterize the complete set of mRNA available in the desired section. However, at present, the technologies of NGS are used to sequence the complete set of mRNA due to some reasons such as high speed and efficiency and low cost. Furthermore, short reads are matched with the short tags produced in these techniques.

Nielsen et al. (2006) introduced a new method of NGS-SAGE; this is known as Deep-SAGE. It works based on the preparation of samples by amplification of emulsion-PCR, eliminating the step of cloning; therefore, this method is simple and significant.

Furthermore, different other sequencing techniques including full-length sequencing of cDNA and generation of ESTs helps the technologies of NGS. The EST technique for sequencing has increased the capacity for profiling of mRNA by the collaboration of NGS. Up to now, libraries of EST by NGS technologies have been assembled in many species of plants such as *Zea mays*, *Arabidopsis thaliana*, and *Medicago truncatula* (Morozova and Marra 2008).

10.13 Small Noncoding RNA Profiling

Small noncoding RNA (sncRNA) profiling has been involved in several crucial mechanisms of plants (Huang et al. 2013b). This group of RNA consists of different other types of RNA, such as siRNA, tRNA, snRNA, rRNA, tasiRNA, and miRNA. All of these have many important roles in gene expression, but siRNA and miRNA are very important and essential in gene expression because these majorly play important roles in post-transcriptional regulation. miRNAs are long RNA, consist of almost 21 nucleotides, and are found in both animals and plants although siRNAs particularly present in plant species only and consist of approximately 21–24 nucleotides. However, both have almost the same size and function, but their specific mode of action and biogenesis are different (Xie et al. 2004).

miRNAs were identified for the first time in *Caenorhabditis elegans*, but at present, they are discovered in many other families such as viruses, animals, fungi, and plants (He and Hannon 2004). Traditionally, micro-RNAs have been recognized by sequencing and cloning of individual miRNAs, which is an unmanageable and slow method. In contrast to conventional methods, the techniques of NGS prevent the step of cloning and make the study of ncRNA easy. Its nature of high efficiency makes it possible to study the sncRNAs in whole genome and produces the small read length by it that is also similar to the size of small interfering RNAs and micro-RNAs (Morin et al. 2008). The profiling of NGS-ncRNA has specific benefits over microarray techniques, such as the potential to identify unique miRNAs and its ability to discover editing of miRNA (Li et al. 2014a). So far, many type of researches have been conducted to identify the role of small RNA profiling in many biological mechanisms of different species of plants such as maize, rice, and wheat (Baldrich and San Segundo 2016; Li et al. 2014a; Lunardon et al. 2016). Gao et al. (2015) studied tobacco to analyze the miRNA expression profiling in various tissues; 50 novel and 165 conserved miRNAs were recognized in this study. Recently, a research has been conducted on *Arabidopsis thaliana*; the function of sRNAs in the regulation of whole-genome methylation of DNA has been exposed by combining data of sequencing of sRNA and bisulfite sequencing of wide genome (Lewsey et al. 2016).

Remarkable studies of sequencing of sncRNAs focused on identifying the Piwi-interacting RNAs, a unique sncRNA present in mammals (Houwing et al. 2007).

10.14 Gene Annotation Using Transcriptome Sequence Data

With advancements in NGS techniques, the cost of sequencing decreases and a large amount of sequencing data of genome are generated every year, whereas multicellular species of eukaryotes is facing difficulty yet in annotation of complete genome. Mainly, the correct exon–intron structure of coding genes in protein and annotation of noncoding genes of RNA lag behind (Brent 2008).

Different methods are used for the annotation of coding protein genes, but recently, the methods of major use has been EST or full-length sequencing of cDNA comprehended by the arrangement of model genome. Using first-generation sequencing techniques, it has been observed that a number of projects of EST sequencing flop to cover very long and rare transcripts by about 20–40 percent. Another major difficulty is the identification of alternative splicing in targets (Brent 2008). The NGS techniques are used to identify the rare transcripts because they provide deeper coverage. Different SGSTs generate short reads, but face many problems during the annotation of various alternating splicing and exact exon–intron structure in coding protein genes. In the future, TGSTs have the ability to produce longer read lengths that will increase genome annotation efficiency.

Interestingly, different studies related to the profiling of gene expression by NGS techniques can also deliver the information of annotation, such as novel genes existence, alternative splicing, and structure of exon and intron (Schneeberger 2014). Furthermore, NGS-based EST data can be utilized in the identification of SNPs and SSRs (Barbazuk et al. 2007).

10.15 Phylogenetic and Ecological Studies

The fundamental concept to study the analysis of phylogenetics depend on sequence of DNA relationship at conserved sites in genome is that the maximum similarity present in any genotype would have maximum resemblance in sequence while the maximum dissimilar genotypes would have many variations from the time when evolution and speciation deviating traits mount up the novel mutations (Patwardhan et al. 2014). For that study, multiple genes sequencing data from populations are needed and the techniques of NGS offers many other possibilities such as high efficiency and cost-effectiveness as compared to classic Sanger sequencing. Targeted sequence apprehends together with the techniques of NGS and provides resources of genomics for non-reference organisms, facilitating many studies such as analysis of phylogeny, variation in population, polyploidy parentage, and various analyses of diversity and gene flow (Grover et al. 2012). Currently, using NGS techniques is promising to study the phylogenetic relationship between closely similar and divergent taxa or species (Steele et al. 2012). In wide-genome studies of phylogenetic the NBS-encoding genes that are disease resistant in cucumbers, it was discovered that the development of NBS genes may have appeared earlier before the divergence of Cucurbitaceae and Arabidopsis. Therefore, these disease-resistant genes in the above-mentioned families are ancient (Zhao et al. 2015). One more phylogenetic analysis founded on SNP data produced through resequencing of the genomes of three thousand various accessions of *Oryza sativa* classified its gene pool into five different groups, namely basmati, temperate and tropical japonica, indica, and aus/boro (RG 2014).

Pease et al. (2016) performed experiments on the different patterns of molecular evolution on wide genome of 13 naturally and reproductively variant species of wild tomato, which are the progenitor of *Solanum lycopersicum*. By relating the variant arrangement of several genes, they discovered many events of rapid successive speciation in the development of *Solanum lycopersicum* in 2.5 million years of evolution and modification of this species.

10.16 Allele Mining

The method that is used to identify the novel and superior alleles of many essential genes founded on the sequence data is called allele mining. Similarly, this method proves the purpose of particular genes in regulating the trait (Ashkani et al. 2015). By using NGS techniques, the mining of alleles becomes the more effective technique over sequencing techniques in the comparison to identify the new alleles based on EcoTIILING. In the first step, identify the targeted gene that controls the trait and then design the primers of various genotypes for PCR to identify the particular trait from a huge diversity. After that, amplicons are sequenced and studied for different diversities. Then, newly identified alleles are integrated with different genotypes that show maximum performance and founded on allele that have particular primers; furthermore, these newly identified alleles can play an important role in plant breeding (Kumar et al. 2010). Currently, Lyu et al. (2013) researched six varieties of rice to observe allelic diversities among them and sequenced their wide genome by using Illumina GA2 platform. By relating these genomes to two standard genomes, a few ETASs were recognized in 9-cis-epoxycarotenoid dioxygenase gene of the IRAT104 upland rice variety. The functional studies of 9-cisepoxycarotenoid dioxygenase gene showed its connotation with the suitability of upland rice (Lyu et al. 2013).

10.17 Genetical Genomics

The basic concept of genetical genomics tells about the study of conventional genetics at maximum profiling expression to clarify the complicated traits at molecular level (Jansen and Nap 2001). One of the key notions in genetical genomics is the presence of hotspots in the genome, where a single variation causes broad downstream alterations in the expression of distant genes that all map to the same genomic locus (Breitling et al. 2008). Therefore, the detention of wide-genome NGS agitations can be used to produce maximum data and has been productively used for mapping different traits (Schneeberger 2014; Zou et al. 2013).

10.18 Epigenetics Studies

Epigenetics is the analysis of phenotypic trait variations without modifying the genomic DNA sequence. These are genetic modifications and extensively classified into two major groups on the basis of epigenetic variations, including DNA methylation and different alterations in post-translation of histone tails (Callinan and Feinberg 2006). At cytosine residues, methylation of DNA occurs by the accumulation of covalent group of methyl at the five carbon of cytosine. Based on NGS techniques, different sequencing methods have been established and applied in the analysis of epigenetics, including WGBS (Cokus et al. 2008), MeDIP-seq (Jacinto et al. 2008), ChIP-seq (Pellegrini and Ferrari 2012), TAB-seq (Yu et al. 2012), and 3C-seq (Dekker et al. 2002). These newly discovered technologies are utilized to discover methylation patterns in DNA, conformation of chromatin, and wide range of nucleic acid and protein integrations.

Whole-genome bisulfite sequencing (WGBS) is associated with the techniques of NGS that facilitate the identification of DNA methylation from whole genome very successfully. Sodium bisulfite is used to treat the genomic DNA that stimulates the transformation of (C) cytosine into (U) uracil through separating from the complex of 5-methylcytosine. Hence, bisulfite-treated DNA sequencing is compared with non-treated DNA sequencing and then it covers up the position of methylated cytosine (Darst et al. 2010). Exact sites of methylation can be detected by TGSTs on DNA template (Flusberg et al. 2010).

DNA sequencing on the basis of NGS in wide-ranging populations have 217 genomes, 144 transcriptomes and 152 methylomes of accessions of wild *Arabidopsis thaliana*, exhibited SMP ranging from 92,646 to 527 (Schmitz et al. 2013). Moreover, Schmitz et al. (2013) analyzed the 152 accessions of phylogeny depending on the newly discovered SMPs and contrasted them by accessions of polygene depending on SNPs and discovered maximum relationship in tree scheme. DNA methylation exhibited less dynamic behavior than gene expression profiles in plants, and only functioned in a single cell type or at a certain developmental stage, according to this study's findings.

Le et al. (2015) contrasted the information produced by WGBS of mutant epigenetic *Arabidopsis thaliana* with their natural accessions and discovered that almost 3 pc of transposable elements are found in the intronic gene sites and have an impact on the expression of genes via RNA-directed DNA methylation-dependent and DNA methylation-independent pathways. The understanding of how intragenic transposable elements influence the transcription of the genes is a unique discovery.

10.19 Regulatory Protein Binding Domain Prediction

Chromatin immunoprecipitation based on NGS is an important approach to analyzing the genome-wide profiling of integration of DNA protein (Varshney et al. 2009). The progress of NGS technologies has surpassed different techniques based on microarrays, including STAGE (sequence tag analysis of genomic enrichment), ChIP-chip, and SAGE previously used in that analysis. ChIP-seq produces a huge amount of data that reveal the understanding of regulation of different gene alterations in epigenetics at genome-wide level, after various major studies of bioinformatics. The techniques of ChIP-seq include DNA precipitation with particular antibodies in response to the selected histone protein and TFs (transcription factors), and then, the DNA precipitated is extracted for further NGS studies. Studies on reads sequence give data about the target regions on the particular histone protein or TFs on a wide-genome level.

Recently, by NGS technologies 1,340 TFBSs and 253 TF–TFBS pairs have been identified in 22 various developing leaf transcriptomic studies in maize (Yu et al. 2015). ChIP-seq reported that in *Brachypodium distachyon*, histone protein, namely H2A.Z, produces a tolerance to heat stress by controlling the packaging of nucleosome in the phase of filling the grain (Boden et al. 2013). In another research, wide-genome detection of DNase-I-hypersensitive (DH) sites in *Arabidopsis thaliana* identified the integration of this site with RNA-II polymerase binding sites. Consequently, in plants, mapping of DH wide-genome site could be utilized for the detection of cis-regulatory DNA elements (Zhang et al. 2012).

10.20 Metagenomic Analysis

Commonly, microorganisms are coupled with plants, so they are located in different forms such as rhizosphere, endophyte, and epiphyte, and these associations have many advantages and a neutral impact on the health of plant. It is essential to consider the dynamics and diversity of microbial populations for getting the basic concept of microbial interaction (Newton et al. 2010). The study of wide genomic microbial population by directly segregating the DNA from the environmental samples is called metagenomics. This includes shotgun preparation, sequencing, and preparing library of metagenomics comprehended by the study of complex data.

Metagenomics offers the basic concepts and understandings to study the microbial communities as various microbe cultures could not be deliberated before applying different traditional techniques (Knief 2014). Different highly efficient NGS techniques allow the metagenome deep sequencing that forces the detection of less abundant microorganisms. The reads are arranged for making the contigs that are assembled in genomes of microbes, which are present in common databases; this process is done after sequencing. A genome of new microbes may be represented by the sequence of the unaligned reads, and after that, the genome of the individual can be reassembled (Albertsen et al. 2013). Roossinck discovered many long-lasting viruses in fungi and plants by the use of metagenomic analysis on the basis of NGS (Roossinck 2015). Moreover, Mhuantong et al. (2015) studied the bagasse metagenome of sugarcane by making a library of fosmids with the use of pyrosequencing. Metagenomic data of relative analysis of genomes detect the conserved and unique decomposing enzymes of biomass in lignocellulolytic microbial population.

10.21 Single-Cell Genomics

Through the study of numerous cells, the perception of the plant genome, its regulation, and expression have originated. Certainly, these studies are informative, but sometimes incompetent to identify any heterogeneity that occurs in the cells of the population (Macaulay and Voet 2014). Overall, the understanding of the genome is that the genome of an individual is persistent in tissues during life. But after sometime, the minor mutation occurs during cell division by the reason that the somatic variation (heterogeneity of genome) is developed within the organisms. These modifications are caused by many diseases and developmental processes (Macaulay and Voet 2014). Because a population of cells is used as starting material in traditional sequencing research, the genomic and transcriptome differences accessible in single cells are lost (Blainey 2013). For that reason, SCGS can deliver the understanding of the single-cell genome. Developments in the separation of individual cells, the proliferation of wide genome, and the technologies of NGS provide the feasibility for sequencing the genome of individuals. While challenges related to the amplification of wide genome are still problematic, advancements in technologies such as TGSTs have the potential to sequence the single cell because it will become a more perfect and fast technology (Macaulay and Voet 2014).

Until now, maximum analyses of SCGS have been conducted in the animal structure; for example, the cells of humans are used to study the tumor. In plants, only minimal research has been conducted, seeing the reality that it is problematic to separate the genome of an individual cell from the cell wall of plants. Diwan et al. (2014) studied the genomic profiles from different tissues of trees such as *Prunus × yedoensis* and *Fagus crenata*. It was determined that they were different systematically from top to bottom of plant. Moreover, Li et al. (2015) sequenced the single microspore culture from a population of tetrad and produced a map having a high resolution for the recombination of meiotic in *Zea mays*. One more research by Farlik et al. (2015) explained the different approaches to analyzing the epigenetics of individuals in wide-genome sequencing of bisulfite and also expressed the bioinformatics tools to study the methylome of the individual cell. Macaulay et al. (2015) introduced the approach of G&T-seq (genome and transcriptome sequencing) of single cell, in which both are sequenced in parallel.

10.22 Exome Sequencing

Commonly, plant genomes are large, repetitive, complex, and polyploid, creating complex conditions for whole-genome sequencing. However, the advancement of NGS technologies for WGS has developed a cost-effective approach, but identifying diverse genotypes by WGS is still costly in exome (complete set of exons present in all genes) situated in genome and shows the short part by 1 to 2%. This sequencing can offer different data that exist in genome's coding region in variant genotypes with meaningful analysis and the cheapest method (Warr et al. 2015).

This sequencing is a two-step method: (i) capturing of exome and (ii) sequencing. There are many approaches to exome capturing that have the following primary steps: genomic DNA isolation and fragmentation, fragment selection holding exons by amplification of PCR, and probe hybridization. Recently, sequencing of single molecule potential of TGSTs has eliminated the step of amplification of PCR of targeted sequence of exon, at the advance decrease the amount of exome sequencing. Exome sequencing depends on the genome annotation accuracy; on the other hand, for crops that have minimal annotated genomes, sequencing data of transcriptome can be utilized.

In wheat, the data of transcriptome were utilized for the improvement of capturing of exome-kit. In the case of barley, the mutation has been identified by exome sequencing such as earlier maturation and detection of gene included in several nodded dwarfism (Mascher et al. 2013). Table 10.3 shows the studies of exome sequencing conducted in different crops.

10.23 Multiple Genome Sequencing and Resequencing

Low cost, high speed, and high efficiency of several techniques of NGS give an opportunity for researchers to understand and make the multiple sequences of genome of the single species of plant. In 2008, the 1001 Genomes Project was introduced, to read wide-genome modifications that occurred in 1,001 project of various accessions of *Arabidopsis thaliana* (http://1001genomes.org/). The genome structure is influenced by various evolutionary issues such as conversion of genes, mutation, polyploidization, mutation, selection, recombination, and several introgressions. The concept of influence on variation of sequence process permits to identify the genesis of genetic diversity and analysis of the variation of alleles that are responsible for phenotypic variation. Cao et al. (2011) sequenced the whole genomes of 80 distinct *A. thaliana* accessions Using the Illumina Genome Analyzer platform. They detected SNPs 4,902,039 and small indels 810,467 within 80 genotypes as well as 1,059 CNVs (copy number variants) dealing with 2.2 Mb of model genome.

Furthermore, Long et al. 2013 conducted a research on multiple sequences of genome in 180 various genotypes of *Arabidopsis thaliana* that were gathered from Sweden and discovered structure variants of about 0.6 million and 4.5 million SNPs. The 3000 rice genome project (3K RGP) was recently completed, with the genomes of 3000 rice accessions gathered from 89 countries being resequenced with an

TABLE 10.3

Exome Sequencing by Using NGS in Some Important Crop Species

Sr. No.#	Species	Platform		Method	References
1	*Triticum aestivum*	Illumina	HiSeq 2000	Array based	(He et al. (2019)
2	*Eucalyptus*		GAIIx		(Dasgupta et al. (2015)
3	*Oryza sativa*		HiSeq 2000		(Henry et al. (2014)
4	*Hordeum vulgare*		HiSeq 2000		(Mascher et al. (2013)
5	*Triticum aestivum*		GAIIx		(Winfield et al. (2012)
6	*Triticum aestivum*		GAIIx		(King et al. (2015)
7	*Pinus taeda*		HiSeq 2000	Solution based	(Neves et al. (2013)
8	*Oryza sativa*		HiSeq 2000		(Udomchalothorn et al. (2014)
9	*Picea rubens*		HiSeq X		(Capblancq et al. (2020)

average sequencing depth of 14X. By this 3K RGP, almost 18.9 million SNPs have been identified in rice that could be worked in the future for the improvement of many crops (RG 2014).

10.24 Accelerating Genetic Gain in Breeding Populations

The rapid identification of cost-effective SNPs and production of genome sequencing have been led by the progression of NGS techniques. It increased the resequencing and sequencing of the wide genome or a minimum portion from many genotypes for the discovery of polymorphism in different species of crops. As a result, in recent years, high-throughput genome-wide SNP genotyping technologies like as genotype-by-sequencing (GBS) have aided in the selection of high breeding value genotypes in breeding populations based on their genotypic composition (Poland et al. 2012) and used earlier in different crops such as legumes for GS (genomic selection) (Beissinger et al. 2013). It has been observed that GBS covers maximum parts of genome and catches many particular genetic variations in population as compared to other genotyping techniques such as SNPs used in different crops now (Bhat et al. 2016). However, coupled NGS–GBS has been found as the standard for genome selection because of its low price, flexibility, and the method used for the prediction of accuracy such as genomic estimated breeding value (GEBV). The accuracy of prediction such as 0.1–0.2 of GEBV has been found maximum by NGS-dependent genotyping compared to other conventional molecular marker methods (Poland et al. 2012). In soybean, the accuracy prediction has been observed to be 0.64 for the production of grain, measured by the cross-validation. This accuracy has significant capability for genome selection (Jarquín et al. 2014). So, GS based on NGS supports increasing the genetic gain in subsequent methods.

- Several numbers of accessions have been promoted by NGS genotyping that assist in discovering maximum genetic variation within the genome. This enhances the intensity of GS. In a single cycle, millions of SNPs can be screened by NGS; however, GS can be used for minimal lines of QTLs having minimum livability (Xu and Crouch 2008). As a result, maximum genetic gain is feasible via GS dependent on NGS.
- Selection is dependent on profiling of marker that supports the detection of genotypes in the population of breeding with maximum standards of breeding by increasing the accuracy and power of selection in the quantitative attributes. This can provide maximum genetic gains every year than the selection of phenotype (Heffner et al. 2010; Zhong et al. 2009).
- Similarly, GS also supports obtaining a higher genetic gain for those attributes that have a permanent time of generation and are challenging to estimate, including tolerance against insects and characteristics for making bread. For this reason, GS has become of low cost. Quantitative genetic traits are significantly affected by the interaction of environment and genotyping. Consequently, the genetic gain and prediction accuracy can be increased by the intermingling of the targeted genomic loci with the environment. It detects the breeding genotypes on the basis of wide-genome marker profiling lacks phenotyping that assists in detecting the tolerant breeding genotype against environmental stress. Hence, identifying millions of SNPs and their role in genotyping in breeding population by the techniques of NGS coupled with GBS can switch the phenotypic GS in the future (Bhat et al. 2016). In chickpea, genomic selection accuracy was calculated for the traits that involved in yield from 0.138 to 0.192 yield of seed to weight of seed, respectively, and the interaction of genotype with environment in the models of GS enhanced the prediction accuracy (Roorkiwal et al. 2018; Wang et al. 2016).

Various models of genomic selection and prediction accuracy have been discovered for constructing the conditions of GS for breeding. GS models comprised of the intermingling of G×E and multiple attributes presented maximum accuracy prediction for the minimal traits of heritability. Furthermore, prediction accuracies in environments and populations were found from minimum to maximum in lentils (Haile et al. 2020). This observation recommended that genomic selection can increase the genetic gain in ecologically and population by genotyping the large breeding population in lentils with large markers (Haile et al. 2020).

Consequently, applications of NGS can assist in enhancing the genetic gain quickly and coupled with GS in plants through genotyping the maximum breeding population with numerous markers.

10.25 Development of Pan-Genome and Super-Pan-Genome

The complete set of genome for the species of individuals is called pan-genome (Tettelin et al. 2005). These genes can be assembled as core and expendable genes. The core genes are preserved in the whole persons, and therefore, these are typically known as housekeeping genes, which are responsible for important functions of cells (Tao et al. 2019). Similarly, the combination of core gene and pan-genome is called closed pan-genome. On the other hand, in pan-genome the dispensable genes exist in either a few or particular individuals, while not present in whole. Functionally, these genes are linked with different adaptive attributes, including signal transduction, activity related to antioxidants and receptors, resistance against abiotic and biotic stresses, and regulation of genes (Gordon et al. 2017; Hurgobin et al. 2018; Li et al. 2014c; McHale et al. 2012; Schatz et al. 2014; Xu et al. 2010). Consequently, these genes participate in species diversity and develop more rapidly as compared to core genes (McHale et al. 2012). In soybean, the substitutions of dispensable genes with non-synonymous and synonymous SNPs have been found maximum (Li et al. 2014c). Furthermore, the substitutions of synonymous and non-synonymous SNPs within coding sites are the basic source of genetic modification in lentils (Sharpe et al. 2013). The understanding of pan-genome dependent on causing the genetic modification, especially variation in structure, present in the substance of gene of the persons is appropriate in the identical species (McHale et al. 2012). It was made possible because of the progression in NGS techniques that allowed the resequencing and sequencing of several accessions that are appropriate to few and many species. These SVs contained the existence or absence of variations, CNVs (copy number variations), and many other types of variations such as transversion, inversion, and inter- or intra-translocation of chromosome (Cook et al. 2012; Feuk et al. 2006; Huang et al. 1991; Qi et al. 2014; Wang et al. 2015). Studies of pan-genome include sequencing of multiple accessions of genome association with cultivated species; on the other hand, genomic sequences of accessions linked with every species present in a genus that is comprised on the studies of the super-pan-genome (Hu et al. 2020; Khan et al. 2020). As wild family members have several particular traits, studies of super-pan-genome can provide a better possibility by using the presence of the variation in genomic structure of a genus within the advancement of genetics by coupling them with the interested traits through the studies of the wide-genome association (Khan et al. 2020). The pan-genome techniques have been used in many crops such as soybean (Li et al. 2014c), cabbage (Xu et al. 2010), mustard (Hurgobin et al. 2018), maize (D'Aoust et al. 1999), rice (Wang et al. 2018), wheat (Montenegro et al. 2017), sesame (Yu et al. 2019), sunflower (Hübner et al. 2019), and tomato (Gao et al. 2019b) for the studies of genetic diversity. In lentils, the technique of NGS has been utilized to sequence RNAs of several accessions associated with the wild and cultivated species and a specific analysis of genome detected SNPs or indels in the genome. On the other hand, only SNPs and indels do not participate in the whole diversity of genes present in species (Saxena et al. 2014; Springer et al. 2009; Wang et al. 2015). Currently, in lentil, the sequence of reference genome is present (Kumar and Gupta 2020). Various structural modifications in chromosomes because of translocations have been observed in and across the species in the past (Jha et al. 2015). In lentil, resequencing of various accessions can support the detection of occurrence of variation in structure at genomic level as found in different crops such as soybean (Lam et al. 2010; Zhou et al. 2015), pulses, and pigeon pea (Varshney et al. 2017). The features mentioned below are used to detect the dispensable genes within the pan-genome, and they will be helpful in crops to study the advancement of the pan-genome.

10.26 Pan-Genome Size

It is distinguished through the number of sequenced individuals included in the pan-genome because it upsurges the percentage of expendable genes and reduces the proportion of primary genes. In rice, a pan-genome with 48,098 genes increased from 3,010 accessions, having 41% dispensable genes; on the other hand, a pan-genome with 40,362 genes had 8% dispensable genes in three accessions (Schatz et al. 2014; Wang et al. 2018).

10.27 Type of Accessions

It has been seen that the utilization of strongly close accessions doesn't finish the extent of the pan-genome. However, it is possible to utilize the different accessions and the combination of accessions of grown and wild species that can support to improve the percentage of dispensable genes in the pan-genome instead of utilizing cultivated species accessions. For instance, in rice the cultivated species *Oryza sativa* has 66 accessions within pan-genome, while *Oryza rufipogon* is a wild species that have 42,580 genes and contains 38% dispensable genes (Zhao et al. 2018). But on the other hand, the accessions of cultivated species in pan-genome have 40,362 genes and 7.83% of dispensable genes (Schatz et al. 2014). Furthermore, Khan et al. (2020) recommended that the utilization of wild species accessions in the advancement of a pan-genome can make it possible to detect the loss of genes in the taming of crop.

10.28 Accelerating the Use of Gene-Based Markers in Breeding

Before the progression of NGS technology, resequencing of ESTs and derivate of unigene amplicons by orthodox method of sequencing has controlled the advancement of markers of SNP dependent on genes that were proofed after PCR (Batley et al. 2003; Wright et al. 2005). Therefore, in lentils, the analysis of transcriptome that depends on NGS technology has recognized the candidate genes that are exposed in the response of environmental stresses (Singh et al. 2019; Singh et al. 2017). Also, SSRs and SNPs are covered by the sequences of these functional unigenes that have the beneficial means for the development of functional markers (Kaur et al. 2011; Pavan et al. 2019; Sharpe et al. 2013; Singh et al. 2019, 2017; Sudheesh et al. 2016b; TEMEL et al. 2015; Verma et al. 2013). The SNPs detected by the studies of transcriptome have been utilized to improve the array that helps genotype and detect their coupling with desired traits (Kaur et al. 2014; Sharpe et al. 2013). Furthermore, the NGS technology makes it better for resequencing of several candidate genes in massive genotypes during the single cycle at the cheapest rate that can be utilized for the coupling of candidate genes of SNPs with the traits of interest (Kaur et al. 2014). Similarly, the sequences of candidate genes can be utilized to generate markers that depend on PCR for breeders, which makes it easy for genotyping of the huge population of breeding completed in other crops (Yang et al. 2012). As well, the utilization of functional markers in mapping can help in the advancement of the flawless markers for breeding.

10.29 Capturing Exome Variation

Identification of the variation in genes through entire genome sequencing of numerous accessions is challenging because of the problems in assembling the complicated and large (4,063 Mb) genome of lentil, which is affected by the duplication of the gene and assembling of the repetitive elements and chromosome (Ogutcen et al. 2018). Therefore, catching the variation in genetics present in the regions of the coding genome is helpful for the advancement of the genetics because it covers the genes that control the numerous attributes of agronomy. However, regions of the coding are more essential as compared to the regions of noncoding for breeding of those plants that have a large complex genome (Bamshad et al. 2011). In lentil research, genomes having a size of 130 Mb (3.2%) are composed of genic area (Ogutcen et al. 2018), which can be common for targeted traits. However, in lentil, possibilities have been generated to catch the variation in genes that exist in genic areas and exome catch arrays containing 85 Mb have been produced in lentil (Ogutcen et al. 2018). These exome capture arrays can be used only to sequence the coding area of protein of the genome instead of the entire genome, so it is the cheapest technique for sequencing (Hodges et al. 2007). Advanced arrays of exome have been utilized to detect the variation in genes among 38 different accessions of lentil, which contain 16 wild species accessions (Ogutcen et al. 2018). They also proposed that the utilization of exome capture arrays in downstream studies comprising the detection of genetic association in the genus of *Lens*, identification of gene, for tracking the advancement of genic markers, selection and detection of useful traits in breeding population, in wild families, and the discovery of genes for adaptive traits is helpful in DNA barcoding in coming years (Ogutcen et al. 2018).

10.30 Future Prospects

The current advancement in techniques of sequencing and NGS computational techniques is composed of proposing significant understandings of different biological mechanisms in plants. The advancement in TGSTs together with several tools of bioinformatics will enhance the precision of all these techniques and facilitate various applications based on NGS. This will make it possible to sequence the massive genomes and can be used for the genomes of crop, which can lead to an enhanced understanding. The techniques of NGS identified the new molecular markers that improve the methods of plant breeding by decoding the best sequences of alleles and genes in unique species of plants in different crops. Different techniques such as exome sequencing and transcriptome will assist in understanding the targeted area of genome and decoding the transcript of cell at the phase of sequencing without understanding the sequence of the genome. Projects related to multiple sequencing of genome will be done by the low-cost techniques, which will produce data about several alleles of various genes in different crops. The information obtained from NGS techniques will be utilized to analyze the function and structure of genome, interaction among proteins and DNA, studies related to phylogenetics, individual genomics and metagenomics by different bioinformatics techniques. By this understanding, multiple sequencing of genome present in biological mechanisms can be read in the future that can never be translated before these techniques.

10.31 Concluding Remarks

In scientific research, the applications of sequencing techniques such as next-generation sequencing have thrown out the advancement and have a significant effect on laboratory experiments, including their cost and scope. The applications of NGS have covered different methods of sequencing that are used to analyze the various biological methods in different plant species. Recently, the Illumina platform, sequencing-by-synthesis technique, is the low-cost sequencing technique despite its smaller read length, but it is prominent and familiar due to its low cost, minimal rate of error, and high efficiency. The major drawback of SGSTs is the formation of smaller read lengths, and it has prevailed over TGSTs. Nanopore TGSTs can generate the longest read length of almost 150 kb, but currently, their usage is limited due to the maximum rate of error. In the future, with the maximum precision of TGSTs, the strategy of genomics will lead to its superior level with maximum sequencing of plant genomes. However, interpreting the massive amounts of data that these studies will create will be a difficult challenge. In different crops, the NGS techniques have been utilized to discover the candidate genes, SNPs, and SSRs shown in abiotic and biotic stresses. The SNPs or SSRs have been utilized to make the markers that define the genetic modification in the collection of germplasm, and they are linked with various phenotypic qualities such as the number of micronutrients, tolerance against diseases, and quality of seeds (Khazaei et al. 2016, 2018, 2017; Lombardi et al. 2014; Sudheesh et al. 2016a). Therefore, the NGS techniques help in identifying a large number of genotypes in the cheapest way as compared to other expensive technologies. While in some other crops, the NGS techniques have been utilized to produce the super-pan-genome and pan-genome through sequencing/resequencing of many accessions. In crops, these challenges have not been developed because of the complexity of genome, which damages the framework and assembling of short reads produced through SGS techniques. TGSS techniques have been developed, which are the cheapest technology that can help in sequencing the long fragments of DNA that produce the framework and assemble the complex genome. This debate will aid the plant science community in terms of developing and utilising genome-wide knowledge of plants in various fundamental and applied research programs using NGS methodologies.

REFERENCES

Albertsen, M., Hugenholtz, P., Skarshewski, A., Nielsen, K. L., Tyson, G. W., and Nielsen, P. H. (2013). Genome sequences of rare, uncultured bacteria obtained by differential coverage binning of multiple metagenomes. *Nature Biotechnology*, 31(6), 533–538.

Allen, A. M., Barker, G. L., Wilkinson, P., Burridge, A., Winfield, M., Coghill, J., Uauy, C., Griffiths, S., Jack, P., and Berry, S. (2013). Discovery and development of exome-based, co-dominant single nucleotide polymorphism markers in hexaploid wheat (Triticum aestivum L.). *Plant Biotechnology Journal*, 11(3), 279–295.

Allen, A. M., Winfield, M. O., Burridge, A. J., Downie, R. C., Benbow, H. R., Barker, G. L. A., Wilkinson, P. A., Coghill, J., Waterfall, C., Davassi, A., Scopes, G., Pirani, A., Webster, T., Brew, F., Bloor, C., Griffiths, S., Bentley, A. R., Alda, M., Jack, P., Phillips, A. L., and Edwards, K. J. (2017). Characterization of a Wheat Breeders' Array suitable for high-throughput SNP genotyping of global accessions of hexaploid bread wheat (Triticum aestivum). *Plant Biotechnology Journal*, 15(3), 390–401.

Arisha, M. H., Aboelnasr, H., Ahmad, M. Q., Liu, Y., Tang, W., Gao, R., Yan, H., Kou, M., Wang, X., Zhang, Y., and Li, Q. (2020). Transcriptome sequencing and whole genome expression profiling of hexaploid sweet potato under salt stress. *BMC Genomics*, 21(1), 197.

Ashkani, S., Yusop, M. R., Shabanimofrad, M., Azadi, A., Ghasemzadeh, A., Azizi, P., and Latif, M. A. (2015). Allele mining strategies: principles and utilisation for blast resistance genes in rice (Oryza sativa L.). *Current Issues in Molecular Biology*, 17, 57–74.

Azam, S., Thakur, V., Ruperao, P., Shah, T., Balaji, J., Amindala, B., Farmer, A. D., Studholme, D. J., May, G. D., and Edwards, D. (2012). Coverage-based consensus calling (CbCC) of short sequence reads and comparison of CbCC results to identify SNPs in chickpea (Cicer arietinum; Fabaceae), a crop species without a reference genome. *American Journal of Botany*, 99(2), 186–192.

Baldrich, P., and San Segundo, B. (2016). MicroRNAs in rice innate immunity. *Rice*, 9(1), 6.

Bamshad, M. J., Ng, S. B., Bigham, A. W., Tabor, H. K., Emond, M. J., Nickerson, D. A., and Shendure, J. (2011). Exome sequencing as a tool for Mendelian disease gene discovery. *Nature Reviews Genetics*, 12(11), 745–755.

Barbazuk, W. B., Emrich, S. J., Chen, H. D., Li, L., and Schnable, P. S. (2007). SNP discovery via 454 transcriptome sequencing. *The Plant Journal*, 51(5), 910–918.

Batley, J., Barker, G., O'Sullivan, H., Edwards, K. J., and Edwards, D. (2003). Mining for single nucleotide polymorphisms and insertions/deletions in maize expressed sequence tag data. *Plant Physiology*, 132 (1), 84–91.

Beissinger, T. M., Hirsch, C. N., Sekhon, R. S., Foerster, J. M., Johnson, J. M., Muttoni, G., Vaillancourt, B., Buell, C. R., Kaeppler, S. M., and de Leon, N. (2013). Marker density and read depth for genotyping populations using genotyping-by-sequencing. *Genetics*, 193(4), 1073–1081.

Bernal, M., Casero, D., Singh, V., Wilson, G. T., Grande, A., Yang, H., Dodani, S. C., Pellegrini, M., Huijser, P., and Connolly, E. L. (2012). Transcriptome sequencing identifies SPL7-regulated copper acquisition genes FRO4/FRO5 and the copper dependence of iron homeostasis in Arabidopsis. *The Plant Cell*, 24 (2), 738–761.

Bhat, J. A., Ali, S., Salgotra, R. K., Mir, Z. A., Dutta, S., Jadon, V., Tyagi, A., Mushtaq, M., Jain, N., and Singh, P. K. (2016). Genomic selection in the era of next generation sequencing for complex traits in plant breeding. *Frontiers in Genetics*, 7, 221.

Blainey, P. C. (2013). The future is now: single-cell genomics of bacteria and archaea. *FEMS Microbiology Reviews*, 37(3), 407–427.

Bleidorn, C. (2016). Third generation sequencing: technology and its potential impact on evolutionary biodiversity research. *Systematics and Biodiversity*, 14(1), 1–8.

Boden, S. A., Kavanová, M., Finnegan, E. J., and Wigge, P. A. (2013). Thermal stress effects on grain yield in Brachypodium distachyon occur via H2A. Z-nucleosomes. *Genome Biology*, 14(6), R65.

Bowers, J., Mitchell, J., Beer, E., Buzby, P. R., Causey, M., Efcavitch, J. W., Jarosz, M., Krzymanska-Olejnik, E., Kung, L., and Lipson, D. (2009). Virtual terminator nucleotides for next-generation DNA sequencing. *Nature Methods*, 6(8), 593–595.

Braslavsky, I., Hebert, B., Kartalov, E., and Quake, S. R. (2003). Sequence information can be obtained from single DNA molecules. *Proceedings of the National Academy of Sciences*, 100(7), 3960–3964.

Breitling, R., Li, Y., Tesson, B. M., Fu, J., Wu, C., Wiltshire, T., Gerrits, A., Bystrykh, L. V., De Haan, G., and Su, A. I. (2008). Genetical genomics: spotlight on QTL hotspots. *PLoS Genetics*, 4(10), e1000232.

Brent, M. R. (2008). Steady progress and recent breakthroughs in the accuracy of automated genome annotation. *Nature Reviews Genetics*, 9(1), 62–73.

Cabeza, R., Koester, B., Liese, R., Lingner, A., Baumgarten, V., Dirks, J., Salinas-Riester, G., Pommerenke, C., Dittert, K., and Schulze, J. (2014). An RNA sequencing transcriptome analysis reveals novel insights into molecular aspects of the nitrate impact on the nodule activity of Medicago truncatula. *Plant Physiology*, 164(1), 400–411.

Callinan, P. A., and Feinberg, A. P. (2006). The emerging science of epigenomics. *Human Molecular Genetics*, 15(suppl_1), R95–R101.

Cao, J., Schneeberger, K., Ossowski, S., Günther, T., Bender, S., Fitz, J., Koenig, D., Lanz, C., Stegle, O., and Lippert, C. (2011). Whole-genome sequencing of multiple Arabidopsis thaliana populations. *Nature Genetics*, 43(10), 956.

Capblancq, T., Butnor, J. R., Deyoung, S., Thibault, E., Munson, H., Nelson, D. M., Fitzpatrick, M. C., and Keller, S. R. (2020). Whole-exome sequencing reveals a long-term decline in effective population size of red spruce (Picea rubens). *Evolutionary Applications*, 13(9), 2190–2205.

Chen, G., Yue, Y., Hua, Y., Hu, D., Shi, T., Chang, Z., Yang, X., and Wang, L. (2013). Comparative transcriptome profiling of a resistant vs. susceptible tomato (Solanum lycopersicum) cultivar in response to infection by tomato yellow leaf curl virus. *PloS One*, 8(11), e80816.

Chen, T., Lv, Y., Zhao, T., Li, N., Yang, Y., Yu, W., He, X., Liu, T., and Zhang, B. (2015). Development and characterization of polymorphic genic-SSR markers in Larix kaempferi. *Molecules*, 20(4), 6060–6067.

Chen, X.-B., Xie, Y.-H., and Sun, X.-M. (2019). SSR marker development in Clerodendrum trichotomum using transcriptome sequencing. *PLOS One*, 14(11), e0225451.

Cokus, S. J., Feng, S., Zhang, X., Chen, Z., Merriman, B., Haudenschild, C. D., Pradhan, S., Nelson, S. F., Pellegrini, M., and Jacobsen, S. E. (2008). Shotgun bisulphite sequencing of the Arabidopsis genome reveals DNA methylation patterning. *Nature*, 452(7184), 215–219.

Cook, D. E., Lee, T. G., Guo, X., Melito, S., Wang, K., Bayless, A. M., Wang, J., Hughes, T. J., Willis, D. K., and Clemente, T. E. (2012). Copy number variation of multiple genes at Rhgl mediates nematode resistance in soybean. *Science*, 338(6111), 1206–1209.

Crossa, J., Burgueno, J., Dreisigacker, S., Vargas, M., Herrera-Foessel, S. A., Lillemo, M., Singh, R. P., Trethowan, R., Warburton, M., and Franco, J. (2007). Association analysis of historical bread wheat germplasm using additive genetic covariance of relatives and population structure. *Genetics*, 177(3), 1889–1913.

Cubizolles, N., Rey, E., Choulet, F., Rimbert, H., Laugier, C., Balfourier, F., Bordes, J., Poncet, C., Jack, P., and James, C. (2016). Exploiting the repetitive fraction of the wheat genome for high-throughput single-nucleotide polymorphism discovery and genotyping. *The Plant Genome*, 9(1), 1–11.

Cui, G., Chai, H., Yin, H., Yang, M., Hu, G., Guo, M., Yi, R., and Zhang, P. (2019). Full-length transcriptome sequencing reveals the low-temperature-tolerance mechanism of Medicago falcata roots. *BMC Plant Biology*, 19(1), 575.

D'Aoust, M.-A., Yelle, S., and Nguyen-Quoc, B. (1999). Antisense inhibition of tomato fruit sucrose synthase decreases fruit setting and the sucrose unloading capacity of young fruit. *The Plant Cell*, 11(12), 2407–2418.

Darst, R. P., Pardo, C. E., Ai, L., Brown, K. D., and Kladde, M. P. (2010). Bisulfite sequencing of DNA. *Current Protocols in Molecular Biology*, 91(1), 7.9.1–7.9.17.

Dasgupta, M. G., Dharanishanthi, V., Agarwal, I., and Krutovsky, K. V. (2015). Development of genetic markers in Eucalyptus species by target enrichment and exome sequencing. *PloS One*, 10(1), e0116528.

Dekker, J., Rippe, K., Dekker, M., and Kleckner, N. (2002). Capturing chromosome conformation. *Science*, 295(5558), 1306–1311.

Diwan, D., Komazaki, S., Suzuki, M., Nemoto, N., Aita, T., Satake, A., and Nishigaki, K. (2014). Systematic genome sequence differences among leaf cells within individual trees. *BMC Genomics*, 15(1), 142.

Egan, A. N., Schlueter, J., and Spooner, D. M. (2012). Applications of next-generation sequencing in plant biology. *American Journal of Botany*, 99(2), 175–185.

Eid, J., Fehr, A., Gray, J., Luong, K., Lyle, J., Otto, G., Peluso, P., Rank, D., Baybayan, P., and Bettman, B. (2009). Real-time DNA sequencing from single polymerase molecules. *Science*, 323(5910), 133–138.

Farlik, M., Sheffield, N. C., Nuzzo, A., Datlinger, P., Schönegger, A., Klughammer, J., and Bock, C. (2015). Single-cell DNA methylome sequencing and bioinformatic inference of epigenomic cell-state dynamics. *Cell Reports*, 10(8), 1386–1397.

Feuk, L., Marshall, C. R., Wintle, R. F., and Scherer, S. W. (2006). Structural variants: changing the landscape of chromosomes and design of disease studies. *Human Molecular Genetics*, 15(suppl_1), R57–R66.

Flusberg, B. A., Webster, D. R., Lee, J. H., Travers, K. J., Olivares, E. C., Clark, T. A., Korlach, J., and Turner, S. W. (2010). Direct detection of DNA methylation during single-molecule, real-time sequencing. *Nature Methods*, 7(6), 461.

Franssen, S. U., Shrestha, R. P., Bräutigam, A., Bornberg-Bauer, E., and Weber, A. P. (2011). Comprehensive transcriptome analysis of the highly complex Pisum sativum genome using next generation sequencing. *BMC Genomics*, 12(1), 227.

Gao, J., Yin, F., Liu, M., Luo, M., Qin, C., Yang, A., Yang, S., Zhang, Z., Shen, Y., and Lin, H. (2015). Identification and characterisation of tobacco micro RNA transcriptome using high-throughput sequencing. *Plant Biology*, 17(3), 591–598.

Gao, J., Li, W.-B., Liu, H.-F., and Chen, F.-B. (2019a). De novo transcriptome sequencing of radish (Raphanus sativus L.) fleshy roots: analysis of major genes involved in the anthocyanin synthesis pathway. *BMC Molecular and Cell Biology*, 20(1), 45.

Gao, L., Gonda, I., Sun, H., Ma, Q., Bao, K., Tieman, D. M., Burzynski-Chang, E. A., Fish, T. L., Stromberg, K. A., and Sacks, G. L. (2019b). The tomato pan-genome uncovers new genes and a rare allele regulating fruit flavor. *Nature Genetics*, 51(6), 1044–1051.

Gordon, S. P., Contreras-Moreira, B., Woods, D. P., Des Marais, D. L., Burgess, D., Shu, S., Stritt, C., Roulin, A. C., Schackwitz, W., and Tyler, L. (2017). Extensive gene content variation in the Brachypodium distachyon pan-genome correlates with population structure. *Nature Communications*, 8(1), 1–13.

Greenleaf, W. J., and Block, S. M. (2006). Single-molecule, motion-based DNA sequencing using RNA polymerase. *Science*, 313(5788), 801–801.

Grover, C. E., Salmon, A., and Wendel, J. F. (2012). Targeted sequence capture as a powerful tool for evolutionary analysis1. *American Journal of Botany*, 99(2), 312–319.

Guo, S., Zheng, Y., Joung, J.-G., Liu, S., Zhang, Z., Crasta, O. R., Sobral, B. W., Xu, Y., Huang, S., and Fei, Z. (2010). Transcriptome sequencing and comparative analysis of cucumber flowers with different sex types. *BMC Genomics*, 11(1), 384.

Hackl, T., Hedrich, R., Schultz, J., and Förster, F. (2014). proovread: large-scale high-accuracy PacBio correction through iterative short read consensus. *Bioinformatics*, 30(21), 3004–3011.

Haile, T. A., Heidecker, T., Wright, D., Neupane, S., Ramsay, L., Vandenberg, A., and Bett, K. E. (2020). Genomic selection for lentil breeding: empirical evidence. *The Plant Genome*, 13(1), e20002.

Hamblin, M. T., and Rabbi, I. Y. (2014). The effects of restriction-enzyme choice on properties of genotyping-by-sequencing libraries: A study in cassava (Manihot esculenta). *Crop Science*, 54(6), 2603–2608.

Hao, W., Wang, S., Liu, H., Zhou, B., Wang, X., and Jiang, T. (2015). Development of SSR markers and genetic diversity in white birch (Betula platyphylla). *PloS One*, 10(4), e0125235.

He, L., and Hannon, G. J. (2004). MicroRNAs: small RNAs with a big role in gene regulation. *Nature Reviews Genetics*, 5(7), 522–531.

He, F., Pasam, R., Shi, F., Kant, S., Keeble-Gagnere, G., Kay, P., Forrest, K., Fritz, A., Hucl, P., Wiebe, K., Knox, R., Cuthbert, R., Pozniak, C., Akhunova, A., Morrell, P. L., Davies, J. P., Webb, S. R., Spangenberg, G., Hayes, B., Daetwyler, H., Tibbits, J., Hayden, M., and Akhunov, E. (2019). Exome sequencing highlights the role of wild-relative introgression in shaping the adaptive landscape of the wheat genome. *Nature Genetics*, 51(5), 896–904.

Head, S. R., Komori, H. K., LaMere, S. A., Whisenant, T., Van Nieuwerburgh, F., Salomon, D. R., and Ordoukhanian, P. (2014). Library construction for next-generation sequencing: overviews and challenges. *Biotechniques*, 56(2), 61–77.

Healy, K. (2007). Nanopore-based single-molecule DNA analysis. *Nanomedicine*, 2(4), 459–481.

Heather, J. M., and Chain, B. (2016). The sequence of sequencers: The history of sequencing DNA. *Genomics*, 107(1), 1–8.

Heffner, E. L., Lorenz, A. J., Jannink, J. L., and Sorrells, M. E. (2010). Plant breeding with genomic selection: gain per unit time and cost. *Crop Science*, 50(5), 1681–1690.

Henry, I. M., Nagalakshmi, U., Lieberman, M. C., Ngo, K. J., Krasileva, K. V., Vasquez-Gross, H., Akhunova, A., Akhunov, E., Dubcovsky, J., and Tai, T. H. (2014). Efficient genome-wide detection and cataloging of EMS-induced mutations using exome capture and next-generation sequencing. *The Plant Cell*, 26(4), 1382–1397.

Hodges, E., Xuan, Z., Balija, V., Kramer, M., Molla, M. N., Smith, S. W., Middle, C. M., Rodesch, M. J., Albert, T. J., and Hannon, G. J. (2007). Genome-wide in situ exon capture for selective resequencing. *Nature Genetics*, 39(12), 1522–1527.

Houwing, S., Kamminga, L., Berezikov, E., Cronembold, D., Girard, A., and Elst, H. (2007). A role for Piwi and piRNAs in germ cell maintenance and transposon silencing in Zebrafish. *Cell*, 129(1), 69–82.

Hu, H., Yuan, Y., Bayer, P. E., Fernandez, C. T., Scheben, A., Golicz, A. A., and Edwards, D. (2020). Legume pangenome construction using an iterative mapping and assembly approach. *Legume Genomics*, 35–47.

Huang, D., Swanson, E. A., Lin, C. P., Schuman, J. S., Stinson, W. G., Chang, W., Hee, M. R., Flotte, T., Gregory, K., and Puliafito, C. A. (1991). Optical coherence tomography. *Science*, 254(5035), 1178–1181.

Huang, S., Deng, L., Guan, M., Li, J., Lu, K., Wang, H., Fu, D., Mason, A. S., Liu, S., and Hua, W. (2013a). Identification of genome-wide single nucleotide polymorphisms in allopolyploid crop Brassica napus. *BMC Genomics*, 14(1), 1–10.

Huang, Y., Zhang, J. L., Yu, X. L., Xu, T. S., Wang, Z. B., and Cheng, X. C. (2013b). Molecular functions of small regulatory noncoding RNA. *Biochemistry (Moscow)*, 78(3), 221–230.

Hübner, S., Bercovich, N., Todesco, M., Mandel, J. R., Odenheimer, J., Ziegler, E., Lee, J. S., Baute, G. J., Owens, G. L., and Grassa, C. J. (2019). Sunflower pan-genome analysis shows that hybridization altered gene content and disease resistance. *Nature Plants*, 5(1), 54–62.

Hurgobin, B., Golicz, A. A., Bayer, P. E., Chan, C. K. K., Tirnaz, S., Dolatabadian, A., Schiessl, S. V., Samans, B., Montenegro, J. D., and Parkin, I. A. (2018). Homoeologous exchange is a major cause of gene presence/absence variation in the amphidiploid Brassica napus. *Plant Biotechnology Journal*, 16(7), 1265–1274.

Hutchison III, C. A. (2007). DNA sequencing: bench to bedside and beyond. *Nucleic Acids Research*, 35(18), 6227–6237.

Iaria, D., Chiappetta, A., and Muzzalupo, I. (2016). De novo transcriptome sequencing of Olea europaea L. to identify genes involved in the development of the pollen tube. *The Scientific World Journal*, 2016.

Ip, C. L., Loose, M., Tyson, J. R., de Cesare, M., Brown, B. L., Jain, M., Leggett, R. M., Eccles, D. A., Zalunin, V., and Urban, J. M. (2015). MinION analysis and reference consortium: Phase 1 data release and analysis. *F1000Research*, 4.

Jaccoud, D., Peng, K., Feinstein, D., and Kilian, A. (2001). Diversity arrays: a solid state technology for sequence information independent genotyping. *Nucleic Acids Research*, 29(4), e25–e25.

Jacinto, F. V., Ballestar, E., and Esteller, M. (2008). Methyl-DNA immunoprecipitation (MeDIP): hunting down the DNA methylome. *Biotechniques*, 44(1), 35–43.

Jansen, R. C., and Nap, J.-P. (2001). Genetical genomics: the added value from segregation. *Trends in Genetics*, 17(7), 388–391.

Jarquín, D., Kocak, K., Posadas, L., Hyma, K., Jedlicka, J., Graef, G., and Lorenz, A. (2014). Genotyping by sequencing for genomic prediction in a soybean breeding population. *BMC Genomics*, 15(1), 740.

Jha, T. b., Mahanti, A., and Ghorai, A. (2015). Karyotype analysis of Indian lentils through EMA based Giemsa staining. *Caryologia*, 68(4), 280–288.

Kaur, S., Cogan, N. O., Pembleton, L. W., Shinozuka, M., Savin, K. W., Materne, M., and Forster, J. W. (2011). Transcriptome sequencing of lentil based on second-generation technology permits large-scale unigene assembly and SSR marker discovery. *BMC Genomics*, 12(1), 265.

Kaur, S., Cogan, N. O., Stephens, A., Noy, D., Butsch, M., Forster, J. W., and Materne, M. (2014). EST-SNP discovery and dense genetic mapping in lentil (Lens culinaris Medik.) enable candidate gene selection for boron tolerance. *Theoretical and Applied Genetics*, 127(3), 703–713.

Khan, A. W., Garg, V., Roorkiwal, M., Golicz, A. A., Edwards, D., and Varshney, R. K. (2020). Super-Pangenome by integrating the wild side of a species for accelerated crop improvement. *Trends in Plant Science*, 25(2), 148–158.

Khazaei, H., Caron, C. T., Fedoruk, M., Diapari, M., Vandenberg, A., Coyne, C. J., McGee, R., and Bett, K. E. (2016). Genetic diversity of cultivated lentil (Lens culinaris Medik.) and its relation to the world's agro-ecological zones. *Frontiers in Plant Science*, 7, 1093.

Khazaei, H., Podder, R., Caron, C. T., Kundu, S. S., Diapari, M., Vandenberg, A., and Bett, K. E. (2017). Marker–trait association analysis of iron and zinc concentration in lentil (Lens culinaris Medik.) seeds. *The Plant Genome*, 10(2), 1–8.

Khazaei, H., Fedoruk, M., Caron, C. T., Vandenberg, A., and Bett, K. E. (2018). Single nucleotide polymorphism markers associated with seed quality characteristics of cultivated lentil. *The Plant Genome*, 11(1), 1–7.

Khera, P., Upadhyaya, H. D., Pandey, M. K., Roorkiwal, M., Sriswathi, M., Janila, P., Guo, Y., McKain, M. R., Nagy, E. D., and Knapp, S. J. (2013). Single nucleotide polymorphism–based genetic diversity in the reference set of peanut (Arachis spp.) by developing and applying cost-effective kompetitive allele specific polymerase chain reaction genotyping assays. *The Plant Genome*, 6(3), 1–11.

King, R., Bird, N., Ramirez-Gonzalez, R., Coghill, J. A., Patil, A., Hassani-Pak, K., Uauy, C., and Phillips, A. L. (2015). Mutation scanning in wheat by exon capture and next-generation sequencing. *PLoS One*, 10(9), e0137549.

Knief, C. (2014). Analysis of plant microbe interactions in the era of next generation sequencing technologies. *Frontiers in Plant Science*, 5, 216.

Kudapa, H., Azam, S., Sharpe, A. G., Taran, B., Li, R., Deonovic, B., Cameron, C., Farmer, A. D., Cannon, S. B., and Varshney, R. K. (2014). Comprehensive transcriptome assembly of chickpea (Cicer arietinum L.) using Sanger and next generation sequencing platforms: development and applications. *PloS One*, 9(1), e86039.

Kumar, S., Banks, T. W., and Cloutier, S. (2012). SNP discovery through next-generation sequencing and its applications. *International Journal of Plant Genomics*, 2012, 1–15.

Kumar, J., and Gupta, D. S. (2020). Prospects of next generation sequencing in lentil breeding. *Molecular Biology Reports*, 1–11.

Kumar, G. R., Sakthivel, K., Sundaram, R. M., Neeraja, C. N., Balachandran, S. M., Rani, N. S., Viraktamath, B. C., and Madhav, M. S. (2010). Allele mining in crops: prospects and potentials. *Biotechnology Advances*, 28(4), 451–461.

Kumar, M., Chauhan, A. S., Kumar, M., Yusuf, M. A., Sanyal, I., and Chauhan, P. S. (2019). Transcriptome sequencing of chickpea (*Cicer arietinum* L.) genotypes for identification of drought-responsive genes under drought stress condition. *Plant Molecular Biology Reporter*, 37(3), 186–203.

Kumari, M., Grover, A., Yadav, P. V., Arif, M., and Ahmed, Z. (2013). Development of EST-SSR markers through data mining and their use for genetic diversity study in Indian accessions of *Jatropha curcas* L.: a potential energy crop. *Genes & Genomics*, 35(5), 661–670.

Lam, H.-M., Xu, X., Liu, X., Chen, W., Yang, G., Wong, F.-L., Li, M.-W., He, W., Qin, N., and Wang, B. (2010). Resequencing of 31 wild and cultivated soybean genomes identifies patterns of genetic diversity and selection. *Nature Genetics*, 42(12), 1053–1059.

Le, T. N., Miyazaki, Y., Takuno, S., and Saze, H. (2015). Epigenetic regulation of intragenic transposable elements impacts gene transcription in Arabidopsis thaliana. *Nucleic Acids Research*, 43(8), 3911–3921.

Levene, M. J., Korlach, J., Turner, S. W., Foquet, M., Craighead, H. G., and Webb, W. W. (2003). Zero-mode waveguides for single-molecule analysis at high concentrations. *Science*, 299(5607), 682–686.

Lewsey, M. G., Hardcastle, T. J., Melnyk, C. W., Molnar, A., Valli, A., Urich, M. A., Nery, J. R., Baulcombe, D. C., and Ecker, J. R. (2016). Mobile small RNAs regulate genome-wide DNA methylation. *Proceedings of the National Academy of Sciences*, 113(6), E801–E810.

Li, X., Acharya, A., Farmer, A. D., Crow, J. A., Bharti, A. K., Kramer, R. S., Wei, Y., Han, Y., Gou, J., and May, G. D. (2012). Prevalence of single nucleotide polymorphism among 27 diverse alfalfa genotypes as assessed by transcriptome sequencing. *BMC Genomics*, 13(1), 568.

Li, A., Liu, D., Wu, J., Zhao, X., Hao, M., Geng, S., Yan, J., Jiang, X., Zhang, L., and Wu, J. (2014a). mRNA and small RNA transcriptomes reveal insights into dynamic homoeolog regulation of allopolyploid heterosis in nascent hexaploid wheat. *The Plant Cell*, 26(5), 1878–1900.

Li, G., Wang, D., Yang, R., Logan, K., Chen, H., Zhang, S., Skaggs, M. I., Lloyd, A., Burnett, W. J., and Laurie, J. D. (2014b). Temporal patterns of gene expression in developing maize endosperm identified through transcriptome sequencing. *Proceedings of the National Academy of Sciences*, 111(21), 7582–7587.

Li, Y. H., Zhou, G., Ma, J., Jiang, W., Jin, L. G., Zhang, Z., Guo, Y., Zhang, J., Sui, Y., Zheng, L., Zhang, S. S., Zuo, Q., Shi, X. H., Li, Y. F., Zhang, W. K., Hu, Y., Kong, G., Hong, H. L., Tan, B., Song, J., Liu, Z. X., Wang, Y., Ruan, H., Yeung, C. K., Liu, J., Wang, H., Zhang, L. J., Guan, R. X., Wang, K. J., Li, W. B., Chen, S. Y., Chang, R. Z., Jiang, Z., Jackson, S. A., Li, R., and Qiu, L. J. (2014c). De novo assembly of soybean wild relatives for pan-genome analysis of diversity and agronomic traits. *Nature Biotechnology*, 32(10), 1045–1052.

Li, X., Li, L., and Yan, J. (2015). Dissecting meiotic recombination based on tetrad analysis by single-microspore sequencing in maize. *Nature Communications*, 6(1), 6648.

Li, C., Chng, K. R., Boey, E. J. H., Ng, A. H. Q., Wilm, A., and Nagarajan, N. (2016). INC-Seq: accurate single molecule reads using nanopore sequencing. *GigaScience*, 5(1), s13742-016-0140-7.

Lim, J.-H., Yang, H.-J., Jung, K.-H., Yoo, S.-C., and Paek, N.-C. (2014). Quantitative trait locus mapping and candidate gene analysis for plant architecture traits using whole genome re-sequencing in rice. *Molecules and Cells*, 37(2), 149.

Lin, Y., Golovnina, K., Chen, Z.-X., Lee, H. N., Negron, Y. L. S., Sultana, H., Oliver, B., and Harbison, S. T. (2016). Comparison of normalization and differential expression analyses using RNA-Seq data from 726 individual Drosophila melanogaster. *BMC Genomics*, 17(1), 1–20.

Liu, L., Li, Y., Li, S., Hu, N., He, Y., Pong, R., Lin, D., Lu, L., and Law, M. (2012). Comparison of next-generation sequencing systems. *BioMed Research International*. 2012, 1–11.

Liu, X., Lu, Y., Yuan, Y., Liu, S., Guan, C., Chen, S., and Liu, Z. (2013). De novo transcriptome of Brassica juncea seed coat and identification of genes for the biosynthesis of flavonoids. *PLoS One*, 8(8), e71110.

Logan-Young, C. J., Yu, J. Z., Verma, S. K., Percy, R. G., and Pepper, A. E. (2015). SNP discovery in complex allotetraploid genomes (Gossypium spp., Malvaceae) using genotyping by sequencing. *Applications in Plant Sciences*, 3(3), 1400077.

Lombardi, M., Materne, M., Cogan, N. O., Rodda, M., Daetwyler, H. D., Slater, A. T., Forster, J. W., and Kaur, S. (2014). Assessment of genetic variation within a global collection of lentil (Lens culinarisMedik.) cultivars and landraces using SNP markers. *BMC Genetics*, 15(1), 150.

Long, Q., Rabanal, F. A., Meng, D., Huber, C. D., Farlow, A., Platzer, A., Zhang, Q., Vilhjálmsson, B. J., Korte, A., and Nizhynska, V. (2013). Massive genomic variation and strong selection in Arabidopsis thaliana lines from Sweden. *Nature Genetics*, 45(8), 884.

Lu, G., Harper, A. L., Trick, M., Morgan, C., Fraser, F., O'Neill, C., and Bancroft, I. (2014). Associative transcriptomics study dissects the genetic architecture of seed glucosinolate content in Brassica napus. *DNA Research*, 21(6), 613–625.

Lunardon, A., Forestan, C., Farinati, S., Axtell, M. J., and Varotto, S. (2016). Genome-wide characterization of maize small RNA loci and their regulation in the required to maintain repression6-1 (rmr6-1) mutant and long-term abiotic stresses. *Plant Physiology*, 170(3), 1535–1548.

Lyu, J., Zhang, S., Dong, Y., He, W., Zhang, J., Deng, X., Zhang, Y., Li, X., Li, B., and Huang, W. (2013). Analysis of elite variety tag SNPs reveals an important allele in upland rice. *Nature Communications*, 4(1), 1–9.

Ma, J.-Q., Huang, L., Ma, C.-L., Jin, J.-Q., Li, C.-F., Wang, R.-K., Zheng, H.-K., Yao, M.-Z., and Chen, L. (2015). Large-scale SNP discovery and genotyping for constructing a high-density genetic map of tea plant using specific-locus amplified fragment sequencing (SLAF-seq). *PloS One*, 10(6), e0128798.

Macaulay, I. C., and Voet, T. (2014). Single cell genomics: advances and future perspectives. *PLoS Genetics*, 10(1), e1004126.

Macaulay, I. C., Haerty, W., Kumar, P., Li, Y. I., Hu, T. X., Teng, M. J., Goolam, M., Saurat, N., Coupland, P., and Shirley, L. M. (2015). G&T-seq: parallel sequencing of single-cell genomes and transcriptomes. *Nature Methods*, 12(6), 519–522.

Mace, E. S., Rami, J.-F., Bouchet, S., Klein, P. E., Klein, R. R., Kilian, A., Wenzl, P., Xia, L., Halloran, K., and Jordan, D. R. (2009). A consensus genetic map of sorghum that integrates multiple component maps and high-throughput Diversity Array Technology (DArT) markers. *BMC Plant Biology*, 9(1), 13.

Madoui, M.-A., Engelen, S., Cruaud, C., Belser, C., Bertrand, L., Alberti, A., Lemainque, A., Wincker, P., and Aury, J.-M. (2015). Genome assembly using Nanopore-guided long and error-free DNA reads. *BMC Genomics*, 16(1), 327.

Mammadov, J., Aggarwal, R., Buyyarapu, R., and Kumpatla, S. (2012). SNP markers and their impact on plant breeding. *International Journal of Plant Genomics*, 2012.

Mardis, E. R. (2013). Next-generation sequencing platforms. *Annual Review of Analytical Chemistry*, 6, 287–303.

Mascher, M., Richmond, T. A., Gerhardt, D. J., Himmelbach, A., Clissold, L., Sampath, D., Ayling, S., Steuernagel, B., Pfeifer, M., and D'Ascenzo, M. (2013). Barley whole exome capture: a tool for genomic research in the genus Hordeum and beyond. *The Plant Journal*, 76(3), 494–505.

McHale, L. K., Haun, W. J., Xu, W. W., Bhaskar, P. B., Anderson, J. E., Hyten, D. L., Gerhardt, D. J., Jeddeloh, J. A., and Stupar, R. M. (2012). Structural variants in the soybean genome localize to clusters of biotic stress-response genes. *Plant Physiology*, 159(4), 1295–1308.

Mhuantong, W., Charoensawan, V., Kanokratana, P., Tangphatsornruang, S., and Champreda, V. (2015). Comparative analysis of sugarcane bagasse metagenome reveals unique and conserved biomass-degrading enzymes among lignocellulolytic microbial communities. *Biotechnology for Biofuels*, 8(1), 16.

Montenegro, J. D., Golicz, A. A., Bayer, P. E., Hurgobin, B., Lee, H., Chan, C. K. K., Visendi, P., Lai, K., Doležel, J., and Batley, J. (2017). The pangenome of hexaploid bread wheat. *The Plant Journal*, 90(5), 1007–1013.

Moon, K.-B., Ahn, D.-J., Park, J.-S., Jung, W. Y., Cho, H. S., Kim, H.-R., Jeon, J.-H., Park, Y.-I., and Kim, H.-S. (2018). Transcriptome profiling and characterization of drought-tolerant potato plant (*Solanum tuberosum* L.). *Molecules and Cells*, 41(11), 979–992.

Morin, R. D., O'Connor, M. D., Griffith, M., Kuchenbauer, F., Delaney, A., Prabhu, A.-L., Zhao, Y., McDonald, H., Zeng, T., and Hirst, M. (2008). Application of massively parallel sequencing to microRNA profiling and discovery in human embryonic stem cells. *Genome Research*, 18(4), 610–621.

Morozova, O., and Marra, M. A. (2008). Applications of next-generation sequencing technologies in functional genomics. *Genomics*, 92(5), 255–264.

Neves, L. G., Davis, J. M., Barbazuk, W. B., and Kirst, M. (2013). Whole-exome targeted sequencing of the uncharacterized pine genome. *The Plant Journal*, 75(1), 146–156.

Newton, A. C., Fitt, B. D., Atkins, S. D., Walters, D. R., and Daniell, T. J. (2010). Pathogenesis, parasitism and mutualism in the trophic space of microbe–plant interactions. *Trends in Microbiology*, 18(8), 365–373.

Nielsen, K. L., Høgh, A. L., and Emmersen, J. (2006). DeepSAGE—digital transcriptomics with high sensitivity, simple experimental protocol and multiplexing of samples. *Nucleic Acids Research*, 34(19), e133–e133.

Ogutcen, E., Ramsay, L., von Wettberg, E. B., and Bett, K. E. (2018). Capturing variation in lens (Fabaceae): development and utility of an exome capture array for lentil. *Applications in Plant Sciences*, 6(7), e01165.

Ophir, R., Sherman, A., Rubinstein, M., Eshed, R., Schwager, M. S., Harel-Beja, R., Bar-Ya'akov, I., and Holland, D. (2014). Single-nucleotide polymorphism markers from de-novo assembly of the pomegranate transcriptome reveal germplasm genetic diversity. *PLoS One*, 9(2), e88998.

Ozsolak, F. (2012). Third-generation sequencing techniques and applications to drug discovery. *Expert Opinion on Drug Discovery*, 7(3), 231–243.

Ozsolak, F., Platt, A. R., Jones, D. R., Reifenberger, J. G., Sass, L. E., McInerney, P., Thompson, J. F., Bowers, J., Jarosz, M., and Milos, P. M. (2009). Direct RNA sequencing. *Nature*, 461(7265), 814–818.

Pabinger, S., Dander, A., Fischer, M., Snajder, R., Sperk, M., Efremova, M., Krabichler, B., Speicher, M. R., Zschocke, J., and Trajanoski, Z. (2014). A survey of tools for variant analysis of next-generation genome sequencing data. *Briefings in Bioinformatics*, 15(2), 256–278.

Pandey, M. K., Agarwal, G., Kale, S. M., Clevenger, J., Nayak, S. N., Sriswathi, M., Chitikineni, A., Chavarro, C., Chen, X., Upadhyaya, H. D., Vishwakarma, M. K., Leal-Bertioli, S., Liang, X., Bertioli, D. J., Guo, B., Jackson, S. A., Ozias-Akins, P., and Varshney, R. K. (2017). Development and evaluation of a high density genotyping 'Axiom_Arachis' array with 58 K SNPs for accelerating genetics and breeding in groundnut. *Scientific Reports*, 7(1), 40577.

Patwardhan, A., Ray, S., and Roy, A. (2014). Molecular markers in phylogenetic studies-a review. *Journal of Phylogenetics & Evolutionary Biology*, 2014.

Pavan, S., Bardaro, N., Fanelli, V., Marcotrigiano, A. R., Mangini, G., Taranto, F., Catalano, D., Montemurro, C., De Giovanni, C., and Lotti, C. (2019). Genotyping by sequencing of cultivated lentil (Lens culinaris Medik.) highlights population structure in the Mediterranean gene pool associated with geographic patterns and phenotypic variables. *Frontiers in Genetics*, 10, 872.

Pease, J. B., Haak, D. C., Hahn, M. W., and Moyle, L. C. (2016). Phylogenomics reveals three sources of adaptive variation during a rapid radiation. *PLoS Biology*, 14(2), e1002379.

Pegadaraju, V., Nipper, R., Hulke, B., Qi, L., and Schultz, Q. (2013). De novo sequencing of sunflower genome for SNP discovery using RAD (Restriction site Associated DNA) approach. *BMC Genomics*, 14(1), 556.

Peleg, Z., Saranga, Y., Suprunova, T., Ronin, Y., Röder, M. S., Kilian, A., Korol, A. B., and Fahima, T. (2008). High-density genetic map of durum wheat× wild emmer wheat based on SSR and DArT markers. *Theoretical and Applied Genetics*, 117(1), 103.

Pellegrini, M., and Ferrari, R. (2012). Epigenetic analysis: ChIP-chip and ChIP-seq. *Next Generation Microarray Bioinformatics*, 802, 377–387.

Poland, J. A., Brown, P. J., Sorrells, M. E., and Jannink, J.-L. (2012). Development of high-density genetic maps for barley and wheat using a novel two-enzyme genotyping-by-sequencing approach. *PloS One*, 7(2), e32253.

Pootakham, W., Jomchai, N., Ruang-areerate, P., Shearman, J. R., Sonthirod, C., Sangsrakru, D., Tragoonrung, S., and Tangphatsornruang, S. (2015). Genome-wide SNP discovery and identification of QTL associated with agronomic traits in oil palm using genotyping-by-sequencing (GBS). *Genomics*, 105(5–6), 288–295.

Qi, X., Li, M.-W., Xie, M., Liu, X., Ni, M., Shao, G., Song, C., Yim, A. K.-Y., Tao, Y., and Wong, F.-L. (2014). Identification of a novel salt tolerance gene in wild soybean by whole-genome sequencing. *Nature Communications*, 5, 4340.

Qiao, Q., Wang, Q., Han, X., Guan, Y., Sun, H., Zhong, Y., Huang, J., and Zhang, T. (2016). Transcriptome sequencing of Crucihimalaya himalaica (Brassicaceae) reveals how Arabidopsis close relative adapt to the Qinghai-Tibet Plateau. *Scientific Reports*, 6(1), 1–8.

Qu, Y., Zhou, A., Zhang, X., Tang, H., Liang, M., Han, H., and Zuo, Y. (2015). De novo transcriptome sequencing of low temperature-treated Phlox subulata and analysis of the genes involved in cold stress. *International journal of molecular sciences*, 16(5), 9732–9748.

RG, P. (2014). The 3,000 rice genomes project. *GigaScience*, 3(1), 2047-217X-3-7.

Roorkiwal, M., Jarquin, D., Singh, M. K., Gaur, P. M., Bharadwaj, C., Rathore, A., Howard, R., Srinivasan, S., Jain, A., and Garg, V. (2018). Genomic-enabled prediction models using multi-environment trials to estimate the effect of genotype× environment interaction on prediction accuracy in chickpea. *Scientific Reports*, 8(1), 1–11.

Roossinck, M. J. (2015). Metagenomics of plant and fungal viruses reveals an abundance of persistent lifestyles. *Frontiers in Microbiology*, 5, 767.

Rusk, N. (2009). Focus on next-generation sequencing data analysis. *Nature Methods*, 6, S1.

Sansaloni, C., Petroli, C., Jaccoud, D., Carling, J., Detering, F., Grattapaglia, D., and Kilian, A. (2011). Diversity Arrays Technology (DArT) and next-generation sequencing combined: genome-wide, high throughput, highly informative genotyping for molecular breeding of Eucalyptus. *Presented at BMC Proceedings.*

Saxena, R. K., Von Wettberg, E., Upadhyaya, H. D., Sanchez, V., Songok, S., Saxena, K., Kimurto, P., and Varshney, R. K. (2014). Genetic diversity and demographic history of Cajanus spp. illustrated from genome-wide SNPs. *PLoS One*, 9(2), e88568.

Schadt, E. E., Turner, S., and Kasarskis, A (2010). A window into third-generation sequencing. *Human Molecular Genetics*, 19(R2), R227–R240.

Schatz, M. C., Maron, L. G., Stein, J. C., Wences, A. H., Gurtowski, J., Biggers, E., Lee, H., Kramer, M., Antoniou, E., and Ghiban, E. (2014). Whole genome de novo assemblies of three divergent strains of rice, Oryza sativa, document novel gene space of aus and indica. *Genome Biology*, 15(11), 506.

Schmitz, R. J., Schultz, M. D., Urich, M. A., Nery, J. R., Pelizzola, M., Libiger, O., Alix, A., McCosh, R. B., Chen, H., and Schork, N. J. (2013). Patterns of population epigenomic diversity. *Nature*, 495(7440), 193–198.

Schneeberger, K. (2014). Using next-generation sequencing to isolate mutant genes from forward genetic screens. *Nature Reviews Genetics*, 15(10), 662–676.

Sharma, R., Mishra, M., Gupta, B., Parsania, C., Singla-Pareek, S. L., and Pareek, A. (2015). De novo assembly and characterization of stress transcriptome in a salinity-tolerant variety CS52 of Brassica juncea. *PloS One*, 10(5), e0126783.

Sharpe, A. G., Ramsay, L., Sanderson, L.-A., Fedoruk, M. J., Clarke, W. E., Li, R., Kagale, S., Vijayan, P., Vandenberg, A., and Bett, K. E. (2013). Ancient orphan crop joins modern era: gene-based SNP discovery and mapping in lentil. *BMC Genomics*, 14(1), 192.

Shu, Y., Li, W., Zhao, J., Liu, Y., and Guo, C. (2018). Transcriptome sequencing and expression profiling of genes involved in the response to abiotic stress in Medicago ruthenica. *Genetics and Molecular Biology*, 41, 638–648.

Singh, B. K., Mishra, D. C., Yadav, S., Ambawat, S., Vaidya, E., Tribhuvan, K. U., Kumar, A., Kumar, S., Kumar, S., and Chaturvedi, K. (2016). Identification, characterization, validation and cross-species amplification of genic-SSRs in Indian Mustard (Brassica juncea). *Journal of Plant Biochemistry and Biotechnology*, 25(4), 410–420.

Singh, D., Singh, C. K., Taunk, J., Tomar, R. S. S., Chaturvedi, A. K., Gaikwad, K., and Pal, M. (2017). Transcriptome analysis of lentil (Lens culinaris Medikus) in response to seedling drought stress. *BMC Genomics*, 18(1), 206.

Singh, D., Singh, C. K., Taunk, J., Jadon, V., Pal, M., and Gaikwad, K. (2019). Genome wide transcriptome analysis reveals vital role of heat responsive genes in regulatory mechanisms of lentil (Lens culinaris Medikus). *Scientific Reports*, 9(1), 1–19.

Springer, N. M., Ying, K., Fu, Y., Ji, T., Yeh, C.-T., Jia, Y., Wu, W., Richmond, T., Kitzman, J., and Rosenbaum, H. (2009). Maize inbreds exhibit high levels of copy number variation (CNV) and presence/absence variation (PAV) in genome content. *PLoS Genet*, 5(11), e1000734.

Steele, P. R., Hertweck, K. L., Mayfield, D., McKain, M. R., Leebens-Mack, J., and Pires, J. C. (2012). Quality and quantity of data recovered from massively parallel sequencing: examples in Asparagales and Poaceae. *American Journal of Botany*, 99(2), 330–348.

Sudheesh, S., Rodda, M. S., Davidson, J., Javid, M., Stephens, A., Slater, A. T., Cogan, N. O., Forster, J. W., and Kaur, S. (2016a). SNP-based linkage mapping for validation of QTLs for resistance to ascochyta blight in lentil. *Frontiers in Plant Science*, 7, 1604.

Sudheesh, S., Verma, P., Forster, J. W., Cogan, N. O., and Kaur, S. (2016b). Generation and characterisation of a reference transcriptome for lentil (Lens culinaris Medik.). *International Journal of Molecular Sciences*, 17(11), 1887.

Sundaresan, S., Philosoph-Hadas, S., Riov, J., Mugasimangalam, R., Kuravadi, N. A., Kochanek, B., Salim, S., Tucker, M. L., and Meir, S. (2016). De novo transcriptome sequencing and development of abscission zone-specific microarray as a new molecular tool for analysis of tomato organ abscission. *Frontiers in Plant Science*, 6, 1258.

Swarbreck, S. M., Lindquist, E. A., Ackerly, D. D., and Andersen, G. L. (2011). Analysis of leaf and root transcriptomes of soil-grown Avena barbata plants. *Plant and Cell Physiology*, 52(2), 317–332.

Tao, Y., Zhao, X., Mace, E., Henry, R., and Jordan, D. (2019). Exploring and exploiting pan-genomics for crop improvement. *Molecular Plant*, 12(2), 156–169.

Temel, H. Y., Göl, D., Akkale, H. B. K., Kahriman, A., and Tanyolac, M. B. (2015). Single nucleotide polymorphism discovery through Illumina-based transcriptome sequencing and mapping in lentil. *Turkish Journal of Agriculture and Forestry*, 39(3), 470–488.

Tettelin, H., Masignani, V., Cieslewicz, M. J., Donati, C., Medini, D., Ward, N. L., Angiuoli, S. V., Crabtree, J., Jones, A. L., and Durkin, A. S. (2005). Genome analysis of multiple pathogenic isolates of Streptococcus agalactiae: implications for the microbial "pan-genome". *Proceedings of the National Academy of Sciences*, 102(39), 13950–13955.

Trick, M., Long, Y., Meng, J., and Bancroft, I. (2009). Single nucleotide polymorphism (SNP) discovery in the polyploid Brassica napus using Solexa transcriptome sequencing. *Plant Biotechnology Journal*, 7(4), 334–346.

Udomchalothorn, T., Plaimas, K., Comai, L., Buaboocha, T., and Chadchawan, S. (2014). Molecular karyotyping and exome analysis of salt-tolerant rice mutant from somaclonal variation. *The Plant Genome*, 7(3), 1–11.

Uitdewilligen, J. G., Wolters, A.-M. A., Bjorn, B., Borm, T. J., Visser, R. G., and Van Eck, H. J. (2013). A next-generation sequencing method for genotyping-by-sequencing of highly heterozygous autotetraploid potato. *PloS One*, 8(5), e62355.

Van Dijk, E. L., Auger, H., Jaszczyszyn, Y., and Thermes, C. (2014). Ten years of next-generation sequencing technology. *Trends in Genetics*, 30(9), 418–426.

Varshney, R. K., Saxena, R. K., Upadhyaya, H. D., Khan, A. W., Yu, Y., Kim, C., Rathore, A., Kim, D., Kim, J., and An, S. (2009). Next-generation sequencing technologies and their implications for crop genetics and breeding. *Trends in Biotechnology*, 27(9), 522–530.

Varshney, R. K., Nayak, S. N., May, G. D., and Jackson, S. A. (2017). Whole-genome resequencing of 292 pigeonpea accessions identifies genomic regions associated with domestication and agronomic traits. *Nature Genetics*, 49(7), 1082.

Verma, P., Shah, N., and Bhatia, S. (2013). Development of an expressed gene catalogue and molecular mark ers from the de novo assembly of short sequence reads of the lentil (L ens culinaris M edik.) transcriptome. *Plant Biotechnology Journal*, 11(7), 894–905.

Wakasa, Y., Oono, Y., Yazawa, T., Hayashi, S., Ozawa, K., Handa, H., Matsumoto, T., and Takaiwa, F. (2014). RNA sequencing-mediated transcriptome analysis of rice plants in endoplasmic reticulum stress conditions. *BMC Plant Biology*, 14(1), 101.

Wang, Y., Xiong, G., Hu, J., Jiang, L., Yu, H., Xu, J., Fang, Y., Zeng, L., Xu, E., and Xu, J. (2015). Copy number variation at the GL7 locus contributes to grain size diversity in rice. *Nature Genetics*, 47(8), 944–948.

Wang, D., Heckathorn, S. A., Mainali, K., and Tripathee, R. (2016). Timing effects of heat-stress on plant ecophysiological characteristics and growth. *Frontiers in Plant Science*, 7, 1629.

Wang, W., Mauleon, R., Hu, Z., Chebotarov, D., Tai, S., Wu, Z., Li, M., Zheng, T., Fuentes, R. R., and Zhang, F. (2018). Genomic variation in 3,010 diverse accessions of Asian cultivated rice. *Nature*, 557(7703), 43–49.

Warr, A., Robert, C., Hume, D., Archibald, A., Deeb, N., and Watson, M. (2015). Exome sequencing: current and future perspectives. *G3: Genes, Genomes, Genetics*, 5(8), 1543–1550.

Winfield, M. O., Wilkinson, P. A., Allen, A. M., Barker, G. L., Coghill, J. A., Burridge, A., Hall, A., Brenchley, R. C., D'Amore, R., and Hall, N. (2012). Targeted re-sequencing of the allohexaploid wheat exome. *Plant Biotechnology Journal*, 10(6), 733–742.

Wolf, J. B. (2013). Principles of transcriptome analysis and gene expression quantification: an RNA-seq tutorial. *Molecular Ecology Resources*, 13(4), 559–572.

Wright, S. I., Bi, I. V., Schroeder, S. G., Yamasaki, M., Doebley, J. F., McMullen, M. D., and Gaut, B. S. (2005). The effects of artificial selection on the maize genome. *Science*, 308(5726), 1310–1314.

Wu, Y., Zhou, Q., Huang, S., Wang, G., and Xu, L.-a. (2019). SNP development and diversity analysis for Ginkgo biloba based on transcriptome sequencing. *Trees*, 33(2), 587–597.

Xie, Z., Johansen, L. K., Gustafson, A. M., Kasschau, K. D., Lellis, A. D., Zilberman, D., Jacobsen, S. E., and Carrington, J. C. (2004). Genetic and functional diversification of small RNA pathways in plants. *PLoS Biology*, 2(5), e104.

Xu, Y., and Crouch, J. H. (2008). Marker-assisted selection in plant breeding: From publications to practice. *Crop Science*, 48(2), 391–407.

Xu, T., Wu, Y.-K., Luo, X., and Guo, L. J. (2010). Plasmonic nanoresonators for high-resolution colour filtering and spectral imaging. *Nature Communications*, 1(1), 1–5.

Xu, Y., Li, X., Lin, J., Wang, Z., Yang, Q., and Chang, Y. (2015). Transcriptome sequencing and analysis of major genes involved in calcium signaling pathways in pear plants (Pyrus calleryana Decne.). *BMC Genomics*, 16(1), 738.

Yadav, N. K., Shukla, P., Omer, A., Pareek, S., and Singh, R. (2014). Next generation sequencing: potential and application in drug discovery. *The Scientific World Journal*, 2014.

Yang, H., Tao, Y., Zheng, Z., Li, C., Sweetingham, M. W., and Howieson, J. G. (2012). Application of next-generation sequencing for rapid marker development in molecular plant breeding: a case study on anthracnose disease resistance in Lupinus angustifolius L. *BMC Genomics*, 13(1), 318.

Yang, Z., Huang, D., Tang, W., Zheng, Y., Liang, K., Cutler, A. J., and Wu, W. (2013). Mapping of quantitative trait loci underlying cold tolerance in rice seedlings via high-throughput sequencing of pooled extremes. *Plos One*, 8(7), e68433.

Yang, P., Han, J., and Huang, J. (2014). Transcriptome sequencing and de novo analysis of cytoplasmic male sterility and maintenance in JA-CMS cotton. *PloS One*, 9(11), e112320.

Yao, L., Wang, J., Li, B., Meng, Y., Ma, X., Si, E., Ren, P., Yang, K., Shang, X., and Wang, H. (2018). Transcriptome sequencing and comparative analysis of differentially-expressed isoforms in the roots of Halogeton glomeratus under salt stress. *Gene*, 646, 159–168.

Yu, M., Hon, G. C., Szulwach, K. E., Song, C.-X., Jin, P., Ren, B., and He, C. (2012). Tet-assisted bisulfite sequencing of 5-hydroxymethylcytosine. *Nature Protocols*, 7(12), 2159.

Yu, C.-P., Chen, S. C.-C., Chang, Y.-M., Liu, W.-Y., Lin, H.-H., Lin, J.-J., Chen, H. J., Lu, Y.-J., Wu, Y.-H., and Lu, M.-Y. J. (2015). Transcriptome dynamics of developing maize leaves and genomewide prediction of cis elements and their cognate transcription factors. *Proceedings of the National Academy of Sciences*, 112(19), E2477–E2486.

Yu, J., Golicz, A. A., Lu, K., Dossa, K., Zhang, Y., Chen, J., Wang, L., You, J., Fan, D., and Edwards, D. (2019). Insight into the evolution and functional characteristics of the pan-genome assembly from sesame landraces and modern cultivars. *Plant Biotechnology Journal*, 17(5), 881–892.

Zakaria, W. N. A. W., Loke, K.-K., Goh, H.-H., and Noor, N. M. (2016). RNA-seq analysis for plant carnivory gene discovery in Nepenthes× ventrata. *Genomics Data*, 7, 18.

Zalapa, J. E., Cuevas, H., Zhu, H., Steffan, S., Senalik, D., Zeldin, E., McCown, B., Harbut, R., and Simon, P. (2012). Using next-generation sequencing approaches to isolate simple sequence repeat (SSR) loci in the plant sciences. *American Journal of Botany*, 99(2), 193–208.

Zhang, W., Zhang, T., Wu, Y., and Jiang, J. (2012). Genome-wide identification of regulatory DNA elements and protein-binding footprints using signatures of open chromatin in Arabidopsis. *The Plant Cell*, 24(-7), 2719–2731.

Zhao, J.-L., Pan, J.-S., Guan, Y., Nie, J.-T., Yang, J.-J., Qu, M.-L., He, H.-L., and Cai, R. (2015). Transcriptome analysis in Cucumis sativus identifies genes involved in multicellular trichome development. *Genomics*, 105(5–6), 296–303.

Zhao, Q., Feng, Q., Lu, H., Li, Y., Wang, A., Tian, Q., Zhan, Q., Lu, Y., Zhang, L., and Huang, T. (2018). Pan-genome analysis highlights the extent of genomic variation in cultivated and wild rice. *Nature Genetics*, 50(2), 278–284.

Zhong, S., Dekkers, J. C., Fernando, R. L., and Jannink, J.-L. (2009). Factors affecting accuracy from genomic selection in populations derived from multiple inbred lines: a barley case study. *Genetics*, 182(1), 355–364.

Zhou, Z., Jiang, Y., Wang, Z., Gou, Z., Lyu, J., Li, W., Yu, Y., Shu, L., Zhao, Y., and Ma, Y. (2015). Resequencing 302 wild and cultivated accessions identifies genes related to domestication and improvement in soybean. *Nature Biotechnology*, 33(4), 408–414.

Zhu, Q.-H., Spriggs, A., Taylor, J. M., Llewellyn, D., and Wilson, I. (2014). Transcriptome and complexity-reduced, DNA-based identification of intraspecies single-nucleotide polymorphisms in the polyploid Gossypium hirsutum L. *G3: Genes, Genomes, Genetics*, 4(10), 1893–1905.

Zhu, H., Wang, H., Zhu, Y., Zou, J., Zhao, F.-J., and Huang, C.-F. (2015). Genome-wide transcriptomic and phylogenetic analyses reveal distinct aluminum-tolerance mechanisms in the aluminum-accumulating species buckwheat (Fagopyrum tataricum). *BMC Plant Biology*, 15(1), 1–13.

Zou, G., Zhai, G., Feng, Q., Yan, S., Wang, A., Zhao, Q., Shao, J., Zhang, Z., Zou, J., and Han, B. (2012). Identification of QTLs for eight agronomically important traits using an ultra-high-density map based on SNPs generated from high-throughput sequencing in sorghum under contrasting photoperiods. *Journal of Experimental Botany*, 63(15), 5451–5462.

Zou, Z., Ishida, M., Li, F., Kakizaki, T., Suzuki, S., Kitashiba, H., and Nishio, T. (2013). QTL analysis using SNP markers developed by next-generation sequencing for identification of candidate genes controlling 4-methylthio-3-butenyl glucosinolate contents in roots of radish, Raphanus sativus L. *PLoS One*, 8(1), e53541.

11

Revisiting Molecular Techniques for Enhancing Sustainable Agriculture

Ankita Chatterjee*
National Chemical Laboratory

CONTENTS

11.1 Introduction

Plants have a natural tendency to form a complex association with varied microbes such as fungi, archaea, bacteria, and protists (Vorholt, 2012). These microbes can reside on the ground (epiphytes), underground (rhizospheric), and inside the plants (endophytes) (Bulgarelli et al., 2013). Nearly all plants adopt microbes as endophytes wherein bacteria survive in a cooperative manner with the host plant tissue and cause no harm to the host cells (Manias et al., 2020). The endophytes, which include bacteria and fungi as microflora, are present in seeds, shoots, roots, or nodules (Mishra et al., 2020). Endophytes can be either gram-positive or gram-negative. Plant–microbe interaction is of paramount importance for maintaining the natural habitat. These interactions impact plant growth and productivity (Lindow and Brandl, 2003; Berendsen et al., 2012) in either a positive or negative way, thereby playing a significant role in sustainable agriculture (Table 11.1). While negative/pathogenic interactions lead to diseases in plants, positive/beneficial interactions with healthy soil microbiota promote plant growth along with an increase in stress tolerance (Reid, 2011; Abhilash et al., 2012). Some well-studied examples of beneficial plant–microbe interactions include symbiosis, where both the host (plant) and the microorganism gain from each other, for example, mutually beneficial relationship between root nodules of leguminous plants (host) and the nitrogen-fixing bacteria (microbe) (Oldroyd et al., 2011). Arbuscular mycorrhizal fungi

* Corresponding author's email: bioinfo.ankita21@gmail.com

DOI: 10.1201/9781003171416-11

TABLE 11.1
Types of Plant–Microbe Interactions

Plant–Microbe Interaction	Description
Plant growth–promoting microorganisms (PGPMs)	Help in nutrient uptake by nitrogen fixation (Taiwo et al., 2019), production of siderophores (Chhabra et al., 2017), and synthesis of growth regulators such as auxins and gibberellins (Parewa et al., 2018)
Mycorrhizae	Assist plants in resource exchange and also nitrogen supply (Roth and Paszkowski, 2017)
Plant disease suppression by rhizobacteria	Protects plants against phytopathogens

(AMF), by residing inside plant roots, help the host plant to absorb phosphate from the soil (Smith and Smith, 2011). Another useful mechanism called quorum sensing wherein bacteria accumulate, detect, and respond to small diffusible communication signals called autoinducers (Waters and Bassler, 2005) facilitates plants' defense mechanism toward both abiotic and biotic environments.

Generally, microbes promote plant growth by fixing nitrogen to enhance nutrients availability, such as sulfur, phosphorous, iron, and copper. They also keep a check on fungal and bacterial diseases and assist in soil bioremediation. State-of-the-art techniques for crop protection also make use of superior organisms, implemented as biocontrol agents and biofertilizers (Parray et al., 2020). Application of microbes for bioremediation is an important survival strategy in plants. Plant growth-promoting bacteria facilitate phytoremediation of heavy metals in polluted soil (Mendoza-Hernández et al. 2019). Cheng et al. (2009) reported the activity of bacteria Pseudomonas putida UW4 toward nickel contamination. A study by Bona et al. (2010) found the key functions of glycolytic enzymes and arsenic (As) transporter (*e.g.*, PgPOR29) during As metabolism. Even though plants possess self-adaptive mechanisms to combat most naturally occurring biotic and abiotic stresses, yet they require the help of microbes to survive and protect themselves against microbial invaders (Turner et al., 2013). In symbiosis, the microorganisms aid the plant in absorption of nutrients or assist in certain biochemical activities that are lacking in plants.

The plant, in return, contributes toward photosynthate for the benefit of the corresponding microbial symbiont in the rhizosphere. By modifying the microflora balance in the rhizosphere, symbiotic associations may also aid in the protection of plants from diseases. The microbial communities associated with plants rely on the microenvironment provided by crops. Based on microenvironments, crops may be classified as carposphere, spermosphere, phyllosphere, and endorhiza. The impact of the microbes on the plant systems is determined by the survival mechanism of the microbial populations (Olanrewaju et al., 2019). Studying plant–microbe interactions is important for several objectives, and the motivation behind much of this research is toward improved and sustainable agricultural productivity. Owing to the interesting and important aspects related to microbes' existence with plants, there's been a growing curiosity to reveal the makeup of microbial communities within the plants. In agriculture practice, microbes are used for the betterment of plant health (Parray et al., 2020). In last few years, there has been a surge of research in identifying novel pathways related to symbiotic benefits (Frantzeskakis et al., 2020; Wille et al., 2019). However, a broader picture of plant microbiome still remains unexplored.

Various factors including chemical, physical, and biological influence the existence and functioning of microbes in the soil. The major abiotic factors include drought and high temperature that affect crop productivity in dry climate. Lack of organic matter in soil further enhances the problem because the non-symbiotic microorganisms rely on organic matter for energy and growth. In recent times, the constant change in natural climate and global warming has severely affected crop production. The present scenario worldwide requires the production from plants to be optimal within the available fertile land. To enhance crop yield and productivity, one of the best ways is to exploit the microbial/biological components. This will also contribute to disease propagation and control (Reid, 2011). The various aspects of plant–microbe interactions that have been studied include the earliest symbiosis between mycorrhizae and plants (Smith and Smith, 2011), pathogenesis (Wirthmueller et al., 2013; Dodds and Rathjen, 2010; Kachroo and Robin, 2013), and nitrogen fixation (Oldroyd et al., 2011). Exploration of such interactions helps in deciphering natural occurrences that influence our daily well-being and could generate applications leading to sustainable resources, less effect on the environment, cleaning of pollution, and impact on atmospheric gases worldwide.

11.2 Molecular Techniques for Deciphering Plant–Microbiome Interactions

Unlike few years back, when majority of microbial communities involved were not culturable (Carvalhais et al., 2013), the present-day molecular techniques have made possible garnering more knowledge about plant–microbe associations and the molecular basis behind it. These advanced techniques help to gauge the effects of perturbations activated by the abiotic and biotic stresses on the diversity of soil microbiome and their interaction with the plants in the constantly changing environment. Majority of these techniques rely on the isolation of microbes for the analysis of diverse microbes present in the soil. The first step in the process involves extraction of DNA and RNA from the soil. The entire microbial genome along with isolated DNA from microbial communities is known as metagenome. Various methods can then be applied for the analysis of DNA extractions depending upon the cloning method, high-throughput sequencing, amplification of polymerase chain reaction (PCR), and microarray hybridization. PCR technique is the most used for quantitatively determining the microbial diversity. For the analysis of functional variations in the microbial community, amplification of certain functional genes may be put to use. The genes used should be specific to certain metabolic processes.

Several next-generation molecular techniques such as high-end next-generation sequencing (NGS), metabolomics, proteomics, and transcriptomics coupled with bioinformatics have proved to be useful and valuable in plant abiotic stress studies (Soda et al., 2015). Proteomics is useful in understanding plant's activity toward abiotic stresses. This might help in the development of new strategies for improving stress tolerance (Gupta et al., 2013). Microbial metabolomics is applied in studying the set of metabolites belonging to microbial communities (Barkal et al., 2016). All of these techniques (Figure 11.1) are discussed in detail later in this chapter.

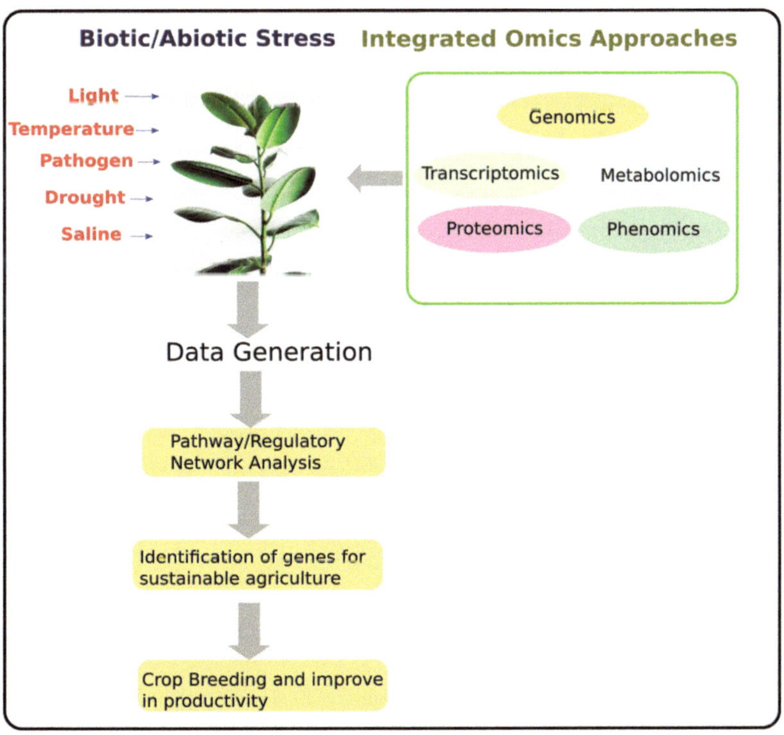

FIGURE 11.1 Various molecular techniques for studying plant–microbiome interaction.

11.3 Why are Plant–Microbe Interactions Important for Crop Improvement?

Crop yield and production is severely affected by important factors such as change in climatic conditions and increasing global warming. One of the most crucial factors in the present-day world is the alarming increase in population and rapid demand for food supply. These challenges have to be overcome by developing ways of efficiently increasing the crop yield. At this end, there's need of exploiting microbial association of plants that play a significant part in disease spreading and check in control (Reid, 2011). The widely researched aspects of plant–microbe interaction has been symbiotic association between plants and mycorrhizae (Smith and Smith, 2011), nitrogen fixation (Oldroyd, 2011), and pathogenesis (Kachroo and Robin, 2013). Crop breeding by using QTL (quantitative trait loci) mapping is an environmentally friendly approach to fighting microbial diseases rather than using pesticides (Gust et al., 2010). For example, several drought resistance QTL have been identified to provide valuable targets in crop breeding (Cui et al., 2018; Li et al., 2018). Another approach to plant disease management also includes the exploitation of biotic and abiotic conditions (Haggag et al., 2015). In fact, crops typically encounter combined attack of biotic and abiotic stresses, which majorly affect their growth and yield (Mahalingam, 2015). All of these stress conditions have an impact on plant–pest interactions by modifying plant defense responses and physiology (Scherm and Coakley, 2003).

NGS is an emerging technology, which has played a prominent role in the complete execution of genome sequencing of several plants and their pathogens producing massive data. As a result, great challenges are faced in translating this massive information for improving crop productivity. Generation of omics data at multiple layers include epitranscriptomes, genomes, transcriptomes, epigenomes, and proteomes (Figure 11.1). The source of these omics data has now extended to micro- and macro-scales (Xu at al., 2019). Omics data are also produced under varying environmental conditions (Groen et al., 2020). Generation of omics data has one application in sustainable agriculture by identifying functional genes that can be engineered to grow new varieties of crops to promote sustainable farming. Table 11.2 lists some of the omics approach-based studies on various plant–pathogen interactions.

TABLE 11.2

Omics Approach-Based Studies on Various Plant–Pathogen Interactions

Pathogen	Host	Reference
Transcriptomics		
Colletotrichum graminicola	*Zea mays*	Bernardo et al. (2012)
Hyaloperonospora arabidopsidis	*Arabidopsis thaliana*	Asai et al. (2014)
Phytoplasma	*Cocos nucifera*	Nejat et al. (2015)
Phytoplasma	*Glycine max*	Jaiswal et al. (2019)
Marssonina coronaria	*Malus domestica*	Feng et al. (2018)
Marssonina coronaria	*Malus baccata*	Zhao et al. (2019)
Diplocarpon rosae	Rose plant	Bhat et al. (2019)
Proteomics		
Pseudomonas syringae pv. tomato	*Solanum lycopersicum*	Margaria et al. (2013)
Verticillium dahliae	Tomato plants	Scandiani et al. (2015)
Mung bean yellow mosaic virus	*Vigna mungo*	Holtappels et al. (2018)
Pseudomonas syringae pv. actinidiae	*Actinidia chinensis*	Meng et al. (2019)
Moniliophthora perniciosa	*Theobroma cacao L.*	Jashni et al. (2019)
Magnaporthe oryzae	*Oryza sativa*	Dos et al. (2020)
Metabolomics		
Pyricularia oryzae	Rice	Kawahara et al. (2012)
Pseudomonas syringae pv. tomato	*Arabidopsis thaliana*	Waskiewicz et al. (2013)
Fusarium tucumaniae	*Glycine max*	Hong et al. (2012)
Verticillium dahliae	*Arabidopsis thaliana*	Azizi et al. (2019)
Fusarium proliferatum	*Asparagus officinalis*	Negrel et al. (2018)
Tuta absoluta	*Solanum lycopersicum*	Zhu et al. (2018)

11.4 Next-Generation Techniques in Studying Plant–Microbe Interactions

11.4.1 Next-Generation Sequencing Analysis of Plant–Microbe Interactions

Plant growth-promoting microorganisms (PGPMs) are intrinsic to plant–microbe interactions. The various processes by which they promote plant growth include surge in nutrient requirement from soil, forbearance to abiotic stress, and pathogen defense. The natural community of plants comprises of diverse types of microorganisms. While some microbes dwell comfortably under stressed conditions and prove beneficial for the plants, others are better taken care of by PGPMs. NGS allows studying plants' behavior with respect to interactome patterns observed in transcriptomics, genomics, and metabolomics, leading to discerning of their survival mechanism (Schlaeppi and Bulgarelli, 2015). Many NGS studies report the role of the bacterial population in plant growth. One of the studies suggested that the microbiota constituents differ from one tissue to another (Rosenberg and Zilber-Rosenberg, 2016).

11.4.2 Transcriptome Scan of Plant–Microbe Interactions

There are instances where microbes concomitantly alter the environment while interacting with plants. For example, saprophytes such as fungi which play a vital role in dead root decomposition and mineralization (Kubartova et al., 2009; Francioli et al., 2021). Another instance is the release of organic acids that cause a decrease in soil pH, thereby causing solubilization of phosphates for easy uptake by roots (Scervino et al., 2010). Several studies have reported on the transcriptional characterization of plant–microbe beneficial interactions, such as *Pseudomonas spp.* in laboratory conditions (Lim et al., 2010; Brotman et al., 2012). Transcriptional profiling of AMF in Petunia hybrida showed a distinctive role for phosphate requiring symbiotically associated genes in the host (Breuillin et al., 2010). Arabidopsis shoots were inoculated with pathogen Pseudomonas syringae, and transcript levels were studied (Verhagen et al., 2004). Studies have shown that variations in the Arabidopsis transcriptome arise because of different traits [FPT9601-T5; MLG45] of *Pseudomonas fluorescens* when inoculated in the roots (Verhagen et al., 2004). Transcriptome analyses of several deleterious plant–microbe relations have been performed mostly using microarrays (Tan et al., 2009; Utsumi et al., 2016; Sham et al., 2017). Xu et al. (2011) applied RNA-Seq to find plant genes that conciliate defense responses against pathogen *Verticillium dahliae*. Some applications of microarrays in plant transcriptomics studies have identified an array of signal transduction pathways that get induced by pathogen-associated molecular patterns (PAMPs) that are related to certain pathogens (Sana et al., 2010; Romero et al., 2017).

Transcriptomics has played a key role in understanding fungi-related diseases in plants. Han et al. (2014) reported genes of some pathogens such as AG1-1A, AG8, and AG3 with functions that are specific to *R. solani*. Similar studies have been carried out on plant-related bacteria using RNA-Seq technology, which revealed that genes' expression differs under some conditions (Coutinho et al., 2015). RNA sequencing technique aids in identifying transcriptome regulations such as small noncoding RNA, antisense RNA, gene operons, and riboswitches (Saberi et al., 2016; Hör et al., 2018). Since the plant transcriptomes are considerably more in number than the bacterial transcriptomes, achieving the desired activity of bacterial mRNA transcripts for differential gene expression and sequencing becomes a challenge.

Metatranscriptomics is an emerging RNA-Seq-based method to detect the expressed transcripts of a microbiome at a particular environmental condition (Shakya et al., 2019). It was used in *Arabidopsis* to identify the bacterial genes in the rhizosphere, which are differently expressed during the development (Chaparro et al., 2014; Chapelle et al., 2016). Application of metatranscriptomics on willow rhizosphere revealed the role of microbial population in phytoremediation (Yergeau et al., 2014). The main advantage of the approach was to infer the possibility of microbial traits appearing in a plant microbiome without requiring the farming of their members. Since its first introduction in 2000s, metatranscriptomics of microbial RNA has increased significantly. It has been applied in the characterization of active microbes (Bashiardes et al., 2016), revealing new microbial relations (Bikel et al., 2015), determining the host–virus relationship (Moniruzzaman et al., 2017), and finding regulatory antisense RNA (Bao et al., 2015). However, despite being a widely accepted method, metatranscriptomics has some of its limitations. Like most of the transcriptomic methods, it also involves critical experimental design, destructive sample collection, and requirement of sufficient material for sequencing. Additionally, it's not always feasible to highlight the complete metatranscriptome owing to the complexity of some microbial communities, the short half-life of RNA, the vast range of transcript expression, and numerous technology-driven limitations (Shakya et al., 2019).

11.4.3 Proteome Analyses of Plant–Microbe Interactions

Proteomics techniques to study plant–pathogen interactions are mostly carried out by liquid chromatography and tandem mass spectrometry. These techniques help in protein identification and any variation during infection. The steps include sample collection, isolation, extraction, and fractionation of proteins by mass spectroscopy. The final step is analysis against a proteome database. Some studies have used proteomics to analyze plant–pathogen relations. Rph15 gene is resistant to *Puccinia hordei*, a causative agent of a disease named leaf rust foliar in barley. A LC–MS/MS analysis of Rph15 discovered various pathogen-associated proteins that are associated with defense mechanism, photosynthesis, carbohydrates metabolism, and protein degradation (Bernardo et al., 2012). Another study performed the proteomic profiling of *Vigna mungo*, when mung bean associated with the yellow mosaic virus (Kundu et al., 2013). They were able to detect the expression of 109 unique proteins. The study also suggested electron transports of the photosystem II as the main attack points in the process of pathogenesis and also during the downregulation of photosynthetic proteins in some genotypes. Similarly, Li et al. (2019) by applying the technique of proteomics studied and observed the mechanisms behind the interaction between poplar plant and the pathogen *Botryosphaeria dothidea*. Santos et al. (2020) performed a comparative analysis of the proteomic changes between Cacao genotypes and pathogen *Moniliophthora perniciosa*.

Metaproteomics and metaproteogenomics also belong to proteomics-based studies. Applications of metaproteomics include measuring the metaproteome of phyllosphere in wild forest trees (Lambais et al., 2017) and identifying the proteins and organisms necessary for nitrogen fixing and oxidation of paddy fields (Bao et al., 2014). Metaproteogenomics uses proteins residing in the microbial population, which are identified from plant microbiota-specific metagenomes. This technique has enhanced protein identification by publicly available databases (Knief et al., 2012). In the past few years, the considerable decrease in expenses involved in next-generation sequencing has strikingly led to enormous metagenome shotgun sequence datasets. In fact, scientific community may start using metagenomics in a manner similar to where 16S rRNA gene fingerprinting methods were used to describe microbial community profiles. However, limitation of proteomics arises due to less protein value, low concentration, and sensitivity owing to host proteins and complex microbes. There's a need to carry out more proteomics studies concerning plant–microbe interactions.

11.4.4 Metabolomics Analyses of Plant–Microbe Interactions

Another novel method used for studying the interactions between plants and microbes is metabolomics. This technique measures the variation in the metabolites' level of the host plant. The maximum utility of this technique is due to its capability in identifying and quantifying numerous compounds simultaneously (Lima et al., 2017). Pathogen attack on plants and consecutive defense response adopted by plants are reflected upon the metabolites. The rapid progress in analytical techniques and data analysis has enabled the analysis of several thousands of compounds in one sample. Researchers are now better equipped to channelize the power of metabolomics to address pivotal questions in the plant–microbe interactions-related study.

When a host's metabolism is affected by the entry of a microbe, several nodulation genes produce Nod factors. In such cases, changes in a metabolite's level can be determined using metabolomics. For example, Negrel et al. (2018) reported the infection of *Plasmopara viticola* in grapevines through MS-based non-targeted metabolomics approach. Another study by Zhu et al. (2018) studied *Phytophthora sojae-mediated disease* in soybean hypocotyls. Several such analyses have been reported using metabolomics to study plant–pathogen interactions (Hong et al., 2012; Su et al., 2018; de Falco et al., 2019). However, the primary concern with many metabolomics approaches is the hurdle to distinguish between plant- and pathogen-specific metabolites. To address this issue, Pang et al. (2018) reported a study wherein by using stable isotope labeling approach, metabolome of a plant bacterial pathogen *Pseudomonas syringae pv.* tomato (Pst) DC3000 was labeled with heavy isotopes. To differentiate plant and bacterial metabolites, the plant metabolites were marked by liquid chromatography–mass spectrometry using multiple reactions monitoring. The study showed that PstDC3000 infection may temper stomatal movement by modulating pathways involved in plant signaling and primary metabolism. Recently, metabolomics of pathogenic microbes that negatively impact crop productivity was performed to generate new biomarkers useful in plant biotechnology (Adeniji et al., 2020). Some very informative works have been reported, such as how sugarcane metabolome and its quality get affected by different soil microbiomes (Huang et al., 2021). The authors performed a comprehensive analysis of microbiome and chemicals present in the soil along with sugarcane metabolome to predict the connection of sugarcane bitterness with soil properties and the host metabolites. Integrating 16S rRNA gene sequencing with liquid chromatography–mass spectrometry (LC–MS) metabolomics, they performed a correlation analysis to examine the interdependence of the host metabolite profile, soil microbiome, and chemicals with sugarcane bitterness.

However, proteomics altogether throws some challenges. First, the expenditure incurred, equipment, technical skill necessary for carrying out metabolic studies make it a less likable one. Second, the availability of not many public metabolomic databases poses difficulty in correlating metabolite and a particular organism. Nevertheless, metabolomics has proved to be an indispensable tool for the detection and quantification of small molecules at the inter-junction of plant–microbe interactions.

11.4.5 Simultaneous Perusal of Interaction between Plants and Related Microorganisms

Interaction transcriptomics, which infers simultaneous perusal of interaction between plants and related microorganisms, is highly useful and recommended because the gene expression of parasitic or symbiotic relationships between plants and microbes is tightly connected. For example, the formation of nodule in legume–rhizobium interactions is strongly associated with developmental events and gene expression in roots (Stancey et al., 2006). Similarly, arbuscular mycorrhiza association has been well specified for gene expression in plant roots (Brechenmacher et al., 2004). A transcriptome analysis of the association between *Magnaporthe oryzae*, a fungal pathogen of the rice, and the host plant, i.e., rice, helped in identifying biotrophy-related effector proteins that might support the plant during hyphal invasion (Mosquera et al., 2009).

Zhang et al. (2020) did perform a transcriptome analysis of *Halomonas* sp. strain MC1, an endophyte. The objective was to decipher the pathogen's plant growth-promoting potential on *Mesembryanthemum crystallinum* (ice plant). The study intended to identify the genes that provide salt tolerance to the plant. The results suggested that under salt stress, MC1 provided ice plant with salt resistance and enabled its survival under stress conditions. Sclerotinia stem rot is a disease of *Brassica napus* (commonly called canola) and is caused by the fungal pathogen *Sclerotinia sclerotiorum.* A transcriptomic analysis of the interaction between *S. sclerotiorum* and canola by Chittem et al. (2020) provided cognizance into the mechanisms of important pathways deployed by *S. sclerotiorum* for the successful infection of canola.

11.4.6 Use of Transcriptomics of Multiple Species in Understanding Plant–Microbe Interactions

Developing disease resistance in crops requires a deeper and clearer understanding of plant and microbiome association. This can be achieved by multi-species transcriptomics analysis. This will in turn require a thorough knowledge of systems biology regarding the processes involved. Abundant data sets derived from proteomics, metabolomics, genomics, and transcriptomics studies may be combined using statistical methodologies and bioinformatics, which will help to identify and integrate key biological processes. Predictive modeling is also used simultaneously. This has enabled a better understanding of ectomycorrhizal interactions within the roots of *Populus tremuloides* (aspen) and *Laccaria bicolor* (Larsen et al., 2011). The study indicated the involvement of transcriptomic data from these complex systems, which facilitated the identification of RNA molecules and the functional genes that cause proteins and metabolites production during the interactions between plants and microbes.

One of the ways of integrating this method is through genome-scale models. Such genome-scale models (GSMs) related to plants differentiate metabolic processes in various compartments such as cytoplasm, chloroplasts, mitochondria, peroxisomes, and vacuoles. The prerequisites for building GSMs are the information collated from the existing proteomics, transcriptomics, genomics, and metabolomics experiments. In recent years, genome-wide research of plants and their related pathogens' interaction has increased. This is primarily due to the progress in genome sequencing and annotations (Shendure et al., 2004). One of the striking features of GSMs is the fact that it doesn't require kinetic information (Pinzon et al., 2010). The very first bacterial genome sequence reported was for Haemophilus influenzae in 1995. This initiated the construction of computational model of any organism. Even solely from genome sequence, it became feasible to visualize the complete functioning of an organism (Seaver et al., 2012). As a first step while developing a GSM of a whole organism, a detailed knowledge of all the genes and their function, their interactions, and the association between different genes is required (Seaver et al., 2012). A metabolic model has been useful in various aspects, such as the interpretation of metabolism and physiology, importance of metabolic reaction processes, gene–protein–reaction (GPR) associations, localization, directionality of reactions, and design of coherent metabolic engineering approach (Murabito et al., 2009; Senger, 2010). Some of the plant genome-scale metabolic models are for Arabidopsis (Poolman et al., 2009), barley (Hordeum vulgare) (Rolletschek et al., 2011), maize (Zea mays) (Dal'Molin et al., 2010), and rice (Poolman et al., 2013; Chatterjee et al., 2015, 2017).

Targeted metabolic reconstructions such as transcriptomic research highlight the relevant information of a specific phenotype network (Pinzon et al., 2010). GSMs have proved to be an essential tool with wide application in exposition of high-throughput data, system metabolic engineering at whole-cell-systems level, discovery and establishment of new hypothesis, understanding the multicellular relationships for the phenotype–genotype gap filling, and exploring the functional evolution of regulatory and metabolic networks (Oberhardt et al., 2009).

11.4.7 Phenomics

Phenomics is a technique for identifying the microbial genotypic–phenotypic properties. Metaphenomics depicts the physiological conditions of microbial metacommunities. More often, while interaction with the environment, plants showcase varying phenotypes.

Plant phenomics has rapidly been emerging as an independent research field. Researchers have categorized phenomics as a holo-omics approach (Xu et al., 2021). Holo-omics is referred to as a new age strategy that combines both host and microbial datasets to enhance approach building for hypothesis development and progress in this area. Revealing the plant–microbe functional relationships should effectively boost plant fitness. The term "holo-omics" has recently been coined by Nyholm et al. (2020) to describe experiments that consolidate multiple omics level data from both host and microbiota domains. Studies have been carried out to prove that these holo-omics approaches hold the capability of unraveling the intricacies of plant–microbiome ecosystem. It is achievable by capturing what is translated, expressed, and formed during plant–microbiome interactions (Xu et al., 2021). As one of the informative tools to understand plant environment interactions, plant phenotyping has applications in crop management practices, studying the outcomes of biostimulants, microbial communities, etc. It has been well established that the phenotype of an organism is influenced by genomic information (Houle et al., 2010). Such a concept is referred to as genotype–phenotype map (Gjuvsland et al., 2013). Integrative phenomics performs a deep analysis of physiological parameters considering the environmental effects, genetic basis, and agricultural practices. It is extremely useful for practical applications in two ways. First, complex crop phenotypes are difficult to quantify. Second, it is a bit difficult to estimate the plant's defense response to environmental variation, specifically using single genetic response. Großkinsky et al. (2018) showed the importance of multi-omics integrated approach in senescence research for a better understanding of the process as a complex system.

11.4.8 Challenges and Future Perspectives

Plants comprise of extensive microbiome. Like humans, plants are also frequently invaded by an array of microorganisms – be it pathogenic or non-pathogenic. Just as humans possess immunity power to get rid of pathogenic infections, plants also utilize their defense mechanisms to stop disease progression by a pathogen. However, pathogens may also escape the solid host innate immunity during evolution (Misra and Chaturvedi, 2015). The ruinous outbreaks of the newly faced pathogens are governed by several factors such as increasing population, human interference, change in world climate, and co-occurrence of host and pathogens (Garrett et al., 2010). Notably, the emergence of new pathogens has caused loss of biodiversity, extermination of several wild species, and decrease in production of crops (Almeida, 2018). Therefore, for the benefit of the mankind, there's an urgent need to identify the emerging pathogens and simultaneously develop strategies to tackle them. The diversity of plant pathogens is severely affected due to trade practices and human-influenced movements of plants. Explication of migration pathways adopted by clever pathogens may be utilized for quarantine approaches and effective disease management (Goss, 2015). Genome sequences of phytopathogens are available. This has provided us with an improved understanding of pathogen's mechanism and adaptive capability while causing plant diseases (Benson et al., 2012; Thynne et al., 2015).

Adaptive evolution of plant pathogens can be studied under population genomics to aid improved disease management strategies. Measures such as restricted movement of plant materials and hence trade will also facilitate the management of emerging plant pathogens. Plant biotechnology holds numerous prospects for studying plant microbiome. Therefore, in order to promote the ongoing research for dealing with emerging phytopathogens, a better collaboration among plant pathologists, epidemiologists, academic researchers, and ecologists should be established. Food is essential for providing nutrients and for balancing human microbiome. Several reports have established the fact that domestic microbiome is heavily influenced by the surrounding vegetation and human inhabitants. These complex communications and connections among microbiomes opens a wide area of mystery and exploration by researchers worldwide.

11.5 Conclusions

The next-generation techniques involving omics have added immensely to our knowledge in studying plant–microbe interactions. However, there still exists several gaps in perceiving the complete picture and answering the assorted questions associated with these interactions for the event of pathogen-resistant crops for human sustainability. The never-ending defiance between plants and pathogens impose several challenges. It includes understanding the biology behind various kinds of stress responses, identification of essential components involved in such interactions, mainly in plant immune responses, understanding the course of disease in plants, abiding response of plant during pathogen attack, successful identification and effective management of newly evolving phytopathogens, and development of crops with pathogen resistance.

In past few years, next-generation biomolecular research has seen significant progress owing to the completion of the genome sequencing project of many plants and microbes (Agrahari et al., 2020). For instance, omics has enabled the characterization of constituents such as inositol monophosphatase, nifH, fixA, nod factors, and ROS scavenging. Pathogens produce different types of toxins that need to be analyzed as it helps to build strategies for increasing the productivity of plants. Omics technology, in several ways, enables the exploration of complex cellular mechanisms involved in the interactions of plants and microbes. Genomics provides a platform to study the evolution of new strains of pathogens and their interactions with plants, thereby contributing to the development of sustainable agriculture strategies. Altogether, it may be said that slowly but steadily, we are reaching a finer understanding of plant immunity and pathogen virulence. This has been made possible by implementing the combined approach of next-generation sequencing technology plus various "omics" technologies along with database development and metabolic modeling. Eventually, with the implementation of new and improved strategies, the problem of world food security may be solved.

REFERENCES

Abhilash et al. 2012. Plant–microbe interactions: novel applications for exploitation in multipurpose remediation technologies. *Trends in Biotechnology*, *30*, 416–420.

Adeniji, A.A. and Babalola, O.O., 2020. Metabolomic applications for understanding complex tripartite plant-microbes interactions: strategies and perspectives. *Biotechnology Reports*, *25*, e00425.

Almeida, R. P. 2018. Emerging plant disease epidemics: Biological research is key but not enough. *PLoS Biology, 16*(8), e2007020.

Asai et al. 2014. Expression profiling during Arabidopsis/downy mildew interaction reveals a highly-expressed effector that attenuates responses to salicylic acid. *PLoS Pathogen, 10*(10), e1004443.

Azizi et al. 2019. Adaptation of the metabolomics profile of rice after Pyricularia oryzae infection. *Plant Physiology and Biochemistry, 144*, 466–479.

Bao et al. 2014. Metaproteomic identification of diazotrophic methanotrophs and their localization in root tissues of field-grown rice plants. *Applied and Environmental Microbiology, 80*(16), 5043–5052.

Bao et al. 2015. Strand-specific community RNA-seq reveals prevalent and dynamic antisense transcription in human gut microbiota. *Frontiers in Microbiology*, *6*, 896.

Barkal et al. 2016. Microbial metabolomics in open microscale platforms. *Nature Communications, 7*(1), 1–11.

Bashiardes et al. 2016. Use of metatranscriptomics in microbiome research. *Bioinformatics Biological Insights*, *10*, 19–25.

Benson et al. 2012. GenBank. *Nucleic Acids Research*, *40*, D48–D53.

Berendsen et al. 2012. The rhizosphere microbiome and plant health. *Trends in Plant Science, 17*(8), 478–486.

Bernardo et al. 2012. Proteomic characterization of the Rph15 barley resistance gene-mediated defence responses to leaf rust. *BMC Genomics, 13*(1), 1–16.

Bhat et al. 2019. Compendium of Colletotrichum graminicola responsive infection-induced transcriptomic shifts in the maize. *Plant Gene, 17*, 100166.

Bikel et al. 2015. Combining metagenomics, metatranscriptomics and viromics to explore novel microbial interactions: towards a systems-level understanding of human microbiome. *Computational and Structural Biotechnology Journal*, *13*, 390–401.

Bona et al. 2010. Proteomic analysis of Pteris vittata fronds: two arbuscular mycorrhizal fungi differentially modulate protein expression under arsenic contamination. *Proteomics, 10*(21), 3811–3834.

Brechenmacher et al. 2004. Expression profiling of up-regulated plant and fungal genes in early and late stages of *Medicago truncatula–Glomus mosseae* interactions *Mycorrhiza, 14*(2004), 253–262.

Breuillin et al. 2010. Phosphate systemically inhibits development of arbuscular mycorrhiza in Petunia hybrida and represses genes involved in mycorrhizal functioning. *The Plant Journal, 64*(6), 1002–1017.

Brotman et al. 2012. Transcript and metabolite analysis of the Trichoderma-induced systemic resistance response to Pseudomonas syringae in Arabidopsis thaliana. *Microbiology, 158*(1), 139–146.

Bulgarelli et al. 2013. Structure and functions of the bacterial microbiota of plants. *Annual Review of Plant Biology*, *64*, 807–838.

Carvalhais et al. 2013. Activation of the jasmonic acid plant defence pathway alters the composition of rhizosphere bacterial communities. *PLoS One, 8*(2), e56457.

Chaparro et al. 2014. Rhizosphere microbiome assemblage is affected by plant development. *The ISME Journal, 8*(4), 790–803.

Chapelle et al. 2016. Fungal invasion of the rhizosphere microbiome. *The ISME Journal, 10*(1), 265–268.

Chatterjee, A. and Kundu, S., 2015. Revisiting the chlorophyll biosynthesis pathway using genome scale metabolic model of Oryza sativa japonica. *Scientific Reports, 5*(1), 1–15.

Chatterjee et al. 2017. Reconstruction of Oryza sativa indica genome scale metabolic model and its responses to varying RuBisCO activity, light intensity, and enzymatic cost conditions. *Frontiers in Plant Science, 8*, 2060.

Cheng et al. 2009. Proteomic analysis of the response of the plant growth-promoting bacterium Pseudomonas putida UW4 to nickel stress. *Proteome Science, 7*(1), 1–8.

Chhabra S. and Dowling D.N. Endophyte-promoted nutrient acquisition: phosphorus and iron. In: Doty S., editor. *Functional Importance of the Plant Microbiome.* Cham: Springer; 2017.

Chittem et al. 2020. Transcriptome analysis of the plant pathogen Sclerotinia sclerotiorum interaction with resistant and susceptible canola (Brassica napus) lines. *PLoS One*, *15*(3), e0229844.

Coutinho et al. 2015. Plant-influenced gene expression in the rice endophyte Burkholderia kururiensis M130. *Molecular Plant-Microbe Interactions, 28*(1), 10–21.

Cui et al. 2018. Simultaneous improvement and genetic dissection of drought tolerance using selected breeding populations of rice. *Frontiers in Plant Science, 9*, 320.

Dal'Molin et al. 2010. C4GEM, a genome-scale metabolic model to study C4 plant metabolism. *Plant Physiology, 154*, 1871–1885.

de Falco et al. 2019. Metabolomics approach based on NMR spectroscopy and multivariate data analysis to explore the interaction between the leafminer Tuta absoluta and tomato (Solanum lycopersicum). *Phytochemical Analysis, 30*(5), 556–563.

Dodds, P. N. and Rathjen, J. P. 2010. Plant immunity: towards an integrated view of plant-pathogen interactions. *Nature Reviews Genetics, 11*, 539–548.

Feng et al. 2019. Responses of Fuji (Malus domestica) and Shandingzi (Malus baccata) apples to Marssonina coronaria infection revealed by comparative transcriptome analysis. *Physiological and Molecular Plant Pathology, 106*, 87–95.

Francioli et al. 2020. Plant functional group drives the community structure of saprophytic fungi in a grassland biodiversity experiment. *Plant and Soil*, 1–15.

Frantzeskakis et al. 2020. Rapid evolution in plant–microbe interactions–a molecular genomics perspective. *New Phytologist, 225*(3), 1134–1142.

Garrett et al. 2010. Emerging plant diseases: what are our best strategies for management? In: E. L. Kleinman, J. A. Delborne et al. editors, *Controversies in Science and Technology from Evolution to Energy, Vol. 3*, New Rochelle, NY: Liebert Publishers.

Gjuvsland et al. 2013. Bridging the genotype–phenotype gap: what does it take? *The Journal of Physiology, 591*(8), 2055–2066.

Goss, E. M. 2015. Genome-enabled analysis of plant pathogen migration. *Annual Review of Phytopathology, 53*, 121–135.

Groen et al. 2020. The strength and pattern of natural selection on gene expression in rice. *Nature, 578*(7796), 572–576.

Griffiths et al. 2011. The bacterial biogeography of British soils. *Environmental Microbiology, 13*(6), 1642–1654

Gupta et al. 2013. Plant abiotic stress: 'Omics' approach. *Journal of Plant Biochemistry & Physiology, 1*(3).

Gust et al. 2010. Biotechnological concepts for improving plant innate immunity. *Current Opinion in Biotechnology, 21*, 204–210.

Hane et al. 2014. Genome sequencing and comparative genomics of the broad host-range pathogen Rhizoctonia solani AG8. *PLoS Genetics, 10*(5), e1004281.

Haggag et al. 2015. Agriculture biotechnology for management of multiple biotic and abiotic environmental stress in crops. *Journal of Chemical and Pharmaceutical Research, 7*, 882–889.

Holtappels et al. 2018. Fire blight host-pathogen interaction: proteome profiles of Erwinia amylovora infecting apple rootstocks. *Scientific Reports, 8*(1), 1–9.

Hong et al. 2012. Metabolomics reveals simultaneous influences of plant defence system and fungal growth in Botrytis cinerea-infected Vitis vinifera cv. Chardonnay berries. *Journal of Experimental Botany, 63*(16), 5773–5785.

Houle et al. 2010. Phenomics: the next challenge. *Nature Reviews Genetics, 11*(12), 855–866.

Hör et al. 2018. Bacterial RNA biology on a genome scale. *Molecular Cell, 70*(5), 785–799.

Huang et al. 2021. Integrative analysis of the microbiome and metabolome in understanding the causes of sugarcane bitterness. *Scientific Reports, 11*(1), 1–11.

Jaiswal et al. 2019. Transcriptomic signature reveals mechanism of flower bud distortion in witches'-broom disease of soybean (Glycine max). *BMC Plant Biology, 19*(1), 1–12.

Jashni et al. 2019. Transcriptome and proteome analyses of proteases in biotroph fungal pathogen Cladosporium fulvum. *Journal of Plant Pathology*, 1–10.

Kachroo, A. and Robin, G. P. 2013. Systemic signaling during plant defense. *Current Opinion in Plant Biology, 16*, 527–533.

Kawahara et al. 2012. Simultaneous RNA-seq analysis of a mixed transcriptome of rice and blast fungus interaction. *PloS One, 7*(11), e49423.

Knief et al. 2012. Metaproteogenomic analysis of microbial communities in the phyllosphere and rhizosphere of rice. *The ISME Journal, 6*(7), 1378–1390.

Kubartová et al. 2009. Diversity and decomposing ability of saprophytic fungi from temperate forest litter. *Microbial Ecology, 58*(1), 98–107.
Kundu et al. 2013. Proteomics approach combined with biochemical attributes to elucidate compatible and incompatible plant-virus interactions between Vigna mungo and Mungbean Yellow Mosaic India Virus. *Proteome Science, 11*(1), 1–14.
Lambais et al. 2017. Phyllosphere metaproteomes of trees from the Brazilian Atlantic forest show high levels of functional redundancy. *Microbial Ecology, 73*(1), 123–134.
Larsen et al. 2011. Using next generation transcriptome sequencing to predict an ectomycorrhizal metabolome. *BMC Systems Biology, 5*(1), 1–14.
Li et al. 2018. Investigating drought tolerance in chickpea using genome-wide association mapping and genomic selection based on whole-genome resequencing data. *Frontiers in Plant Science, 9*, 190.
Li et al. 2019. Comparative proteomic analysis of plant–pathogen interactions in resistant and susceptible poplar ecotypes infected with *Botryosphaeria dothidea. Phytopathology, 109*(12), 2009–2021.
Lim et al. 2010. Transcriptomic changes in the root of oil palm (Elaeis guineensis Jacq.) upon inoculation with Bacillus sphaericus UPMB10. *Tree Genetics & Genomes, 6*(5), 793–800.
Lima et al. 2017. Metabolomic study of chardonnay grapevines double stressed with esca-associated fungi and drought. *Phytopathology, 107*(6), 669–680.
Lindow, S.E. and Brandl, M.T., 2003. Microbiology of the phyllosphere. *Applied and Environmental Microbiology, 69*(4), 1875–1883.
Mahalingam, R. 2015. Consideration of combined stress: a crucial paradigm for improving multiple stress tolerance in plants. *Combined Stresses in Plants* (pp. 1–25). Springer, Cham.
Manias et al. Isolation and characterization of endophytes: biochemical and molecular approach. Prospects for Sustainable Agriculture, Woodhead Publishing Series in Food Science, Technology and Nutrition, 2020. 1–14.
Margaria et al. 2013. Novel aspects of grapevine response to phytoplasma infection investigated by a proteomic and phospho-proteomic approach with data integration into functional networks. *BMC Genomics, 14*(1), 1–15.
Meng et al. 2019. Proteomics of rice—*Magnaporthe oryzae* interaction: what have we learned so far? *Frontiers in Plant Science, 10*, 1383.
Mendoza-Hernández et al. 2019. Phytoremediation of mine tailings by Brassica juncea inoculated with plant growth-promoting bacteria. *Microbiological Research, 228*, 126308.
Misra, B.B. and Chaturvedi, R. 2015. When plants braces for the emerging pathogens. *Physiological and Molecular Plant Pathology, 92*, 181–185.
Mishra et al. Plant-microbe interactions for bioremediation and phytoremediation of environmental pollutants and agro-ecosystem development. In: Bharagava, R. and Saxena, G., editors. *Bioremediation of Industrial Waste for Environmental Safety*. Singapore: Springer, 2020.
Mosquera et al. 2009. Interaction transcriptome analysis identifies Magnaporthe oryzae BAS1–4 as biotrophy-associated secreted proteins in rice blast disease. *The Plant Cell, 21*(4), 1273–1290.
Murabito et al. 2009. Capturing the essence of a metabolic network: a flux balance analysis approach. *Journal of Theoretical Biology, 260*, 445–452.
Moniruzzaman et al. 2017. Virus-host relationships of marine single-celled eukaryotes resolved from metatranscriptomics. *Nature Communications 8*, 16054.
Negrel et al. 2018. Identification of lipid markers of Plasmopara viticola infection in grapevine using a non-targeted metabolomic approach. *Frontiers in Plant Science, 9*, 360.
Nejat et al. 2015. Transcriptomics-based analysis using RNA-Seq of the coconut (Cocos nucifera) leaf in response to yellow decline phytoplasma infection. *Molecular Genetics and Genomics, 290*(5), 1899–1910.
Nyholm et al. 2020. Holo-omics: integrated host-microbiota multi-omics for basic and applied biological research. *Iscience*, 101414.
Olanrewaju et al. 2019. Plant health: feedback effect of root exudates-rhizobiome interactions. *Applied Microbiology and Biotechnology, 103*(3), 1155–1166.
Oldroyd et al. 2011. The rules of engagement in the legume-rhizobial symbiosis. *Annual Review of Genetics, 45*, 119–144.
Pang et al. 2018. Metabolomics of early stage plant cell–microbe interaction using stable isotope labeling. *Frontiers in Plant Science, 9*, 760.

Parewa et al. Sustainable crop production and soil health management through plant growth-promoting rhizobacteria. In: Meena, V. (eds) *Role of Rhizospheric Microbes in Soil* (pp. 299–329). Springer, Singapore, 2018
Parray, J.A. and Shameem, N. 2019. *Sustainable Agriculture: Advances in Plant Metabolome and Microbiome*. Academic Press.
Pinzon et al. 2010. Trageted metabolic reconstruction: a novel approach for the characterization of plant-pathogen interactions. *Briefings in Bioinformatics, 12*, 151–162.
Poolman et al. 2013. Responses to light intensity in a genome-scale model of rice metabolism. *Plant Physiology, 162*(2), 1060–1072.
Reid, A. 2011. Microbes helping to improve crop productivity. *Microbe 6*, 435–439.
Rolletschek et al. 2011. Combined noninvasive imaging and modeling approaches reveal metabolic compartmentation in the barley endosperm. *The Plant Cell, 23*(8), 3041–3054
Romero et al. Gaining insight into plant responses to beneficial and pathogenic microorganisms using metabolomic and transcriptomic approaches. In: Kalia, V. and Saini, A. editors. *Metabolic Engineering for Bioactive Compounds*. Singapore: Springer, 2017.
Rosenberg, E. and Zilber-Rosenberg, I., 2016. Microbes drive evolution of animals and plants: the hologenome concept. *MBio, 7*(2).
Roth, R. and Paszkowski, U., 2017. Plant carbon nourishment of arbuscular mycorrhizal fungi. *Current Opinion in Plant Biology, 39*, 50–56.
Saberi et al. 2016. Natural antisense RNAs as mRNA regulatory elements in bacteria: a review on function and applications. *Cellular & Molecular Biology Letters, 21*(1), 1–17.
Sana et al. 2010. Metabolomic and transcriptomic analysis of the rice response to the bacterial blight pathogen Xanthomonas oryzae pv. oryzae. *Metabolomics, 6*(3), 451–465.
Santos et al. 2019. The pathogen Moniliophthora perniciosa promotes differential proteomic modulation of cacao genotypes with contrasting resistance to witches´ broom disease. *BMC Plant Biology,* 20, 1. https://doi.org/10.1186/s12870-019-2170-7
Scandiani et al. 2015. Metabolic profiles of soybean roots during early stages of Fusarium tucumaniae infection. *Journal of Experimental Botany, 66*(1), 391–402.
Scervino et al. 2010. Soil fungal isolates produce different organic acid patterns involved in phosphate salts solubilization. *Biology and Fertility of Soils, 46*(7), 755–763.
Scherm, H. and Coakley, S. M. 2003. Plant pathogens in a changing world. *Australasian Plant Pathology, 32*, 157–165.
Schlaeppi, K. and Bulgarelli, D. 2015. The plant microbiome at work. *Molecular Plant-Microbe Interactions, 28*(3), 212–217.
Seaver et al. 2012. Frontiers in metabolic reconstruction and modeling of plant genomes. *Journal of Experimental Botany, 63*, 2247–2258.
Senger, R. S. 2010. Biofuel production improvement with genome-scale models: the role of cell composition. *Biotechnology Journal, 5*, 671–685.
Sham et al. 2017. Microarray analysis of Arabidopsis WRKY33 mutants in response to the necrotrophic fungus Botrytis cinerea. *PLoS One*, *12*(2), e0172343.
Shakya et al. 2019. Advances and challenges in metatranscriptomic analysis. *Frontiers in Genetics, 10*, 904.
Shendure et al. 2004. Advanced sequencing technologies: methods and goals. *Nature Reviews Genetics, 5*, 335–344.
Smith, S.E. and Smith, F.A. 2011. Roles of arbuscular mycorrhizas in plant nutrition and growth: new paradigms from cellular to ecosystem scales. *Annual Review of Plant Biology, 62*, 227–250.
Soda et al. 2015. Omics study for abiotic stress responses in plants. *Advances in Plants & Agriculture Research, 2*(1), 00037.
Stancey G. et al. 2006. Genetics and functional genomics of legume nodulation. Current Opinion in Plant Biology, *9*(2006), 110–121.
Su et al. 2018. The dynamic transcriptome and metabolomics profiling in Verticillium dahliae inoculated Arabidopsis thaliana. *Scientific Reports, 8*(1), 1–11.
Taiwo et al. Plant-microbes relationships in soil ecological system and benefits accruable to food health. Field Crops: Sustainable Management by PGPR. In: Maheshwari, D. and Dheeman, S., editors. *Sustainable Development and Biodiversity*, Vol. 23 Cham: Springer, 2019.
Tan et al. 2009. Assessing the impact of transcriptomics, proteomics and metabolomics on fungal phytopathology. *Molecular Plant Pathology, 10*(5), 703–715.

Thynne et al. 2015. Phytopathogen emergence in the genomics era. *Trends in Plant Science, 20*, 246–255.

Turner et al. 2013. The plant microbiome. *Genome Biology, 14*(6), 1–10.

Utsumi et al. 2016. Cassava (Manihot esculenta) transcriptome analysis in response to infection by the fungus Colletotrichum gloeosporioides using an oligonucleotide-DNA microarray. *Journal of Plant Research, 129*(4), 711–726.

Verhagen et al. 2004. The transcriptome of rhizobacteria-induced systemic resistance in arabidopsis. *Molecular Plant-Microbe Interactions*, *17*(8), 895–908.

Vorholt, J. A. 2012. Microbial life in the phyllosphere. *Nature Reviews Microbiology, 10*, 828–840.

Waters, C.M. and Bassler, B.L. 2005. Quorum sensing: cell-to-cell communication in bacteria. *Annual Review of Cell and Developmental Biology, 21*, 319–346.

Waśkiewicz et al. 2013. Plant-pathogen interactions during infection process of asparagus with Fusarium spp. *Open Life Sciences, 8*(11), 1065–1076.

Wille et al. 2019. Insights to plant–microbe interactions provide opportunities to improve resistance breeding against root diseases in grain legumes. *Plant, Cell & Environment, 42*(1), 20–40.

Wirthmueller et al. 2013. On the front line: structural insights into plant–pathogen interactions. *Nature Reviews Microbiology, 11*, 761–776.

Xu et al. 2011. Lignin metabolism has a central role in the resistance of cotton to the wilt fungus Verticillium dahliae as revealed by RNA-Seq-dependent transcriptional analysis and histochemistry. *Journal of Experimental Botany, 62*(15), 5607–5621.

Xu et al. 2019. Population-level analysis reveals the widespread occurrence and phenotypic consequence of DNA methylation variation not tagged by genetic variation in maize. *Genome Biology, 20*(1), 1–16.

Xu et al. 2021. Holo-omics for deciphering plant-microbiome interactions. *Microbiome*, *9*(1), 1–11.

Yergeau et al. 2014. Microbial expression profiles in the rhizosphere of willows depend on soil contamination. *The ISME Journal, 8*(2), 344–358.

Zhao et al. 2019. Transcriptome profiling reveals the response process of tomato carrying Cf-19 and Cladosporium fulvum interaction. *BMC Plant Biology, 19*(1), 1–12.

Zhu et al. 2018. Metabolomics analysis of soybean hypocotyls in response to Phytophthora sojae infection. *Frontiers in Plant Science*, 9, 1530.

12

Nanotechnology in Plant Pathology: An Overview

Ratul Moni Ram*
KVK Saran

Prakash Jyoti Kalita*
Myongji University

H.B. Singh
Somvanshi Research Foundation

CONTENTS

12.1 Introduction

Nanotechnology may be regarded as one of the emergent fields of science and technology that holds a pivotal position in the current technological group. Owing to its interdisciplinary nature, it can be well incorporated into various scientific fields such as physics, chemistry, material sciences, medicine,

* Corresponding author's email: ratulbhu006@gmail.com

DOI: 10.1201/9781003171416-12

pharmaceuticals, and agriculture. The concept of nanotechnology has first come to light during the classical talk on "There's Plenty of Room at the Bottom" by American physicist Richard Feynman (1959). However, the term "nanotechnology" was first used by Norio Taniguchi (1974). In general language, nanotechnology may be regarded as the science of manipulation of matter at the scale of 1–100 nm (Sekhon, 2014). The term "nano" usually denotes one billionth (10^{-9}); thus, nanotechnology is the science dealing with particles having the size of one billionth of a meter. In simpler terms, we can say that a nanometer is 1/80,000 the diameter of a human hair. When a material is compacted to nano-form, it expresses some novel properties that remain absent in the original form. In an attempt to define nanotechnology, Joseph and Morrison (2006) stated it as "the manipulation or self-assembly of individual atoms, molecules or molecular clusters into structures to create materials devices with new or vastly different properties." NPs have proved to be a boon for the scientific world due to their unique property, higher growth, and wider applicability. The potential of NPs has been explored in different sectors, *viz.* medicine, agriculture, cosmetics, electronics, and environmental remediation for societal benefits (Feiner, 2006; Patolsky et al., 2006; Hu and Chen, 2007). This tremendous potential of NPs has made nanotechnology-based researches more prevalent among the scientific community over the past few years.

12.2 Need for Nanotechnology in Agriculture

Currently, the overall world population has been estimated to be around 7.7 billion and is projected to touch 9.6 billion by 2050 and 11.2 billion by end of this century (United Nations Economic and Social Council, 2018). This indicates that to ensure food security for such an enormous population, new systems for gaining energy will be required. However, producing more quantities of food requires more natural resources, land consumption, water supply, and energy (Glenn and Florescu, 2015). The current population demands an increase in crop production through traditional and advanced technologies simultaneously by maintaining the ecosystem. Thus, very soon, scientific farming through the need-based utilization of inputs will pave the way for enhanced production.

Agriculture is always considered the heart of most developing nations. In addition to a source of food and fiber, it also contributes to a country's economy, i.e., gross domestic product (GDP). With the existence and widespread promotion of the green revolution during the 1970s, the main thrust was laid upon increasing the overall agricultural production. To achieve the target, chemical pesticides were applied non-judiciously for the management of biotic stresses. However, much later, it was realized that apart from managing the plant pathogens, these chemicals resulted in a reduction in soil fertility, development of pesticide-resistant pathogens, accumulation of pesticide residue in the food chain, and several environmental and health hazards (Tilman et al., 2002). These negative effects of chemicals lead researchers to shift their attention toward alternative means and exploiting their potential in disease management.

Over the past few decades, nanotechnology has gained widespread attention in agriculture by deciphering robust utility (Kah and Hofmann, 2014). Modification of agriculture using this technology has captivated the thoughts of scientists and researchers on a global basis (Adholeya et al., 2017; Jogaiah et al., 2019; Joshi et al., 2019). Nanotechnology has been termed as one of the six "key enabling technologies" that contribute to sustainable growth and development in areas devoting toward a greener economy (European Commission, 2017). As a result of their outsmart performance in disease resistance and plant growth promotion, nanotechnology has occupied prime space in the agricultural sector (Ghormade et al., 2011; Mishra et al., 2014b). The various roles of nanotechnology in different agricultural sectors are presented in Figure 12.1. The unique properties of NPs, viz., shape and size, high surface area, and reactivity, have the tremendous ability in changing the scenario of agriculture in developing countries.

Prior to on-field application of NPs, their toxicological effects on plant and soil microbes should be studied (Ahmad et al., 2020). However, in general, the toxic effect of NPs on soil–plant interrelations is still not resolved completely (Antisari et al., 2011; Lee et al., 2012). Nanoformulations are regarded as an eco-friendly alternative to chemicals in combating pest attack, although their uncontrolled release in the atmosphere might pose threats to flora and fauna (Banik and Sharma, 2011). In recent times, engineered NPs have proved their potential for using them as a prime disease management tool (Ocsoy et al., 2013).

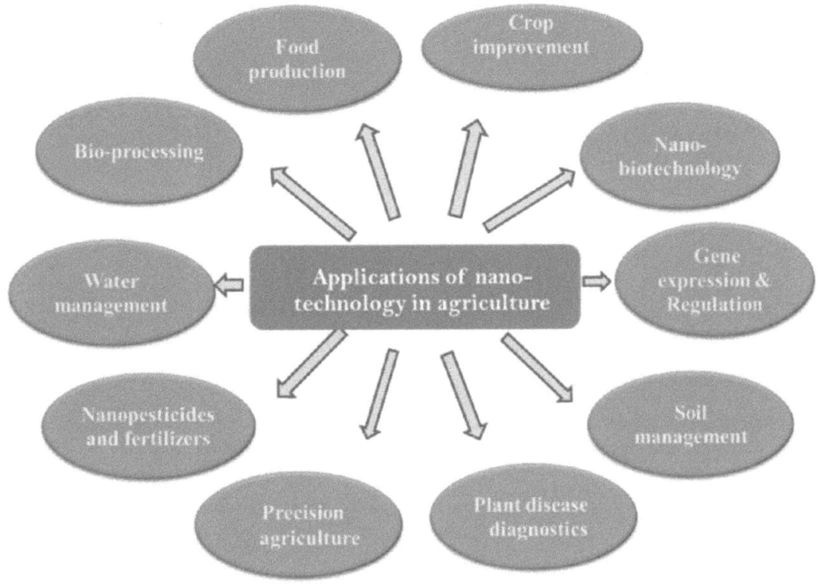

FIGURE 12.1 Multifarious role of nanotechnology in agriculture.

12.3 Nanoformulations of Agrochemicals for Crop Improvement

The role of nanotechnology in crop improvement has received accolades focusing on the advancement in target delivery of nanofertilizers and nanopesticides (Rodrigues et al., 2017; Yin et al., 2018) or as sensors for checking the status of water or nutrient in an area (Kah et al., 2018). The conventional inputs, i.e., land, water, fertilizers, and pesticides, incorporated in agriculture take place inefficiently as a large portion of chemicals applied per year did not reach the target as they may be washed away or not properly placed (de Rosa et al., 2010; de Oliveira et al., 2014). Hence, the development of advanced agricultural technology has become the need of the hour to confront such challenges. The past decade has witnessed above 3,000 patent applications on nanopesticides being filed (Kah et al., 2012). The nano-based products has been gaining widespread popularity in agriculture over the years owing to their reduced rate of application and greater effectiveness in a short period (Xu et al., 2010; Mishra et al., 2014b; Rajput et al., 2018). As a result, they are applied in various fields to enhance crop production (Table 12.1). Nanoformulations facilitate the precise release of chemicals under specific time and circumstances, thereby preventing delivery loss of particles through runoff, leaching, infiltration, volatilization, etc. (Luque and Hermosin, 2013; Sangeetha et al., 2017). The nanopesticide formulations provide a large surface area and thus enhance affinity to the target (Bergeson, 2010).

The two principal factors that make nanomaterials different from their counterparts are higher surface area and quantum effects. Morphology, aspect ratio/size, structure/composition, surface area, solubility, release of toxic ions, and behavior toward ROS (O_2/H_2O) are few peculiar features of NPs (Somasundaran et al., 2010).

Nanotechnology can be used irrespective of the type of crop. It promotes biosource use and enhances carbon assimilation and soil aggregation through eco-friendly means. Nanofertilizers enhance the nutrient use efficiency (NUE) of plants by three times and also impart stress tolerance (Corradini et al., 2010). Such nanofertilizers containing nutrients or growth promoters encapsulated in NPs facilitate slow and efficient release to target. In comparison with chemical fertilizers based on dose and cost, the former is economically cheaper and require a small dose. Nano-encapsulation is also being done to regulate the release of target chemicals. Nano-encapsulation comprises nanosized particles of the active ingredients being coated by a protective layer. Nano-encapsulation of pesticides helps in pest management along with checking residue accumulation in soil. Nanoformulations of agrochemicals are a complex entity as

TABLE 12.1

Various Applications of NP-Based Formulations in Agriculture

Sl. No.	NPs	Applications	Reference
1.	Nano-encapsulated agrochemicals	Control of phytopathogens	Wani et al. (2019); Dhillon and Mukopadhyay (2015)
2.	Nano-encapsulated herbicides	Control of parasitic weeds	Perez-de-Luque and Diego (2009)
3.	Silver NPs (AgNPs)	Control of phytopathogens	Park et al. (2006), Min et al. (2009)
4.	Titanium dioxide (TiO_2) NPs	Antifungal and antibacterial activity, delivery vehicle for NP transportation	Siddiqui et al. (2018)
5.	Chitosan polymer NPs	Facilitate controlled and targeted NPK release	Corradini et al. (2010)
6.	Essential oil of *Artemisia arborescens* incorporated into solid lipid NPs	Insecticidal potential against young ones of *Bemisia tabaci* and *Aphis gossipy* adults	Lai et al. (2006)
7.	NP polymers	Insecticidal activity against *Tribolium castaneum*	Yang et al. (2009)
8.	Silica NPs	Insecticidal activity	Barik et al. (2008)

they contain several surfactant polymers (organic) and metal NPs (inorganic) in the nanometer size range (Sekhon, 2014; Jogiah et al., 2020).

In addition to the above-mentioned facts, nanotechnology also plays a promising role in seed technology. In the case of wind-pollinated crops, nanobiosensors specific to contaminating pollen are used to detect contamination and minimize it. Similarly, it can also be used to select a particular target crop and avoid unnecessary cross-pollination. The recent era has witnessed the transfer of novel genes into seeds and its commercial success. Nano-barcodes may also facilitate tracking of such seeds that are encodable, readable, and durable (Nicewarner Pena et al., 2001). Nano-coating of seeds with non-metals will not only protect seeds from pests and pathogens, but also reduce their rate of application. Similarly, Su and Li (2004) developed a quantum dot (QD) technique that facilitates diagnosis of unviable and infected seeds.

12.4 Synthesis of Metal NPs from Plant Extract

The majority of NPs are manufactured physically and chemically, but nowadays, biologically produced NPs are gaining more popularity due to their eco-friendly nature. Hence, much focus is emphasized on the synthesis of NPs through natural methods instead of using chemical methods because natural methods are efficient, cost-effective, and environment-friendly. Biologically, NPs are synthesized by in vitro reduction of metal ions (Table 12.2) in the presence of microorganisms, whole-plant extract, and leaf or fruit extract (Armendariz et al., 2004; Ankamwar, 2010; Ali et al., 2011, Banerjee, 2011). The process of natural biological reduction of metal NPs is quite rapid as it can be easily carried out at room temperature and under moderate pressure. Besides this, it is so effective and fast that it can be readily scaled up. The most commonly found reducing agents in plant extracts are alkaloids, terpenoids, and phenolic compounds as represented in Figure 12.2, which in turn play a crucial role in the synthesis of NPs.

There have been several reports that suggest that the NPs synthesized from these natural methods are efficient in the treatment of cancer, diabetes, and wounds, show anti-inflammatory activity, and are used in the making of antibacterial and antiphytofungal agents. In pharmaceutical industries, the biologically synthesized NPs are used in the production of drugs (Wagner et al., 2006), which in turn confers many advantages; for example, it increases the half-life of the drug, and also it enhances the targeted drug delivery in an efficient way (Sahoo et al., 2007).

TABLE 12.2

Metal NPs Synthesized from Various Range of Plant Extracts with Their Applications

Sl. No.	Plant Species Used	Metal Used	Use of the Synthesized Nanoparticle	Average Size of Synthesized Nanoparticle (nm)	Reference
1.	*Ruellia tuberosa* leaf	Iron oxide	Antimicrobial properties	52.78	Vasantharaj et al. (2018)
2.	*Cassia absus*	Silver	Antibacterial activity	18–25	Jehan et al. (2018)
3.	*Albizia lebbeck* stem bark	Zinc oxide	Antimicrobial and antioxidant action	66.25	Umar et al. (2018)
4.	*Tridax procumbens* leaf	Copper oxide	Mosquito larvicidal activity against dengue, Zika, and chikungunya	16	Muthamil et al. (2018)
5.	*Zingiber officinale* and *Thymus vulgaris*	Silver	Antifungal activity against *C. albicans*	12 and 18	Mohammadi et al. (2019)
6.	*Kalanchoe daigremontiana* leaf	Silver	Antibacterial activities	25	Molina et al. (2019)
7.	*Cucurbita pepo* leaf	Zinc oxide	Treatment of femoral fracture	8	Hu et al. (2019)
8.	*Carpesium cernuum*	Silver	Antioxidant activity	13.0±0.2	Ahn et al. (2019)
9.	Rangoon creeper leaves	Silver	Antibacterial action against *S. aureus* and *E. coli*	12	Birusanti et al. (2019)
10.	*Cinnamomum camphora* fruit	Silver	Inhibition of *F. oxysporum*	20.3	Huang et al. (2019)
11.	*Artemisia scoparia*	Silver	Enhancing shelf life of fresh fruits	12.0–23.3	Hanif et al. (2019)
12.	*Panax notoginseng* leaf	Gold	Anticancer activity	100	Wang et al. (2019)
13.	*Coptis chinensis*	Silver	Anticancer activity	6–45	Pei et al. (2019)
14.	*Albizia procera* leaf	Silver	Antibacterial activities against *E. coli* and *S. aureus*	6.18	Rafique et al. (2019)
15.	*Manilkara zapota* leaf	Silver	Induce apoptosis in human colorectal carcinoma cells	24	Shaniba et al. (2019)
16.	*Euphorbia helioscopia* leaf	Iron and copper oxide	Antifungal activities	7–10	Henam et al. (2019)
17.	*Matricaria chamomilla L, Olea europaea, Lycopersicon esculentum M.*	Zinc oxide	Antibacterial activity against *Xanthomonas oryzae pv. oryzae*	40.5–124.0	Ogunyemi et al. (2019)
18.	*Ricinus communis*	Zinc oxide	Antioxidant, antifungal, and anticancer activity	20	Shobha et al. (2019)
19.	*Fumariae herba*	Platinum	Catalytic properties	30	Dobrucka (2019)
20.	*Rhazya stricta* root	Silver	Antibacterial activity	20	Shehzad et al. (2018)

After the synthesis of NPs, the next important step is their characterization, which is mainly carried out based on their size and shape as in many cases uniform-sized NPs are needed (Jiang et al., 2009). The morphology of the synthesized NPs is determined by several techniques, *viz.* X-ray diffraction (XRD), UV–visible spectroscopy, field emission scanning electron microscopy (FE-SEM), high-resolution transmission electron microscopy (HR-TEM), atomic force microscopy (AFM), and Fourier

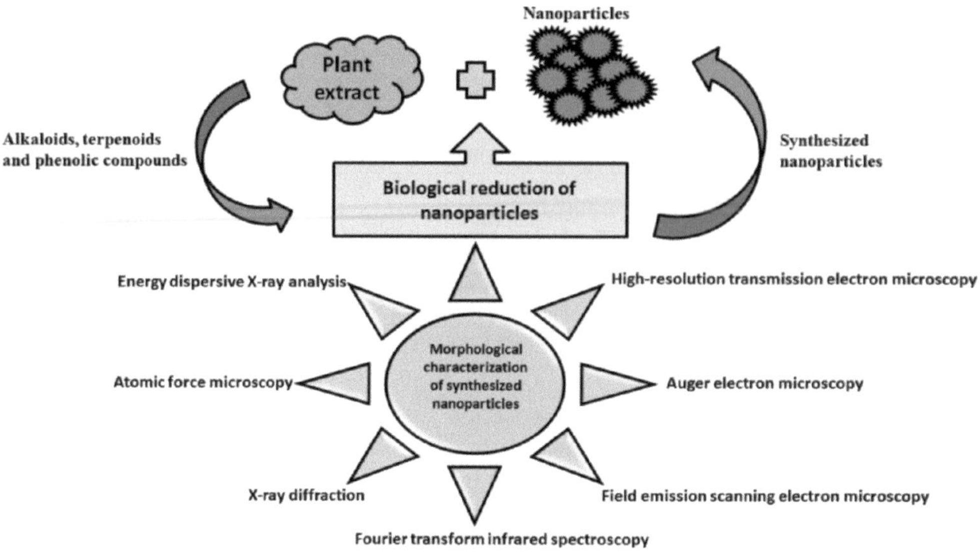

FIGURE 12.2 Pictorial representation of the process of NPs synthesis from plant extracts.

transform infrared spectroscopy (FTIR) (Sepeur, 2008). In the current section, we will discuss in detail the synthesis of metal NPs from various natural sources.

12.5 Synthesis of Silver NPs from Plant Extract

Silver NPs (AgNPs) are used for several purposes, among which one is the treatment of human hepatocellular carcinoma (HepG2), but the chemical synthesis of AgNPs is toxic and it affects the environment. So, nowadays, plant extracts are used for the synthesis of AgNPs. Recently, anthocyanin extract from callus (AE-AgNPs) and purple basil (BC-AgNPs) have been used for this purpose, and it was reported that the AE-AgNPs show considerable cytotoxicity toward human hepatocellular carcinoma (HepG2) cell line in comparison with BC-AgNPs. These NPs are spherical with BC-AgNPs having an average size of 50.97 ± 0.10 nm and AE-AgNPs 42.73 ± 1.24 nm measured with XRD and SEM (Abbasi et al., 2019). Aritonang et al. (2019) developed the method for synthesizing AgNPs using the plant extracts of *Impatiens balsamina* and *Lantana camara* leaves. The synthesis of AgNPs was confirmed through ultraviolet–visible spectrophotometry and TEM., It is visually shown that when $AgNO_3$ is treated with plant extracts, it turns into greyish-brown color in the case of *L. camara* extract and brownish-yellow upon treatment with *I. balsamina* extract. TEM analysis revealed that the average size of the AgNPs synthesized was less than 24 nm. These AgNPs show considerable antimicrobial activity against *Staphylococcus aureus* and *Escherichia coli*, which was comparable to the ciprofloxacin, a well-known antibiotic. Hue et al. (2019) synthesized AgNPs from the extract of *Bauhinia acuminate* plant flower, which helps in both reduction and stabilization of AgNPs. The formation of NPs was confirmed with FTIR and XRD. Their average size was reported to be around 17 nm. They further reported that these AgNPs can efficiently induce the proliferation of mesenchymal stem cells (MSCs); hence, they can be used for curing bone fracture. Shelar et al. (2019) synthesized AgNPs from the fruit peel extract of *Momordica charantia* and confirmed it through FTIR and XRD. The synthesized AgNPs caused the complete death of worms at 2 mg/ml concentration, whereas at 20 ppm concentration, it showed 85% mortality in larvae of mosquito *Aedes albopictus*. Hence, the report suggests that the AgNPs synthesized from *M. charantia* show considerable helminthicidal and larvicidal activities. Another group synthesized AgNPs from the extracts of *Galega officinalis*, and the synthesized AgNPs were characterized by using TEM and XRD. Results revealed that the NPs showed high antimicrobial activity against *E. coli*, *S. aureus*, and *P. syringae* (Manosalva et al., 2019).

AgNPs synthesized from the fruit extract of *Prosopis farcta* were reported to be bearing considerable antimicrobial activities against *Streptococcus pneumonia*, *S. aureus*, and *E. coli*. The average size of the synthesized AgNPs was found to be around 12.68 nm when analyzed with XRD and TEM. Hence, this report suggests that the AgNPs synthesized from the *Prosopis farcta* fruit extract can be used as a potent antimicrobial agent (Salari et al., 2019). AgNPs synthesized from the aqueous extract of Piper betle showed considerable antiphytofungal activity (Khan et al. 2019). The synthesized AgNPs were analyzed by XRD, FTIR, and FE-SEM and were reported to be bearing an average size of 6–14 nm. These AgNPs showed a considerable antiphytofungal activity against *F. solani* and *Alternaria brassicae*. These AgNPs act on the fungal mycelium cell wall, thereby causing cellular permeability. In a nutshell, it can be assumed that the biologically synthesized AgNPs are highly efficient and can hence be used to replace the nanoparticles synthesized using the much expensive and toxic chemical and physical methods.

12.6 Synthesis of Gold NPs from Plant Extract

Besides AgNPs, many researchers are using gold NPs (AuNPs) also for the treatment of cancer. One group has developed AuNPs from *Magnolia officinalis* and has determined its size through dynamic light scattering and reported it to be around 128 nm. The shape of the synthesized AuNPs was determined by AFM, EDX, and HR-TEM. Reports suggest that the AuNPs are capable of inducing the expression of the apoptotic genes in the A549 cells, thereby carrying out the cell cytotoxicity and apoptosis in a significant way (Zheng et al., 2019). Similarly, Li et al. (2019) used *Marsdenia tenacissima* (MT) for the synthesis of AuNPs. The synthesized AuNPs size was characterized by HR-TEM. Further, these AuNPs were tested for their potential ability toward inducing cytotoxicity in HepG2 cells. Reports suggest that the AuNPs significantly induce the formation of reactive oxygen species along with a change in the mitochondrial membrane potential. They reported a significant induction of Bax, caspase-9, and caspase-3 genes; besides this, some anti-apoptotic genes expression was reported to be repressed in the treated samples. Hence, the results indicate that MT-AuNPs have significant potential for carrying out anticancer activities. Moustafa et al. (2019) synthesized monodispersed and polydispersed AuNPs from the leaf and seed extracts of *Peganum harmala L.* They found that the average size of the monodispersed and polydispersed AuNPs was 43.44 and 52.04 nm, respectively, as observed with XRD, FE-SEM, EDX, and TEM. Further, it was found that the isotropic AuNPs bear a significant antibacterial effect against *E. coli* and *S. aureus*.

The AuNPs synthesized from the extract of *Ephedra sinica Stapf* were reported to possess anti-neuroinflammatory activities. The size of the synthesized NPs was around 57.6 ± 3.07 nm as determined by the dynamic light scattering technique. They also confirmed the crystalline structure of the AuNPs with XRD and FTIR. These NPs were reported to be showing anti-neuroinflammatory activities in microglia cells, which was confirmed through ELISA and flow cytometry. These NPs have shown a significant reduction in the lipopolysaccharide-mediated induction of mediators such as reactive oxygen species, prostaglandin E2, and nitric oxide. Altogether, these findings suggest the potential use of the AuNPs synthesized from *E. sinica Stapf* in curing neurodegenerative diseases (Park et al., 2019).

12.7 Synthesis of Copper NPs from Plant Extract

Zangeneh et al. (2019) synthesized CuNPs from the leaf extract of *Falcaria vulgaris*. The synthesis of NPs was confirmed through XRD, TEM, FTIR, and FE-SEM. These NPs were reported to be efficient in cutaneous wound healing in rats. Treatment of the rats with CuNPs significantly induced the wound contracture and fibrocytes/fibroblast rate. Besides this, the synthesized CuNPs also exhibit antifungal properties as they effectively inhibit the fungal growth at a concentration of 2–4 mg/ml and removed them completely at 4–8 mg/ml. Similarly, the synthesized CuNPs also showed some antibacterial properties by inhibiting the bacterial growth at 2–8 mg/ml and removing it completely at a concentration of 4–16 mg/ml. Hence, the reports suggest that the CuNPs synthesized from the leaf extract of *F. vulgaris* show potential cutaneous wound healing, antibacterial, and antifungal properties. Rajeshkumar et al. (2019) used the medicinal plant *Cissus arnotiana* for the synthesis of CuNPs. The morphology of the CuNPs was characterized by

UV–visible spectroscopy and XRD. The synthesized CuNPs were reported to be spherical with an average size ranging between 60 and 90 nm as analyzed by TEM. The CuNPs showed significant antibacterial activities against gram-negative bacteria with an inhibition zone of 22.20 ± 0.16 mm at 75 μg/ml.

12.8 Synthesis of Iron and Zinc Oxide NPs from Plant Extract

Karade et al. (2019) used Gardenia, a traditional medicinal plant used for the synthesis of superparamagnetic α-Fe_2O_3 NPs. The phenolic compounds present in plant extract carried out the reduction of the NPs. The synthesized NPs were spherical with an average size of around 5 nm. Further, their measurement of magnetic ability has revealed its superparamagnetic character with a non-saturating MS value of 8.5 emu/g at room temperature. They have also tested the cytotoxic impact on human mesenchymal cells and reported it to have a promising role in inducing cytotoxic activities, which can be used in the pharmaceuticals field in near future. Vinotha et al. (2019) synthesized ZnO NPs from the leaf extract of *Costus igneus*. The presence of Ci-ZnO NPs was confirmed with GC–MS and proton nuclear magnetic resonance (1H NMR) spectroscopy. They found that Ci-ZnO NPs showed significant antidiabetic activity. Besides this, Ci-ZnO NPs also exhibited considerable antioxidant and antibacterial activities against some pathogenic bacteria such as *Vibrio parahaemolyticus*, *Proteus vulgaris*, and *Streptococcus mutans*. Chandra et al. (2019) synthesized ZnO NPs from the leaf extract of *Berberis aristata*. They reported that the synthesized ZnO NPs exhibit considerable antibacterial activities against *Bacillus subtilis*, *Klebsiella pneumoniae*, *E. coli*, and *B. cereus*; besides this, they also bear some antioxidant properties. Bayrami et al. (2019) synthesized ZnO NPs from the leaf extract of whortleberry. The synthesized ZnO NPs were injected into the intraperitoneal region of alloxan-induced diabetic rats for analyzing the level of insulin, HDL, triglyceride, and blood glucose level. Reports suggest that the injected rats have shown significant improvement in health status upon treatment in comparison with untreated rats. Besides this, the ZnO NPs were also effective in inhibiting gram-negative and gram-positive bacteria. Hence, the above works demonstrate that the ZnO and iron oxide NPs synthesized from the leaf extract of different plants can be used as an antidiabetic, antibacterial, or cytotoxic agent.

12.9 Application of Nanotechnology in Plant Pathology

In this current era, plant pathology utilizes several forms of NPs, such as metallic oxides, metalloids, non-metals, carbon nanomaterials, dendrimers, liposomes, and quantum dots. These diverse NPs have got tremendous potential, and they function differently. Nano-barcodes serve as an important tool in detecting pathogen DNA and distinguish each strain precisely and determine its stage of application. NPs have also been utilized as biomarkers for the identification of distinctive compounds produced in different stages of the diseased plant as compared to the healthy ones (Chartuprayoon et al., 2010). Thus, nano-based diagnostic kits and sensors increase speed, power, and limit of detection (Chinnamuthu and Boopathi, 2009; Yao et al., 2009). NP-based sensors might prove to be promising for the better detection of viral pathogens in the plant (Baac et al., 2006; Mishra et al., 2015a). The enzyme-based biosensors treated with NPs of Ag, Au, Co, Ti, etc., not only facilitate accurate and fast diagnosis of plant infection, but also aid in the detection of pesticide residues (Khan and Rizvi, 2014).

12.10 Silver NPs

Silver NPs hold a pioneer position in the list of NPs explored for pathogen control on account of their historical antimicrobial behavior (Richards, 1991), and a lot of research has been done on their effectiveness in managing several phytopathogens (Prasad 2014; Jogaiah et al., 2019). Not only in plant pathology, they play vivid roles in other fields such as insect management, food preservation, biomedicine, and plant growth promotion (Table 12.4). Paret et al. (2006) reported that the combined application of NPs of Ag and Si with a hydrophilic polymer to cucumber foliage @0.3 µg/ml considerably reduces powdery mildew infection. Kim et al. (2008) tested the efficacy of NPs of colloidal Ag against *Sphaerotheca pannosa*, the pathogen causing powdery mildew of rose. Applications @10 µg/ml on plants infected with powdery mildew lead to the slow disappearance of pathogen colonies on leaves, which did not reappear until seven days. Lamsal et al. (2011) did further exploration of the concept and applied NPs at varying rates, i.e., 10, 30, 50, and 100 µg/ml, to leaves of pumpkin and cucumber 3–4 weeks pre- and post-pathogen infection. At both times of application, plants treated @100 µg/ml had only 20% disease incidence compared to 80% in untreated plants. Jagana et al. (2017) used nano-Ag in varying concentrations to treat banana fruits for the management of anthracnose disease. The disease severity was found to be minimum (6.7%) when nano-Ag was applied @2,000 µg/ml, while the untreated fruits resulted in 75.6% severity. In another experiment, Moussa et al. (2013) tested the effectiveness of AgNPs against pathogens attacking cereals. They synthesized AgNPs in the supernatant culture of *Serratia* spp. and reported that even lower concentrations, i.e., 2 µg/ml, effectively prevent conidial germination of *Bipolaris sorokiniana,* the wheat spot blotch pathogen. Similarly, Mishra et al. (2014a) reported the ability of AgNPs in managing *B. sorokiniana* on wheat. The altered expression of genes involved in melanin biosynthesis on treatment with AgNPs is the main reason behind their antifungal activity.

Apart from foliar pathogens, nano-Ag is also effective against soilborne pathogens and nematodes. The soilborne diseases reported to be suppressed are those caused by *Fusarium* spp., *Phytophthora parasitica*, and *Meloidogyne* spp. Even low doses, i.e., ≤100 µg/ml, of nano-Ag prove to be extremely toxic. The formulations of nano-Ag were also found effective in controlling pathogens such as *Sclerotium cepivorum* and *Colletotrichum gloeosporioides* (Aguilar-Mendez et al., 2011; Jung et al., 2010). Moreover, it is also effective in managing plant-parasitic nematodes, viz. *Meloidogyne* spp. (Abdellatif et al., 2016; Ardakani, 2013). It was found that the application of the same on second-stage juveniles of root-knot nematode inactivated them entirely within 6 h (Cromwell et al., 2014). Also, in combination with conventional nematicides, the LC50 dose to suppress nematode gets reduced significantly (Nassar, 2016).

The potential of nano-Ag in combination with a biocontrol agent has been found promising. Mallaiah (2015) reported that a consortium of Ag with biocontrol agents such as *Bacillus subtilis*, *Pseudomonas fluorescens*, and *Trichoderma viride* leads to an increase in Fusarium wilt suppression on *Crossandra* plants. A single application of Ag particles yielded 5% more flowers compared to control, while the consortium led to enhancements of 12%, 14%, and 15% with *B. subtilis*, *P. fluorescens*, and *T. viride*, respectively. Studies on combining nano-Ag with fungicides have also been done by researchers. The fungicidal activity of fluconazole enhanced by 0.35–0.37 times against *Fusarium oxysporum, Alternaria alternata*, and *Cladosporium herbarum* (Bholay et al., 2013). Similarly, Gajbhiye et al. (2009) biosynthesized nano-Ag from *A. alternata* and reported that in combination with fluconazole, the antifungal activity was enhanced against *Phoma glomerata*. Fungi proved to be a rich source for the biosynthesis of most of the AgNPs. Moussa et al. (2013) documented that more than 48 fungal species can produce nano-Ag, even including plant-pathogenic genera, viz. *Aspergillus, Penicillium, Verticillium, and Fusarium* (Table 12.3).

TABLE 12.3

Multifarious Role of Biosynthesized Silver NPs

Sl. No.	NPs	Applications	Reference
1.	Silver (Ag) NPs	Antifungal action against *Bipolaris sorokiniana*	Mousa et al. (2013), Mishra et al. (2014a; 2015b)
2.	AgNPs	Antifungal action against *Fusarium oxysporum*	Gopinath and Velusamy (2013)
3.	AgNPs	Antifungal and antibacterial effect against a range of phytopathogens	Jaidev and Narasimha (2010), Mala et al. (2012), Lee et al. (2013)
4.	AgNPs	Biomedical application	Priyadarshini et al. (2013)
5.	AgNPs	Antimicrobial and anticoagulation activity	Jeyaraj et al. (2013)
6.	AgNPs	Fruit and vegetable preservation	Fayaz et al. (2009)

12.11 Copper NPs

Copper is known to play a key role in the biological activities of all organisms and is considered an essential micronutrient. Cu ions may be toxic to microorganisms, humans, and the environment, so this peculiar property has attracted researchers to use CuNPs in a controlled manner for agricultural benefits. The first experiment on bactericidal/fungicidal ability of nano-Cu in the field was done by Giannousi et al. (2013). Elmer et al. (2016, 2018) applied CuNPs as a nanofertilizer/supplement to augment disease resistance against the wilt pathogens (Evans et al., 2007). CuNPs were found to be very efficient in suppressing the wilt of tomato and brinjal. Elmer et al. (2018) did a foliar application of CuO NPs on watermelon seedlings and transplanted them in pot soil infected with *F. oxysporum* f. sp. *niveum*. Significant results of disease suppression were obtained. Similarly, Ponmurugan et al. (2016) used CuNPs biosynthesized from *Streptomyces griseus* cultures to suppress *Poria hypolateritia*, the tea red root fungus. The nano-form of copper was found to exhibit fourfold higher activity against bacterial blight on pomegranate at 10^{-4} times the concentration of recommended copper dose (Mondal and Mani, 2012). The CuO NPs exhibiting antibacterial activity against *Xanthomonas* spp. were reported by several researchers (Ocsoy et al., 2013; Strayer-Scherer et al., 2018); however, their antifungal activity against fungi such as *Curvularia lunata*, *A. alternata*, *F. oxysporum*, and *Botrytis cinerea* has also been documented (Kaushik and Dutta, 2017; van Piet et al., 2016; Bramhanwade et al., 2016).

12.12 Zinc NPs

ZnNPs exhibit antimicrobial behavior against several phytopathogens (Graham et al., 2016; Hafeez et al., 2014; Rajiv et al., 2013). *In vitro* analyses showed that nano-Zn is antagonistic to fungi, viz. *F. oxysporum*, *A. alternata*, *B. cinerea*, *Mucor plumbeus*, *Penicillium expansum*, *Rhizoctonia solani*, *Rhizopus stolonifer*, and *Sclerotinia sclerotiorum* (He et al., 2011; Sardella et al., 2017). It is also effective against bacteria (Graham et al., 2016; Indhumathy and Mala, 2013) and *M. incognita* (Kaushik and Dutta, 2017). Moreover, zinc nanoparticles also act as antiviral agents for the management of cucumber mosaic virus (CMV) in brinjal (El-Sawy et al., 2017).

12.13 Silicon NPs

SiNPs have been used for a long time in optical materials, biosensors, electrochemistry, and biocatalysts. However, the initial report of nano-Si in disease management was the use of an Ag-Si complex by Park et al. (2006) who constructed a complex that suppresses cucumber powdery mildew @0.3 μg/ml. However, the study did not reveal the original element showcasing the suppression ability. Elmer et al. (2018) did a foliar application of nano-SiO to evaluate its potential in managing Fusarium wilt of watermelon and found no significant difference in disease control and plant growth promotion between treated and untreated plants.

12.14 Magnesium NPs

The NPs of MgO were also examined for their ability to control disease after the discovery of their antimicrobial activity (Huang et al., 2005). Imada et al. (2016) demonstrated that nano-MgO enhances systemic disease resistance against *Ralstonia solanacearum* on tomatoes when applied at preinfection stage. In another experiment, Wani and Shag (2012) reported that nano-MgO inhibits conidial germination of *F. oxysporum*, *R. stolonifer*, *M. plumbeus*, and *A. alternata* more efficiently than nano-ZnO when applied @50 μg/ml.

12.15 Sulfur NPs

Choudhury et al. (2010) conducted *in vitro* experiments and reported that nano-S can control *F. oxysporum*, *F. solani*, *Aspergillus niger*, and *Venturia inaequalis*. However, Rao and Paria (2013) also stated the ability of nano-S in controlling *F. solani* and *V. inaequalis*.

12.16 Gold NPs

A lot of data are available on the biosynthesis of nano-Au, but their ability to control disease has been documented very less. However, Jayaseelan et al. (2013) generated nano-Au in extracts of *Abelmoschus esculentus* and reported their antifungal ability to *A. niger*, *A. flavus*, *Puccinia graminis*, and *Candida albicans*.

12.17 Carbon NPs

Carbon nanomaterials are quite miscellaneous in their shape and function. Various types have been synthesized so far, but among them, three groups are very popular, i.e., carbon nanotubes (single-walled or multi-walled), fullerenes (C60; buckyballs), and graphene oxides (both oxidized and reduced forms). The use of carbon nanomaterials has widely been exhibited for decades in engineering, textiles, paints, medicine, electronics, and other sectors. However, their disease suppression activity was realized much later against fungi (Sarlak et al., 2014; Wang et al., 2014) and bacteria (Liu et al., 2009; Wang et al., 2013). Wang et al. (2014) reported that single-walled carbon nanotubes inhibit conidial germination of *F. graminearum* and *F. poae* followed by multi-walled carbon nanotubes and graphene oxide. Ocsoy et al. (2013) incorporated a dsDNA-directed nano-Ag onto graphene oxide to suppress the tomato bacterial wilt caused by *X. perforans* in the greenhouse and found significant results. Berry et al. (2014) used carboxylated forms of single-walled carbon nanotubes in enzyme degradation of *Trametes versicolor* and *Phlebia tremellosa* causing white rot. Moreover, the plant growth activities of carbon nanotubes were reported by several researchers (Khodakovskaya et al., 2009; Tripathi et al., 2011; Wang et al., 2014).

12.18 Titanium Dioxide (TiO_2) NPs

The antimicrobial property of TiO_2 NPs is attributed to its crystal shape, size, and structure. These particles have been receiving incredible attention in recent years due to their potent ability as antimicrobial agents (Lyu et al., 2017). The photocatalytic activities of nano-TiO_2 add to its antifungal property (Boxi et al., 2016; De Filpo et al., 2013). The antifungal activity of TiO_2 NPs has been reported in several cases (Haghighi et al., 2013; Morteza et al., 2013), but owing to their toxicity (Shah et al., 2017), TiO_2 nanoparticles can be easily combined with eco-friendly polymers for efficient delivery (Allahverdiyev et al., 2011). Paret et al. (2013) designed a light-activated nano-TiO_2/Zn complex that can suppress rose bacterial leaf spot caused by *Xanthomonas* sp. Since the complex also contained nano-ZnO, the individual effect of particles could not be assessed.

12.19 Molybdenum NPs

The growth of chickpea seedlings turned out to be two- to threefold higher upon treatment with MoNPs with nitrogen-fixing bacteria in comparison with water (Taran et al., 2014); however, its applicability in managing pathogens is still under trial.

12.20 Liposomes

Liposome molecules contain a spherical vesicle having at least a phospholipid bilayer. They are of various sizes and have a vacant interior that can be incorporated with antimicrobial substances. Matouskova et al. (2016) reported that liposome-encapsulated antimicrobial components, viz. chitosan and herbal extracts, significantly inhibit gram-negative bacteria. The liposomes are stable in water for longer periods, which facilitates their usage in irrigation treatments. Perez-de-Luque et al. (2012) reported that amphotericin bound to liposomes forms nanodisks, which are effective at a concentration of 10 μg/ml in irrigation water and suppress the incidence of chickpea wilt.

12.21 Dendrimers

Dendrimers are branched, tree-like NP structure comprising a central core and contain different functional groups. They play multiple roles and are mostly used as delivery agents. They are more popular in medical sciences, but have few applications in plant pathology. Dendrimers can facilitate basipetal transportation of chemicals to the roots and can be used for rescue treatments (Khairnar et al., 2010) (Table 12.4).

12.22 Conclusions

Nanotechnology has emerged as a potential tool in crop improvement through higher nutrient use efficiency and enhanced disease diagnostics measures and management practices. It also facilitates the development of sensitive molecular tools for the manipulation of plants and pathogens. The use of NPs enables the controlled release of agrochemicals to their target and thus minimizes the delivery losses. The application of nanoparticles to stabilize biocontrol operations has been anticipated, which will further help in reducing the environmental hazards. The nanosensors can detect pathogens at a very low range, i.e., parts per billion or even more. However, nanotechnology has begun to make genuine advances in plant disease identification and management as most of its full utility has just been explored in recent years. The new tools such as biosensors and quantum dots have shown a promising effect, and thus, there is the likelihood of them replacing conventional assays such as ELISA and other preliminary

TABLE 12.4

Potential Use of Various Types of NPs in Plant Pathology

Sl. No.	Nanoparticle Type	Potential Use	Reference
1.	Metalloids, metallic oxides, non-metals, and their composites	Nanopesticides and nanofertilizers, delivery agent for transportation of NPs	Park et al. (2006), Lamsal et al. (2011)
2.	Quantum dots	Disease detection and diagnostics tool	Rispail et al. (2014)
3.	Dendrimers	Delivery agent for transportation of NPs	Khairnar et al. (2010)
4.	Fullerenes (buckyballs)	Antimicrobial agents, delivery agent for transportation of NPs	Hao et al. (2017)
5.	Single-walled/ multi-walled nanotubes	Antimicrobial agents, delivery agent for transportation of NPs	Wang et al. (2014)
6.	Graphene oxide sheets (oxidized or reduced types)	Delivery agent for transportation of NPs	Chen et al. (2013), Ocsoy et al. (2013)
7.	Nanobiosensors	Disease detection and diagnostics tool	James, (2013), Lin et al., (2014)
8.	Liposomes	Delivery vehicle for genetic or antimicrobial products	Matouskova et al. (2016), Perez de-Luque et al. (2012)

detection tools. To enhance crop production with maintaining ecological balance, agricultural science has to draw benefits from nanotechnology. However, the main limitation in integrating nanotechnology into phytopathology is that presently, it is being sought after at a limited scale and by a few laboratories only. So, in near future, more research should be carried out to discover, adapt, and apply nanotechnology on a vast scale so that the global concern of food security could be minimized to a certain extent.

REFERENCES

Abbasi et al. 2019. A comparative evaluation of the antiproliferative activity against HepG2 liver carcinoma cells of plant-derived silver nps from basil extracts with contrasting anthocyanin contents. *Biomolecules* 9, 320.

Abdellatif et al. 2016. Green NPs engineering on root-knot nematode infecting eggplants and their effect on plant DNA modification. *Iran. J. Biotechnol.* 14(4), 250– 59.

Adholeya et al. 2017. Zero draft policy on regulation of nanoproducts in agriculture. The Energy and Resource Institute, TERI. www.teriin.org.

Aguilar-Mendez et al. 2011. Synthesis and characterization of silver NPs: effect on phytopathogen *Colletotrichum gloesporioides. J. Nanopart Res.* 13, 2525–2532.

Ahmad et al. 2020. Green synthesis and characterization of zinc oxide nanoparticles using *Eucalyptus globules* and their fungicidal ability against pathogenic fungi of apple orchards. *Biomolecules* 10, 425. Available from: https://doi.org/10.3390/biom10030425.

Ahn et al. 2019. Green synthesis and biological activities of silver NPs prepared by *Carpesium cernuum* extract. *Arch. Pharm. Res.* 10, 1007.

Ali et al. 2011. Plant extract mediated synthesis of silver and gold NPs and its antibacterial activity against clinically isolated pathogens. *Colloids Surf B Biointerfaces* 85, 360–365.

Allahverdiyev et al. 2011. Antimicrobial effects of TiO_2 and Ag_2O nanoparticles against drug-resistant bacteria and leishmania parasites. *Future Microbiol.* 6(8), 933–940.

Ankamwar, B., 2010. Biosynthesis of gold NPs (green-gold) using leaf extract of *Terminalia catappa. Eur. J. Chem.* 7, 1334–1339.

Antisari et al. 2011. Response of soil microbial biomass to CeO_2 nanoparticles. *EQA-Int. J. Environ. Qual.* 7, 1–16.

Ardakani, A.S., 2013. Toxicity of silver, titanium and silicon NPs on the root-knot nematode, *Meloidogyne incognita*, and growth parameters of tomato. *Nematol* 15(6), 671–677.

Aritonang et al. 2019. Synthesis of silver NPs using aqueous extract of medicinal plants' (*Impatiens balsamina* and *Lantana camara*) fresh leaves and analysis of antimicrobial activity. *Int. J. Microbiol.* 10, 1155.

Armendariz et al. 2004. Size controlled gold nanoparticle formation by *Avena sativa* biomass: use of plants in nanobiotechnology. *J. Nanopart. Res.* 6, 377–382.

Baac et al. 2006. Antibody-based surface plasmon resonance detection of intact viral pathogen. *Biotechnol. Bioeng.* 94(4), 815–819.

Banerjee, J. and Narendhirakannan, R. 2011. Biosynthesis of silver NPs from *Syzygium cumini* (L.) seed extract and evaluation of their in vitro antioxidant activities. *Dig. J. Nanomater. Biostruct.* 6, 961–968.

Banik, S. and Sharma, P. 2011. Plant pathology in the era of nanotechnology. *Indian. Phytopathol.* 64, 120–127.

Barik et al. 2008. Nanosilica—from medicine to pest control. *Parasitol. Res.* 103, 253–258.

Bayrami et al. 2019. A facile ultrasonic-aided biosynthesis of ZnO NPs using *Vaccinium arctostaphylos* L. leaf extract and its antidiabetic, antibacterial, and oxidative activity evaluation. *Ultrason. Sonochem.* 55, 57–66.

Bergeson, L.L. 2010. Nanosilver pesticide products: What does the future hold? Environ. *Qual. Manage.* 19, 73–82.

Berry et al. 2014. Oxidative enzymatic response of white-rot fungi to singlewalled carbon nanotubes. *Environ. Pollut.* 193, 197–204.

Bholay et al. 2013. Fungicides potential of biosynthesized silver NPs against phytopathogens and potentiation of Fluconazol. *World Res. J. Pharm. Res.* 1, 12–15.

Birusanti et al. 2019. Sustainable green synthesis of silver NPs by using Rangoon creeper leaves extract and their spectral analysis and anti-bacterial studies. *IET Nanobiotechnol.* 13, 71–76.

Bouwmeester et al. 2009. Review of health safety aspects of nanotechnologies in food production. *Regul. Toxicol. Pharmacol.* 53(1), 52–62.

Boxi et al. 2016. Ag doped hollow TiO_2 nanoparticles as an effective green fungicide against *Fusarium solani* and *Venturia inaequalis* phytopathogens. *Nanotechnology* 27, 085103.
Bramhanwade et al. 2016. Fungicidal activity of Cu NPs against *Fusarium* causing crop diseases. *Environ. Chem. Lett.* 14(2), 229–235.
Chandra et al. 2019. Phyto-mediated synthesis of zinc oxide NPs of *Berberis aristata*: Characterization, anti-oxidant activity and antibacterial activity with special reference to urinary tract pathogens. *Matter Sci. Eng. C Mater. Biol. Appl.* 102, 212–220.
Chartuprayoon et al. 2010. Detection of Plant Pathogen using LPNE Grown Single Conducting Polymer Nanoribbon. Abstract #2278, 218th ECS Meeting.
Chen et al. 2013. A new function of graphene oxide emerges: inactivating phytopathogenic bacterium, *Xanthomonas oryzae* pv. *oryzae*. *J. Nanopart. Res.* 15(5), 1658.
Chinnamuthu, C.R. and Boopathi, P.M. 2009. Nanotechnology and agroecosystem. *Madras Agric. J.* 96(1–6), 17–31.
Choudhury et al. 2010. Nanosulfur: a potent fungicide against food pathogen, *Aspergillus niger. AIP Conf. Proc.* 1276, 154–157.
Corradini et al. 2010. A preliminary study of the incorporation of NPK fertilizer into chitosan NPs. *Exp. Polymer Lett.* 4(8), 509–515.
Cromwell et al. 2014. Nematicidal effects of silver NPs on root-knot nematode in bermudagrass. *J. Nematol.* 46(3), 261–266.
De Filpo et al. 2013. Preventing fungal growth in wood by TiO_2 nanoparticles. *Int. Biodeterior. Biodegrad.* 85, 217–222.
de Oliveira et al. 2014. Application of nanotechnology for the encapsulation of botanical insecticides for sustainable agriculture: Prospects and promises. *Biotechnol. Adv.* 32, 1550–1561.
de Rosa et al. 2010. Nanotechnology in fertilizers. *Nat. Nanotechnol.* 5, 91.
Dhillon, N.K. and Mukhopadhyay, S.S. 2015. Nanotechnology and allelopathy: synergism in action. *J. Crop. Weed.* 11(2), 187–191.
Dobrucka, R. 2019. Biofabrication of platinum NPs using *Fumariae herba* extract and their catalytic properties. *Saudi J. Biol. Sci.* 26, 31–37.
EC, Communication from the Commission to the European Parliament, the Council, the European Economic and Social Committee and the Committee of the Regions. A European strategy for Key Enabling Technologies - A bridge to growth and jobs.
Elmer, W.H. and White, J. 2016. NPs of CuO improves growth of eggplant and tomato in disease infested soils. *Environ. Sci. Nano.* 3, 1072–1079.
Elmer et al. 2018. Effect of metalloid and metallic oxide NPs on Fusarium wilt of watermelon. *Plant Dis.* 11, 89–98.
El-Sawy et al. 2017. Antiviral activity of 2 nitromethyl phenol, zinc nanoparticles and seaweed extract against cucumber mosaic virus (CMV) in Eggplant. *J. Virol. Antivir. Res.* 6, 2.
Evans et al. 2007. Copper and plant disease. 27, 177–188.
Fayaz et al. 2009. Biogenic synthesis of silver NPs and their synergistic effect with antibiotics: a study against Gram-positive and Gram-negative bacteria. *Nanomed. NBM* 6, 103–109.
Feiner, L.F. 2006. Nanoelectronics: crossing boundaries and borders. *Nat. Nanotechnol.* 1, 91–92.
Gajbhiye et al. 2009. Fungus-mediated synthesis of silver NPs and their activity against pathogenic fungi in combination with fluconazole. *Nanomed. Nanotechnol. Bio. Med.* 5(4), 382–386.
Galbraith, D.W. 2007. Nanobiotechnology: silica breaks through in plants. *Nat. Nanotechnol.* 2(5), 272–273.
Ghormade et al. 2011. Perspectives for nano-biotechnology enabled protection and nutrition of plants. *Biotechnol. Adv.* 29, 792–803.
Giannousi et al. 2013. Synthesis, characterization and evaluation of copper based nanoparticles as agrochemicals against *Phytophthora infestans*. *RSC Adv.* 3(44), 21743–52.
Glenn, J.C. and Florescu, E. 2016. Millennium project team. 2015–16 state of the future. *J. Socialomics* 5(3), 1–6.
Gopinath, V. and Velusamy, P. 2013. Extracellular biosynthesis of silver NPs using *Bacillus* sp. GP-23 and evaluation of their antifungal activity towards *Fusarium oxysporum*. *Spectrochimica Acta A: Mol. Biomol. Spectr.* 106, 170–174.
Graham et al. 2016. Potential of nano-formulated zinc oxide for control of citrus canker on grapefruit trees. *Plant Dis.* 100(12), 2442–2447.

Hafez et al. 2014. Assessment of antibacterial activity for synthesized zinc oxide nanorods against plant pathogenic strains. *Int. J. Sci. Technol. Res.* 3(9), 318–324.

Haghighi et al. 2013. Antifungal activity of TiO_2 nanoparticles and EDTA on Candida albicans biofilms. *Infection, Epidemiol. Microbiol.* 1 (1), 33–38.

Hanif et al. 2019. Formulation of active packaging system using *Artemisia scoparia* for enhancing shelf life of fresh fruits. *Mater. Sci. Eng. C Mater. Biol. Appl.* 100, 82–93.

Hao et al. 2017. Potential applications and antifungal activities of engineered nanomaterials against gray mold disease agent *Botrytis cinerea* on rose petals. *Front. Plant Sci.* 8, 1332.

He et al. 2011. Antifungal activity of zinc oxide NPs against *Botrytis cinerea* and *Penicillium expansum. Microbiol. Res.* 166(3), 207–215.

Henam et al. 2019. Microwave synthesis of NPs and their antifungal activities. *Spectrochim Acta. A: Mol. Biomol. Spectrosc.* 213, 337–341.

Hu, L. and Chen, G., 2007. Analysis of optical absorption in silicon nanowire arrays for photovoltaic applications. *Nano. Lett.* 7, 3249–3252.

Hu et al. 2019. *Cucurbita pepo* leaf extract induced synthesis of zinc oxide NPs, characterization for the treatment of femoral fracture. *J Photochem. Photobiol. B* 195,12–16.

Huang et al. 2005. Controllable preparation of nano-MgO and investigation of its bactericidal properties. *J. Inorg. Biochem.* 99(5), 986–993.

Huang et al. 2019. Inhibition of *Fusarium oxysporum* by AgNPs biosynthesised using *Cinnamomum camphora* fruit extract. *IET Nanobiotechnol.* 13, 42–45.

Hue et al. 2019. Biosynthesis of Silver NPs using *Bauhinia acuminate* flower extract and their effect to promote osteogenesis of MSCs and improve meniscus injury healing. *J. Photochem. Photobiol. B* 197, 111536.

Imada et al. 2016. Magnesium oxide nanoparticles induce systemic resistance in tomato against bacterial wilt disease. *Plant Pathol.* 65(4), 551–560.

Indhumathy, M. and Mala, R. 2013. Photocatalytic activity of zinc sulphate nano material on phytopathogens. *Int. J. Agric. Environ. Biotechnol.* 6(4S), 737–743.

Jagana et al. 2017. Green NPs: a novel approach for the management of banana anthracnose caused by *Colletotrichum musae. Int. J. Curr. Microbiol. Appl. Sci.* 6(10), 1749–1756.

Jaidev, L.R. and Narasimha, G. 2010. Fungal mediated biosynthesis of silver NPs, characterization and antimicrobial activity. *Colloids Surf. B Biointerfaces* 81(2), 430–433.

James, C. 2013. Polypyrrole nanoribbon based chemiresistive immunosensors for viral plant pathogen detection. *Anal. Methods* 5, 3497–3502.

Jayaseelan et al. 2013. Green synthesis of gold nanoparticles using seed aqueous extract of *Abelmoschus esculentus* and its antifungal activity. *Ind. Crops Prod.* 45, 423–29.

Jehan et al. 2018. Investigative evaluation of *Cassia absus* for antibacterial capacity and biomimetic synthesis of silver NPs. *IET Nanobiotechnol.* 12, 1062–1066.

Jiang et al. 2009. Characterization of size, surface charge, and agglomeration state of nanoparticle dispersions for toxicological studies. *J. Nanopart. Res.* 11,77–89.

Jogaiah et al. 2019. *Ganoderma applanatum*-mediated green synthesis of silver nanoparticles: structural characterization and in vitro and in vivo biomedical and agrochemical properties. *Arab. J. Chem.* 12, 1108–1120.

Jogaiah et al. (eds.). 2020. *Advances in Nano-Fertilizers and Nano-Pesticides in Agriculture: A Smart Delivery System for Crop Improvement.* Woodhead Publishing.

Joseph, T. and Morrison, M. 2006. Nanotechnology in Agriculture and Food Institute of Nanotechnology. A Nanoforum report, retrieved from http://www. nanoforum.org/dateien/temp/ nanotechnology.

Joshi et al. 2019. Mycogenic selenium nanoparticles as potential new generation broad-spectrum antifungal molecules. *Biomolecules* 9 (9), 419.

Jung et al. 2010. The effect of nano-silver liquid against the white-rot of the green onion caused by *Sclerotium cepivorum. Mycobiology* 38(1), 39–45.

Kah, M. and Hofmann, T. 2014. Nanopesticide research: current trends and future priorities. *Environ. Int.* 63, 224–235.

Kah et al. 2012. Nano-pesticides: state of knowledge, environmental fate and exposure modelling Nano-pesticides: state of knowledge. *Crit. Rev. Environ. Sci. Technol.* 43:16, 1823–1867.

Kah et al. 2018. A critical evaluation of nanopesticides and nanofertilizers against their conventional analogues. *Nat. Nanotechnol.* 13, 677–684.

Karade et al. 2019. A green approach for the synthesis of α-Fe2O3 NPs from *Gardenia resinifera* plant and it's *In vitro* hyperthermia application. *Heliyon* 5, 2044.

Kaushik, H. and Dutta, P., 2017. Chemical synthesis of zinc oxide nanoparticle: its application for antimicrobial activity and plant health management. *Paper Presented at the 109th Annual Meeting of the American Phytopathological Society*, San Antonio, TX, August.

Khan et al. 2019. Optimization of process parameters for the synthesis of silver NPs from *Piper betle* leaf aqueous extract, and evaluation of their antiphytofungal activity. *Environ. Sci. Pollut. Res. Int.* 10, 1007.

Khairnar et al. 2010. Dendrimers: potential tool for enhancement of antifungal activity. *Int. J. PharmTech. Res.* 2(1), 736–739.

Khodakovskaya et al. 2009. Carbon nanotubes are able to penetrate plant seed coat and dramatically affect seed germination and plant growth. *ACS Nano.* 3(10), 3221–3227.

Kim et al. 2008. Antifungal effectiveness of nanosilver colloid against rose powdery mildew in greenhouses. *Solid State Phenom.* 135, 15–18.

Lai et al. 2006. *Artemisia arborescens* L Essential Oil-Loaded Solid Lipid Nanoparticles for potential agricultural application: Preparation and characterization. *AAPS Pharm. Sci. Tech.* 7, 1–9.

Lamsal et al. 2011. Inhibition effect of silver NPs against powdery mildew on cucumber and pumpkin. *Mycobiology* 39, 26–32.

Lee et al. 2012. Effects of soil-plant interactive system on response to exposure to ZnO nanoparticles. *J. Microbiol. Biotechnol.* 22, 1264–1270.

Lee et al. 2013. Synthesis of silver NPs using cow milk and their antifungal activity against phytopathogens. *Mater. Lett.* 105, 128–131.

Li et al. 2005. Multiplexed detection of pathogen DNA with DNA-based fluorescence nanobarcodes. *Nat. Nanotechnol.* 23(7), 885–889.

Li et al. 2019. Synthesis and characterization of gold NPs from *Marsdenia tenacissima* and its anticancer activity of liver cancer HepG2 cells. *Artif. Cells Nanomed. Biotechnol.* 47, 3029–3036.

Lin et al. 2014. Direct detection of orchid viruses using nanorod-based fiber optic particle plasmon resonance immunosensor. *Biosens. Bioelectron.* 51, 371–378.

Liu et al. 2008. Preparation of fluorescence starch-nanoparticle and its application as plant transgenic vehicle. *J. Cent. South. Univ.* 15, 768–773.

Liu et al. 2009. Sharper and faster "nano darts" kill more bacteria: a study of antibacterial activity of individually dispersed pristine single-walled carbon nanotube. *ACS Nano.* 3(12), 3891–3902.

Lu et al. 2003. Virus-induced gene silencing in plants. *Methods* 30(4), 296–303.

Lyons, K. and Scrinis, G. 2009. *Under the regulatory radar? Nanotechnologies and their impacts for rural Australia. Tracking Rural Change: Community, Policy and Technology in Australia, New Zealand and Europe.* Australian National University E Press, Canberra, 151–171.

Lyu et al. 2017. Titanium as a beneficial element for crop production. *Front. Plant. Sci.* 8, 597.

Mala et al. 2012. Synergistic bactericidal activity of silver NPs and ciprofloxacin against phytopathogens. *J. Cell Tissue Res.* 12, 3249–3254.

Mallaiah B. 2015. Integrate approaches for the management of Crossandra (*Crossandra infundibuliformis* L. nees) wilt caused by *Fusarium incarnatum* (Desm.) Sacc. *PhD Thesis*, Tamil Nadu Agric. Univ., Madurai, India.

Manosalva et al. 2019. Green synthesis of silver NPs: effect of synthesis reaction parameters on antimicrobial activity. *World J. Microbiol. Biotechnol.* 35, 88.

Matouskova et al. 2016. Effect of encapsulation on antimicrobial activity of herbal extracts with lysozyme. *Food Technol. Biotechnol.* 54(3), 304–316.

Mishra, S. and Singh, H.B. 2015a. Biosynthesized silver nanoparticles as a nanoweapon against phytopathogens: exploring their scope and potential in agriculture. *Appl. Microbiol. Biotechnol.* 99, 1097–1107.

Mishra, S. and Singh, H.B. 2015b. Silver nanoparticles mediated altered gene expression of melanin biosynthesis genes in *Bipolaris sorokiniana. Microbiol. Res.* 172, 16–18.

Mishra et al. 2014a. Biofabricated silver NPs act as a strong fungicide against *Bipolaris sorokiniana* causing spot blotch disease in wheat. *PLoS One* 9(5), e97881.

Mishra et al. 2014b. Nanotechnology: exploring potential application in agriculture and its opportunities and constraints. *Biotech. Today* 4, 9–14.

Mohammadi et al. 2019. Green synthesis of silver NPs using *Zingiber officinale* and *Thymus vulgaris* extracts: characterisation, cell cytotoxicity, and its antifungal activity against *Candida albicans* in comparison to fluconazole. *IET Nanobiotechnol.* 13, 114–119.

Molina et al. 2019. Green synthesis of Ag nanoflowers using *Kalanchoe daigremontiana* extract for enhanced photocatalytic and antibacterial activities. *Colloids Surf. B Biointerfaces* 180, 141–149.

Morteza et al. 2013. Study of photosynthetic pigments changes of maize (*Zea mays* L.) under nano TiO_2 spraying at various growth stages. *Springer Plus* 2(1), 247.

Moussa et al. 2013. Botryticidal activity of nanosized silver-chitosan composite and its application for the control of gray mold in strawberry. *J. Food Sci.* 78(10), 589–594.

Moustafa, N.E. and Alomari, A.A. 2019. Green synthesis and bactericidal activities of isotropic and anisotropic spherical gold NPs produced using *Peganum harmala* L leaf and seed extracts. *Biotechnol. Appl. Biochem.* 10, 1002.

Muthamil Selvan et al. 2018. Green synthesis of copper oxide NPs and mosquito larvicidal activity against dengue, zika and chikungunya causing vector *Aedes aegypti*. *IET Nanobiotechnol.* 12, 1042–1046.

Nassar, A.M. 2016. Effectiveness of silver nano-particles of extracts of *Urtica urens* (Urticaceae) against root-knot nematode *Meloidogyne incognita*. *Asian J. Nematol.* 5, 14–19.

Nicewarner-Pena et al. 2001. Submicrometer metallic barcodes. *Science* 294, 137–141.

Ocsoy et al. 2013. Nanotechnology in plant disease management: DNA-directed silver NPs on graphene oxide as an antibacterial against *Xanthomonas perforans*. *ACS Nano.* 7(10), 8972–8980.

Ogunyemi et al. 2019. Green synthesis of zinc oxide NPs using different plant extracts and their antibacterial activity against *Xanthomonas oryzae* pv. *oryzae*. *Artif. Cells Nanomed. Biotechnol.* 47, 341–352.

Paret et al. 2013a. Evaluation of a light-activated nanoparticle formulation of titanium dioxide with zinc for management of bacterial leaf spot on rosa 'Noare'. *HortScience* 48(2), 189–192.

Park et al. 2006. A new composition of nanosized silica-silver for control of various plant diseases. *Plant Pathol. J.* 22, 295–302.

Park et al. 2019. Anti-neuroinflammatory effects of *Ephedra sinica* Stapf extract-capped gold NPs in microglia. *Int. J. Nanomed.* 14, 2861–2877.

Patolsky et al. 2006. Nanowire sensors for medicine and life sciences. *Nanomedicine* 1, 51–65.

Pei et al. 2019. Biosynthesis, characterization, and anticancer effect of plant-mediated silver NPs using *Coptis chinensis*. *Int. J. Nanomed.* 14, 1969–1978.

Perez-de-Luque, A. and Rubiales, D. 2009. Nanotechnology for parasitic plant control. *Pest. Manag. Sci.* 65(-5), 540–545.

Perez-de-Luque et al. 2012. Effect of amphotericin B ´ nanodisks on plant fungal diseases. *Pest. Manag. Sci.* 68(1), 67–74.

Ponmurugan et al. 2016. Antifungal activity of biosynthesised copper NPs evaluated against red root-rot disease in tea plants. *J. Exp. Nanosci.* 11(13), 1019–1031.

Prasad, R. 2014. Synthesis of silver nanoparticles in photosynthetic plants. *J. Nanopart.* 2014, 963961.

Rafique et al. 2019. Novel and facile synthesis of silver NPs using *Albizia procera* leaf extract for dye degradation and antibacterial applications. *Mater. Sci. Eng. C Mater. Biol. Appl.* 99,1313–1324.

Rajeshkumar et al. 2019. Antibacterial and antioxidant potential of biosynthesized copper NPs mediated through *Cissus arnotiana* plant extract. *J. Photochem. Photobiol. B* 197, 111531.

Rajiv et al. 2013. Bio-fabrication of zinc oxide NPs using leaf extract of *Parthenium hysterophorus* L. and its size-dependent antifungal activity against plant fungal pathogens. *Spectrochim. Acta Part A* 112, 384–387.

Rajput et al. 2018. Nanotechnology: comprehensive understanding of interaction, toxicity, and fate of biosynthesized nanoparticles in agroecosystem. In: Singh, H.B., Mishra, S., Fraceto, L.F., de lima, R., (Eds.) *Emerging Trends in Agri-Nanotechnology*. CAB International, UK pp. 135–163.

Rao, K.J. and Paria, S. 2013. Use of sulfur nanoparticles as a green pesticide on *Fusarium solani* and *Venturia inaequalis* phytopathogens. *RSc Adv.* 3 (26), 10471–10478.

Richards, R.M. 1981. Antimicrobial action of silver nitrate. *Microbios* 31, 83–91.

Rispail et al. 2014. Quantum dot and superparamagnetic nanoparticle interaction with pathogenic fungi: internalization and toxicity profile. ACS Appl. *Mater. Interfaces* 6(12), 9100–9110.

Rodrigues et al. 2017. Nanotechnology for sustainable food production: promising opportunities and scientific challenges. *Environ. Sci. Nano.* 4, 767–781.

Sahoo et al. 2007. The present and future of nanotechnology in human health care. *Nanomed. Nanotechnol. Biol. Med.* 3, 20–31.

Salari et al. 2019. In-vitro evaluation of antioxidant and antibacterial potential of greensynthesized silver NPs using *Prosopis farcta* fruit extract. *Iran. J. Pharm. Res.* 18, 430–455.

Sangeetha et al. 2017. Production of bionanomaterials from agricultural wastes. In: Prasad, R., Kumar, M., Kumar, V. (Eds.), *Nanotechnology: An Agricultural Paradigm*. Springer Nature, Singapore, pp. 33–58.

Sardella et al. 2017. Physiological effects and mode of action of ZnO NPs against postharvest fungal contaminants. *Paper Presented at the 109th Annual Meeting of the American Phytopathological Society*, San Antonio, TX, August.

Sarlak et al. 2014. Synthesis of nanopesticides by encapsulating pesticide nanoparticles using functionalized carbon nanotubes and application of new nanocomposite for plant disease treatment. *J. Agric. Food Chem.* 62(21), 4833–4838.

Sekhon, B.S. 2014. Nanotechnology in agri-food production: an overview. *Nanotechnol. Sci. Appl.* 7(2), 31–53.

Sepeur S. 2008. *Nanotechnology: Technical Basics and Applications.* Hannover: Vincentz.

Shah et al. 2017. Hazardous effects of titanium dioxide nanoparticles in ecosystem. *Bioinorg. Chem. Appl.* 2017, 1.

Shaniba et al. 2019. *Manilkara zapota* (L.) P. Royen leaf extract derived silver NPs induce apoptosis in human colorectal carcinoma cells without affecting human lymphocytes or erythrocytes. *Biol. Trace Elem. Res.* 10, 1007.

Shehzad et al. 2018. Synthesis characterization and antibacterial activity of silver nanoparticles using *Rhazya stricta*. *Peer J.* 6:c6086.

Shelar et al. 2019. Helminthicidal and larvicidal potentials of biogenic silver NPs synthesized from medicinal plant *Momordica charantia*. *Med. Chem.* 10, 2174.

Shobha et al. 2019. Synthesis and characterization of zinc oxide NPs utilizing seed source of *Ricinus communis* and study of its antioxidant, antifungal and anticancer activity. *Mater. Sci. Eng. C Mater. Biol. Appl.* 97, 842–850.

Siddiqui et al. 2018. Titanium dioxide and zinc oxide nanoparticles affect some bacterial diseases, and growth and physiological changes of beetroot. *Int. J. Veg. Sci.* 1–22.

Somasundaran et al. 2010, Nanoparticles: Characteristics, mechanisms and modulation of biotoxicity. *Kona Powder Particle J.* 28, 38–49.

Strayer-Scherer et al. 2018. Advanced copper composites against copper-tolerant *Xanthomonas perforans* and tomato bacterial spot. *Phytopathology* 108, 196–205.

Su, X.L. and Li, Y. 2004. Quantum dot biolabelling coupled with immunomagnetic separation for detection of *Escherichia coli* 0157H. *Anal. Chem.*, 76, 4806–4810.

Taniguchi, N. 1974. On the basic concept of nano-technology. *Proceedings of the International Conference on Production Engineering Tokyo Part II Japan Society of Precision Engineering.*

Taran et al. 2014. The effect of colloidal solution of molybdenum nanoparticles on the microbial composition in rhizosphere of *Cicer arietinum* L. *Nanoscale Res. Lett.* 9(1), 289.

Tilman et al. 2002. Plant diversity and composition: effects on productivity and nutrient dynamics of experimental grasslands. In: Loreau, M., Naeem, S., Inchausti, P. (eds) *Biodiversity and Ecosystem Functioning.* Oxford University Press, Oxford, pp. 21–35.

Torney et al. 2007. Mesoporous silica NPs deliver DNA and chemicals into plants. *Nat. Nanotechnol.* 2(5), 295–300.

Tripathi et al. 2011. Growth stimulation of gram (*Cicer arietinum*) plant by water soluble carbon nanotubes. *Nanoscale* 3(3), 1176–1181.

Umar et al. 2018. Biosynthesis of zinc oxide NPs using *Albizia lebbeck* stem bark, and evaluation of its antimicrobial, antioxidant, and cytotoxic activities on human breast cancer cell lines. *Int. J. Nanomed.* 14, 87–100.

United Nations Economic and Social Council, 2018. Programme implementation and future programme of work of the Secretariat in the field of population in 2017. E/CN.9/2018/6.

van Viet et al. 2016. Fusarium antifungal activities of copper nanoparticles synthesized by a chemical reduction method. *Nanomaterials* 2016, e1957612.

Vasantharaj et al. 2019. Biosynthesis of iron oxide NPs using leaf extract of *Ruellia tuberosa*: Antimicrobial properties and their applications in photocatalytic degradation. *J. Photochem. Photobiol. B* 192, 74–82.

Vinotha et al. 2019. Synthesis of ZnO NPs using insulin-rich leaf extract: anti-diabetic, antibiofilm and antioxidant properties. *J. Photochem. Photobiol. B* 197, 111541.

Wagner et al. 2006. The emerging nanomedicine landscape. *Nat. Biotechnol.* 24, 1211–1218.

Wang et al. 2013. Evaluation of antibacterial effects of carbon nanomaterials against copper-resistant *Ralstonia solanacearum*. *Colloids Surf. B Biointerfaces* 103(1), 136–142.

Wang et al. 2014. Evaluation and mechanism of antifungal effects of carbon nanomaterials in controlling plant fungal pathogen. *Carbon* 68, 798–806.

Wang et al. 2019. Synthesis of gold NPs from leaf *Panax notoginseng* and its anticancer activity in pancreatic cancer PANC-1 cell lines. *Artif. Cells Nanomed. Biotechnol.* 47, 1216–1223.

Wani, A.H. and Shah, M.A. 2012. A unique and profound effect of MgO and ZnO nanoparticles on some plant pathogenic fungi. *J. Appl. Pharm. Sci.* 2(3), 40–44.

Wani et al. 2019. Nanoencapsulation of agrochemicals, fertilizers, and pesticides for improved plant production. In: Ghorbanpour, M., Wani, S.H. (eds) *Advances in Phytonanotechnology*, Academic Press, pp. 279–298.

Wiesman et al. 2007. Novel cationic vesicle platform derived from vernonia oil for efficient delivery of DNA through plant cuticle membranes. *J. Biotechnol.* 130(1), 85–94.

Xu et al. 2010. Applications and toxicological issues surrounding nanotechnology in the food industry. *Pure Appl. Chem.* 82(2), 349–372.

Yang, L. and Watts, D.J. 2005. Particle surface characteristics may play an important role in phytotoxicity of alumina NPs. *Toxicol. Lett.* 158(2), 122–132.

Yao et al. 2009. Fluorescence silica nanoprobe as a biomarker for rapid detection of plant pathogens. In: Yin, Y., Wang, X. (eds) *Advanced Materials Research*, Vol. 79. Trans Tech Publications, pp. 513–516.

Yin et al. 2018. Opportunities to advance sustainable design of nano-enabled agriculture identified through a literature review. *Environ. Sci. Nano.* 5, 11–26.

Zangeneh et al. 2019. Novel synthesis of *Falcaria vulgaris* leaf extract conjugated copper NPs with potent cytotoxicity, antioxidant, antifungal, antibacterial, and cutaneous wound healing activities under in vitro and in vivo condition. *J. Photochem. Photobiol. B* 197, 111556.

Zheng et al. 2019. Gold nanoparticles synthesized from *Magnolia officinalis* and anticancer activity in A549 lung cancer cells. *Artif. Cells Nanomed. Biotechnol.* 47, 3101–3109.

13

An Overview of CRISPR and Gene Chip Technology to Study Plant–Microbe Interaction

Prassan Choudhary*
Amity University

CONTENTS

13.1 Introduction

The interplay between plants and microbes has been subject to great amount of attention. The interactions can be neutral, beneficial, or harmful to the plants. Plant microbe interactions are intriguing as they involve a maze of underlying mechanisms. Plants generally respond to pathogen attack via the pathogen-associated molecular patterns (PAMPs) and danger-associated molecular patterns (DAMPs) (Lodha and Basak 2012). Also, pathogen effectors are recognized, resulting in effector-triggered immunity or PAMP-triggered immunity. These plant strategies kick-start the signaling pathways, resulting in the production of antimicrobial compounds that combat the pathogen. Similarly, certain stains, which are known to be endophytic in nature, impact the nutrient uptake and impart tolerance against environment stress. The impact of abiotic stresses such as drought, salinity, temperature, soil acidity, and waterlogging also impact such plant–microbe interactions (Berg 2009). Despite their importance, such microbes are still an understudy and need much attention to be characterized properly.

Omics tools are those that follow a top-down approach to understand the association and effects of plants and microbes. This approach uses genomics, proteomics, metabolomics, and transcriptomics to uncover the molecular interactions resulting from such plant–microbe face-offs (Swarupa et al. 2016). Nowadays, a collective term "holo-omics," coined by Nyholm et al., is being used to refer to studies involving host–microbe interactions (Nyholm et al. 2020; Xu et al. 2021). Xu et al. (2018) studied the root microbiome associated with drought tolerance in the life cycle of drought struck sorghum plant. CRISPR/Cas 9 is another powerful tool being used to reveal the crosstalk among microbial communities and plant response (Rubin et al. 2022). All such efforts are being done to unravel the structural and functional relationship between plants and their partners. In order to draw a more holistic perspective on the matter, there is a need to discuss various experimental studies and designs that are being framed. Gene chips

* Corresponding author's email: prassan.pratik@gmail.com

DOI: 10.1201/9781003171416-13

are employed to screen hundreds of genes, which are an essentiality underlying various cascade pathways. Recently, AtMAD (database for *Arabidopsis thaliana* multi-omics association) has been launched to throw better light on the association studies of the plant (Lan et al. 2021). This would help in understanding plant genomics and would make it easy to understand such association studies more effectively.

In this chapter, a more recent picture of the omics approaches has been focused in association with recent technologies being used to develop a better understanding on the matter (Sharma et al. 2020). In order to exemplify host–microbiome interface, opportunities for widening system-level comprehensions need to be shaped, which would extensively inform microbial advances to improve fitness and health of the host.

13.2 Holo-Omics for Plant Biology: Advantages and Challenges

Plants are associated with diverse microbiota due to their survival in different biogeochemical diverse soils (Franzosa et al. 2015; Castrillo et al. 2017). A number of studies have focused on –holo-omics approaches to decipher the plant biology in association with microbiota (Table 13.1). The most frequently used tool is transcriptomics as they present a suitably annotated outlook on host operations (Salas-González et al. 2021). The main benefit of such data type is the availability of well-developed tools, which strengthens the analytics presenting a cascade of host-specific expression analyses in plants. Castrillo et al. deciphered the association between microbiome composition and phosphate starvation response (PSR) function appearing in *Arabidopsis*. It was the consequence of this design that the research community unraveled that there is a coordination between the plant immune system and microbial identification

TABLE 13.1

Recent Studies in the Application of Holo-Omics and Microarray Technologies to Understand Plant–Microbe Interactions

Host	Sample Type(s)	Methodology Used	References
Holo-Omics Platforms			
Sorghum	Rhizosphere	16S, host genomic	Deng et al. (2021)
Rice	Sheath	Sheath infection-based RNA sequencing (RNA-Seq) analysis	Jeon et al. (2020)
Brassica rapa	Soil, rhizosphere, root	16S, ionomics, metabolomics, phenome	Ichihashi et al. (2020)
Strawberry	Phyllosphere	RNA-Seq analysis, RT-qPCR	Puławska et al. (2020)
Brassica napus	Petioles	Transcriptome, Illumina HiSeq 2500	Chittem et al. (2020)
Maize	Soil, root, rhizosphere, shoot	16S, plant extracts	Kudjordjie et al. (2019)
Peanut	Rhizosphere, root	Metagenome, RNA-Seq of host, metatranscriptome	Li et al. (2019)
Arabidopsis, wheat, rice, and *N. benthamiana*	Root, leaf	16S, metabolites	Huang et al. (2019)
Microarray and High-Throughput Platforms			
Arabidopsis thaliana	Mixed infection	Microarrays	Tahmasebi et al. (2021)
Arabidopsis thaliana	Leaves	Microarrays, differential gene expression	Pathak et al. (2020)
Spinach	Root	Genome sequencing	Holmes et al. (2020)
Nicotiana tabacum	Stem, leaf	Glycan microarrays	Weiller et al. (2020)
N. benthamiana	Root	Oligonucleotide microarrays, microRNAs	Liu et al. (2020)
Tomato	-	Peptide microarrays	Yeo and Devarenne (2020)
Cycas debaoensis and *C. fairylakea*	Root	Microarray (GeoChip 5.0)	Chang et al. (2020)
N. benthamiana	Epidermal cells	Deep sequencing, transcriptome	Fan et al. (2014)

with nutritional indications during microbiome assemblage. As the 16S rRNA compositional reports suggested that the microbiomes of PSR mutants were dissimilar from those of wild-type *Arabidopsis*, it was confirmed that this design demonstrated the direct connection of host's root microbiome with PSR (Castrillo et al. 2017). Such studies also encounter certain challenges during their execution. Successful collaborations among plant and microbiologists are required as such studies typically require a broad range of expertise to implement. In order to bridge the gaps of kingdom-specific pathways, statisticians and computational biologists should be there to recognize and implement the approaches with suitable statistical justifications (Joyce and Palsson 2006).

There is an absolute need for sustained development of compatible bioinformatics tools as it can be a daunting task to determine which to pick, as some are specific to the entire data type and experiments, while others center on the specific questions under purview (Franzosa et al. 2015). Integration of orthogonal data sets within one framework is also a challenge as the studies discussed here fail to spot principle associations among manifold omics layers as the focus is on separate omics analyses initially and then the integration of results from separate layers is done (Hasin et al. 2017). Studies have suggested newly developed strategies, for instance network-free non-Bayesian approaches, network- and kernel-based approaches, and network-free Bayesian approaches, which can suitably unearth the non-linear associations in host–microbe interaction cases (Bersanelli et al. 2016). Direct integration approaches such as transkingdom network (TransNet) analysis integrate and analyze holo-omics data set as they allocate the assembly of networks among differentially expressed elements integrated through correlation analysis (Rodrigues et al. 2018).

Metabolites form a vital component of the working border of host–microbe interactions. The importance of adding metabolomics data while carrying out holo-omics studies is that much of the data assimilation requirements with experimental, analytical that are needed for metabolomics studies are fully attuned with the addition of transcriptomics, genomics, and proteomics research and hence metabolomics is able to endow with a common denominator to the design and analysis of future holo-omics experiments (Schloss 2018; Xu et al. 2021). Hu et al. (2018) combined primary metabolite determination, microbiome profiling, host-specific defense gene expression, microbial complementation analysis, and herbivore growth assays to find that benzoxazinoids (BXs) are exudates of the roots of maize and wheat, which lead to reaction rings with soil microbial communities that impact the coming generations of crop performance (Hu et al. 2018). The host-specific phenomics and gene expression analysis in the study confirmed that conditioned shifts in soil microbiota drives changed host defense in the immediate next origination and modified the levels of phytohormones established to regulate herbivory. Some challenges related to this approach include that numerous peaks are not exclusive as corresponding to specific metabolites and due to noised data, replications play a crucial role in such studies (Krumsiek et al. 2012). Also, the amount of identifiable metabolites is quite less as compared to that of recognizable transcripts and genes from the genome or transcriptome layer, which hinders the use of metabolomics data resulting in impaired interpretation of the final results.

13.3 Recent Advancements in Plant–Microbe Crosstalk Studies

13.3.1 CRISPR/Cas9 Manipulation

Targeted genetic manipulation has led to a new paradigm in the area of studying microbial communities. DART (DNA-editing All-in-one RNA-guided CRISPR/Cas Transposase) systems are being successfully used to understand the microbial communities, both free living and associated with plants (Rubin et al. 2022). Gene loss-of-function mutants in association with gain-of-function germplasms are vital reserves for gene utility studies along with crop genetic improvisations (Boorisjuk et al. 2019). Nowadays, the CRISPR/Cas approach is receiving much attention as it establishes the roles of individual microbes and helps in defining and assigning particular roles to microbes in a dynamic environment (Prabhukarthikeyan et al. 2020). CRISPR/Cas systems have primarily been divided into Class 1, which includes types I, III, and IV, and Class 2, which consists of types II, V, and VI, based on structural, evolutionary, and functional characteristics with discrete machinery of guide RNA biogenesis and target interference (Garcia.-Doval and Jinek 2017).

The best characterized CRISPR/SpCas9 (type II) system is primarily applicable in plant (host) aimed at genome targeting. It comes originally from *Streptococcus pyogenes* consisting of three components: a trans-activating small RNA (tracrRNA), the SpCas9 nuclease, and a mature CRISPR RNA (crRNA) (Bhaya et al. 2011; Cong et al. 2013). A major limitation of the CRISPR/SpCas9 application system is that it requires PAM sequence for DNA target cleavage (Hua et al. 2019). Consequently, numerous Cas9 variants and orthologues with distinctive PAM specificities, such as SaCas9, ScCas9, SpCas9 (VQR), SpCas9-NG, xCas9, SpCas9-NRCH, and SpCas9-NRRH, have been isolated, thereby enhancing the extent of genome-editing gizmos in plants (Hua et al. 2018, 2019; Wang et al. 2020). CRISPR/Cas12 comes as the second most well-recognized CRISPR system. It has been classified as belonging to type V and has separate evolutionary genesis and structural architecture when compared to Cas9. Among the known RuvC-resembling domain-having Cas12 have been explained along with wide-ranging functions, and both CRISPR/Cas12b (C2c1) and CRISPR/Cas12a (Cpf1) have effectively been applied for genome editing in plants (Zaidi et al. 2017; Ming et al. 2020). Cas14, also a type V protein, is a model method for imparting resistance against high-fidelity SNP genotyping and ssDNA plant viruses due to its sequence-neutrality and unobstructed cleavage (Harrington et al. 2018; Khan et al. 2019).

In comparison with traditional methods for genetic manipulation of the microbial genome, which have been related to inefficient homologous recombination, the CRISPR/SpCas9 tools are more efficient and quite simple in many cases. Moreover, they provide a high-throughput experimental platform to dissect gene function at the whole-genome level in plant pathogens (Zhang et al. 2018). Hence, they can be used to confer plant with disease resistance. Type III effectors have been reported to be secreted into the plant cell during the infection process in bacteria (Büttner and He 2009). These effectors mainly disrupt the host's defense pathways, thereby activating the *S* genes for disease development (Zaidi et al. 2018). Therefore, both *S* genes and negative regulators of plant innate immune response are important target sites for CRISPR/Cas9-mediated gene editing to improve plant resistance.

A study reported the knockout of the *OsSWEET11* gene, which is directed for TALEs in rice, resulting in improving defiance to *X. oryzae* pv. *oryzae* (Xoo) noticeably without changing the development of pollen (Kim et al. 2019). TALE-binding elements (EBEs) present in the promoter zone of both *OsSWEET14* and *OsSWEET11* were also aimed at, which led to the rice lines hauling indels conferring robust resistance to most *Xoo* strains (Xu et al. 2019). Similarly, fungal pathogens have a complex colonization mechanism with the host. CRISPR/SpCas9-mediated knockout imparted challenge in opposition to the powdery mildew fungal pathogen *Blumeria graminis* f. sp. *tritici* involving its homologs in wheat (*TaMLOs*) (Wang et al. 2014). Similarly, *OsERF922* encoding transcription factor belonging to the APETALA2/ethylene response factor (AP2/ERF) superfamily in rice has been reported to act as both positive and negative regulators in plant defense against different pathogens (Langner et al. 2018). In 2016, Wang et al. applied CRISPR/SpCas9 to alter *OsERF922* resulting in rice lines carrying different frameshift indels, which showed enhanced resistance to the rice blast fungus *M. oryzae* without upsetting plant enhancement and similar agronomic qualities tested. Plant–virus interactions have also been studied using these genome-editing tools (Macovei et al. 2018; Bastet et al. 2019; Gomez et al. 2019). CRISPR technologies are expected to widen the studies related to plant–pathogen relationship and imparting broad resistance against pathogens in future.

13.3.2 Role of Microarrays and Gene Chip Technologies

The previous two decades have successfully used microarrays and such parallel technologies to characterize multi-genes at the same time. The technology primarily uses a glass platform needed to spot oligonucleotides or cDNAs. Microarrays are used to measure and outline the expression of numerous genes with the aim of gene discovery and global expression profiling (Lodha and Basak 2012). Microarray technology has extensively and effectively been used to identify regulatory genes and key defense genes and to elucidate signal transduction pathways involved in disease resistance and their links to intermittent pathways (Table 13.1). Hence, this technology can be used as a tool to encompass the developmental stages of a pathogen entry and infection caused to the host. Development of disease progression has also been analyzed in different environmental conditions for a clear understanding of the process.

The basic protocol employed in this technique is the extraction of mRNA from control and experimental samples. Using reverse transcriptase polymerase chain reaction (RT-PCR), the mRNAs is converted to cDNA. Further, the two cDNA pools are fluorescently labeled with the help of two different fluorochromes, mixed and hybridized for a specified period of time. Specific gene sequences are present as individual spots where the hybridization takes place, after which the surplus cDNA is cleaned off. A laser scanner is used to identify the intensity of the fluorophores formed during hybridization. Spots which fluoresce with a particular label are analyzed for its upregulation or downregulation with respect to the conditions of the study (Tan et al. 2009).

The applications of microarray gene chips to understand plant–pathogen interactions are immense. Cellier et al. (2017) used ArrayTube microarray technology to refurbish a protocol that differentiates 17 major groups of *Ralstonia solanacearum* species complex (Rssc) in a lone multiplex reaction. Rssc is considered a priority plant pathogen, and the study helped in providing phylogenetic incongruence of Rssc strains, which may be used to signal emerging potent pathogenic strains. Microarray-based analysis targeted *WRKY33* transcription factor to alter susceptibility against *Botrytis cinerea* repressing 1,054 genes at the early inoculation stages (Sham et al. 2017). The study emphasized that novel oxylipin signal transduction pathways can be elucidated using the identification of molecular components involved in cyclopentanone signaling (Sham et al. 2017). Similarly, Ertani et al. (2017) used alfalfa-based protein hydrolysate (EM)-based microarray to understand the differential expression of genes entailed in carbon (C) and nitrogen (N) metabolism, nutrient uptake, photosynthesis, and developmental processes in tomato. The study also reported the upregulation of genes involved in secondary metabolism, enhancing phenol content in EM-treated plants (Ertani et al. 2017). Previously, a microarray-based transcriptomic study reported the instigation of both sulfur and nitrogen assimilation pathways treated with an *Ascophyllum nodosum* seaweed extract in *Brassica napus* plants (Jannin et al. 2013). In a later study, the same extract was applied to B. napus and the upregulation of expression of transport-specific genes specifically nitrate, iron, sulfate, copper, and NRAMP3 was reported, which led to increased mineral transport in the plant (Billard et al. 2014).

Although the microarray technology has its advantages, it has certain pitfalls that can limit its use globally. The cost of reading gene chips developed using microarray technology is quite high. Also, such studies require a team of well-equipped biologists and analytical computational biologists along with the clear understanding of software packages to be used for data analysis (Cellier et al. 2017).

13.4 Conclusions

The complex nature of plant–microbe interactions is being subject to immense scrutiny. Inter- and intradisciplinary team of researchers is required to understand the complex layers of mechanisms being triggered as a result of crosstalk between host and microbes. Such studies can not only help in understanding the interaction dynamics, but also help in other areas of study, such as molecular diagnostics and crop protection. A sustainable growth in crop production can thus be achieved to manage the ever-growing population.

Acknowledgment

The corresponding author heartily thanks the Head of Department, Amity Institute of Biotechnology, Amity University.

Conflict of Interest

The authors declare no conflict of interest.

REFERENCES

Bastet et al. Mimicking natural polymorphism in eIF 4E by CRISPR/Cas9 base editing is associated with resistance to potyviruses. *Plant Biotechnol J.* 2019;17(9):1736–1750.

Berg G. Plant–microbe interactions promoting plant growth and health: perspectives for controlled use of microorganisms in agriculture. *Appl Microbiol Biotechnol.* 2009;84(1):11–18.

Bersanelli et al. Methods for the integration of multi-omics data: mathematical aspects. *BMC Bioinformatics.* 2016;17(2):167–177.

Bhaya et al. CRISPR-Cas systems in bacteria and archaea: versatile small RNAs for adaptive defense and regulation. *Annu Rev Genet.* 2011;45:273–297.

Billard et al. Two biostimulants derived from algae or humic acid induce similar responses in the mineral content and gene expression of winter oilseed rape (Brassica napus L.). *J Plant Growth Regul.* 2014;33(2):305–316.

Borisjuk et al. Genetic modification for wheat improvement: from transgenesis to genome editing. *Biomed Res Int.* 2019; 2019:6216304.

Büttner D and He SY. Type III protein secretion in plant pathogenic bacteria. *Plant Physiol.* 2009;150(4): 1656–1664.

Castrillo et al. Root microbiota drive direct integration of phosphate stress and immunity. *Nature.* 2017; 543(7646):513–518.

Cellier et al. Tube-wise diagnostic microarray for the multiplex characterization of the complex plant pathogen *Ralstonia solanacearum. Front Plant Sci.* 2017;8:821.

Chang et al. GeoChip 5.0 microarray: a descriptive overview of genes present in coralloid root microbiomes of Cycas debaoensis and Cycas fairylakea. *Extrem Life, Biospeology Astrobiol.* 2020;12(1):5–20.

Chittem et al. Transcriptome analysis of the plant pathogen Sclerotinia sclerotiorum interaction with resistant and susceptible canola (Brassica napus) lines. *PLoS One.* 2020;15(3):e0229844.

Cong et al. Multiplex genome engineering using CRISPR/Cas systems. *Science* (80-). 2013;339(6121): 819–823.

Deng et al. Genome wide association study reveals plant loci controlling heritability of the rhizosphere microbiome. *ISME J.* 2021;1–14.

Ertani et al. Transcriptome-wide identification of differentially expressed genes in *Solanum lycopersicon L.* in response to an alfalfa-protein hydrolysate using microarrays. *Front Plant Sci.* 2017;8:1159.

Fan et al. Deep sequencing–based transcriptome profiling reveals comprehensive insights into the responses of Nicotiana benthamiana to Beet necrotic yellow vein virus infections containing or lacking RNA4. *PLoS One.* 2014;9(1):e85284.

Franzosa et al. Sequencing and beyond: integrating molecular 'omics' for microbial community profiling. *Nat Rev Microbiol.* 2015;13(6):360–372.

Garcia-Doval C and Jinek M. Molecular architectures and mechanisms of Class 2 CRISPR-associated nucleases. *Curr Opin Struct Biol.* 2017;47:157–166.

Gomez et al. Simultaneous CRISPR/Cas9 mediated editing of cassava eIF 4E isoforms nCBP.1 and nCBP.2 reduces cassava brown streak disease symptom severity and incidence. *Plant Biotechnol J.* 2019;17(2): 421–434.

Harrington et al. Programmed DNA destruction by miniature CRISPR-Cas14 enzymes. *Science* (80-). 2018;362(6416):839–842.

Hasin et al. Multi-omics approaches to disease. *Genome Biol.* 2017;18(1):1–15.

Holmes et al. A high-throughput genomic screen identifies a role for the plasmid-borne type II secretion system of Escherichia coli O157: H7 (Sakai) in plant-microbe interactions. *Genomics.* 2020;112(6):4242–4253.

Huang et al. A specialized metabolic network selectively modulates Arabidopsis root microbiota. *Science* (80-). 2019;364(6440):6389.

Hu et al. Root exudate metabolites drive plant-soil feedbacks on growth and defense by shaping the rhizosphere microbiota. *Nat Commun.* 2018;9(1):1–13.

Hua et al. Precise A·T to G·C base editing in the rice genome. *Mol Plant.* 2018;11(4):627–630.

Hua et al. Genome engineering in rice using Cas9 variants that recognize NG PAM sequences. *Mol Plant.* 2019;12(7):1003–1014.

Ichihashi et al. Multi-omics analysis on an agroecosystem reveals the significant role of organic nitrogen to increase agricultural crop yield. *Proc Natl Acad Sci.* 2020;117(25):14552–14560.

Jannin et al. *Brassica napus* growth is promoted by *Ascophyllum nodosum* (L.) Le Jol. seaweed extract: microarray analysis and physiological characterization of N, C, and S metabolisms. *J Plant Growth Regul.* 2013;32(1):31–52.

Jeon et al. Transcriptome profiling of the rice blast fungus Magnaporthe oryzae and its host Oryza sativa during infection. *Mol Plant-Microbe Interact.* 2020;33(2):141–144.

Joyce AR and Palsson BØ. The model organism as a system: integrating 'omics' data sets. *Nat Rev Mol cell Biol.* 2006;7(3):198–210.

Khan et al. Targeting plant ssDNA viruses with engineered miniature CRISPR-Cas14a. *Trends Biotechnol.* 2019;37(8):800–804.

Kim et al. CRISPR/Cas9-targeted mutagenesis of Os8N3 in rice to confer resistance to *Xanthomonas oryzae pv. oryzae*. *Rice.* 2019;12(1):1–13.

Krumsiek et al. Mining the unknown: a systems approach to metabolite identification combining genetic and metabolic information. *PLoS Genet.* 2012;8(10):e1003005.

Kudjordjie et al. Maize synthesized benzoxazinoids affect the host associated microbiome. *Microbiome.* 2019;7(1):1–17.

Lan et al. AtMAD: *Arabidopsis thaliana* multi-omics association database. *Nucleic Acids Res.* 2021;49(D1):D1445–D1151.

Langner et al. CRISPR crops: plant genome editing toward disease resistance. *Annu Rev Phytopathol.* 2018;56:479–512.

Li et al. Legacy of land use history determines reprogramming of plant physiology by soil microbiome. *ISME J.* 2019;13(3):738–751.

Liu et al. Genome-wide microRNA profiling using oligonucleotide microarray reveals regulatory networks of microRNAs in Nicotiana benthamiana during Beet necrotic yellow vein virus infection. *Viruses.* 2020;12(3):310.

Lodha TD and Basak J. Plant–pathogen interactions: what microarray tells about it? *Mol Biotechnol.* 2012; 50(1):87–97.

Macovei et al. Novel alleles of rice eIF4G generated by CRISPR/Cas9 targeted mutagenesis confer resistance to Rice tungro spherical virus. *Plant Biotechnol J.* 2018;16(11):1918–1927.

Ming et al. CRISPR–Cas12b enables efficient plant genome engineering. *Nat Plants.* 2020;6(3):202–208.

Nyholm et al. Holo-omics: integrated host-microbiota multi-omics for basic and applied biological research. *Iscience.* 2020;101414.

Pathak et al. Computational analysis of microarray data of Arabidopsis thaliana challenged with Alternaria brassicicola for identification of key genes in Brassica. *J Genet Eng Biotechnol.* 2020;18(1):1–20.

Prabhukarthikeyan et al. Understanding the plant-microbe interactions in CRISPR/Cas9 era: indeed a sprinting start in marathon. *Curr Genomics.* 2020;21(6):429–443.

Puławska et al. Transcriptome analysis of Xanthomonas fragariae in strawberry leaves. *Sci Rep.* 2020;10(1):1–10.

Rodrigues et al. Transkingdom networks: a systems biology approach to identify causal members of host–microbiota interactions. In: Robert G. Beiko, Will Hsiao, John Parkinson (eds), *Microbiome Analysis.* New York: Humana Press, Springer; 2018. pp. 227–242.

Rubin et al. Species-and site-specific genome editing in complex bacterial communities. *Nature microbiology.* 2022; 7(1):34–47.

Salas-González et al. Coordination between microbiota and root endodermis supports plant mineral nutrient homeostasis. *Science* (80–). 2021;371(6525):0695.

Schloss PD. Identifying and overcoming threats to reproducibility, replicability, robustness, and generalizability in microbiome research. *MBio.* 2018;9(3):e00525–18.

Sham et al. Microarray analysis of Arabidopsis WRKY33 mutants in response to the necrotrophic fungus *Botrytis cinerea*. *PLoS One.* 2017;12(2):e0172343.

Sharma et al. Deciphering the omics of plant-microbe interaction: perspectives and new insights. *Curr Genomics.* 2020;21(5):343–362.

Swarupa et al. Omics-driven approaches in plant–microbe interaction. In: Singh, D., Singh, H., Prabha, R. (eds), *Microbial Inoculants in Sustainable Agricultural Productivity.* New Delhi: Springer; 2016. pp. 61–84.

Tan et al. Assessing the impact of transcriptomics, proteomics and metabolomics on fungal phytopathology. *Mol Plant Pathol.* 2009;10(5):703–715.

Tahmasebi et al. Microarray analysis of Arabidopsis thaliana exposed to single and mixed infections with Cucumber mosaic virus and turnip viruses. *Physiol Mol Biol Plants*. 2021;27(1):11–27.

Wang et al. Simultaneous editing of three homoeoalleles in hexaploid bread wheat confers heritable resistance to powdery mildew. *Nat Biotechnol*. 2014;32(9):947–951.

Wang et al. Enhanced rice blast resistance by CRISPR/Cas9-targeted mutagenesis of the ERF transcription factor gene OsERF922. *PLoS One*. 2016;11(4):e0154027.

Wang et al. Targeted base editing in rice with CRISPR/ScCas9 system. *Plant Biotechnol J*. 2020;18(8):1645

Weiller et al. Overexpression of VvPGIP1 and NtCAD14 in tobacco screened using glycan microarrays reveals cell wall reorganisation in the absence of fungal infection. *Vaccines*. 2020;8(3):388.

Xu et al. Drought delays development of the sorghum root microbiome and enriches for monoderm bacteria. *Proc Natl Acad Sci*. 2018;115(18):E4284–E4293.

Xu et al. Engineering broad-spectrum bacterial blight resistance by simultaneously disrupting variable TALE-binding elements of multiple susceptibility genes in rice. *Mol Plant*. 2019;12(11):1434–1446.

Xu et al. Holo-omics for deciphering plant-microbiome interactions. *Microbiome*. 2021;9(1):1–11.

Yeo I-C and Devarenne TP. Screening for potential nuclear substrates for the plant cell death suppressor kinase Adi3 using peptide microarrays. *PLoS One*. 2020;15(6):e0234011.

Zaidi et al. CRISPR-Cpf1: a new tool for plant genome editing. *Trends Plant Sci*. 2017;22(7):550–553.

Zaidi et al. Genome editing: targeting susceptibility genes for plant disease resistance. *Trends Biotechnol*. 2018;36(9):898–906.

Zhang et al. Bacterial genome editing with CRISPR-Cas9: taking *Clostridium beijerinckii* as an example. In: Braman J. (ed), *Synthetic Biology*. Methods in Molecular Biology, vol. 1772. New York: Humana Press, Springer; 2018. pp. 297–325.

14

Functional Genomic Approaches to Improve Rice Productivity through Leaf Architecture

Robert Nepomuceno, Cristine Marie B. Brown, and Marilyn B. Brown*
University of the Philippines

CONTENTS

14.1 Introduction

Rice (*Oryza sativa* L.) serves as a staple food for half of the world's population and is considered to be one of the most important crops in the world. It belongs to the grass family Poaceae and has a number of different cultivars and varieties that are grown all over the world. It is a primary source of carbohydrates in people's diet, particularly in economically disadvantaged countries, where it accounts for 50%–80% of the daily calorie intake. Moreover, it possesses a minute amount of other minerals such as calcium and iron (Gopalan et al., 2007).

The Philippines has a long history of rice cultivation, where Austronesian people from South China and Taiwan brought them in 3,200 BC (Alojado, 2014). Currently, it is a major agricultural commodity in the Philippines, generating 4.8B USD (international price) in 2014. Locally, it stands second behind sugar cane in terms of production in tonnage (FAOSTAT, 2014). Although the Philippines is considered to be the eighth largest producer of rice in the world, the country still imports rice. This inadequacy will be a major theme in the world as the world's population continues to exponentially expand subsequently increasing the demand for food. The world population is projected to increase from 7.2B to 9.6B by 2050, and in order to support the population, it is estimated that an increase in yield of 60% is necessary (Sheehy et al., 2008).

The genus of *Oryza* is comprised of 24 species subdivided into species complexes based on similarity in morphology and cytology. It has $n=12$ chromosomes with genomes ranging from AA to KKLL (Vaughan, 1994, 1989). This diversity has been the basis of traditional breeding for a wide range of

* Corresponding author's email: drmbbrown@yahoo.com.ph

DOI: 10.1201/9781003171416-14

important characteristics (Vaughan et al., 2003). However, traditional breeding techniques alone are not expected to contribute to an increase in yield of 60%. A radical approach to crop improvement is required in order to accomplish this goal.

There are three types of photosynthesis in plants: C_3, C_4, and CAM. C_3 photosynthesis is employed by ~300,000 species, representing 95% of all the plant species. This type of photosynthesis is inherently inefficient as ribulose-1,5-bisphosphate carboxylase/oxygenase (rubisco, EC 4.1.1.39) also uses O_2 as a substrate, leading to photorespiration (Bowes et al., 1971). Photorespiration is responsible for the loss of 25% of carbon assimilated in photosynthesis (Andrews and Lorimer, 1987). It is a relic of evolution caused by the fact that rubisco evolved at a time of high CO_2 and low atmospheric O_2 concentration 400 MYA. It reduces the rate of carbon fixation in plants employing C_3 photosynthesis. Adaptation for photorespiration involves the high expression of rubisco in the leaf, making the enzyme a major sink of nitrogen in C_3 species. This also explains the relative abundance of the enzyme in the biosphere (Jin et al., 2006). Moreover, C_3 species are required to maintain high conductance in order to absorb more CO_2, leading to water loss. C_4 species, on the other hand, employ spatial separation between rubisco and O_2 by pumping CO_2 in bundle sheath cells—an area devoid of O_2 and where carbon fixation occurs eliminating the effects of photorespiration. Consequently, C_4 species have a higher water and nitrogen use efficiency (Mathews and Van Holde, 1990; Hopkins, 1999; Moore et al., 2003; Simpson, 2010). Crassulacean acid metabolism (CAM) species, on the other hand, employs temporal separation by absorbing CO_2 at night and then conducting carbon fixation during the day. This minimizes water loss as stomates are closed during this period (Lee, 2010).

One radical approach to rice productivity improvement is through the conversion of the C_3 photosynthesis of rice into C_4. The major enzymes required for the pathway have already been elucidated; however, there is still a need to identify novel genes and regulatory factors essential for the process. Functional analysis of rice genome through the use of gain of function mutants such as full-length cDNA overexpression can be used to determine gene function. Gain-of-function mutants offer an advantage over loss-of-function mutants since rice has a large portion of redundant genes, and the function of such genes is impossible to identify using a gene knockout approach (Higuchi-Takeuchi et al., 2013).

14.2 Review of Literature

14.2.1 Rice Taxonomy and Origin

Rice is under the genus *Oryza*, of the family Poaceae (formerly Gramineae) of the order Poales (Plants Database USDA, 2014). There are two major eco-geographical races of rice—the indica and the japonica. Race in this context refers to a subset of a population geographically and reproductively isolated from each other. Japonica is primarily found in temperate areas, while indica ranges from the tropics to subtropics (Garris et al., 2005).

Rice domestication is generally believed to have occurred 9,000 years ago from its progenitor, *O. rufipogon* (Oka, 1988). It is widely believed that the center of origin and diversity of rice is China, specifically the river valleys of Yangtze, where Mekong River is considered to be the primary center of origin for *O. sativa* (Porteres, 1956; Higham and Lu, 1998).

One of the earliest archaeological evidences of rice domestication is found in China, where plant microstructures such as leaf blade silica phytoliths were found dating back to 12,000 BC (Zhao et al., 1998). Several oddly shaped bones assumed to have been used as tools for rice cultivation (Fuller, 2007) dating back to 5000–4000 BC were also found. As early as 100 AD, there was already a Chinese record about the morphological and interspecific distinction between the two *O. sativa* subspecies, then known as the "Hsien" and "Keng," but now commonly referred to as indica and japonica (Matsuo et al., 1997). Isozyme analysis using 15 polymorphic loci on 1,700 *O. sativa* varieties revealed six different groupings (Glaszman, 1987). Aside from indica and japonica, aus, aromatic, rayada, and ashina were identified as distinct groups. RFLP analysis was only able to distinguish between indica and japonica (Wang and Tanksley, 1989), while SSR analysis using 234 accessions and looking at 169 nuclear loci has identified five groups—aus and indica are subgrouped under indica and tropical and temperate japonica and aromatic subgrouped under japonica (Garris et al., 2005). Intronic sequence comparison and patterns of retrotransposon insertion have revealed that indica and japonica diverged from each other 200,000 to 400,000 years ago (Zhu and Ge, 2005).

There are several models to explain the origin of the indica and japonica rice subspecies, suggesting either single or multiple origins. Single-origin models suggest that rice originated from wild rice and divergence between indica and japonica occurred after domestication (Oka and Morishima, 1982). This is supported by multiple analyses on genes that show high sequence similarity (Li and Sang, 2006; Tan et al., 2008). Multiple-origin models, on the other hand, suggest that indica and japonica are products of separate and independent domestication (Fuller et al., 2010) from *O. rufipogon* and that the sequence similarity observed in some genes of indica and japonica was a product of introgression. This model can explain the differences in allelic frequencies between the two subspecies (Second, 1982) and is further supported by phylogenetic analyses where japonica and indica are associated in different clades (Cheng et al., 2003). Moreover, isozyme and RFLP comparison revealed that indica and japonica are closer to *O. rufipogon* varieties than they are to each other, suggesting that indica and japonica descended or were domesticated independently from subgroups or subspecies of *O. rufipogon* (Garris et al., 2005).

The origin and center of diversity of *O. glaberrima,* on the other hand, are thought to have been on the inner delta of the Niger River and Guinean coast in Africa 4,000 years ago (Chang and Bardenas, 1965). Several pieces of evidence supporting this assertion exist in the form of linguistic and archaeological evidence (Blench, 2006; Klee and Neumann, 2000). Molecular marker (SNP and SSR) and isozyme analyses revealed that *O. barthii* is closely related and most possibly a progenitor to *O. glaberrima* (Semon et al., 2005).

14.2.2 Rice Genetic Resources

Rice breeding is highly dependent on the amount and quality of its genetic resources, which represents a wide array of rice relatives that are rich in agronomically important trait variations. Due to the importance of rice, there has been a major effort in collecting and conserving its genetic resources. Germplasm collections around the world represent the different cultivars and varieties of the 25 *Oryza* species. IRRI maintains 108,000 accessions of rice (Jackson and Lettington, 2003). In addition, rice accessions are also maintained in other CGIAR centers such as West Africa Rice Development Association (WARDA), International Center for Tropical Agriculture (CIAT), and International Institute of Tropical Agriculture (IITA), Nigeria. There are also rice germplasms maintained in other Asian countries such as China and Indonesia.

Although rice is rich in terms of genetic diversity, as indicated by the number of its wild relatives and the variation in chromosomal configurations, there have been reports that the widely utilized varieties and cultivars in different countries represent a narrow genetic base (Mishra, 2002). Sun et al. (2001), on the other hand, suggested that a wide genetic variability is still present particularly in wild species.

Having a rich genetic resource makes it possible to develop varieties that have the agronomically important traits of the wild relatives such as resistance to biotic and abiotic stresses (Brar and Khush, 2003). For instance, *O. nivara* has long been established to be a source of resistance to grassy stunt virus and a potential reservoir for resistance against rice sheath blight disease (Prasad and Eizena, 2008).

14.2.3 Improving Rice Productivity

Human population has been projected to reach nine billion by 2050, and to support the burgeoning population's demand for food, it has been estimated that there is a need for an increase of 60% in rice yield (Sheehy et al., 2008). Multiple strategies can be employed to meet this demand: (i) reducing the losses incurred during harvesting, (ii) increasing the area allotted to rice cultivation, and (iii) increasing the yield per hectare. Accompanied by the increasing demand for food is the increase in the demand for housing and shelter facilities. Thus, it is common to observe arable land being converted to commercial lots for housing. Such is the case in Nueva Ecija, wherein the province's Provincial Physical Framework Plan explicitly states: "lands for agricultural population will have to be sacrificed for urban developments." Lands allotted to rice cultivation are thus decreasing. Losses in post-harvest processing are estimated to be about 26% (Ren-Yong et al., 1990), and this value increases in areas afflicted with challenging weather. If post-harvest losses were to be eliminated completely, it would still not suffice to remedy the increase in demand brought about by the increase in population.

Yield is a function of the proportion of the total biomass allocated to grains. This proportion is also called the harvest index of a crop. Biomass is a function of the intercepted photosynthetically active radiation (PAR) and the radiation use efficiency (RUE) or its efficiency to utilize PAR (Monteith, 1977). Given this, an equation for yield has been formulated by Mitchell and Sheehy (2006).

$$Y = HI\varepsilon\sum_{i=1}^{n} Qifi$$

where Y is the grain yield (g/m^2), HI is the harvest index, ε is the radiation use efficiency or RUE expressed in g/MJ, n is the duration of growth in days, Qi is the PAR on the crop on the ith day expressed in MJ/m^2, and fi is the fraction of incident PAR intercepted on the ith day. Yield can potentially be enhanced by increasing any of the variables in the equation.

Carbon fixation in C_3 species starts at the uptake of inorganic carbon in the atmosphere, where its dissolution in aqueous cellular components to form bicarbonate is catalyzed by carbonic anhydrase. The carbon is then released in the mesophyll cells and fixed by rubisco into a five-carbon molecule ribulose-1,5-bisphosphate (RuBP), forming an unstable six-carbon compound that spontaneously dissociates to form two three-carbon compound phosphoglycerate. This is then converted to glyceraldehyde 3-phosphate and then reduced by ATP and NADPH to form RuBP. Despite its reputation as the most important enzyme on the planet, rubisco is extremely inefficient. It is a very slow-acting enzyme, fixing only 3–10 molecules of $CO_2 s^{-1}$ per molecule of rubisco. To compensate, rubisco is highly expressed in the leaf (Ellis, 2010), explaining the relative abundance (1.68 g/m^2) of the enzyme and also making the enzyme a major sink of nitrogen in plants (Jin et al., 2006). In addition, rubisco has an oxygenase activity, making it possible for the enzyme to fix O_2 instead of CO_2 to form phosphoglycolate. This compound is later on recycled, in a process known as photorespiration requiring the expenditure of ATP to release RuBP (Bowes et al., 1971).

Oxygenase activity is favored at high-temperature and high-O_2 environments. Since rice is a C_3 plant grown most commonly under tropical conditions coupled with the increasing trend in global temperature, it can be assumed that rice is severely afflicted by photorespiration. Thirty-three percent of the total rubisco in rice may be functioning as an oxygenase at 30°C and current CO_2 concentration, resulting in a 30% reduction in productivity. Rubisco evolved 400 MYA, a time wherein atmospheric conditions had a much higher concentration of CO_2 than O_2 (Berner and Kothovala, 2001) favoring carboxylase over oxygenase activity. However, the present conditions are reversed and photorespiration is partially favored with rubisco performing one oxygenation in every 80–100 carboxylation.

It is therefore widely regarded that the rate-limiting step in photosynthesis is rubisco itself. Attempts to increase productivity naturally target this enzyme (Martin et al., 2012). One approach is to increase rubisco in the leaves. However, further increasing the expression of the enzyme will require an increase in the rate of fertilizer application not to mention changes in leaf microcellular structures to accommodate the increase in the enzyme. This approach does not favor economic and environmental sustainability as an increase in the input will inevitably burden the farmers (Mitchell and Sheehy, 2006). Zhu et al. (2003) attempted to simulate the expression of foreign rubisco with a high affinity to CO_2 in C_3 species. Based on the kinetic properties of the enzyme and canopy photosynthesis, it is concluded that at most 25% increase in productivity is expected from this endeavor.

Most C_4 species possess a unique leaf wreath-like cellular arrangement known as Kranz anatomy. The arrangement consists of mesophyll cells separating bundle sheath cells, which in turn surround the vascular bundles. Initial carboxylation occurs in mesophyll cells and is transported to bundle sheath cells, where rubisco resides and the actual carbon fixation occurs.

C_4 and CAM photosynthesis forms are adaptations to an arid, hot, and dry environment (Edwards et al., 2010) evolving independently 45 times in 19 lineages. A form of C_4 photosynthesis occurs in single-celled organisms such as in the case of *Hydrilla verticillata* (Bowes et al., 2002). The absence of Kranz anatomy and its relative simplicity leads to an attempt to incorporate this system in terrestrial crops (Hausler et al., 2002). However, the low rate of assimilation led to the conclusion that it is unsuitable for use in increasing C_3 crop productivity.

14.2.4 C_4 Leaf Anatomy

C_4 species generally possess Kranz anatomy—a unique arrangement of mesophyll and bundle sheath cells in the leaf. This is characterized by at most two mesophyll cells in between the bundle sheath cells radially arranged around the vascular bundles. The effect is a decreased distance in between veins, leading to a higher vein density per unit area and allowing for rapid exchange in photosynthates. An increase in vein density is considered to be a prerequisite in the development of C_4 photosynthesis (McKown and Dengler, 2009).

In contrast, C_3 species have multiple mesophyll cells separating the bundle sheath cells. Mesophyll cells are where carbon fixation occurs in C_3 species, and that multiplicity of the cell is a form of adaptation to the inefficiency of the main carboxylating enzyme. Mesophyll cells in C_3 species appear green due to the presence of chloroplast, while in C_4, it appears colorless since chloroplast mainly resides in the bundle sheath cells. Chloroplast arrangement inside the bundle sheath and mesophyll cells differs in C_3 and C_4 species. In C_4 species, chloroplasts are concentrated along the bundle sheath cell wall adjacent to the neighboring mesophyll cells. In C_3 species, chloroplasts are arranged along the border of the mesophyll cells. The feature places the carbon fixing cells in areas with high concentrations of initially fixed CO_2 (Leegod et al., 2000). Since the vein in C_4 species are narrowly spaced, there is an increased number of veins per given area as well as a decreased distance in between veins. Various leaf anatomical differences between a C_3 and C_4 are shown in Table 14.1.

Cells of the mesophyll cell make up the bulk of internal leaf tissue and are the major site of photosynthesis in a C_3 plant. One of the main differences in the mesophyll cells of C_4 and C_3 plants is that in most C_3 plants, mesophyll cells are highly lobed (Figure 14.1). Structural convolutions increase surface area, and as such, highly developed mesophyll cell lobes are implicated in high mesophyll cell conductance with respect to CO_2 flux (Adachi, 2013) and have been suggested to be involved with CO_2 recycling.

14.2.5 Leaf Development

Leaves are the site of photosynthesis and photoperception in plants and an integral portion of the shoot system. Despite its apparent simplicity, the genetic and molecular processes involved in leaf development are complex, and although studied extensively, the mechanisms underlying the process need further elucidation (Tsukaya, 2013).

Leaves develop from leaf primordia or founder cells located flanking the shoot apical meristem (Reinhardt et al., 2000). Leaf primordia are regions of localized high auxin concentration or auxin maxima. The arrangement of auxin maxima gives rise to a pattern in leaf arrangement or phyllotaxy.

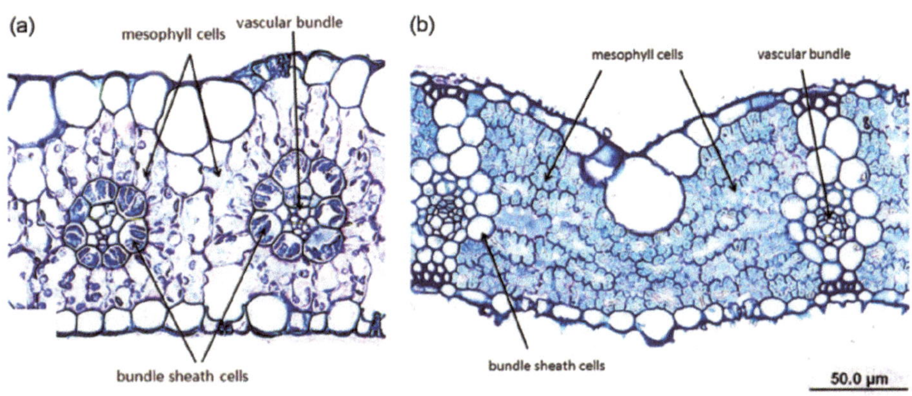

FIGURE 14.1 Leaf anatomical differences between a C_4 (a: sorghum) and a C_3 (b: rice) species. Magnification 400×.

TABLE 14.1
Summary of the Leaf Anatomical Differences between C_3 and C_4 Species

	Maize		Sorghum		Rice	
Traits	**Mean±SE**	**Range**	**Mean±SE**	**Range**	**Mean±SE**	**Range**
Vein density (no. mm^{-1})	9±0.65	42,226	8±0.71	42,226	5±0.26	42,130
Interveinal distance (µm)	115±8.62	102–127	121±8.42	99–135	192±9.00	164–207
Vein width (µm)						
Small	83±3.44	60–99	74±2.32	53–86	43±1.47	31–50
Large	130±6.42	114–145	181±12.86	143–213	105±7.11	93–127
Lateral number of mesophyll cells (MCs)	2±0.2	42,006	2±0.3	42,038	7±0.64	42,163
MC width (µm)	24±0.76	13–38	28±0.87	15–44	24±0.44	13–42
Number of Bundle Sheath Cells (BSCs) in Vein						
Small vein	6±0.52	42,162	6±0.73	42,131	8±0.67	42,226
Large vein	14±1.39	13–16	14±1.26	42324	16±1.52	14–18
BSC Area (µm^2)						
Small vein	681±80.02	579–808	406±54.33	323–473	119±16.54	97–155
Large vein	314±59.52	205–387	266±29.07	222–302	180±22.77	151–218
Small vein	301±97.67	131–409	239±39.26	201–333	7±1.68	42,133
Large vein	149±20.95	109–173	148±24.87	120–205	7±3.48	42,108
Small vein	48±8	31–59	59±6	52–73	8±2	42,137
Large vein	47±5	41–52	55±6	49–68	6±3	42,075

Arabidopsis has a spiral phyllotaxy, which is the effect of the unequal distribution of auxin maxima in the left and right sides (Chitwood et al., 2012). Rice, in contrast, has an alternate phyllotaxy resulting from auxin maxima localization on opposing sides (O'Connor et al., 2014).

The establishment of dorsoventrality occurs concurrent with the bulging of the leaf primordium. The process is important for the determination of abaxial and adaxial surface and flat outgrowth of lamina. It is dependent on a hypothetical factor from shoot apical meristem termed as "anlagen factor" (Efroni et al., 2010). The factor is theorized to be gamma-aminobutyric acid (GABA) shunt metabolite; deficiency in the factor results in the disruption in leaf adaxial–abaxial polarity of *Arabidopsis* (Tatematsu et al., 2011). *Arabidopsis* class III homeodomain leucine zipper (HD-Zip III) proteins are responsible for adaxial side determination. Members of the family include PHABULOSA (PHB), PHAVOLUTA (PHV), and REVOLUTA (REV; AT5G60690). KANADI (KAN) and ETTIN (ETT)/AUXIN RESPONSE FACTOR 3 (ARF3) (AT2G33860) and ARF4 (Pekker et al., 2005) family, on the other hand, are responsible for the abaxial side determination. Members of the KAN family include KAN1 (AT5G16560) and KAN2 (AT1G32240) (Mc Connel et al., 2001).

Cell proliferation in the early leaf primordium initially decelerates and resumes after differentiation of apical and basal regions. ANGUSTIFOLIA3 (AN3)/AtGRF-INTERACTING FACTOR1 (At-GIF1; AT5G28640) positively regulates cell proliferation in the inner tissue based on mRNA expression. Further analysis, however, showed that the inner tissue and the epidermis are under the control of AN3/At-GIF1 (Kawade et al., 2013). Leaf primordium grows toward the apex and the opposite side forming a crescent shape. Rapid cell division transforms the leaf primordium into a hood-shaped structure. A procambial strand is formed during this phase. The leaf primordium eventually envelopes the shoot apical meristem and is now conical in shape, leading to the formation of the blade sheath boundary. Periclinal division in the adaxial surface of epidermal cells in the blade sheath boundary leads to the protrusion of a ligule primordium. Vascular bundles start to develop at this stage; xylem and phloem are recognizable in the midrib. Stomata formation at the distal region also occurs at this stage. Following ligule primordium differentiation, leaf blade elongation occurs, reaching its maximum length. Leaf sheath elongation remains suppressed until leaf blade elongation reaches completion. During leaf sheath elongation, structures of the epidermis such as stomata and bulliform cells initially become apparent from the apex and then vascular bundles mature. The tip of the leaf blade starts to emerge from the sheath, and at this point, leaf epidermal structures are almost complete. Unequal elongation in abaxial and adaxial cells of the leaf blade results in the curving of the leaf as it emerges from the sheath (Itoh et al., 2005).

14.2.6 Rice Genome Functional Analysis

It has been estimated that rice has 32,000 genes based on the genome sequencing conducted in 2005 (IRGSP, 2005). Around 80% of those genes are putative protein-coding genes. A limited number of loci have had its function annotated (Rice Annotation Project, 2007). In order to increase productivity, it is a must to make use of the resources provided by this project.

Functional analysis of the genome of rice can make use of forward and reverse genetics. Reverse genetics attempts to determine gene function by determining the phenotypic effects of a gene sequence, which was derived from sequencing, while forward genetics attempts to determine the genetic basis of a particular trait. Resources for such an approach have been developed, such as t-DNA (An et al., 2005), Tos17 (Miyao et al., 2003), and transposon and retrotransposon (Upadhaya et al., 2002) insertion lines.

t-DNA insertion mutagenesis makes use of a transposon (Parinov et al., 2004) or t-DNA to disrupt gene function by inserting within the gene. A priori knowledge of the t-DNA or transposon allows the insert to act as a marker for downstream analysis. t-DNA insertion lines have an average of 1.4 copies per line. Insertion lines have been developed for japonica cultivars such as Dongjin, Hwayoung, Nipponbare, Zhonghua 11, Zhonghua 15, and Tainung 67 with the pertinent information such as denotation, phenotype, and flanking sequence tags (FSTs) available publicly online (Zhang et al., 2006).

Transposon mutagenesis makes use of Activator-Dissociation (Ac/Ds) system (Chin et al., 1999) and Enhancer/Suppressor-Mutator (En/Spm-dSpm) system (van Enckevort et al., 2005). Ac/Ds system alters maize genome by replication and transposition across maize genomes in preference to gene-rich regions.

Retrotransposon mutagenesis makes use of Tos17, which is an endogenous rice retrotransposon. It was identified using RT-PCR degenerate primers based on LTR retrotransposons, particularly the conserved amino acid sequence for the copia superfamily. Tos17 is induced by tissue culture, and it is the only type of retrotransposon to cause insertional mutagenesis during tissue culture compared to other retrotransposon families (Hirochika et al., 1996). A population of 47,000 Tos17 insertion mutants using Nipponbare have been developed by Miyao et al. (2003).

Gene knockout can be induced to identify the resulting phenotypes from the mutation. However, 29% of rice genes have multiple copies (IRGSP, 2005) and some genes are only required for stress response. A loss of function of a particular gene will not necessarily produce a phenotype if its function is replaced by a gene of identical function or its function is only required in a particular situation. Moreover, in order to utilize the genetic resource from insertional mutagenesis, FSTs for each mutant line must be isolated and identified, which is time-consuming, laborious, and costly. An alternative method is a gain-of-function approach through gene activation. An example of this would be activation tagging, which makes use of transcriptional enhancers from the cauliflower mosaic virus 35S promoter. The enhancer is positioned in the left border of the t-DNA, the insertion of which would randomly enhance the transcription of a particular gene in the vicinity of the insertion site (Jeong et al., 2002). The optimum distance for transcription enhancement is 7 kb of the insertion site (Wan et al., 2009). Another method is the full-length cDNA overexpressor gene hunting system.

The full-length cDNA overexpression system (FOX gene hunting system) (Ichikawa et al., 2006; Kondou et al., 2009) makes use of the full-length cDNA as it contains all the information for the production of functional mRNA and proteins and it is constitutively overexpressed in a host system. *Arabidopsis* FOX lines were first developed using this approach using 10,000 fl-cDNA derived from *Arabidopsis thaliana* (Seki et al., 2002). An equal molar ratio of each fl-cDNA was used to generate plant expression vector, and an *Agrobacterium* library containing the vector is generated. The *Agrobacterium* library is in turn used to introduce the fl-cDNA into *Arabidopsis* to generate T_1 seeds. The T_1 seeds are selfed to generate T_2 seeds, which in turn are used for screening. Once a plant or a line of a phenotype of interest is identified, the responsible fl-cDNA can be easily isolated and its identity determined. More than 23,000 of *Arabidopsis* FOX lines were generated using this approach (Kondou et al., 2009).

Rice FOX was developed to analyze rice gene function. It uses fl-cDNA derived from rice, which is then constitutively overexpressed in the rice host. Approximately 12,000 independent rice FOX lines were generated for rice FOX utilizing 13,980 rice fl-cDNA. Approximately 16.6% of the rice FOX lines manifest an apparent phenotypic change. Nakamura et al. (2007) observed an association between extreme dwarfism and overexpression of gibberellin 2-oxidase or RING finger-containing protein, dwarfism with overexpression of eukaryotic translation-initiation factor 4G and green callus and overexpression of golden2-like transcription factor (GLK1). Verification was done through retransformation and overexpression of the above-mentioned fl-cDNA into the host, and it was shown that the same phenotypic changes were exhibited (Nakamura et al., 2007).

Functional genome analysis can also be done through the use of heterologous gene overexpression system or by using fl-cDNA library derived from one organism and then using another plant species as a host system. Such is the case in the rice FOX *Arabidopsis* lines wherein rice fl-cDNA is overexpressed in an *Arabidopsis* host using pRiceFox expression vector. More than 33,000 rice FOX *Arabidopsis* lines were generated using 13,000 independent rice fl-cDNAs. A broad-spectrum disease resistance (*BSR1*) to both bacterial (*P. syringae* and *X. oryzae*) and fungal pathogens (*C. higginsianum* and *M. oryzae*) in dicotyledonous and monocotyledonous plants was identified using this type of population (Dubouzet et al., 2011).

14.2.7 Semi-Quantitative PCR

Semi-quantitative PCR compares the relative abundance (quantitation) of target templates. PCR-based methods of nucleic acid quantitation have the advantage of greater sensitivity in target detection. Comparison with traditional methods of nucleic acid quantitation revealed a good correlation with PCR-based method by comparing multidrug resistance (MDR1) gene expression in MDRKB cell lines (Noonan et al., 1990). Park and Mayo (1991) studied temporal expression and localization of the progesterone receptor (PR) mRNA in the rat ovary and utilized PCR-based quantitation and *in situ* hybridization. The proportion of RNA of the compared templates is equal in both methods.

14.2.8 Gas Exchange Measurements

The response of C_4 species to varying atmospheric CO_2 concentrations is also different from that of C_3. C_4 species can exhibit a higher rate of photosynthesis at a lower atmospheric CO_2 concentration. This is due to its CO_2 concentrating mechanism resulting in an elevated CO_2 concentration around rubisco. C_3 species, on the other hand, is dependent on the atmospheric CO_2 concentration, resulting in a much lower rate of photosynthesis at lower concentrations of CO_2. The measurement of photosynthesis is conducted through gas exchange measurements using an infrared gas analyzer (IRGA) apparatus. It is a common technique used in measuring the net rate of photosynthetic carbon assimilation in individual leaves, a whole plant or canopy. Gas exchange measurement of a particular leaf is conducted by enclosing the leaf in a chamber and then determining the change in CO_2 concentration in the air supplied within the chamber. Measurement of the concentration of gases inside the chamber is conducted by using infrared gas analysis.

Figure 14.2 shows the comparison of the responses of four different plants to varying concentrations of atmospheric CO_2. Maize, sorghum, and green foxtail millet (*Setaria viridis*) represent the C_4 species, while rice represents C_3. Green foxtail millet (*S. viridis*) of the grass family is currently gaining much interest as a new model for studying C_4 photosynthesis. The difference in response is often used as an indicator of the type of photosynthesis (Downton and Tregunna, 1968). The point in which the curve intersects the X-axis is called the CO_2 compensation point. This corresponds to the concentration of CO_2 wherein the net carbon assimilation rate is zero or when the rate of carbon assimilation is equal to respiration. CO_2 compensation points of C_4 species are very low generally at around 0–10, while C_3 species has a compensation point of around 50 at 30°C rising up to 70 at 35°C. The initial slope of the curve, on the other hand, represents the carboxylation efficiency. C_4 species generally have a steeper slope and consequently higher carboxylation efficiency as compared to C_3. The difference is primarily due to the higher affinity of PEP carboxylase to CO_2 as compared to rubisco at cellular pH (Edwards and Walker, 1983).

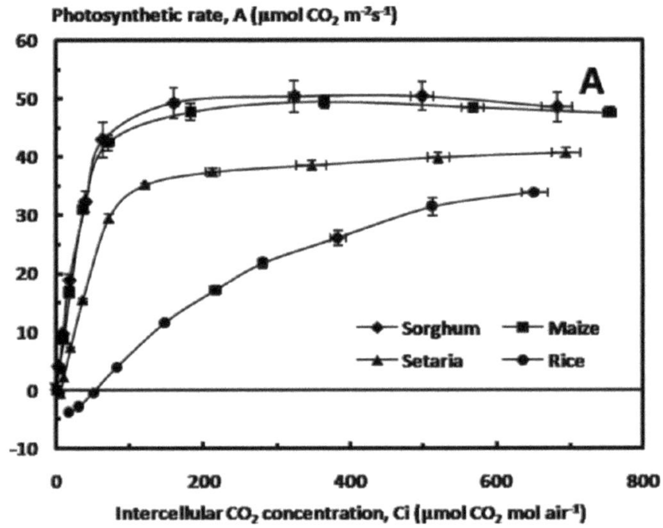

FIGURE 14.2 Photosynthetic CO_2 assimilation as a function of internal CO_2 concentration of sorghum, maize, *Setaria*, and rice. Measurement conditions were 400 μbar CO_2, 30°C leaf chamber temperature, and 1500 μmol photons m^{-2}/s.

14.3 Conclusions

Rice functional genomics to identify novel associations in gene and gene function is an essential field of study especially today that sequencing has become common and cheap. There is an increasing need to make sense of the data generated from sequencing endeavors, the resources of which are invaluable in the field of plant breeding. This is especially true for cutting-edge development of lines with limited known genetic resources, where the goal is not just focused on one trait, but on the total overhaul of an organism in the noble goal of attaining food security.

REFERENCES

Adachi S. 2013. The mesophyll anatomy enhancing CO_2 diffusion is a key trait for improving rice photosynthesis. *Journal of Experimental Botany*. 64(4): 1061–1072.

Alojado JM. 2014. History of Philippine Cuisine. (http://www.philippine-islands.ph/cn/history_of_philippine_cuisine-aid_8.html). Alojado Publishing International. Retrieved: 2014 May 6.

Andrews TJ and GH Lorimer. 1987. *The Biochemistry of Plants: A Comprehensive Treatise, Vol. 10, Photosynthesis* (eds Hatch, M. D. & Boardman, N. K.) 131–218. New York: Academic Press.

Berner RA and Z Kothavala. 2001. GEOCARB III: a revised model of atmospheric CO2 over Phanerozoic time. *American Journal of Science* 301: 182–204.

Blench R. 2006. *Archaeology, Languages, and the African Past*. Lanham: Alta Mira Press, 217–219.

Bowes et al. 1971. Phosphoglycolate production catalyzed by ribulose diphosphate carboxylase. *Biochemical and Biophysical Research Communications* 45: 716–722.

Bowes et al. 2002. C4 mechanisms in aquatic angiosperms: comparisons with terrestrial C4 systems. *Functional Plant Biology* 29: 379–392.

Brar DS and G Khush. 2003. Utilization of wild species of genus Oryza in rice improvement. In: Nanda, J.S. and Sharma, S. D. (eds.), *Monograph on Genus Oryza*, pp. 283–309. Science Publishers. Enfield, New Hampshire and Plymouth, UK, pp. 283–309.

Cheng et al. 2003. Polyphyletic origin of cultivated rice: Based on the interspersion pattern of SINEs. *Molecular Biology and Evolution*. 20: 67–75.

Chitwood et al. 2012. Leaf asymmetry as a developmental constraint imposed by auxin-dependent phyllotactic patterning. *Plant Cell* 24: 2318–2327.

Downton WJS and EB Tregunna. 1968. Carbon dioxide compensation – its relation to photosynthesis carboxylation reactions, systematic of the Graminae and leaf anatomy. *Canadian Journal of Botany* 46: 207–215.
Dubouzet et al. 2011. Screening for resistance against Pseudomonas syringae in riceFOX Arabidopsis lines identified a putative receptorlike cytoplasmic kinase gene that confers resistance to major bacterial and fungal pathogens in Arabidopsis and rice. *Plant Biotechnology Journal* 9: 466–485.
Edwards GE and DA Walker. 1983. *C3, C4: Mechanism and Cellular and Environmental Regulation of Photosynthesis.* Oxford: Blackwell Scientific Publications.
Efroni et al. 2010. Morphogenesis of simple and compound leaves: A critical review. *Plant Cell* 22: 1019–1032.
Ellis JR. 2010. Tackling unintelligent design. *Nature* 463(7278): 164–165.
Food and Agriculture Organization (fao.org) accessed 27 October 2014.
Fuller DQ. 2007. Contrasting patterns in crop domestication and domestication rates: recent archaeo-botanical insights from the Old World. *Annals of Botany* 100(5): 903–924.
Fuller et al. 2010. Consilience of genetics and archaeobotany in the entangled history of rice. *Archaeological and Anthropology Sciences* 2: 115–131.
Garris et al. 2005. Genetic structure and diversity in Oryza sativa L. *Genetics*169:1631–1638.
Glaszmann JC. 1987. Isozymes and classification of Asian rice varieties. *Theoretical and Applied Genetics* 74: 21– 30.
Gopalan et al. 2007. *Nutritive Value of Indian Foods.* National Institute of Nutrition (NIN), ICMR.
Hausler et al. 2002. Overexpression of C4-cycle enzymes in transgenic C3 plants: a biotechnological approach to improve C3-photosynthesis. *Journal of Experimental Botany* 53: 591–607.
Higham C and TLD Lu. 1998. The origins and dispersal of rice cultivation. *Antiquity* 72: 867–877.
Higuchi-Takeuchi et al. 2013. High-throughput analysis of rice genes by means of the heterologous full-length cDNA overexpressor (FOX)-hunting system. *The International Journal of Developmental Biology.* 57:517–523.
Hirochika et al. 1996. Retrotransposons of rice involved in mutations induced by tissue culture. *Proceedings of the National Academy of Sciences* 93: 7783–7788.
Ichikawa et al. 2006. The FOX hunting system: an alternative gain-of-function gene hunting technique. *The Plant Journal: For Cell and Molecular Biology* 48(6): 974–985.
International Rice Genome Sequencing Project (IRGSP). 2005. The map-based sequence of the rice genome. *Nature* 436: 793–800.
Itoh et al. 2005. Rice plant development: from zygote to spikelet. *Plant Cell Physiology* 46: 23–47.
Jackson MT and RJL Lettington. 2003. Conservation and use of rice germplasm, an evolving paradigm under the International Treaty on Plant Genetic Resources for Food and Agriculture. In: Sustainable rice production for food security. Proceedings of the 20th session of the International Rice Commission. Bangkok, Thailand, 23–26 July 2002, pp. 75–88.
Jin et al. 2006. Antisense inhibition of Rubisco activase increases Rubisco content and alters the proportion of Rubisco activase in stroma and thylakoids in chloroplasts of rice leaves. *Annals of Botany* 97: 739–744.
Kawade et al. 2013. Non-cell-autonomously coordinated organ-size regulation in leaf development. *Development* 137: 4221–4227.
Klee et al. 2000. Four thousand years of plant exploitation in the Chad Basin of northeast Nigeria. I. The archaeobotany of Kursakata. *Vegetation History and Archaeobotany* 9: 223–237.
Kondou et al. 2009. Systematic approaches to using the FOX hunting system to identify useful rice genes. The Plant Journal: For Cell and Molecular Biology. *The Plant Journal: For Cell and Molecular Biology.* 57(5): 883–894.
Lee JS. 2010. Stomatal opening mechanism of CAM plants. *Journal of Plant Biology* 53: 19–23.
Leegod et al. 2000. *Photosynthesis: Physiology and Metabolism.* Amsterdam: Elsevier Science Publishing.
Li et al. 2006. Rice domestication by reducing shattering. *Science* 311: 1936–1939.
Martin et al. 2012. Rubisco activity and regulation as targets for crop improvement. *Journal of Experimental Botany.* 64(3): 717–730.
Mathews CK and KE Van Holde. 1990. *Biochemistry.* Redwood City, CA: The Benjamin/Cummings Publishing Co., Inc. pp. 643–669.
Matsuo et al. 1997. *Science of the Rice Plant.* Tokyo: Food and Agriculture Policy Research Center.
Mcconnell et al. 2001. Role of PHBULOSA and PHAVOLUTA in determining radial patterning in shoots. *Nature* 411: 709–713.

Mckown AD and NG Dengler. 2007. Key innovations in the evolution of Kranz anatomy and C4 vein pattern in Flaveria (Asteraceae). *American Journal of Botany* 94(3): 382–399.

Mishra B. 2002. Varietal improvement for rice production in India. In: Nguyen, VN. (Ed.) *Genetic diversity in rice production, Case studies from Brazil, India and Nigeria*. Food and Agriculture Organization of the United Nations (FAO), Rome, Italy, pp. 37–91.

Mitchell PL and JE Sheehy. 2006. Supercharging rice photosynthesis to increase yield. *New Phytologist* 171: 688–693.

Miyao et al. 2003. Target site specificity of the Tos17 retrotransposon shows a preference for insertion within genes and against insertion in retrotransposon-rich regions of the genome.*The Plant Cell* 15:1771–1780.

Monteith JL. 1977. Climate and the efficiency of crop production in Britain. *Philosophical Transactions of the Royal Society* 281: 277–294.

Nakamura et al. 2007. A genome wide gain of function analysis of rice genes using the FOX hunting system. *Plant Molecular Biology* 65: 357–371.

Noonan et al. 1990. Quantitative analysis of MDR1 (multidrug resistance) gene expression in human tumors by polymerase chain reaction. *Proceedings of the National Academy of Sciences* 87: 7160–7164.

O'Connor et al. 2014. A Division in PIN-Mediated Auxin Patterning during Organ Initiation in Grasses. *PLoS Computational Biology* 10(1): e1003447.

Oka HI. 1988. *Origin of Cultivated Rice*. Japan Science Society. Press/Elsevier, Tokyo/Amsterdam.

Parinov et al. 2004, Tol2 transposon-mediated enhancer trap to identify developmentally regulated zebrafish genes in vivo. *Developmental Dynamics*. 231: 449–459.

Park OK and KE Mayo. 1991. Transient expression of progesterone receptor messenger RNA in ovarian granulosa cells after the preovulatory luteinizing hormone surge. *Molecular Endocrinology* 5: 967–978.

Plants Database USDA, NRCS. 2014. The PLANTS Database (http://plants.usda.gov, 6 May 2014). National Plant Data Center, Baton Rouge, LA 70874-4490 USA, and ITIS Retrieved May 6, 2014.

Prasad B and GC Eizenga. 2008. Rice sheath blight disease resistance identified in Oryza spp. accessions. *Plant Diseases*.92: 1503–1509.

Porteres R. 1956. Taxonomic agrobotanique der riz cultives O. sativa Linne. et O. glaberrima Steudelo. *Journal d'agriculture traditionnelle et de botanique appliquée*. 3: 341–384.

Reinhardt et al. 2000. Auxin regulates the initiation and radial position of plant lateral organs. *Plant Cell* 12: 507 518.

Ren-Yong et al. 1990. Loss assessments and factor-finding analysis of grain post-production systems in China. In *Proceedings of the Thirteenth ASIAN Seminar on Grain Postharvest Technology, Brunei Darussalam*, 4–7 Sept. 1990. ed. Naewbanij, J.O. ASIAN grain post-harvest programme, Bangkok. pp. 370–392.

Rice Annotation Project. 2007. Curated genome annotation of Oryza sativa ssp. japonica and comparative genome analysis with Arabidopsis thaliana. *Genome Research*. 17: 175–183.

Second G. 1982. Origin of the genetic diversity of cultivated rice (Oryza spp): study of the polymorphism scored at 40 isozyme loci. *The Japanese Journal of Genetics*. 57: 25–57.

Seki et al. 2002. Functional annotation of a full-length Arabidopsis cDNA collection. *Science* 296: 141145.

Semon et al. 2005. The population structure of African cultivated rice Oryza glaberrima (Steud.): evidence for elevated levels of LD caused by admixture with O. sativa and ecological adaptation. *Genetics* 169: 1639–1647.

Sheehy et al. 2008. Harnessing photosynthesis in tomorrow's world: Humans, crop production and poverty alleviation. In: Allen JF, Gantt E, Golbeck JH, Osmond B, eds. *Photosynthesis. Energy from the Sun. 14th International Congress on Photosynthesis*, Springer, pp. 1243–1248.

Sun et al. 2001. Comparison of the genetic diversity of common wild rice (Oryza rufipogon Griff.) and cultivated rice (O. sativa L.) using RFLP markers. *Theoretical and Applied Genetics* 102:57 162.

Tan et al. 2008. Control of a key transition from prostrate to erect growth in rice domestication. *Nature Genetics* 40: 1360–1364.

Tatematsu et al. 2011. Succinic semialdehyde dehydrogenase is involved in the robust patterning of Arabidopsis leaves along the adaxial-abaxial axis. *Plant Cell Physiology*. 52:1340–1353.

Tsukaya H. 2013. Leaf development. *Arabidopsis Book* 11:e0163.

Upadhyaya et al. 2002. An iAc/Ds gene and enhancer trapping system for insertional mutagenesis in rice. *Functional Plant Biology* 29: 547–559

Van Enckevort et al. 2005. EU-OSTID: a collection of transposon insertional mutants for functional genomics in rice. *Plant Molecular Biology* 59: 99–110.

Vaughan DA. 1989. The genus Oryza L. Current status of taxonomy. IRRI Research Paper Serial No. 138.

Vaughan DA. 1994. *The Wild Relatives of Rice: A Genetic Resources Handbook*. IRRI, Manila, Philippines.

Vaughan et al. 2003. Diversity in the Oryza genus. *Current Opinion in Plant Molecular Biology* 6: 139–146.

Wang ZY and SD Tanksley. 1989. Restriction fragment length polymorphism in Oryza sativa L. *Genome* 32: 1113–1118.

Zhang et al. 2006. RMD: a rice mutant database for functional analysis of the rice genome. *Nucleic Acids Research* 34: D745–748.

Zhao et al. 1998. Distinguishing rice (Oryza sativa, Poaceae) from wild Oryza species through phytolith analysis 2: Finalized method. *Economic Botany* 52:134–145.

Zhu Q and S Ge. 2005. Phylogenetic relationships among A-genome species of the genus Oryza revealed by intron sequences of four nuclear genes. *New Phytologist* 167: 249–2.

Zhu et al. 2003. Would transformation of C3 crop plants with foreign Rubisco increase productivity? A computational analysis extrapolating from kinetic properties to canopy photosynthesis. *Plant, Cell and Environment* 27: 155–165.

15

Tapping the Role of Plant Volatiles Inducing Multi-Trophic Interactions for Sustainable Agricultural Production

Ratul Moni Ram
KVK Saran

Ashim Debnath
Acharya Narendra Deva University of Agriculture & Technology, Kumarganj

Khumbarti Debbarma
University of Agricultural Sciences, GKVK

Yengkhom Linthoingambi Devi
College of Agriculture, IGKV

Jayakumar Pathma*
Lovely Professional University

CONTENTS

15.1 Introduction

Plant volatiles are secondary metabolites produced by the plants and include an array of low molecular weight (<300 Da) organic compounds with elevated vapor pressure (0.01 kPa) at temperatures as low as 20°C, varied biochemistry, and specific bioactivity. Plants are the primary producers that fix

* Corresponding author's email: pathmajayakumar@gmail.com

DOI: 10.1201/9781003171416-15

atmospheric carbon dioxide and synthesize required carbohydrates, which provide energy for all life forms on earth. Baldwin (2010) reported that nearly one-fifth of the atmospheric carbon dioxide fixed by plants is released back into the atmosphere as volatile organic compounds (VOCs). The plants use these compounds to protect themselves and their neighbors from biotic and abiotic stresses. The concept of "talking trees" postulated in the early 1980s was subjected to criticism until more scientific evidence on plant volatiles triggering defense-related response in neighboring plants was recorded. Rhoades (1983) reported that surprisingly, the willow trees in close vicinity to those infested by tent caterpillars showed resistance to the caterpillar attack as compared to willow trees that grew far apart. It was hypothesized that the neighbors were alerted by the airborne signals from infested trees and hence produced metabolites to resist infestation. Years later, it became evident that plants, when damaged either mechanically or infested by microbes or arthropods, produce distinct organic compounds, of which the majority are volatile and prime the defense mechanisms of neighboring plants to evade biotic stress (Heil & Karban, 2010; Meents & Mithofer, 2020). But the bitter fact is that in most cases, the receiver benefits at the cost of the emitter. This is true in many cases, be it intraspecific communication between plants of the same species or interspecific communication between plants of different species or between plants and organisms belonging to different kingdoms, including insects and microbes.

Plants being sessile are bestowed with the ability to produce an array of secondary metabolites, of which few are volatile. Many secondary metabolites are produced regularly, while a few compounds are stress-induced and have varied functions (Figure 15.1). Nearly 57.31% of the VOCs identified were reported to be produced by both healthy and affected plants (Gulati et al., 2020). These compounds have high vapor pressure, diffuse in the atmosphere, and are available as olfactory cues for the receiver within a specific distance, which could be neighboring plants, insects, and microbes in a defined distance. The majority of these infochemicals are beneficial to the producing plants, while a few are detrimental. Beneficial volatiles produced by the plant favor the emitter by (i) inducing "allelopathic effects" hampering weed seed germination and growth (Jabran et al., 2015); (ii) acting as "allomones" that repel insect herbivores and provide a direct advantage to the emitter (Tlak Gajger & Dar, 2021; Wahengbam et al., 2021); and (iii) acting as "synomones," which attract arthropod carnivores that devour arthropod herbivore pests damaging the host plant thereby indirectly benefitting the emitter or attract pollinators benefitting both producer and receiver. However, certain volatiles produced by plants attract their herbivore pest or phytopathogen and are thus beneficial to the receivers, but detrimental to the producer and are classified as "kairomones" (Pathma et al.

FIGURE 15.1 Bioactivity of plant volatiles and their role in sustainable agriculture.

2021a, b). Few plant volatiles recruit beneficial microbes and engineer healthy plant microbiomes that improve plant growth and immunity against various biotic and abiotic stresses (Pathma et al. 2021a, b). Plant-derived volatile signaling molecules belong to various chemical classes and induce specific communication. The effectiveness of these compounds depends upon the existing climatic conditions, especially temperature, humidity, and wind velocity (Heil & Karban, 2010). This chapter will provide us with a better understanding of the biochemistry, biosynthesis, and bioefficacy spectrum of these VOCs. This knowledge will help us tap these biomolecules and synthesize their mimics, which will act as highly promising, eco-friendly, and cost-effective substitutes for crop protection chemicals and synthetic fertilizers and help in sustaining crop productivity in an eco-friendly and cost-effective manner.

15.2 Volatile Organic Compounds (VOCs): Classification and Biosynthesis

Volatile organic compounds from plants are of low molecular weight and represent nearly one percent of secondary metabolites produced by plants. VOCs are highly diverse in terms of their biochemistry, biosynthetic pathways involved, site of synthesis, and bioactivity. However, they possess certain unique features; for example, their lipophilic nature makes them bypass cellular barriers and diffuse into the atmosphere. They are synthesized by methylation, acylation, or reduction of hydrophilic precursors. Molecular studies revealed that genes controlling their biosynthetic pathways have undergone duplications and several alterations at specific sites, thereby changing the substrate-specificity of enzymes leading to diverse products. These biomolecules either alone or in combination induce specific reactions either within the organism or between other organisms, including plants, insects, and microbes that had coevolved and are in close vicinity. Different VOCs are produced both by aerial plant parts (leaves, stem, flowers, and seeds) and by subterranean root cells, and blends of these volatiles induce multi-trophic interactions among organisms dwelling both above and below ground (Bouwmeester et al., 2019). The majority of the VOCs are released during stress when plant cells are subjected to damage and play an important role in biosignaling between plants, microbes, and herbivores, thereby altering their behavior and physiology to survive harsh conditions (Holopainen, 2004). They help plants evade pathogen infections and arthropod infestations by warning the neighboring plants as well as priming their immunity (Erb & Kliebenstein, 2020). They arrest pest infestations and improve productivity by inviting predators, parasitoids, and beneficial pollinators toward them (De Boer et al, 2004; McCormick et al., 2012). They also recruit phyllosphere and rhizosphere microbes that improve plant health (Junker & Tholl, 2013). Nevertheless, many volatiles are also attractive to plant herbivores and invite the insects to prefer the emitters as a host plant for insect nutrition and oviposition (Derstine et al., 2020).

VOCs from plant could be broadly categorized into four major classes, namely (i) terpenoids, (ii) phenylpropanoids/benzenoids, (iii) fatty acid derivatives, and (iv) compounds from amino acids excluding L-phenylalanine (Baldwin, 2010).

15.2.1 Terpenoids

The majority of plant secondary metabolites are terpenoids with nearly 40,000 compounds recorded to date (Yu & Utsumi, 2009). Terpenoids are synthesized from the condensation of five-carbon compounds, namely isopentenyl pyrophosphate (IPP) and its allyl isomer dimethylallyl pyrophosphate (DMAPP) (Karunanithi & Zerbe, 2019). Their production is both species-specific and site-specific. For instance, MEP pathway is used for the formation of cytosolic sesquiterpene in snapdragon flowers (Dudareva et al., 2005), whereas both MEP and MVA pathways are used in carrot leaves and roots for the formation of sesquiterpenes (Hampel et al., 2005). In higher plants, terpenoids are produced in two specific sites, namely cytosol and plastids, via independent pathways, viz. mevalonic acid (MVA) and 2-methylerythritol 4-phosphate/1-deoxy-xylulose 5-phosphate (MEP/DOXP), respectively (Figure 15.2).

In the cytosolic mevalonic acid pathway, the process is initiated by condensation of acetyl-CoA into 3-hydroxy-3-methylglutaryl-CoA, followed by reduction to mevalonic acid trailed by phosphorylation and decarboxylation leading to the formation of the final product isopentenyl pyrophosphate (IPP) (Lange et al., 2000). The plastidial methylerythritol phosphate (MEP) pathway starts with the condensation

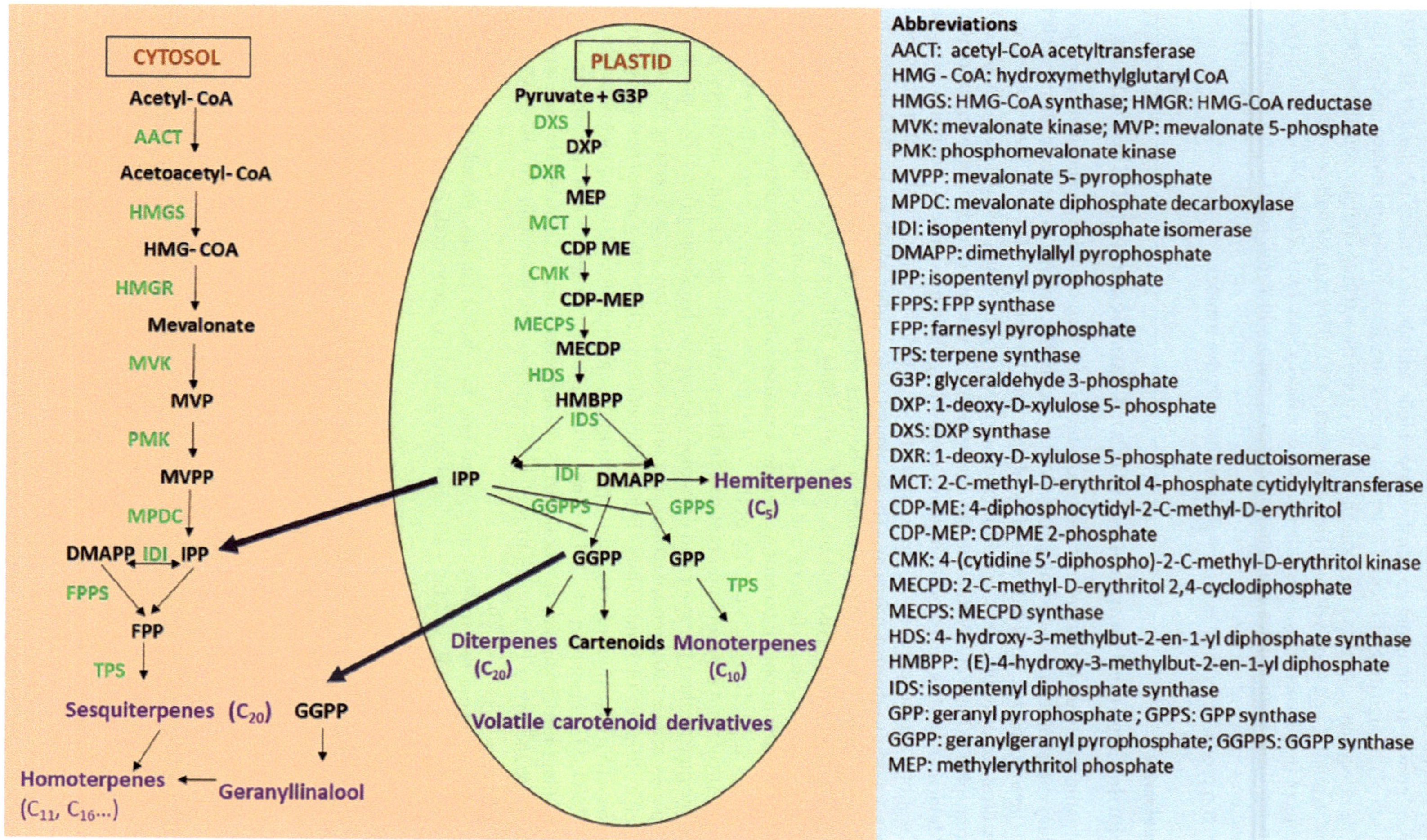

FIGURE 15.2 Biosynthesis of volatile terpenoids in plants.

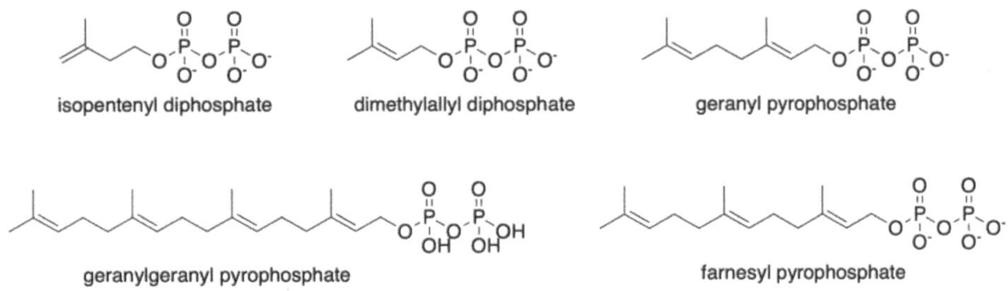

FIGURE 15.3 Structure of terpene precursors.

of D-glyceraldehyde 3-phosphate (G3P) along with pyruvate for the production of 1-deoxy-D-xylulose 5-phosphate (DOXP). Isomerization and reduction take place for the formation of intermediates. Different substrates such as DMAPP and IPP are produced, which further leads to the production of geranyl pyrophosphate (GPP), farnesyl pyrophosphate (FPP), and geranylgeranyl pyrophosphate (GGPP), the precursors for terpene synthesis (Figure 15.3). The mevalonic acid (MVA) pathway synthesizes only IPP, whereas both IPP and DMAPP are produced in the methylerythritol phosphate (MEP) pathway with a 6:1 ratio (Rohdich et al., 2003). For maintaining a balance between both the pathways, isopentenyl diphosphate isomerase (IDI) plays a very important role, which converts IPP to DMAPP. IPP, DMAPP, GPP, and FPP act as connecting metabolites between both MVA and MEP pathways, and they transport through the plastid (inner envelope membrane) with transporter (Wang et al., 2019). Three molecules of IPP condense with one molecule of DMAPP in plastid to form GGPP (C20), while two molecules of IPP condense with one molecule of DMAPP in the cytosol to form FPP. Terpenoid precursors, GGPP and FPP, are acted by enzymes terpene synthases/cyclases (TPSs) in an interestingly specific way owing to the TPS gene, which ultimately decides the nature and diversity of the terpenoid produced (Baldwin, 2010). Thus, terpenoid diversity results from the ability of enzymes to modify TPS products via a different process such as dehydrogenation, hydroxylation, and acylation (Dudareva et al., 2004). Volatile terpenoids are mainly produced in plants from carotenoids by dioxygenase cleavage, enzymatic transformation, and acid-catalyzed conversion (Winterhalter & Rouseff, 2001). Products such as sesquiterpenes are derived in the mevalonic acid (MVA) pathway, whereas monoterpenes, hemiterpenes, and diterpenes are produced in the plastidial methylerythritol phosphate (MEP) pathway (Hsieh et al., 2008).

Terpenoids are classified based on the number of carbon units (Table 15.1). The simplest terpenoid is isoprene, which is referred to as hemiterpene with one 5-carbon unit. Monoterpenes include C_{10} compounds [(*E*)-β-ocimene, linalool], while sesquiterpenes include C_{15} compounds [*(E, E)*-β-farnesene, *(E)*-β-caryophyllene, α-humulene], and homoterpenes contain structures with varying number of carbons such as *(E)*-4,8-dimethylnona-1,3,7-triene (DMNT) with eleven carbons and *(E, E)*-4,8,12-trimethyltrideca-1,3,7,11-tetraene (TMTT) with 16 carbons, while diterpenes have 20 carbons (Boncan et al., 2020). Generally, leaves produce monoterpenes, while flowers produce sesquiterpenes; however, glandular trichomes and resin ducts on leaves can produce both (Dudareva et al., 2006). When tissue damage occurs due to herbivore or pathogen attack, more predominantly monoterpenes and sesquiterpenes and, in some cases, certain diterpenes are released, which act as phytoalexins that alarm and mediate the defense mechanisms in neighboring plants apart from resisting the attack themselves (Dudareva et al., 2004; Holopainen & Blande, 2012).

15.2.2 Phenylpropanoid/Benzenoid Compounds

Phenylpropanoids and benzenoids derived from amino acid L-phenylalanine (Phe) form the second largest class of VOCs engaged in reproduction and defense (Knudsen et al., 2006). Phenylalanine (Phe) is synthesized through the shikimate biosynthetic pathway in plastid, but it is further converted to volatile compounds outside this organelle (Maeda & Dudareva, 2012). L-Phenylalanine ammonia-lyase (PAL) catalyzes the biosynthesis of phenylpropanoids/benzenoids in most cases and deaminates phenylalanine to trans-cinnamic acid (Schnepp & Dudareva, 2007). Cinnamic acid (CA) is later converted into benzenoids by shortening the propyl side chain via the oxidative or non-oxidative pathway, or sometimes both

TABLE 15.1
Diverse Classes of Volatile Terpenoids

S. No.	VOC Class	Biosynthesis Pathway	Nomenclature	Structure
1.	Hemiterpene	MEP	Isoprene	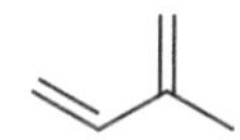
2.	Monoterpene	MEP	β-Ocimene	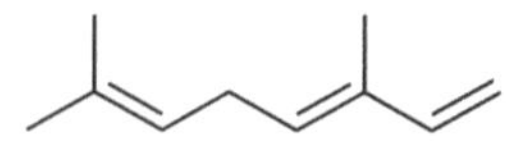
3.	Monoterpene	MEP	Linalool	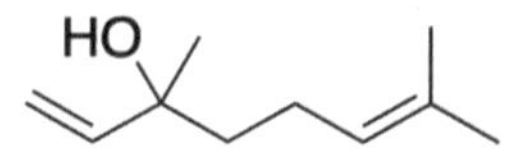
4.	Sesquiterpene	MVA	β-Farnesene	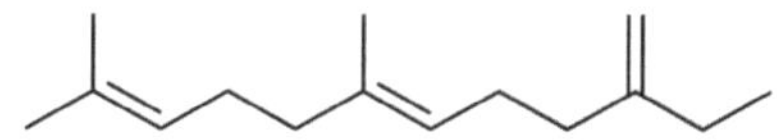
5.	Sesquiterpene	MVA	β-Caryophyllene	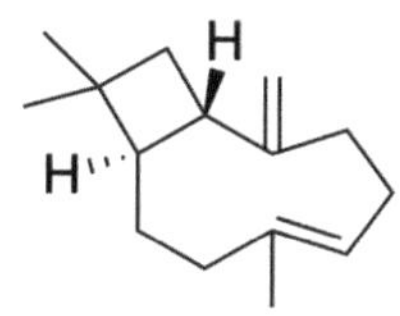
6.	Sesquiterpene	MVA	α-Humulene	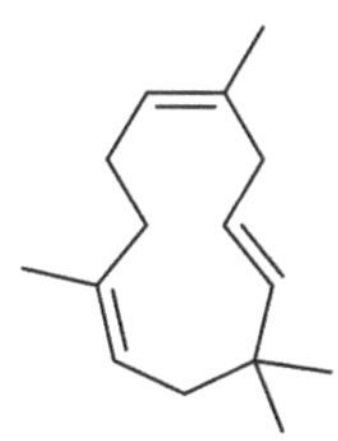
7.	Homoterpene	MEP and MVA	4,8-Dimethylnona-1,3,7-triene (DMNT)	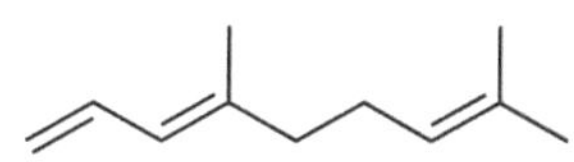
8.	Homoterpene	MEP and MVA	*(E, E)*-4,8,12-trimethyltrideca-1,3,7,11-tetraene (TMTT)	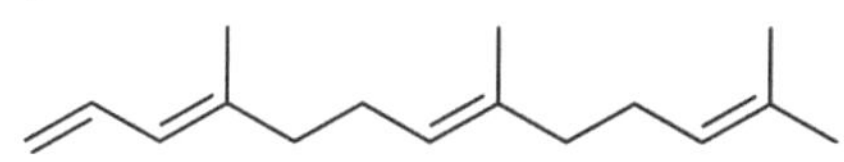

(Orlova et al., 2006). The oxidative pathway begins with the activation of CA to its CoA thioester and ultimately forms benzoyl-CoA after oxidation, hydration, and cleavage of the b-keto thioester (Klempien et al., 2012). In non-oxidative pathway, NAD+ dependent benzaldehyde dehydrogenase converts benzaldehyde to benzoic acid (Qualley et al., 2012). In biosynthetic pathways of volatile benzenoids, two enzymes acyltransferases and methyltransferases were found to be effective (D'Auria, 2006; D'Auria et al., 2003). Figure 15.4 is the schematic representation of the biosynthesis of phenylpropanoids.

Phenylpropanoids usually contain an aromatic phenyl group (six carbons) with a propane (three carbons) side chain. The diverse biological activity is attributed to the location of the propenyl double bond and the substituent on benzene ring. Based on the structure, phenylpropanoids can be classified into five major groups, i.e., monolignols, flavonoids, phenolic acids, coumarins, and stilbenes, of which the first three are common among all terrestrial plants (Sharma et al., 2019).

Flavonoids contain a C_{15} backbone depicted as C_6–C_3–C_6 comprising of two phenyl rings along with a heterocyclic pyran ring, based on which they are further categorized as flavones (apigenin, baicalein, and luteolin), isoflavones (rotenone, purerarin, pterocarpin, and trifolirhizin), flavonols (quercetin, kaempferol, and myricetin), flavanonols (dihydroquercetin, dihydromyricetin, and dihydrokaempferol), flavanones (hesperidin, naringenin, and liquiritigenin), proanthocyanidins, anthocyanidins (cyanidin), and aurones (sulfuretin) (Nabavi et al., 2020). Figure 15.5 shows the representative structures of few flavonoids.

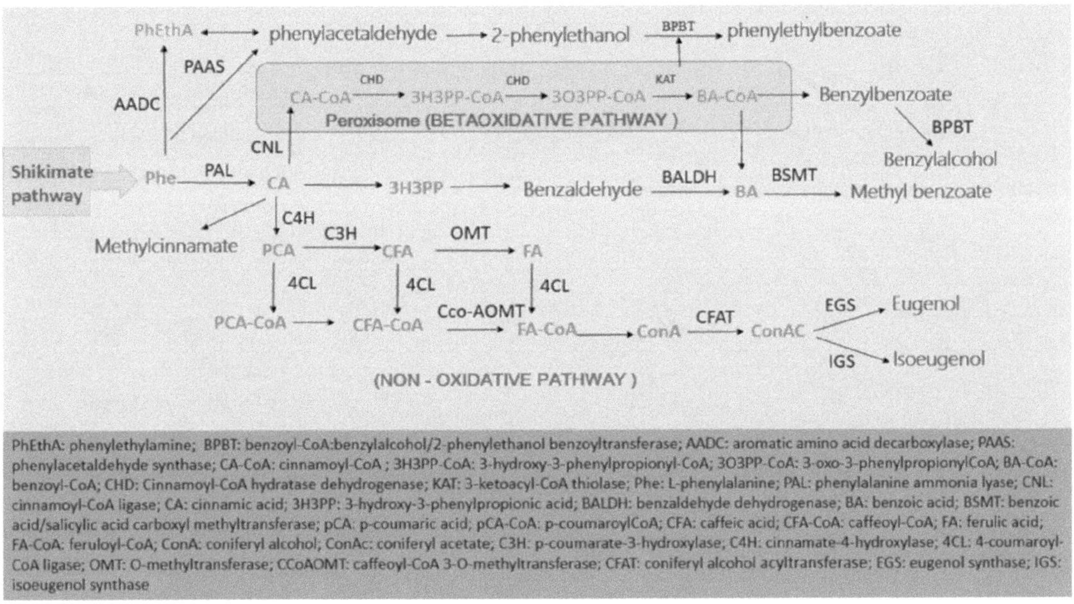

FIGURE 15.4 Biosynthesis of volatile benzenoid/phenylpropanoid compounds.

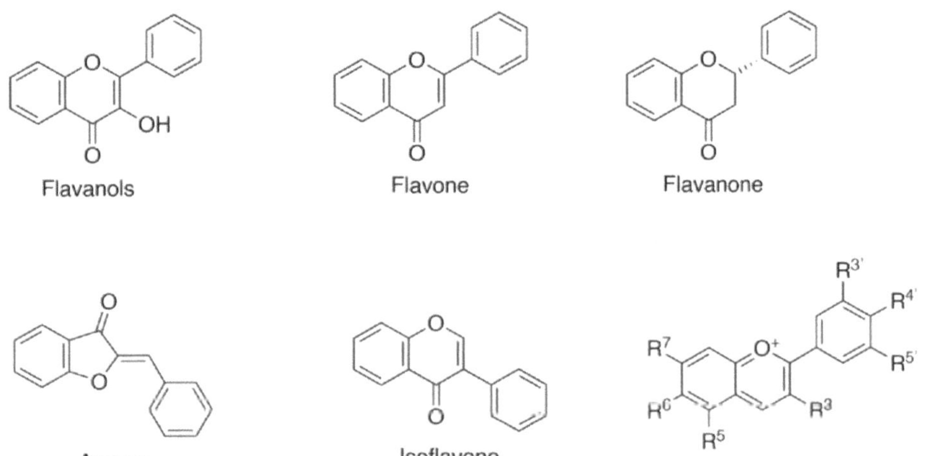

FIGURE 15.5 Representative structures of few flavonoids.

Monolignols are monomers of lignin and lignan and are also referred to as hydroxycinnamyl alcohol. Coniferyl alcohol, sinapyl alcohol, and p-coumaryl alcohol are common monolignols (Figure 15.6), which have similar C6–C3 skeleton, but show varying methoxylation degree in C3 and C5 sites of the aromatic ring. Coniferyl aldehyde via two discrete pathways yields coniferyl alcohol and sinapyl alcohol. Sinapyl aldehyde is reduced to sinapyl alcohol by sinapyl alcohol dehydrogenase; p-coumaryl alcohol is produced from its analogue in the same way. Lignin {$(C6–C3)_n$} is a polymer mainly synthesized by oxidative polymerization of the above three monolignols to form monomeric lignin units, viz. guaiacyl (G), siringyl (S), and p-hydroxyphenyl (H), and the reaction is catalyzed by peroxidases and laccases (Wang et al., 2013; Marchiosi et al., 2020).

Phenolic acids (phenolcarboxylic acids) contain a benzene ring connected with at least one or more hydroxyl (OH) or methoxy (OCH_3) groups. Based on the structure, phenolic acids can be further classified

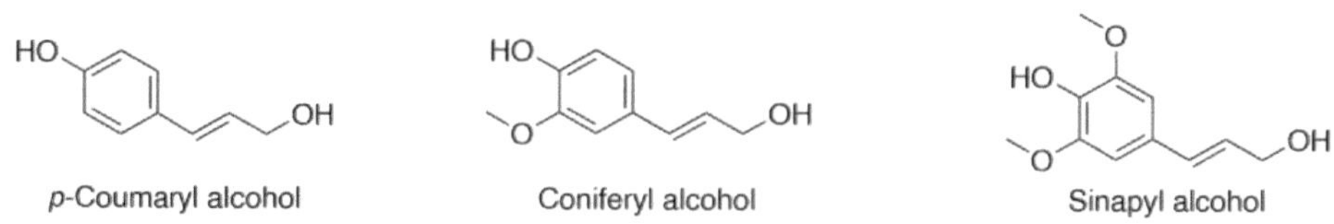

FIGURE 15.6 Representative structures of few monolignols.

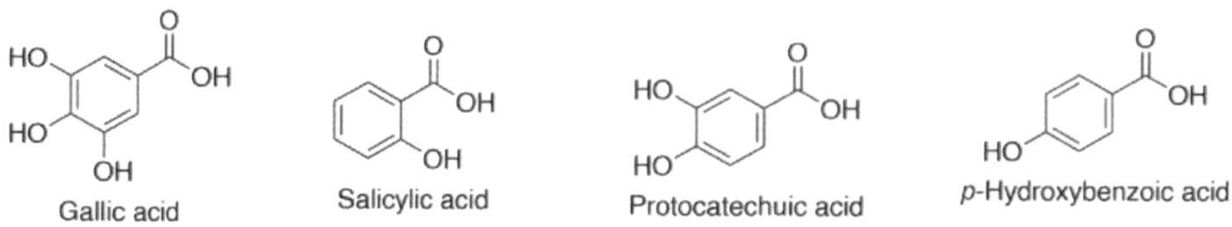

FIGURE 15.7 Representative structures of few members of hydroxybenzoic acid group.

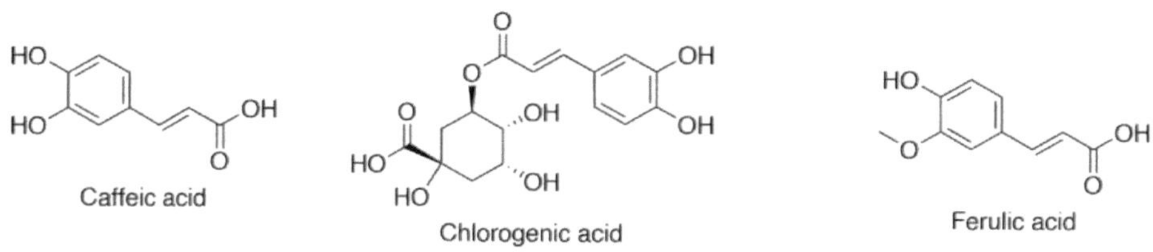

FIGURE 15.8 Representative structures of few members of hydroxycinnamic acid group.

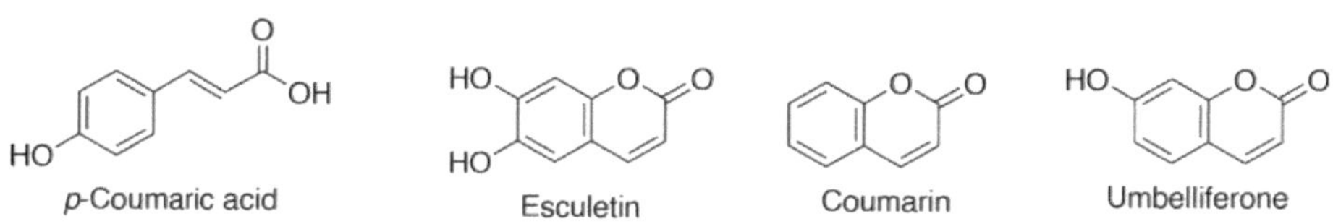

FIGURE 15.9 Representative structures of few members of coumarin.

into two subgroups, namely hydroxybenzoic acid group (C_6–C_1) (Figure 15.7) and hydroxycinnamic acid group (C_6–C_3) (Figure 15.8) (Khadem and Marles, 2010; Widhalm & Dudareva, 2015). Salicylic acid, gallic acid, p-hydroxybenzoic acid, protocatechuic acid, syringic acid, gentisic acid, vanillic acid, isovanillic acid, and veratric acid are members of hydroxybenzoic acid group, while caffeic acid, chlorogenic acid, ferulic acid, isoferulic acid, p-coumaric acid, p-coumaroylquinic acid, sinapic acid, rosmarinic acid, etc., are members of the hydroxycinnamic acid group (Robbins, 2003; Marchiosi et al., 2020).

Coumarins are benzopyrones (C_6–C_3). The term was derived from "Coumarou," a French word referring to tonka beans (*Coumarouna odorata*), from which they were first reported. Based on structure, coumarins can be further divided into (i) simple coumarins, (ii) pyranocoumarins, (iii) furanocoumarins, and (iv) coumarins with substituent group in the pyrone ring (Keating & O'kennedy, 1997). Simple coumarins include coumarin, esculetin, scopoletin, and umbelliferone (Figure 15.9) and contain no furan or pyran group in the basic structure, while pyranocoumarins and furanocoumarins have a pyran or furan group, respectively, on the benzene ring of benzopyrone backbone (Shimizu, 2014).

Stilbenes are small phenylpropanoids with molecular weight approximately ranging between 200–300 g/mol and made of diphenylethylene backbone (C_6–C_2–C_6). Resveratrol is the most commonly found stilbene in plants and occurs in both *cis* and *trans* isomeric forms (Figure 15.10). Roupe et al. (2006)

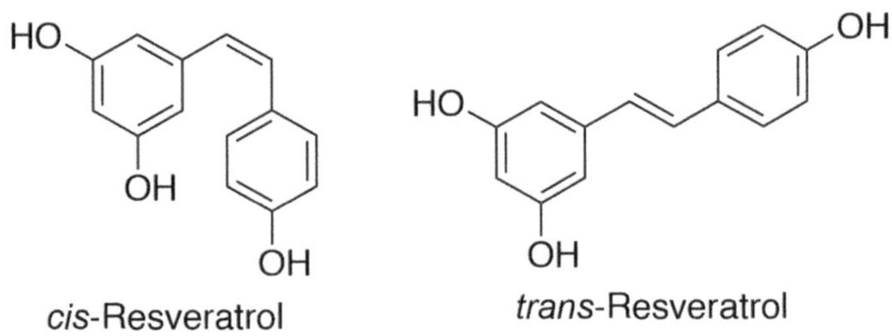

FIGURE 15.10 Representative structures of members of silbene.

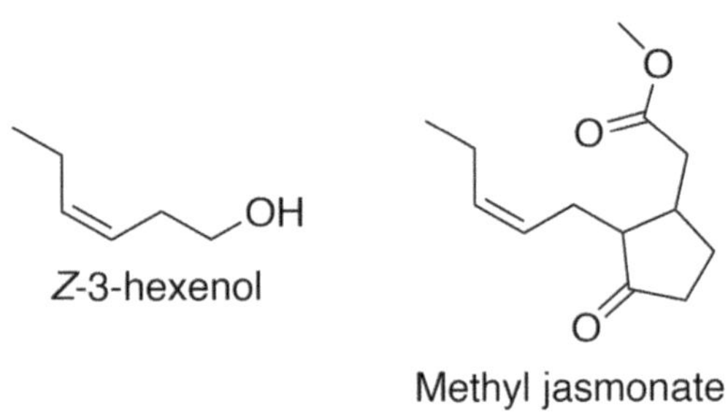

FIGURE 15.11 Representative structures of few fatty acid derivatives.

reported that stilbene production is induced in plants subjected to environmental stress and protects them from wounds, microbes, insect herbivores, and harmful ultraviolet radiations.

15.2.3 Fatty Acid Derivatives

Methyl jasmonates, 1-hexanal, trans-2-hexenal, cis-3-hexenol, nonanal, etc., are few fatty acid derivatives (Figure 15.11) that form an important class of plant VOCs derived from precursors such as linoleic acids or linolenic acids, which are C_{18} unsaturated fatty acids. In the "lipoxygenase (LOX) pathway," linoleic or linolenic acid undergoes oxygenation to form 9-hydroperoxy and 13-hydroperoxy intermediates, which further metabolize to form volatile compounds and operate as two branches (Figure 15.12) (Feussner & Wasternack, 2002). For the formation of jasmonic acid (JA), 13-hydroperoxy intermediate is used in the allene oxide synthase branch where jasmonic acid is then converted into methyl jasmonate in the presence of JA carboxyl methyltransferase (Seo et al, 2001; Song et al, 2005). The hydroperoxide lyase (HPL) branch converts both 9- and 13-hydroperoxy intermediates into C_6 and C_9 aldehydes (3-hexenal or 3,6-nonadienal) and corresponding C12- or C9-ω fatty acids (12-oxo-dodecenoic acid or 9-oxononanoic acid), which act as substrates for alcohol dehydrogenases and undergo reduction reaction giving rise to volatile alcohols and their esters (Figure 15.3) (Grechkin, 1998; Prestage et al., 1999; Akacha et al., 2005; D'Auria et al., 2007; Gigot et al., 2010). These saturated and unsaturated C6/C9 aldehydes and alcohols are synthesized in green organs of plants and are responsible for the "fresh green" aroma of vegetables and fruits apart from being produced at the site of injury and are called GLVs (green leaf volatiles). In addition, these fatty acids are also components of floral volatile blends of several species of plants, including wild snapdragon, oil-secreting *Lysimachia*, and carnation.

15.2.4 Amino Acid Derivatives

The floral scents and fruit aroma contain a blend of organic volatiles containing alcohols, aldehydes, acids, esters, nitrogen- and sulfur-containing compounds, etc., with the majority of them derived from various amino acids such as leucine, isoleucine, valine, alanine, and methionine and their intermediates

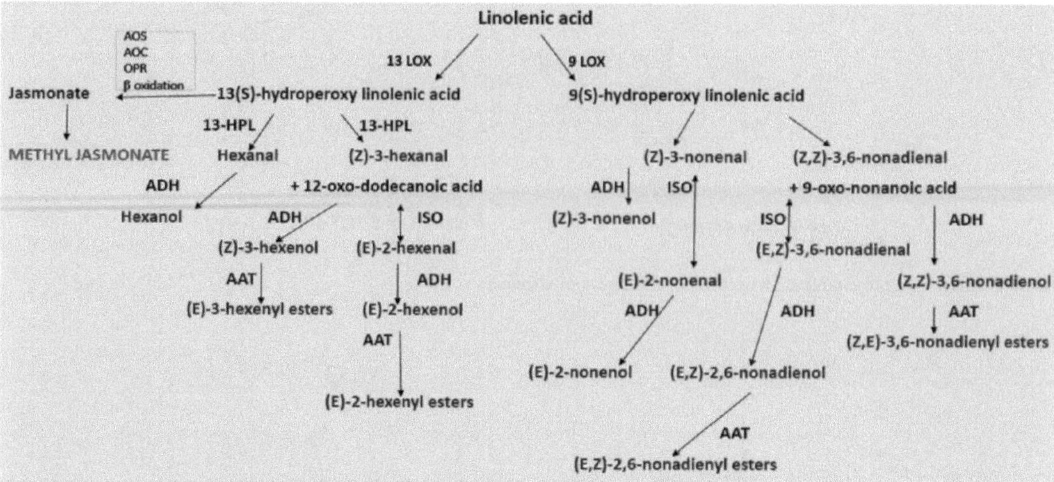

Abbreviations: alcohol acyltransferase (AAT); alcohol dehydrogenase (ADH); allene oxide cyclase (AOC); allene oxide synthase(AOS); 9-hydroperoxide lyase (9-HPL); 13- hydroperoxide lyase (13-HPL); isomerase (ISO); 9-lipoxygenase (9-LOX); 13-lipoxygenase (13-LOX); 12-oxophytodienoate reductase (OPR).

FIGURE 15.12 Biosynthesis of volatile fatty acid derivatives.

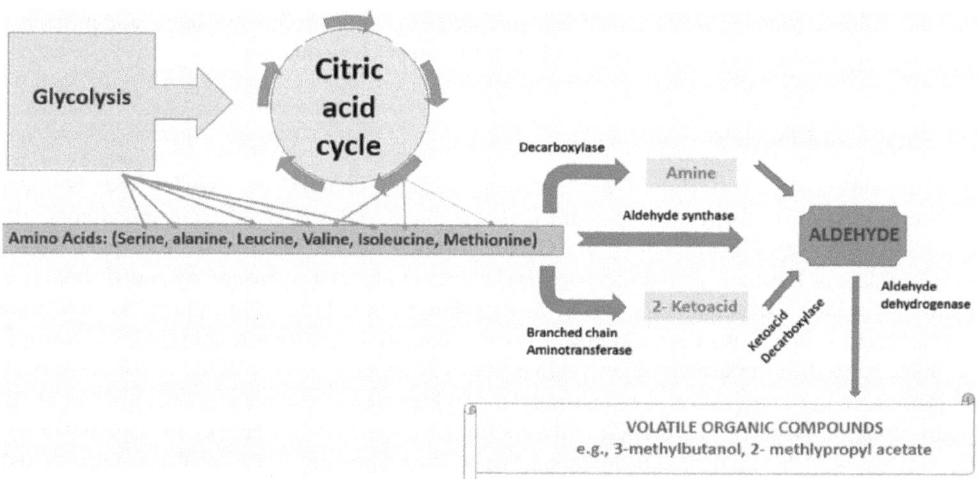

FIGURE 15.13 Biosynthesis of volatiles derived from branched-chain amino acids.

(Knudsen and Gershenzon, 2006). Branched-chain volatile aldehydes, alcohols, and esters in fruits such as strawberry, apple, banana, and tomato originate from branched-chain amino acids, viz. valine, isoleucine, and leucine (El Hadi et al., 2013). Synthesis of amino acid-derived plant volatiles was studied using radiolabeled or isotope-labeled precursors. Amino acids undergo transamination or deamination leading to the production of α-keto acid. The process continues with decarboxylation, reductions, oxidations, and/or esterifications, which lead to the formation of acids, aldehydes, alcohols, and esters (Figures 15.13 and 15.14) (Reineccius, 2006). Methionine is reported to be the precursor for sulfur-containing volatiles, namely dimethyl disulfide and volatile thioesters. In strawberry, the precursor alanine in the presence

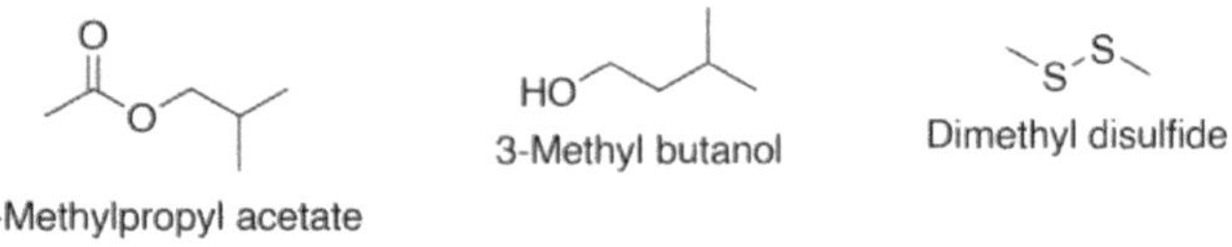

FIGURE 15.14 Representative structures of few amino acid derivatives.

of strawberry alcohol acyltransferase (SAAT) forms volatile ethyl esters that impart characteristic flavor (El Hadi et al., 2013).

15.3 Role of Volatile Organic Compounds (VOCs) in Plant Growth

VOCs are released constitutively from different plant parts such as leaves, flowers, and fruits into the soil through the root system and also into the atmosphere. Both healthy and plants under stress release volatiles of different nature and intensity that function to suit the purpose. The higher vapor pressure of the volatile compounds make them reach far distances, and their complex blends initiate not only intra-species, but also inter-kingdom signaling provoking changes in the receiving organisms. Many of these interactions, on proper speculation, could be tapped efficiently and used to improve crop growth and yield directly by improving their nutrition and indirectly by reducing stress.

Comparatively less research has been done to find out the effect of VOCs on plant growth and productivity. VOCs emitted from certain plant species were found to possess allelopathic effects and inhibited growth of other plant species, which would otherwise compete for similar resources (Abd-ElGawad et al., 2020). Plant VOCs, namely isoprenoids including a few mono- and sesquiterpenes and hexenal, were evidenced to inhibit weed seed germination and root growth and help in weed control, thereby enabling the producer plant in the better acquisition of water, nutrients, and light, thereby increasing crop productivity (Puig et al., 2018; Brilli et al., 2019). Aging of the plants is associated with changes in the hormonal concentration and is characterized by increased accumulation of ROS followed by cellular organ damage (Stark, 2005; Campisi and Vijg, 2009; Woo et al., 2013). Certain isoprenoids synergize cytokinins biosynthesis, thereby increasing the antioxidant activity resulting in the prevention of cell degradation and cell death, which extend the longevity of flowers and leaves with a favorable influence on the overall crop production (Dani et al., 2016).

Endophytic and epiphytic microbes colonizing the plants' tissues and their rhizosphere and phyllosphere regions play an important role in overall plant nutrition, health, and productivity (Pathma et al., 2021a). Root exudates from healthy and damaged plants have significant blends of compounds with a specific function. Flavonoids from legume were known to invite nitrogen-fixing *Rhizobium* (Peters et al., 1986), while benzoxazinoids from maize roots were documented to recruit plant growth-promoting rhizobacteria such as *Pseudomonas putida* (Neal et al., 2012). All these beneficial microbes associated with plants improved plant growth tremendously by mobilizing plant nutrients and by producing plant growth-promoting hormones. They also protect plants from pathogenic microbes and herbivore attack by producing antibiotics, hydrolytic enzymes, and insecticidal toxins. These beneficial microbes recruited by the plant volatiles also produce their microbial volatiles, which in turn affect the plant growth beneficially. VOCs from beneficial fungi and bacteria were known to modulate gene expression responsible for cell wall rigidity and expansion, thereby increasing the associated plant size (Zhang et al., 2007). Under tissue culture condition, volatiles from *Bacillus badius* M12 induced morphogenesis in tobacco callus and increased the browning of callus by antibody biosynthesis (Gopinath et al., 2015). Also, some of the PGPR volatiles such as indole and dimethyl hexadecylamine (DMHDA) were found to increase the density of root hairs as well as the length of primary and lateral roots, ultimately increasing the root volume and surface area (Castulo-Rubio et al., 2015). Increased root volume increases the rhizosphere region, ultimately increasing the population and diversity of plant growth-promoting microbes. This improves nutrient availability, especially iron and phosphorus, leading to better crop productivity (Sharifi and Ryu, 2017). Certain microbial volatiles apart from improving the quantity of crop biomass in terms of fruit or seed or tuber yield also improve the quality of the crop by increasing the number of sugars, secondary metabolites, and essential oils. Volatiles from *Bacillus subtilis* GB03 along with benzaldehyde were shown to increase the essential oil content and biomass in medicinal plants such as *Atractylodes lancea* and *Codonopsis pilosula* (Zhou et al., 2016). Volatile compounds also help in the stockpiling of sugars such as glucose, sucrose, and starch, which is required to increase the quality of crops such as potato and sugar beet. Thus, plant volatiles play a significant part in improving plant growth both quantitatively and qualitatively.

15.4 Role of Plant Volatiles in Stress Management

As mentioned earlier, out of the total known plant secondary metabolites, only one percent represent plant volatile compounds and have been reported from over ninety plant families. Both internal and external factors (damage of plant parts) help in the production process (Laothawornkitkul et al., 2008). These organic plant volatiles act as infochemicals kindling multi-trophic interactions and specific behavioral response in the receivers and thus have a tremendous potential to enhance crop protection and productivity. Although healthy plants produce volatiles, plants subjected to biotic (Dicke and Baldwin, 2010) and abiotic stresses (Loreto and Schnitzler, 2010) produce compounds to protect themselves from further damage. Stress induces a cascade of events in plants resulting in the activation of defense-related genes and various biosynthetic pathways to produce compounds to evade or minimize plant damage. Thus, plants under stress release bioactive compounds and improve their resistance either by stabilizing cell membranes indirectly or by direct quenching of reactive oxygen species (ROS). Plants on being exposed to stress, just like humans, retain the memory of the stress, which helps them to respond better and protect themselves from forthcoming stressful situations (Hilker & Schmulling, 2019). Certain factors help in shaping this plant's stress memory and are referred to as "priming stimuli" and include a large number of VOCs. Volatile nature of these compounds not only aids in quick dispersal in atmosphere and diffusion through soil pores reaching neighboring plants, but also helps them diffuse through cell membranes to quickly reach distant plant parts. A "primed" plant shows an early, quick, and comparatively intense response to further stressful conditions (Conrath et al., 2015; Mauch-Mani et al., 2017). Numerous studies evidence the role of VOCs in priming plants' defense against environmental stresses (Cofer et al., 2018), phytopathogens (Ameye et al., 2015), and insect pests (Kim and Felton, 2013).

15.4.1 VOCs Repelling Insect Herbivores

Plant volatiles can influence the behavior and physiology of the dependent herbivore and either attract them (Guerin and Ryan, 1984) or repel them by acting as feeding repellents and ovipositional deterrents (Bruce et al. 2005). McCormick et al. (2012) described that a unique blend of volatile compounds is released by plants damaged by insect herbivores, which provide indirect defense and are referred to as herbivore-induced plant volatiles (HIPVs). Naturally occurring wild plant species were found to possess a high degree of resistance to pests and diseases as compared to cultivated species. This might be due to the upregulation of defense-related genes and their products in the natural coevolution process. A study conducted to investigate the level of defense provided by plant volatile compounds against cotton pests, using one natural and five commercial cotton varieties, showed that the rate of volatile emission from the leaves damaged by beet armyworm in the natural variety was much higher in comparison with the emission from commercial varieties damaged by the same insect. The volatiles produced included lipoxygenase products, monoterpenes, and sesquiterpenes (Loughrin et al., 1995). Williams et al. (1980) reported that certain wild tomato species produced medium-length methyl ketones, which offered protection against pests. Transferring such desirable traits from wild species to tomato cultivars through crossing would lead to the development of novel tomato cultivars with pest resistance suitable for sustainable development. Rasmann et al. (2005) documented that roots of few maize varieties when damaged by western corn rootworm larvae produced many volatile compounds, of which (E)-β-caryophyllene, a sesquiterpene, attracted entomopathogenic nematodes that efficiently parasitized and killed the corn rootworm, thereby protecting the plants from further damage. However, North American maize lines when stressed could not emit (E)-β-caryophyllene from roots and hence suffered heavy damage on infestation by western corn rootworm. Understanding the production mechanisms of plant volatiles and their bioactivity provides plant breeders with the opportunity to develop new maize genotypes resistant to the pest by reposition (E)-β-caryophyllene emission by insect-damaged roots.

15.4.2 VOCs Suppressing Phytopathogens

VOCs can suppress pathogen growth and development. Recently, a study has been conducted to screen twenty-two different VOCs for their bioefficacy against three fungal pathogens, *viz. Fusarium oxysporum, Colletotrichum lindemuthianum*, and *Botrytis cinerea*. It was observed that eugenol showed a high degree of fungal growth suppression followed by other volatiles, namely trans-2-decenal, (+) carvone, nonanal, citral, *L*-linalool, nerolidol, against all three fungal pathogens (Quintana-Rodriguez et al., 2018). Several other studies also documented the capacity of VOCs emitted from plant leaves to suppress pathogen growth and development. Cuminaldehyde and *p*-cymene showed antifungal activity against *Verticillium dahliae, B. cinerea, Alternaria mali*, and *F. oxysporum* (Sekine et al., 2007). *In vitro* studies showed that germination and growth of *Monilinia laxa* causing brown rot of stone fruit were hampered by volatile compounds such as trans-2-hexenal, citral, and carvacrol (Neri et al., 2007). When wheat plant leaves are subjected to mechanical damage, they released a volatile compound Z-3-hexenyl acetate, which activated resistance against the fungal pathogen *Fusarium graminearum* (Ameye *et al.,* 2015). Z-3-hexenyl acetate was also found to reduce the damage of maize plant during cold stress (Brilli et al., 2011; Cofer et al., 2018). Damaged plants release certain monoterpenes, among which pinene was found to trigger accumulation of ROS and activate genes related to salicylic acid production and those inducing systemic acquired resistance (SAR) in the emitter plants apart from inducing defense in neighboring plants (Riedlmeier et al., 2017). Laura et al. (2018) reported that methyl jasmonate (MeJA) in low concentrations primed plant defense by triggering wound-inducible gene in paddy. Post-bloom fruit drop in citrus caused by *Colletotrichum acutatum* was suppressed when exposed to linalool under *in vitro* conditions (Marques et al., 2015). *Botrytis cinerea* was also found to be sensitive to the application of monoterpenes such as limonene *in vitro*. However, (+)-limonene induced the growth of another fungal pathogen *Penicillium digitatum*. In such instances, the blend of another volatile compound citral served the purpose, which indicates that at times two or more VOCs may be required for crop protection from diverse pathogens (Simas et al., 2017). Ahmad et al. (2015) showed that plant volatiles can interfere with pathogenic bacterial quorum sensing molecules, thereby protecting the emitting plants from pathogenic infection. Studies on the profile and activity of volatiles from healthy and pathogen-infested plants proved that significant difference occurred in the volatile blends. Gulati et al. (2020) showed that tomato plants infected by *Fusarium oxysporum* produced volatiles that efficiently attracted beneficial antagonists *Bacillus* sp. toward their rhizosphere as compared to the healthy plants.

Besides the production of VOCs from the plants against several biotic and abiotic stresses, the VOCs can also be emitted by soil-dwelling, beneficial, and plant growth-promoting rhizobacteria (PGPR). PGPR apart from promoting plant growth can increase the tolerance of the plants to stress. PGPR can produce both non-volatile and volatile substances. Ryu et al. (2003) reported microbial VOC emission in the early 21st century, which is now understood to play an important role in microbe–microbe and plant–microbe interactions. Volatiles from *Bacillus subtilis* GBO3 were found to increase the growth of *Arabidopsis. Pseudomonas* species isolated from soybean and canola was found to inhibit the mycelial growth of the fungal phytopathogen *Sclerotinia sclerotiorum*. Likewise, the expression of VOCs from *Bacillus subtilis* protected the plants treated with them from *Clavibacter michiganensis*, causing bacterial ring rot of potato (Rajer et al., 2017). VOCs from *Bacillus* and *Acinetobacter* inhibited the growth of *Phytophthora capsici* (Syed-Ab-Rahman et al., 2019).

15.4.3 VOCs Evading Abiotic Stress

Apart from tackling biotic stress, plant volatiles have also been reported to protect plants from abiotic stress. Drought, salinity, temperature extremities, ultraviolet radiation, and metal toxicity were found to interfere with the volatile mixtures produced by the affected plant (Loreto et al. 2014a, b; Forieri et al.

2016; Ameye et al. 2017). In drought conditions, *Arabidopsis* plants exposed to 2,3-butanediol showed high-stress tolerance by increasing stomatal closure, thereby reducing water loss (Cho et al., 2008). Several studies using *Arabidopsis, Ocimum, Vicia faba*, etc., documented that the volatiles from a plant subjected to salt stress can induce salt tolerance or prime the neighboring plants to deal with salinity effectively without suffering yield loss (Lee and Seo, 2014; Landi et al., 2020). Generally, volatile isoprenoids are said to act as antioxidants protecting plants from abiotic stress. In plants exposed to a higher temperature, isoprene and monoterpenes prevent hydrogen peroxide accumulation and lipid peroxidation in cell membranes (Vickers et al., 2009; Loreto et al., 2014a, b). Also microbial volatiles help in evading abiotic stress in plants. *Arabidopsis* plants exposed to VOCs from *Bacillus subtilis* GB03 were found to accumulate higher levels of glycine, betaine, and choline as compared to those plants without VOC treatment. Choline and glycine act as osmoprotectants and thus help plants survive water-stressed conditions. Further, molecular analysis revealed that the production of PEAMT, an essential enzyme, was higher under dehydration or water-stressed condition (Zhang et al., 2010). Proline is an important osmolyte that increases under stress condition. Potato plants treated with PGPR showed higher proline accumulation and ROS-scavenging enzymes expression and were tolerant to abiotic stresses such as salinity, drought, and heavy metal toxicity (Gururani et al., 2013). Dimethyl disulfide (DMDS) is another sulfur-containing compound produced by many soil fungi and bacteria, and their emission protected the sulfur-deficient plants from growth retardation (Meldau et al., 2013; Kanchiswamy et al., 2015).

15.5 Application in Crop Improvement

Plant volatiles, as well as microbial volatile compounds, can act as intra- and interspecies signals and induce systemic defense response in plants against abiotic and biotic stresses. This provides us with an opportunity to utilize them appropriately to manage pests, diseases, deficiency, drought, and temperature extremities. As compared to the direct activation of defense responses in plants, priming does not require a costly activation of metabolic pathways (Martinez-Medina et al., 2016) and therefore provides a sustainable method to develop novel crop protection strategies. Scientific investigations evidence that VOCs prime plant defenses against biotic and abiotic stress factors (Landi et al., 2020; Ninkovic et al., 2019). Owing to their similarity in mode of action to vaccines, VOCs are referred to as "green vaccines" (Luna-Diez, 2016). This offers huge scope for the biologists and chemists to understand the variation in volatile profiles and their activity and use them in crop improvement and develop varieties with specific volatile signatures that are detrimental for arthropod pests and phytopathogens and appealing to the beneficial arthropods and microbes.

15.6 Conclusions and Future Prospects

Plants produce an array of metabolites required for normal growth and development as well as for protecting themselves from abiotic and biotic stresses. All these metabolites are designed to be produced with specific spatial and temporal patterns, and their biosynthesis is regulated by innate and induced gene expression and molecular signaling. Among the secondary metabolites are various volatile compounds that play a specific role in plant physiology and defense as well as influence the other floral and faunal communities in the vicinity to generate ecological sustainability. Advancements in metabolomics and the advent of chemical characterization techniques help us understand gene expression, gene product purification, and metabolite characterization. Bioactivity evaluation of the volatile metabolites provides us with opportunities to explore different biomolecules available in nature's repository and to utilize them efficiently in sustainable crop production in harmony with nature. In-depth studies to understand the role of these volatile semiochemicals inducing inter- and intraspecific interactions are essential to tap their benefit to the fullest. This knowledge will help us replace toxic agrochemicals used for crop production and pest management with environmentally safer green molecules with negligible toxicity to non-target organisms and make profits uplifting farmer's living standards while assuring consumers' interest for safe food and, above all, preserving mother earth and its biodiversity, thereby ensuring sustainability.

Acknowledgments

The authors wish to acknowledge the support offered by Dr. R. Nagarajaprakash, Assistant Professor (Chemistry), School of Chemical Engineering and Physical Sciences, Lovely Professional University, Phagwara, Punjab, India, to draw chemical structures supporting the chapter content.

REFERENCES

Abd-ElGawad et al. (2020). Essential oil of Bassia muricata: Chemical characterization, antioxidant activity, and allelopathic effect on the weed Chenopodium murale. *Saudi Journal of Biological Sciences, 27*(7), 1900–1906.

Aharoni et al. (2000). Identification of the SAAT gene involved in strawberry flavor biogenesis by use of DNA microarrays. *Plant Cell, 12*, 647–661.

Ahmad et al. (2015). The impact of plant volatiles on bacterial quorum sensing. *Letters in Applied Microbiology, 60*(1), 8–19.

Akacha et al. (2005). Production of hexenol in a two-enzyme system: Kinetic study and modelling. *Biotechnology Letters, 27*(23), 1875–1878.

Ameye et al. (2015). Priming of wheat with the green leaf volatile Z-3-hexenyl acetate enhances defense against *Fusarium graminearum* but boosts deoxynivalenol production. *Plant Physiology, 167*(4), 1671–1684.

Ameye et al. (2017) Green leaf volatile production by plants: A meta-analysis. *New Phytologist, 220*, 666–683.

Arimura et al. (2010). Acquired immunity to herbivory and allelopathy caused by airborne plant emissions. *Phytochemistry, 71*(14–15), 1642–1649.

Baldwin, I. T. (2010). Plant volatiles. *Current Biology, 20*(9), R392–R397.

Boncan et al. (2020). Terpenes and terpenoids in plants: Interactions with environment and insects. *International Journal of Molecular Sciences, 21*(19), 7382.

Bouwmeester et al. (2019). The role of volatiles in plant communication. *The Plant Journal, 100*(5), 892.

Brilli et al. (2011). Detection of plant volatiles after leaf wounding and darkening by proton transfer reaction "time-of-flight" mass spectrometry (PTR-TOF). *PLoS One, 6*(5), e20419.

Brilli et al. (2019). Exploiting plant volatile organic compounds (VOCs) in agriculture to improve sustainable defense strategies and productivity of crops. *Frontiers in Plant Science, 10*, 264.

Bruce et al. (2005). Insect host location: A volatile situation. *Trends in Plant Science, 10*(6), 269–274.

Campisi, J. and Vijg, J. (2009). Does damage to DNA and other macromolecules play a role in ageing? If so, how?. *Journals of Gerontology Series A: Biomedical Sciences and Medical Sciences, 64*(2), 175–178.

Castulo-Rubio et al. (2015). Volatile organic compounds produced by the rhizobacterium Arthrobacter agilis UMCV2 modulate Sorghum bicolor (strategy II plant) morphogenesis and SbFRO1 transcription in vitro. *Journal of Plant Growth Regulation, 34*(3), 611–623.

Cho et al. (2008). 2R, 3R-butanediol, a bacterial volatile produced by Pseudomonas chlororaphis O6, is involved in induction of systemic tolerance to drought in Arabidopsis thaliana. *Molecular Plant-Microbe Interactions, 21*(8), 1067–1075.

Cofer et al. (2018). Green leaf volatiles protect maize (Zea mays) seedlings against damage from cold stress. *Plant, Cell & Environment, 41*(7), 1673–1682.

Conrath et al. (2015). Priming for enhanced defense. *Annual Review of Phytopathology, 53*, 97–119.

D'Auria, J. C. (2006). Acyltransferases in plants: A good time to be BAHD. *Current Opinion in Plant Biology, 9*(3), 331–340.

D'Auria et al. (2003). The SABATH family of MTs in Arabidopsis thaliana and other plant species. *Recent Advances in Phytochemistry, 37*, 253–284.

D'Auria et al. (2007). Characterization of a BAHD acyltransferase responsible for producing the green leaf volatile (Z)-3-hexen-1-yl acetate in *Arabidopsis thaliana. The Plant Journal, 49*(2), 194–207.

Dani et al. (2016). Do cytokinins, volatile isoprenoids and carotenoids synergically delay leaf senescence?. *Plant, Cell & Environment, 39*(5), 1103–1111.

De Boer et al. (2004). Identification of volatiles that are used in discrimination between plants infested with prey or nonprey herbivores by a predatory mite. *Journal of Chemical Ecology*, *30*(11), 2215–2230.

Derstine et al. (2020). Plant volatiles help mediate host plant selection and attraction of the spotted lanternfly (Hemiptera: Fulgoridae): A generalist with a preferred host. *Environmental Entomology*, *49*(5), 1049–1062.

Dicke, M., and Baldwin, I. T. (2010). The evolutionary context for herbivore-induced plant volatiles: Beyond the 'cry for help'. *Trends in Plant Science*, *15*(3), 167–175.

Dudareva et al. (2004). Biochemistry of plant volatiles. *Plant Physiology*, *135*, 1893–1902.

Dudareva et al. (2005). The nonmevalonate pathway supports both monoterpene and sesquiterpene formation in snapdragon flowers. *Proceedings of the National Academy of Sciences*, *102*(3), 933–938.

Dudareva et al. (2006). Plant volatiles: Recent advances and future perspectives. *Critical Reviews in Plant Sciences*, *25*(5), 417–440.

Effah et al. (2019). Potential roles of volatile organic compounds in plant competition. *Perspectives in Plant Ecology, Evolution and Systematics*, *38*, 58–63.

El Hadi et al. (2013). Advances in fruit aroma volatile research. *Molecules*, *18*(7), 8200–8229.

Erb, M. and Kliebenstein, D. J. (2020). Plant secondary metabolites as defenses, regulators, and primary metabolites: The blurred functional trichotomy. *Plant Physiology*, *184*(1), 39.

Feussner, I. and Wasternack, C. (2002). The lipoxygenase pathway. *Annual Review of Plant Biology*, *53*(1), 275–297.

Forieri et al. (2016). Salinity stress effects on direct and indirect defence metabolites in maize. *Environmental and Experimental Botany*, *122*, 68–77.

Gigot et al. (2010). The lipoxygenase metabolic pathway in plants: Potential for industrial production of natural green leaf volatiles. *Biotechnology, Agronomy and Society and Environment*, *14*, 451–460.

Goff, S. A. and Klee, H. J. (2006). Plant volatile compounds: Sensory cues for health and nutritional value?. *Science*, *311*(5762), 815–819.

Gopinath et al. (2015). A new initiative in micropropagation: Airborne bacterial volatiles modulate organogenesis and antioxidant activity in tobacco (Nicotiana tabacum L.) callus. *In Vitro Cellular & Developmental Biology-Plant*, *51*(5), 514–523.

Grechkin, A. (1998). Recent developments in biochemistry of the plant lipoxygenase pathway. *Progress in Lipid Research*, *37*(5), 317–352.

Guerin, P. M. and Ryan, M. F. (1984). Relationship between root volatiles of some carrot cultivars and their resistance to the carrot fly, *Psila rosae. Entomologia experimentalis et Applicata*, *36*(3), 217–224.

Gulati et al. (2020). A non-invasive soil-based setup to study tomato root volatiles released by healthy and infected roots. *Scientific Reports*, *10*(1), 12704.

Gururani et al. (2013). Plant growth-promoting rhizobacteria enhance abiotic stress tolerance in *Solanum tuberosum* through inducing changes in the expression of ROS-scavenging enzymes and improved photosynthetic performance. *Journal of Plant Growth Regulation*, *32*(2), 245–258.

Hampel et al. (2005). Biosynthesis of mono-and sesquiterpenes in carrot roots and leaves (*Daucus carota* L.): Metabolic cross-talk of cytosolic mevalonate and plastidial methylerythritol phosphate pathways. *Phytochemistry*, *66*(3), 305–311.

Heil, M. and Karban, R. (2010). Explaining evolution of plant communication by airborne signals. *Trends in Ecology & Evolution*, *25*(3), 137–144.

Hilker, M. and Schmülling, T. (2019). Stress priming, memory, and signalling in plants. *Plant, Cell & Environment*, *42*(3), 753–761.

Holopainen, J. K. (2004). Multiple functions of inducible plant volatiles. *Trends in Plant Science*, *9*, 529–533.

Holopainen, J. K. and Blande, J. D. (2012). Molecular plant volatile communication. *Sensing in Nature*, 17–31.

Hsieh et al. (2008). Chloroplast localization of methylerythritol 4-phosphate pathway enzymes and regulation of mitochondrial genes in ispD and ispE albino mutants in Arabidopsis. *Plant Molecular Biology*, *66*(6), 663–673.

Jabran et al. (2015). Allelopathy for weed control in agricultural systems. *Crop Protection*, *72*, 57–65.

Junker, R. R. and Tholl, D. (2013). Volatile organic compound mediated interactions at the plant-microbe interface. *Journal of Chemical Ecology*, *39*, 810–825.

Kanchiswamy et al. (2015). Chemical diversity of microbial volatiles and their potential for plant growth and productivity. *Frontiers in Plant Science*, *6*, 151.

Karunanithi, P. S. and Zerbe, P. (2019). Terpene synthases as metabolic gatekeepers in the evolution of plant terpenoid chemical diversity. *Frontiers in Plant Science*, *10*.

Keating, G. J. and O'kennedy, R. (1997). The chemistry and occurrence of coumarins. *Coumarins: Biology, Applications and Mode of Action*, 23–66.

Khadem, S. and Marles, R. J. (2010). Monocyclic phenolic acids; hydroxy- and polyhydroxybenzoic acids: Occurrence and recent bioactivity studies. *Molecules*, *15*, 7985–8005.

Kim, J. and Felton, G. W. (2013). Priming of antiherbivore defensive responses in plants. *Insect Science*, *20*(3), 273–285.

Klempien et al. (2012). Contribution of CoA ligases to benzenoid biosynthesis in petunia flowers. *The Plant Cell*, *24*(5), 2015–2030.

Knudsen, J. T. and Gershenzon, J. (2006). The chemical diversity of floral scent. Teoksessa N. Dudareva & E. Pichersky (toim.) *Biology of Floral Scent*, 27–52.

Knudsen et al. (2006). Diversity and distribution of floral scent. *The Botanical Review*, *72*(1), 1.

Landi et al. (2020). "Help is in the air": Volatiles from salt-stressed plants increase the reproductive success of receivers under salinity. *Planta*, *251*(2), 48.

Lange et al. (2000). Isoprenoid biosynthesis: The evolution of two ancient and distinct pathways across genomes. *Proceedings of the National Academy of Sciences*, *97*(24), 13172–13177.

Laothawornkitkul et al. (2008). Discrimination of plant volatile signatures by an electronic nose: A potential technology for plant pest and disease monitoring. *Environmental Science & Technology*, *42*(22), 8433–8439.

Laura et al. (2018). Epigenetic control of defense genes following MeJA-induced priming in rice (*O. sativa*). *Journal of Plant Physiology*, *228*, 166–177.

Lee K. and Seo P.J. (2014). Airborne signals from salt-stressed *Arabidopsis* plants trigger salinity tolerance in neighboring plants. *Plant Signaling & Behavior*, *9*, e28392.

Liu et al. (2015). MYB transcription factors as regulators of phenylpropanoid metabolism in plants. *Molecular Plant*, *8*(5), 689–708.

Loreto, F. and Schnitzler, J. P. (2010). Abiotic stresses and induced BVOCs. *Trends in Plant Science*, *15*(3), 154–166.

Loreto et al. (2014a). Plant volatiles and the environment. *Plant, Cell & Environment*, *37*, 1905–1908.

Loreto et al. (2014b). Volatile isoprenoids and their importance for protection against environmental constraints in the Mediterranean area. *Environmental and Experimental Botany*, *103*, 99–106.

Loughrin et al. (1995). Volatiles emitted by different cotton varieties damaged by feeding beet armyworm larvae. *Journal of Chemical Ecology*, *21*(8), 1217–1227.

Luna, E. (2016). Using green vaccination to brighten the agronomic future. *Outlooks on Pest Management*, *27*(3), 136–140.

Maeda, H. and Dudareva, N. (2012). The shikimate pathway and aromatic amino acid biosynthesis in plants. *Annual Review of Plant Biology*, *63*, 73–105.

Marchiosi et al. (2020). Biosynthesis and metabolic actions of simple phenolic acids in plants. *Phytochemistry Reviews*, *19*(4), 865–906.

Marques et al. (2015). Structural and biochemical characteristics of citrus flowers associated with defence against a fungal pathogen. *AoB Plants*, *7*, 1–10.

Martinez-Medina et al. (2016). Recognizing plant defense priming. *Trends in Plant Science*, *21*(10), 818–822.

Mauch-Mani et al. (2017). Defense priming: An adaptive part of induced resistance. *Annual Review of Plant Biology*, *68*, 485–512.

McCormick et al. (2012). The specificity of herbivore-induced plant volatiles in attracting herbivore enemies. *Trends in Plant Science*, *17*(5), 303–310.

Meents, A. K. and Mithöfer, A. (2020). Plant–plant communication: Is there a role for volatile damage-associated molecular patterns? *Frontiers in Plant Science*, *11*, 1–13. doi: 10.3389/fpls.2020.583275

Meldau et al. (2013). Dimethyl disulfide produced by the naturally associated bacterium Bacillus sp B55 promotes *Nicotiana attenuata* growth by enhancing sulfur nutrition. *The Plant Cell*, *25*(7), 2731–2747.

Nabavi et al. (2020). Flavonoid biosynthetic pathways in plants: Versatile targets for metabolic engineering. *Biotechnology Advances*, *38*, 107316.

Nakamura et al. (2001). Two distinct isopentenyl diphosphate isomerases in cytosol and plastid are differentially induced by environmental stresses in tobacco. *FEBS Letters*, *506* 61–64.

Neal et al. (2012). Benzoxazinoids in root exudates of maize attract *Pseudomonas putida* to the rhizosphere. *PLoS ONE*, *7*, e35498.

Neri et al. (2007). Fungicidal activity of plant volatile compounds for controlling *Monilinia laxa* in stone fruit. *Plant Disease*, *91*(1), 30–35.

Ninkovic et al. (2019). Who is my neighbor? Volatile cues in plant interactions. *Plant Signaling & Behavior, 14*(9), 1634993.

Orlova et al. (2006). Reduction of benzenoid synthesis in petunia flowers reveals multiple pathways to benzoic acid and enhancement in auxin transport. *The Plant Cell, 18*(12), 3458–3475.

Pathma et al. (2021a). Deciphering and harnessing plant microbiomes: Detangling the patterns and process—a clean, green road to sustainable agriculture. In: Yadav A.N., Singh J., Singh C., Yadav N. (eds) *Current Trends in Microbial Biotechnology for Sustainable Agriculture. Environmental and Microbial Biotechnology.* Springer, Singapore.

Pathma et al. (2021b). Microbial biofertilizers and biopesticides: Nature's assets fostering sustainable agriculture. In: Prasad R., Kumar V., Singh J., Upadhyaya C.P. (eds) *Recent Developments in Microbial Technologies. Environmental and Microbial Biotechnology.* Springer, Singapore.

Peters et al. (1986). A plant flavone, luteolin, induces expression of *Rhizobium meliloti* nodulation genes. *Science, 233*, 977–980.

Ponzio et al. (2013). Ecological and phytohormonal aspects of plant volatile emission in response to single and dual infestations with herbivores and phytopathogens. *Functional Ecology, 27*, 587–598.

Prestage et al. (1999). Volatile production in tomato fruit with modified alcohol dehydrogenase activity. *Journal of the Science of Food and Agriculture, 79*(1), 131–136.

Puig et al. (2018). The consistency between phytotoxic effects and the dynamics of allelochemicals release from *Eucalyptus globulus* leaves used as bioherbicide green manure. *Journal of Chemical Ecology, 44*(7), 658–670.

Qualley et al. (2012). Completion of the core β-oxidative pathway of benzoic acid biosynthesis in plants. *Proceedings of the National Academy of Sciences, 109*(40), 16383–16388.

Quintana-Rodriguez et al. (2018). Shared weapons in fungus-fungus and fungus-plant interactions? Volatile organic compounds of plant or fungal origin exert direct antifungal activity in vitro. *Fungal Ecology, 33*, 115–121.

Rajer et al. (2017). Volatile organic compounds produced by a soil-isolate, Bacillus subtilis FA26 induce adverse ultra-structural changes to the cells of *Clavibacter michiganensis* ssp. *sepedonicus*, the causal agent of bacterial ring rot of potato. *Microbiology, 163*(4), 523–530.

Rasmann et al. (2005). Recruitment of entomopathogenic nematodes by insect-damaged maize roots. *Nature, 434*(7034), 732–737.

Reineccius, G. (2006). *Flavour Chemistry and Technology.*CRC Press, Boca Raton, FL.

Rhoades, D.F. (1983). Responses of alder and willow to attack by tent caterpillars and webworms: Evidence for pheromonal sensitivity of willows. In: Hedin, PA, (ed.) *Plant Resistance to Insects*, pp. 55–68, American Chemical Society, Washington, DC.

Riedlmeier et al. (2017). Monoterpenes support systemic acquired resistance within and between plants. *The Plant Cell, 29*(6), 1440–1459.

Robbins, R. J. (2003). Phenolic acids in foods: An overview of analytical methodology. *Journal of Agriculture and Food Chemistry, 51*, 2866–2887.

Rohdich et al. (2003). The deoxyxylulose phosphate pathway of isoprenoid biosynthesis: Studies on the mechanisms of the reactions catalyzed by IspG and IspH protein. *Proceedings of the National Academy of Sciences, 100*(4), 1586–1591.

Roupe et al. (2006). Pharmacometrics of stilbenes: Seguing towards the clinic. *Current Clinical Pharmacology, 1*(1), 81–101.

Ryu, C. M. and Farag, M. (2003). a, Hu, C. *H., Reddy, MS, Wei, H.-X., Paré, PW, & Kloepper, JW*, 4927–4932.

Schnepp, J. and Dudareva, N. (2018). Floral scent: Biosynthesis, regulation and genetic modifications. *Annual Plant Reviews Online*, 240–257.

Sekine et al. (2007). Antifungal effects of volatile compounds from black zira (*Bunium persicum*) and other spices and herbs. *Journal of Chemical Ecology, 33*(11), 2123–2132.

Seo et al. (2001). Jasmonic acid carboxyl methyltransferase: A key enzyme for jasmonate-regulated plant responses. *Proceedings of the National Academy of Sciences, 98*(8), 4788–4793.

Sharifi R. and Ryu C.M. (2017). Chatting with a tiny belowground member of the holobiome: Communication between plants and growth-promoting rhizobacteria. *Advances in Botanical Research, 82*, 135–160.

Sharma et al. (2019). Response of phenylpropanoid pathway and the role of polyphenols in plants under abiotic stress. *Molecules, 24*(13), 2452.

Shimizu, B. (2014). 2-Oxoglutarate-dependent dioxygenases in the biosynthesis of simple coumarins. *Frontiers in Plant Science, 5,* 549.

Simas et al. (2017). Citrus species essential oils and their components can inhibit or stimulate fungal growth in fruit. *Industrial Crops and Products, 98,* 108–115.

Song et al. (2005). Isolation and characterization of a jasmonic acid carboxyl methyltransferase gene from hot pepper (*Capsicum annuum* L.). *Journal of Plant Biology, 48,* 292–297.

Stark, G. J. (2005). Functional consequences of oxidative membrane damage. *The Journal of Membrane Biology, 205*(1), 1–16.

Syed-Ab-Rahman et al. (2019). Suppression of *Phytophthora capsici* infection and promotion of tomato growth by soil bacteria. *Rhizosphere 9,* 72–75.

Tlak Gajger, I. and Dar, S. A. (2021). Plant Allelochemicals as Sources of Insecticides. *Insects, 12*(3), 189.

Vickers et al. (2009). A unified mechanism of action for volatile isoprenoids in plant abiotic stress. *Nature Chemical Biology, 5,* 283–291.

Wahengbam et al. (2021). Insecticides derived from natural products: Diversity and potential applications. In: Yadav A.N., Singh J., Singh C., Yadav N. (eds) *Current Trends in Microbial Biotechnology for Sustainable Agriculture. Environmental and Microbial Biotechnology.* Springer, Singapore.

Wang et al. (2013). Plant cell wall lignification and monolignol metabolism. *Frontiers in Plant Science, 4,* 220.

Wang et al. (2019). Towards efficient terpenoid biosynthesis: Manipulating IPP and DMAPP supply. *Bioresources and Bioprocessing, 6*(1), 6.

Widhalm, J. R. and Dudareva, N. (2015). A familiar ring to it: Biosynthesis of plant benzoic acids. *Molecular Plant, 8,* 83–97.

Williams et al. (1980). 2-Tridecanone: A naturally occurring insecticide from the wild tomato *Lycopersicon hirsutum* f. *glabratum. Science, 207*(4433), 888–889.

Winterhalter, P. and Rouseff, R. L. (Eds.). (2001). *Carotenoid-Derived Aroma Compounds.* .ACS Publications, Washington, DC, 1–7.

Woo et al. (2013). Plant leaf senescence and death–regulation by multiple layers of control and implications for aging in general multiple layers of control and implications for ageing in general. *Journal of Cell Science, 126,* 4823–4833.

Yu, F. and Utsumi, R. (2009). Diversity, regulation, and genetic manipulation of plant mono-and sesquiterpenoid biosynthesis. *Cellular and Molecular Life Sciences, 66*(18), 3043–3052.

Zhang et al. (2007). Rhizobacterial volatile emissions regulate auxin homeostasis and cell expansion in Arabidopsis. *Planta, 226*(4), 839–851.

Zhang et al. (2010). Choline and osmotic-stress tolerance induced in Arabidopsis by the soil microbe Bacillus subtilis (GB03). *Molecular Plant-Microbe Interactions, 23*(8), 1097–1104.

Zhou et al. (2016). Volatiles released by endophytic *Pseudomonas fluorescens* promoting the growth and volatile oil accumulation in *Atractylodes lancea. Plant Physiology and Biochemistry, 101,* 132–140.

16

Desiccation Tolerance in Orthodox and Recalcitrant Seeds

Bahman Fazeli-Nasab*
University of Zabol

Raheba Jami
Herat University

Naser Vahabi
University of Zabol

Abdolhosein Rezaei
Gorgan University of Agriculture Science and Natural Resources, Golestan University

Ramin Piri
University of Tehran

Moharam Valizadeh
University of Sistan and Baluchestan

CONTENTS

* Corresponding author's email: bfazeli@uoz.ac.ir

DOI: 10.1201/9781003171416-16

16.1 Introduction

16.1.1 Desiccation Tolerance as the Basis of Long-Term Seed Viability

Terrestrial organisms, such as plants, are constantly faced with the problem of drying up due to dry air. Therefore, during the evolution of plants on the planet, they have acquired adaptations that allow them to survive or prevent drying out (Oliver, Velten, and Mishler 2005). One of the mechanisms that can be important in the survival of orthodox plant seeds is its tolerance to desiccation. By definition, desiccation means the loss of free water from the protoplast, which is a matter of physiology and intracellularity. In addition to low plants (many lichens, ferns, and mosses), desiccation has also been found in seeds and pollen grains. In general, the desiccation in seeds is different from the desiccation of vegetative tissues in plants during drought (Wood 2007). The desiccation due to its destructive oxidative processes can damage the integrity of the seed cell membrane and thus cause the deterioration of the cell membrane and loss of germination and storage capacity of the seed. To prevent these damages in the seed, the mechanism of tolerance to its effluent, with the help of various cellular processes, protects the seed and improves the seed quality (Zhao et al. 2020).

Desiccation tolerance (DT) refers to an organism's ability to shield cell membranes from the detrimental effects of water removal while still maintaining a bilayer membrane structure in the absence of an aqueous medium. Desiccation resistance, on the other hand, is described as a seed's ability to survive and germinate after being exposed to water (Leprince and Buitink 2010; Alpert 2005).

Only orthodox seeds tolerate desiccation in higher plants. This feature allows orthodox seeds to withstand extreme climatic conditions. Orthodox seeds in dry form can be preserved for a long time and retain their quality, depending on the type of seed, storage temperature, and humidity (Berjak and Pammenter 2008). Tolerance to orthodox seeds is one of their most important and basic characteristics. It is also a prerequisite for completing the plant life cycle as a suitable and compatible solution to enable the seed to survive, during its discharge on the mother plant, drying, storage, or environmental stresses, and to ensure the better distribution of the species (Menezes-Blackburn et al. 2018; Ballesteros, Fanega-Sleziak, and Davies 2021). In fact, orthodox seeds acquire drought tolerance during development and lose it during germination (Kermode and Finch-Savage 2002). Recital citrate seeds do not experience the desiccation stage during their growth on the mother plant and their water content remains high at all stages of their development until germination, so they are not able to grow for long periods of time. They cannot be stored (Ballesteros, Fanega-Sleziak, and Davies 2021). These seeds include many species of tropical and subtropical trees, including rubber and cocoa, mango and cherries. Also includes woody species belonging to the family Dipterocarpaceae, water grasses and some tree species belonging to temperate climates such as sycamore and oak (Kijowska-Oberc et al. 2020; Chandra, Parkhey, and Keshavkant 2018).

In orthodox seeds, tolerance to desiccation depends on their developmental stage. But in general, this mechanism occurs when the seeds reach their maximum dry weight (before the desiccation phase) (Chandra, Parkhey, and Keshavkant 2018). As a result, the commencement of desiccation tolerance in different plant seeds occurs at different times throughout seed development and is dependent on the plant species, rate of water loss from the seed, and final moisture content following desiccation. For example, in cereals, the ability of seeds to tolerate desiccation occurs in the early stages of nutrient accumulation, as it has been observed that the embryos of corn, wheat, and barley are completely resistant to rapid desiccation in the early stages of accumulation (Lehner et al. 2006). Tolerance to desiccation in legume seeds occurs at a later time during development, i.e., at maturity (Ellis, Hong, and Roberts 1987).

Effort tolerance in orthodox seeds is associated with several cellular and biochemical processes involving various compounds such as LEA proteins, the accumulation of high amounts of non-reducing soluble sugars (such as sucrose and raffinose family oligosaccharides (RFOs) (raffinose, galacosinol, and Terhalose) (Angelovici et al., 2010), ABA (Kermode and Finch-Savage 2002), RFO (Hincha, Zuther, and Heyer 2003), and PEG (Buitink, Hoekstra, and Leprince 2002).

One of their most significant and fundamental traits is their tolerance for conventional seeds. It is also a prerequisite for completing the plant life cycle, as a suitable and compatible solution to enable the seed to survive, during its discharge on the mother plant, drying, storage, or environmental stresses, and to ensure better species distribution (Menezes-Blackburn et al. 2018; Ballesteros, Fanega-Sleziak, and Davies 2021). Drought tolerance is acquired throughout development and lost during germination in orthodox seeds (Kermode and Finch-Savage 2002).

Desiccation tolerance is most likely due to a complex trait with each factor being equally relevant and decisive. As a result, all of these processes must be involved in order to achieve maximum desiccation tolerance (Leprince et al. 2004). However, different compounds are synthesized during seed development or in the desiccation stage to increase seed resistance to desiccation resistance. Some are required for DT, some only for consumption during germination and membrane protection during water uptake, and some for both DT and germination (Angelovici et al. 2010). For example, RFOs have been suggested to protect cellular cohesion during desiccation by stabilizing membranes in water loss, providing energy-producing substrates during germination, and inhibiting some of the hydroxyl radical scavengers (Hincha, Zuther, and Heyer 2003). An increase in the amount of trehalose is also associated with gaining tolerance to desiccation, as well as with regulating germination (Panda, Rangani, and Parida 2021). Flavonoids also restrict permeability to solutes, minimizing damage during early germination (Nobakht et al. 2017; Tagousop et al. 2018). They also create a chemical barrier against fungal infections due to their antibacterial capabilities (Biharee et al. 2020). Tocopherols are lipophilic antioxidants that prevent non-enzymatic lipid oxidation in seed desiccation, storage, and germination period (Latef et al. 2020).

Several amino acids increase in desiccation-tolerant seedlings, including the non-protein amino acid GABA, whose levels fall early in seed germination. By changing the GABA route, the quick drop in GABA may aid to supply the energy required by the TCA cycle (Hijaz and Killiny 2019). Reactive oxygen species (ROS) (hydrogen peroxide, superoxide radicals, and hydroxyl radicals) are commonly thought of as toxic molecules whose accumulation causes fat peroxidation, protein oxidation, enzyme inactivation, and DNA damage. Eventually, these production factors will help cell death to occur (Kibinza et al. 2011). During germination, active oxygen species are released, which are produced in the seeds during the desiccation. This increase in the production and release of ROS causes the cell membrane structure to be disrupted and the membranes to lose their health, resulting in an increase in the leakage of electrolytes from the cell (Goel and Sheoran 2003).

Seeds' resistance to desiccation may be linked to their ability to eliminate reactive oxygen species (ROS) in order to avoid damaging actions produced by these molecules, such as lipid peroxidation. Enzymes such as superoxide dismutase, catalase, and glutathione ascorbate cycle enzymes, as well as antioxidant substances such as glutathione or ascorbate, vitamin E, and other molecules that remove and inactivate ROS, are examples of these systems (Bailly et al. 2000). The accumulation of non-enzymatic antioxidant components during drying may also help to protect cells from ROS. Enzyme activity is, in fact, inversely proportional to the amount of water in the cell. Water binds to macromolecular structures under low humidity, decreasing enzyme molecular mobility and accessibility to their substrates. This condition suggests that the AOS of antioxidant chemicals is most likely connected to the protection of oxidative damage in environments with low water content (Bernardy et al. 2020).

Researchers, from the ecological aspects of desiccation sensitivity, came to three general conclusions, which are as follows: (i) The wetter the plant habitat and the higher the temperature, the more sensitive it is to desiccation; (ii) sensitivity to desiccation in non-dormant seeds is more and shows higher frequency than in dormant seeds; and (iii) desiccation-sensitive seeds are larger than desiccation-resistant seeds (Tweddle et al. 2003). The ability to survive in dried state is the result of adaptation that prevents cell damage caused by physiological and biochemical changes before the seeds begin to dry. It is not yet clear what major changes occur in the cell over a short period of time (4–5 days) that change from seed susceptibility to its DT stage. Initial activation of protective compounds in the seed probably occurs during this period. Some chemicals (hormones, proteins, enzymes, and carbohydrates) and cellular structures (membranes) have been implicated in this process (from sensitivity to drought tolerance). The buildup of carbohydrates and proteins such as dehydrins is linked to seed protection against harm (Oliver et al. 2020; Kurek, Plitta-Michalak, and Ratajczak 2019; de Abreu et al. 2008; Fazeli-Nasab, Rahmani, and Khajeh 2021).

Desiccation tolerance is the ability to survive severe water loss and the biological role of DT, which depends on dry storage capacity (seed longevity). The role of ABA signaling in tolerance, the role of sugar signaling in seed tolerance, LEA protein accumulation, the involvement of LEA proteins and HSPs in seed tolerance, the role of carbohydrates in seed tolerance, the role of sugars in tolerance, the role of RFO in seed DT, and the role of enzymatic and non-enzymatic antioxidants in seed DT determine the desiccation (Nakashima et al. 2009; Morales et al. 2017).

16.2 Mechanisms of Desiccation Tolerance in Developing Seeds

The ability of seeds to survive in the dry state is a result of their adaptation to prevent cell degradation during water loss. Before the seeds begin to dry, some physiological and biochemical changes occur that prevent damage to cellular components due to water loss. It is not yet clear what major changes in the cell occur in a short period (4–5 days). Primary induction of protective components is likely to occur when the seeds change from susceptible to susceptible to effluent. It has been suggested that some compounds (hormones, proteins, enzymes, and sugars) or cellular structures (membranes) play a key role in this transfer (from sensitivity to drought tolerance) (Kumar et al. 2017; Badhan et al. 2018).

The buildup of carbohydrates and proteins is linked to seed protection against harm (dehydrins). Sugars can protect cells from drying out by generating a glassy phase, changing the characteristics of the cell membrane, or substituting water molecules at membrane junctions. The presence of disaccharides (trehalose or sucrose) helps to keep cell membranes stable during drying. Sugars may help to stabilize protein structure by forming hydrogen bridges between hydroxyl carbohydrate groups and polar protein regions. High concentrations of sugars inhibit molecular mobility and limit biochemical reactions. Accumulation of sugars is not the only way that plants tolerate drought, but also enzymes of sugar metabolism, such as sucrose-phosphate synthase, sucrose synthase are also very important to tolerate its effluent (Liu et al. 2019; Yu et al. 2021; Solís-Guzmán et al. 2017). Proteolytic enzymes can play an important role in the degradation of abnormal proteins and in protein repair by drying. When protein reproduction is low, enzymes involved in the removal of toxic mediators produced during oxygen metabolism and cell membrane lipid peroxidation (e.g., glutathione reductase, superoxide dismutase) are likely to be very important in DT (Zhang et al. 2020a; Rodriguez-Concepcion, D'Andrea, and Pulido 2019).

Positive regulation (rearrangement) of genes, encoding enzymes that detoxify ROS, has been confirmed under stress. The cell membrane is considered to be the main site of damage caused by drying. Changes in the structure or composition of the cell membrane are related to its tolerance to desiccation. These changes may provide resistance to disruption of membrane lamellar formation. High levels of unsaturated membrane fatty acids facilitate their oxidation and lead to membrane damage. Increasing the number of fat-soluble antioxidants, on the other hand, may protect membrane lipids during drying. It is evident that there are systems in orthodox seed cells and drought-tolerant plants that may limit drying damage, retain physiological integrity in the dry state, and subsequently activate repair systems after rehydration (Bartels and Sunkar 2005; Smolikova et al. 2021; Abbas et al. 2018; Meher, Koundal, and Gajbhiye 2010).

16.3 Master Transcription Factors and Regulatory Mechanisms of Desiccation Tolerance

Oxidative stress is caused by an excess of ROS and a lack of antioxidant activity. Many antioxidants, including ascorbate, glutathione, tocopherols, quinones, flavonoids, and phenolics, are activated by plants to lower it (Kranner and Birtić 2005). However, photosynthesis must be altered to successfully restrict ROS production (Rasouli et al. 2020; Fazeli-Nasab and Sayyed 2019). At maturity, the photosynthetic system in seeds is normally destroyed (Bewley, Bradford, and Hilhorst 2012). Both early and late seed maturation show expression of such defensive mechanisms. A transcription network known as the LAFL network is required for proper seed maturation program implementation. The key regulators in this transcription network are LEAFY COTYLEDON1 (LEC1), LEC2, FUSCA (FUS) 3, and ABI3 from the CAPAAT box-linked factors of the HAP3 family, and they interact in a complicated way. B3 transcription factors have the domains LEC2, FUS3, and ABI3 (Horstman et al. 2017).

At seed maturity, mutations in each of these genes cause severe abnormalities. ABI3-5 mutants are implicated in chromatin density and nucleus size reduction (van Zanten et al. 2011), whereas MtABI3 mutants have been shown to have decreased LEA protein expression (Delahaie et al. 2013). Re-induction of DT was studied in alfalfa roots using transcriptum method, and it was concluded that 3-millimeter-long roots are sensitive to rapid drying, but pretreatment with PEG can make them tolerate it. During a period of gene expression, it was shown that about 1,300 different genes were expressed during DT re-induction in PEG-treated alfalfa roots (Boudet et al. 2006). Many of these genes (720) are involved in the cell cycle and early metabolism, which are controlled there. Sucrose accumulates in drought-tolerant roots, and transcription and metabolite tests have revealed that sucrose is generated with a variety of lipids and carbohydrates (Boudet et al. 2006). Furthermore, transcription rapidly induces LEA proteins, and proteomic characterization of the heat shock protein causes their buildup (Boudet et al. 2006). Genes that are expressed during seed maturation (between 14 and 20 days after flowering) and genes that are expressed at the root during DT re-induction have been found to have a high overlap (Terrasson et al. 2013). It appears that after DT re-induction, germinated (partial) seeds revert to their initial developmental stage and restore their original structure, according to transcription data (Boudet et al. 2006).

Fifty genes upstream of the DT re-induction genes are regulated during germination in Arabidopsis, whereas more than 50 genes downstream of the DT re-induction genes exhibit greater expression during germination, supporting the "reversion" theory (Maia et al. 2011). The biggest number of distinct genes expressed occurred in the DT rearrangement for alfalfa (2,829 genes downstream vs 740 genes upstream), according to transcription data (Terrasson et al. 2013). Arabidopsis is a plant that belongs to the genus *Arabidopsis* (414 genes downstream versus 263 genes upstream) (Maia et al. 2011). Both plant species' downstream genes are engaged in cellular metabolic activities, biogenesis, and growth. In Arabidopsis, photosynthetic genes are also controlled downstream. Stress response, reaction to abiotic stimulus, response to dehydration, reaction to ABA stimulus, lipid localization, seed development, and fetal development at the end have shown more overlap between GO in alfalfa and Arabidopsis. They are the seedlings (Terrasson et al. 2013; Maia et al. 2011).

MtABI3 (Barreto et al. 2019), MtABI4 (Ochatt and Abirached-Darmency 2019), MtABI5 (Ochatt and Abirached-Darmency 2019), and MtAP2 EREBP gene are all closely connected with the DT gene, making them ideal candidates for DT regulators. MtABI3 is one of the most essential transcription factors; in Arabidopsis, a huge number of MtABI3-related genes have been discovered as direct targets of ABI3 (Le Signor et al. 2018; Verdier et al. 2013).

16.4 The Role of Various Factors and Signaling in Developing Seeds

16.4.1 The Role of ABA Signaling in DT

ABA is an abbreviation/acronym for Abscisic acid. The principal regulator of plant development in response to environmental challenges is phytohormones. In the field of ABA detection and downstream signaling, more than 100 gene loci have been discovered so far (Cutler et al. 2010). Seed growth processes, including nutrient accumulation, dormancy, and DT, are all regulated by ABA (Kermode and Finch-Savage 2002).

The ability to fully and adequately tolerate drought was an important evolutionary step that played a key role in arid lands. In fact, the genes responsible for the synthesis and signaling of ABA hormone are found during dehydration in terrestrial plants, which may have been important in achieving DT and drought tolerance during plant development (Umezawa et al. 2010). In this regard, *P. patens* has shown that deletion of ABA INSENSITIVE 3 (ABI3) or class A PP2Cs (Arabidopsis PP2Cs orthologs that regulate ABA signaling) or both will affect DT acquisition (Figure 16.1) (Komatsu et al. 2013). For example, pretreatment of *Marchantia polymorpha* with ABA results in morphological changes in plant survival after drying (Eklund et al. 2018), or in Arabidopsis, DT is obtained shortly after seed maturity (in ABA mutants compared to wild type) (Abley et al. 2020). ABA mutants may not have mutated completely in Arabidopsis, and several mutant ABA1 alleles have been reported, as well as double-mutant analysis with other ABA-sensitive mutants (ABA2 and ABA3) (Feitosa-Araujo et al. 2020; Barrero et al. 2005).

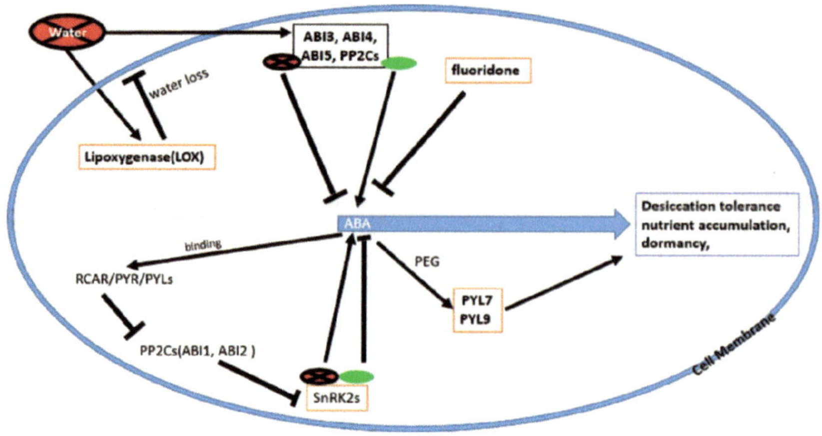

FIGURE 16.1 The molecular mechanism of ABA signaling in DT.

As a result, it's been claimed that a different mechanism could produce modest amounts of ABA (Barrero et al. 2005). Low levels of ABA are most likely enough to cause DT. The fact that the particular expression of ABA antibodies in tobacco seed (Nicotiana tabacum) results in substantially stronger seed phenotypes, including seeds sensitive to desiccation, supports this theory (Kim et al. 2020). The triple mutants of the SnRK2 gene (snrk2.2/3.6) in Arabidopsis have significant ABA signaling defects (Katsuta et al. 2020; Lu et al. 2020).

ABA is required for the induction of DT in germinating seedlings. By treating alfalfa rootstocks and employing the mutant ABA2-1 Arabidopsis (disrupted in ABA biosynthesis), fluridone (an inhibitor of ABA production) stimulates re-induction of DT (Maia et al. 2014). The amount of ABA in Arabidopsis germinated seeds does not appear to be altered, but there is likely considerable sensitivity to ABA. PEG treatment stimulated the expression of two ABA receptors (PYL7 and PYL9) (Maia et al. 2014).

In terms of ABA signaling, numerous mutations have been discovered, which play a role in the re-induction of DT in germinated seeds. Two mutants of ABI5, MtABI5-1 and MtABI5-2, for example, were unable to re-induce DT in alfalfa roots when exposed to osmotic stress (Terrasson et al. 2013). Several ABA signaling mutants, including as ABI3-8, ABI3-9, ABI4-3, and ABI5-7, were shown to have phenotypes in the re-induction of DT in germinated seeds in a study. In germinated seeds, all mutants had a lower ability to re-induce DT. These findings point to a distinction between the acquisition of DT during seed development and the re-induction of DT in germinated seeds. To explain these disparities, two hypotheses have been proposed. First, the well-defined pathways are involved in the induction of DT during seed development, followed by germination, with the exception of ABI3 (which is definitely engaged in both situations). Second, ABA, ABI4, and ABI5 are among the mechanisms that cause DT (Maia et al. 2014; Khan, Ali, Khan, et al. 2020).

ABI3, ABI4, and ABI5 have all been linked to the DT gene, implying that the second hypothesis is correct (Khan, Ali, Khan, et al. 2020; Verdier et al. 2013). A limited developmental window of ABA sensitivity has been observed in the post-germination stage and before plant growth, and it has been suggested that immature plants show ambient osmotic state in this time. When early seedlings are exposed to water loss, ABA stimulates the growth of germinating embryos via ABI3 and ABI5, shielding them from water loss (Lopez-Molina, Mongrand, and Chua 2001).

16.4.2 The Role of LEA Protein and HSP Signaling in Seed DT

In cotton and wheat, LEA proteins were first discovered in the late stages of seed maturity (Rasouli and Fazeli-Nasab 2014). LEA proteins are protective molecules that appear to work by substituting water, ion sequencing, reducing ROS, or stabilizing protein and membrane structures in response to drought stress (Figure 16.2)(Rasouli and Fazeli-Nasab 2014). Drought tolerance proteome research frequently uncovers LEA proteins. Heat-induced proteomic changes in *M. truncatula* seeds have been seen during

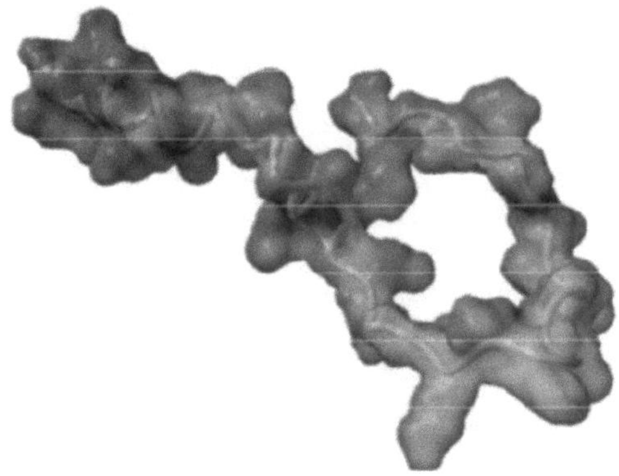

FIGURE 16.2 LEA2 protein model of wheat generated using YASARA.

germination and following re-establishment of drought tolerance (Boudet et al. 2006). In *M. truncatula*, six LEA proteins from four gene groups were discovered, including Em6, MP2, PM18 isoform, six SBP65 isoforms, PM25, and one DHN3 isoform, all of which were linked to DT (Boudet et al. 2006). LEA proteins of Em6, PM18, and PM25 have also been shown to accumulate in considerable amounts in *M. truncatula* following seed development and drying (Chatelain et al. 2012).

This further confirms the association between the lack of LEA protein accumulation and susceptibility to seed drying. This study also showed that many drought-tolerant LEA proteins are upregulated with ABI3. Other preservatives whose presence is important in gaining tolerance to desiccation during seed development are important proteins such as LEA proteins or maturation proteins and HSPs that accumulate during the desiccation stage of growth. LEA proteins, as their name implies, go back to the end stage of embryogenesis. LEA proteins are small, hydrophilic, and largely unstructured and heat-resistant proteins. LEA proteins have a wide range of desiccation protection functions with different capabilities including ion bonding, antioxidant activity, preservation and stabilization of membrane and protein structures (Smolikova et al. 2021).

So far, 31 of the 81 genes in the LEA family have been identified as being associated with abiotic stresses such as cold, osmotic stress, salinity, and drought. On the other hand, LEA genes are one of the different genes that increase their expression in all desiccation-resistant organisms, which has been proven by transcriptomic studies (Leprince and Buitink 2010). Some of these maturation proteins are associated with the seed's ability to advance seedling growth, while others are associated with DT (Campobenedetto et al. 2020). Maturity proteins may be an essential part of the primary response system that protects against stresses at the onset of desiccation before saccharides reach high levels, or they may act in conjunction with oligosaccharides to play a role in the development of DT (Han et al. 2020; Leprince et al. 2017).

One class of these proteins are HSPs (Figure 16.3), which are expressed under stress in the cell. The role of these proteins is to prevent the conformation of proteins under stress factors, to repair denatured proteins, to accelerate the folding of proteins, to destroy badly folded proteins, and to locate them. These proteins are present in all living cells in the attached or non-attached state to proteins and are located in the nucleus and cytoplasm of the cell (Liu et al. 2021). In this regard, there are reports of HSPs in seed development before the events related to drying (Zinsmeister et al. 2020). The interaction of tiny HSPs with other proteins as molecular chaperones has also been proposed as a way to prevent proteins from malfunctioning in the dry state (Buitink, Hoekstra, and Leprince 2002) and to allow appropriate re-folding in the event of water loss (Mtwisha et al. 2006). It was reported that when soybean seeds were isolated and exposed to a relative humidity of 61%, HSPs were synthesized in the whole seed. These proteins are probably compatible with a parent factor that prevents them from accumulating before the seed vascular connection is severed from the mother plant. Such a maternal factor has already been implicated in inducing the expression of LEA protein genes in cotton (Shu et al. 2020).

There is evidence to suggest that seed proteins, which are commonly used in storage, may also play a role in protecting against drought stress. A similar protein of 53 kDa LEA (ASP53) has been reported in

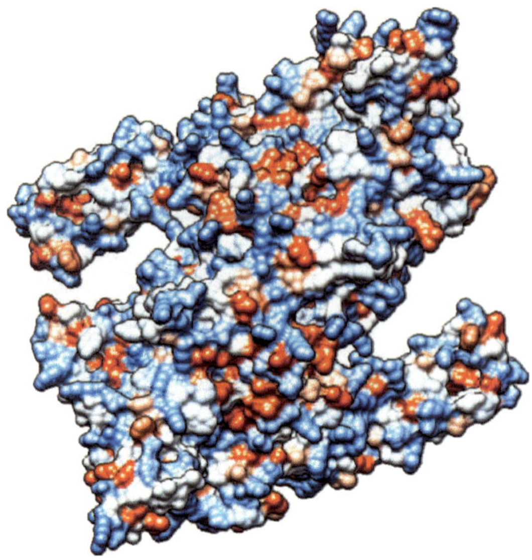

FIGURE 16.3 HSP protein model of *Arabidopsis thaliana* generated using SWISS-MODEL workspace.

the seeds of Acacia erioloba, which has the properties of stored protein as well as LEA. These proteins are thermally stable and are able to prevent the denaturation of enzymes and reduce the rate of loss of secondary structure of temperature-dependent hemoglobin, thus acting as a chaperone (Mtwisha et al. 2007).

16.4.3 The Role of Carbohydrate Signaling in Seed DT

In the 1980s, information was published about growing soluble carbohydrates and dry seeds that were associated with the accumulation of non-reducing sugars, especially sucrose, with specific oligosaccharides (Vertucci and Farrant 1995). Soluble saccharides can protect cytosolic components from desiccation damage, in vitro osmotic stress, frostbite, and heat. The protective mechanisms by which carbohydrates stabilize membranes under desiccation conditions have been proposed in two ways. One hypothesis is the replacement of glucose hydroxyl groups with hydration sites of sensitive intracellular components (i.e., at the membrane level) (Chen et al. 2020; Alfei 2020), and the second is the presence of glucose solutions between the cell membrane bilayers that prevent the membrane fluid phase change. Both of these mechanisms lead to the formation of a high-viscosity glassy phase (glassy phase) where harmful reactions are suppressed (Hasanuzzaman et al. 2020; Tokarz et al. 2021).

Raffinose oligosaccharides help prevent sucrose from crystallizing at low water concentrations and keep dry membranes stable (Schill and Hengherr 2018). Trehalose, as a disaccharide, has the potential to protect membranes against the negative consequences of water loss due to its ability to create an aqueous glass at low water concentrations (Zhang et al 2020b; Lee et al. 2017). Trehalose has been proposed as a candidate for maintaining the composition of storage and structural proteins in Arabidopsis seeds following water loss (Salem and Giavalisco 2019). Sucrose, which is found in high amounts in all dried orthodox seeds , has long been regarded a trehalose alternative. In theory, it is progressively implicated in the development of intracellular glass phases in the field of safeguarding the membrane in the dry condition, as was conceivable for trehalose (the "water substitution hypothesis"). Accumulation of sucrose and especially series of oligosaccharides of raffinose family and galactosyl cyclitols for tolerance to dry desiccation has been recorded in a wide range of seeds (Buitink, Hoekstra, and Leprince 2002; Kermode and Finch-Savage 2002).

The presence of huge amounts of sugars, especially disaccharides and oligosaccharides, is one of the features of desiccation-resistant species, including orthodox seeds. In fact, the presence of specific carbohydrates causes severe physical repercussions in a live cell due to the massive scarcity of intracellular water. In explaining this phenomenon, we must say that the loss of moisture leads to excessive proximity of the membrane surfaces to each other, and if this phenomenon occurs, the membrane compounds decompose and the membrane experiences a phase change from fluidity to gelatin, and this leads to cell death. However, the presence of small sugar solutions between the membranes can limit the excessive proximity of the two sides of the cell membrane and lead to a reduction in physical stress that causes the fluid phase to change to gelatin (Kermode and Finch-Savage 2002; Meier et al. 2017; Sun, Irving, and Leopold 1994).

To understand how sugars and other solutions can affect the behavior of the membrane phase, first, we need to understand how moisture loss physically affects membrane lipids. It has been reported that general physical conditions such as temperature, imposed mechanical loads, osmotic pressures, composition distribution, and velocity or equilibrium influence the phase behavior of membranes and solutions. So temperature, as an important factor, is always considered. Regarding how this change of state occurs, a temperature has been considered for it, which is known as transition temperature of lipid fluid to gel phase and is abbreviated as Tm (Bryant, Koster, and Wolfe 2001; Liu, Chang, and Lin 2006; She et al. 2016).

If glucose solutions are present between cell membranes, this change in temperature (T_m) does not increase as much as in their absence. The state in which these small sugar solutions are between the cell membranes is called the glass phase, which keeps the membrane fluid. Regarding why this phase of replacing water with sugar is known as the glass phase, we must say that when the solution of a hydrated body dries, the viscosity of the solution increases. If the viscosity increases to approximately 10–14 pascals (Pa), it becomes a glassy solution and the resulting solid state is called the glassy phase. Meanwhile, the temperature change of T_m mode is below T_0 temperature (this temperature is considered as the basis for comparing the effects of moisture loss against the change temperature (T_m) of the lipid phase) (Oliver et al. 2020; Koster and Anderson 1995). However, due to the gradual decrease in water content, there is a gradual increase in the change temperature, which raises the temperature of T_m above the temperature of T_0 and enters the cell into the gelatin phase. This increase in phospholipid T_m during moisture loss can be quantified by the sealing forces and the physical stresses exerted on the membrane and macromolecules at low moisture content (Bryant and Wolfe 1992; Pomeisl et al. 2020).

In addition to the role of sugars as a protective substance, high amounts of sugars are able to establish seedlings in an ABA-dependent manner (Alferez, de Carvalho, and Boakye 2021), and seedlings that contain both ABA and sugar are more resistant to drying (Alferez, de Carvalho, and Boakye 2021; Dekkers, Schuurmans, and Smeekens 2008). Over a period of time, generally 2 to 3 days following germination, germinated seeds become sensitive to sugars (Merchante and Stepanova 2017). Sugars form embryonic marker genes, including ABI3, ABI5, and numerous LEA genes, during this developmental stage (Dekkers, Schuurmans, and Smeekens 2008). The presence of sugars after ABA-induced drought treatment, for example, helped *M. polymorpha* seeds to survive (Eklund et al. 2018). *C. sativus* has a harmful impact on the environment (Leprince et al. 2004). In Arabidopsis, the hexokinase-1 protein serves as both a signaling molecule and an enzyme (Karve, Xia, and d Moore 2012). On the other hand, it has not been proven that glucosamine's influence on DT re-induction is attributable to enzymatic activity inhibition or glucose signaling. As a result, sugars are likely to serve as a structural defender as well as a signaling mediator. However, all of the genetic resources accessible in Arabidopsis can be employed as a model plant to examine the re-induction of DT in order to solve such issues and ambiguities (Wang et al. 2015; Guimaraes et al. 2016).

16.4.3.1 The Role of Sugars, Especially Sucrose, in DT

Preventing the passage of the liquid phase to the lipid jelly phase and the related leaking through the membrane is one component of seed stability in the dry state (Schill and Hengherr 2018; Koshland and Tapia 2019). Many investigations have shown that soluble sugars can impede the temperature transfer of phospholipids from the gel phase to the liquid phase (T_m) during post-emergence. Other sugars, such as raffinose, sucrose, or the ratio of sucrose to raffinose, have been implicated in seed desiccation tolerance in studies (Stachura, Malajczuk, and Mancera 2019). Sucrose, in particular, is important in preventing the crystalline phase from transferring to the jelly or glassy phase in the membrane's lipid bilayer and so tends to form a crystalline state (Stachura, Malajczuk, and Mancera 2019). This impact is comparable to the preceding theory in many ways (Bryant and Wolfe 1992), which explains how sugars' osmotic and volumetric effects prevent the two neighboring layers from re-approaching and, as a result, the mechanical forces that converge as the two layers approach. Another idea emphasizes the significance of a particular hydrogen connection between sugars and lipids. DT has also been linked to a rise in soluble carbs (raffinose and sucrose) as well as a higher raffinose-to-sucrose ratio (Black et al. 1999).

16.4.3.2 The Role of RFO in Seed DT

Various compounds are synthesized during seed development or in the desiccation stage to increase the seed's ability to resist desiccation. Some are for DT, and some only for consumption during germination and for protecting the membrane during water uptake, and some for both DT and germination (Angelovici et al. 2010). Raffinose family of oligosaccharides are used to protect cellular integrity during desiccation by stabilizing membranes in water loss, providing energy-producing substrates during germination, and inhibiting some of the hydroxyl radical (OH) scavengers (Hincha, Zuther, and Heyer 2003). Other sugars, such as raffinose, sucrose, or the ratio of sucrose to raffinose, prevent moisture loss and affect the distance between the two layers of the membrane, so they are essential and play an important role in tolerance to its desiccation. Loss of moisture causes the membrane surfaces to get too close to each other, and if this phenomenon occurs, the membrane compounds decompose and the membrane changes phase from fluidity to gelatin, resulting in the death of the living cell. However, the presence of small sugar solutions between the membranes can limit the excessive proximity of the two sides of the cell membrane, and this leads to a reduction in physical stress (Williams and Leopold 1989; Hibshman, Clegg, and Goldstein 2020).

Tolerance to desiccation has been demonstrated by examining changes in the amount of soluble carbohydrates (oligosaccharides) and the activity of antioxidant enzymes during different harvest dates (32–57 days after flowering) and drying at 15°C and a relative humidity of 70% in beans. Acquisition of effluent tolerance at 38 days after flowering and at the same time with the accumulation of soluble sugars (raffinose, stachyose and sucrose) begins and reaches its maximum at 46 days after flowering and at the same time with the maximum accumulation of these sugars (Bailly et al. 2000). Seed-resistant desiccation, on the other hand, had minimal monosaccharides (glucose and fructose) and this amount remained unchanged during development, while the amount of soluble sugars such as raffinose, stachyose and the ratio of these oligosaccharides to sucrose increased. Seed-resistant seeds also showed high activity of catalase and glutathione reductase and low activity of ascorbate peroxidase and superoxide dismutase, while the opposite situation was observed in dried immature seeds (Bailly et al. 2000).

In the study of changes in germination ability of durum wheat seeds, soluble carbohydrates and activity of antioxidant enzymes during different stages of maturation and also artificial drying (for two weeks in clusters at 20°C), it was reported that the DT accumulated high levels of soluble sugars (raffinose and sucrose) and an increase in the ratio of raffinose to sucrose has been observed (Lehner et al. 2006).

16.5 Antioxidants, Both Enzymatic and Non-Enzymatic, Play a Role in Seed Desiccation Tolerance

Carbohydrate metabolism and ROS elimination abilities appear to be favorable activities in terms of seed germination potential (Davari, Solouki, and Fazeli-Nasab 2018). ROS such as hydrogen peroxide, superoxide radicals, and hydroxyl radicals are engaged in different areas of seed physiology and are hazardous chemicals whose accumulation causes fat peroxidation, degradation, oxidation, and oxidation. Cell death is caused by a mutation in DNA (Kibinza et al. 2011). The structure of cell membranes is disrupted, and the membranes lose their normal function as a result of increased ROS generation and release, resulting in an increased electrolyte leakage from the cell (Goel and Sheoran 2003). The antioxidant cellular system, which includes the elimination of enzymatic poisoning, detoxification, and antioxidant chemicals, is responsible for 'ROS' dual role in plants. Such processes have the ability to either remove ROS, which are commonly created during stressful situations, or carefully control ROS concentrations to regulate signaling pathways. Seed tolerance to developing seeds, germination completion, and seed storage capacity are all influenced by detoxification processes (Bailly et al. 2000; Puač et al. 2018).

Seeds' resistance to desiccation may be linked to their ability to eliminate ROS in order to avoid damaging actions produced by these molecules, such as lipid peroxidation. Superoxide dismutase, catalase, and glutathione ascorbate cycle enzymes are examples of these mechanisms, as are antioxidant substances such glutathione or ascorbate, vitamin E, and other substances that remove and inactivate ROS (Bailly et al. 2000). The accumulation of non-enzymatic antioxidant components could also help protect cells from ROS during drying. Enzyme activity is, in fact, inversely proportional to the amount of water in the cell. Water binds firmly to macromolecular structures at low humidity, decreasing enzyme molecular mobility and accessibility to its substrates. This condition shows that antioxidant chemicals' protection of oxidative damage in low water content is more likely to be connected to ROS. Peroxy-rhodopsin (Prxs) are thiol-dependent antioxidants that have the ability to decrease hydrogen peroxide and hydroxyl. They build up in the seeds as they mature and dry (Reczek and Chandel 2015). Cisperoxy-rhodopsin appears to be expressed only in dry fbv (flower buds visible) tissues. Furthermore, peroxyroxins serve a critical function in preserving the nucleus' integrity and thus genetic information after drying (Lu et al. 2019).

Acquisition and loss of DT close to the capacity of cells is to inhibit ROS. Loss of viability after drying of the seeds of *Quercus robur*, *Shorea robusta*, and *Theobroma cacao* is connected with loss of cellular antioxidant potential or formation of free radicals (Kurek, Plitta-Michalak, and Ratajczak 2019; Chandra, Parkhey, and Keshavkant 2018).

16.6 Tolerance to Desiccation in Germinated Seeds

DT occurs at the end of seed maturity, just before drying, and then fades away during germination (Bewley, Bradford, and Hilhorst 2012). Seeds that have been subjected to rapid drying treatments (water loss for more than 2 hours) perish before or shortly after germination (Maia et al. 2011). Germinated seeds, on the other hand, have a wider time window during which they may tolerate drying. When a modest osmotic stress (using a PEG technique) is applied before quick drying, the presence of this window can be shown. *Cucumis sativus*, *Impatiens walleriana*, *Medicago*, *Tabebuia impetiginosa* (Brazilian tree species), and more recently, *Arabidopsis* have all demonstrated this (Maia et al. 2014). As a result, modest osmotic stress is thought to induce tolerance to desiccation in germinating seedlings. Therefore, tolerance to desiccation depends totally on the growth stage of germinated seeds (Delahaie et al. 2013; Maia et al. 2011). For example, alfalfa seeds with a root length of 1 mm survive in the fast-drying state. When the roots are up to 2.7 mm long, the treatment of mild osmotic stress with PEG determines their ability to tolerate desiccation (Buitink et al. 2003).

Four distinct growth stages were identified to determine the developmental window in which germinated Arabidopsis seeds can be induced to tolerate desiccation (Maia et al. 2011): (i) seed coat splitting, (ii) root expansion, (iii) primary root 0.5–3.5 mm, and (iv) initial root hair appearance. The seeds were able to endure drying following PEG treatment in the first three phases, and the plant survival rate was close to 100%. This percentage has been reduced to 20%–40% in the fourth stage, indicating that this skill has been substantially gone at this point.

Different segments of the seed show a variable level of re-induction of tolerance to desiccation. In Arabidopsis, cotyledons are the most resistant tissue, followed by hypocotyl and roots (Buitink et al. 2003).

The cell cycle is linked to the loss of tolerance to desiccation. The transition from DT to desiccation-sensitive state occurs when root cells enter the G2 phase, during which DNA is doubled (Faria et al. 2005; Li et al. 2018), which could alter DT induction. Pre-germination DNA synthesis in tomato seeds does not support such a connection. As a result, it's unclear whether DNA replication or cell cycle activation is the primary cause of DT loss. Seed priming, for example, limits seed life by synthesizing DNA in germinated seeds in tomatoes. However, the loss of tolerance to desiccation is not the end of the story. Furthermore, DNA repair has damaged Arabidopsis mutants (DNA ligase enzymes—AtLIG4 and AtLIG6), resulting in a reduction in seed life (Waterworth et al. 2010). Water loss also affects the dynamics and integrity of microtubules, which can be linked to a lack of tolerance to desiccation. DT has been associated with structural alterations in chromatin density and core size (van Zanten et al. 2011), and these structural modifications go beyond re-dewatering the seeds. Dormant seeds that have been rehydrated have a tiny nucleus. The return to the size of the bigger nucleus in non-dormant seeds appears to be related to germination, with the highest rise occurring between 2 and 3 days after sowing (van Zanten et al. 2011). Despite the fact that various processes are linked to the loss of DT in germinated seeds, the genetic control and molecular mechanisms involved are yet unknown.

16.7 Experimental Approaches to DT

Drought tolerance is acquired throughout development and lost during germination in orthodox seeds; hence, evaluating the response to seed water loss during development or germination is a standard way for determining DT. This approach has been used on rice seeds in proteomics investigations. However, understanding the many events relating to seed drought tolerance of seed growth and germination is difficult (Leprince and Buitink 2010). To address this issue, some physiological models have been constructed. One of the models employed in proteomics study of *M. truncatula* seeds is re-establishment of drought tolerance in germinated seeds utilizing mild osmotic stress with PEG solution (Boudet et al. 2006). CaCl2 and methyl viologen (MV) are chemical reagents that can increase and decrease the drying tolerance of germinated chickpea seeds, respectively (Wang, Møller, and Song 2012). Seed drying and germination tolerance are thus inextricably linked. Chickpea drying tolerance proteins have been identified using this method. It's tough to develop a model for evaluating recital citrate seedlings as they grow and germinate. However, chemical reagents such as NO and H_2O_2 purification can be used to modify their drying tolerance (Bai et al. 2011).

16.8 Modification in Desiccation: Removal of Cytoplasmic Water

When water is lost from a cell, the volume of the cell shrinks, causing the cytoplasmic components to decrease and the cell contents to become significantly viscous, potentially triggering protein deformation and membrane fusion. A wide range of chemicals have been identified that can prevent such unwanted molecular interactions in membrane models and protein systems.

Proline, glutamate, glycine betaine, carnitine, mannitol, sorbitol, fructan, polyols, trehalose, sucrose, and oligosaccharides are only a few examples. Despite their chemical differences, these chemicals prefer to leave the surface of proteins and hydrate them. Regardless of whether dehydration is caused by dryness, frostbite, or osmotic shock, many plants and microbes collect suitable salts in response to cellular water loss. Because they don't interfere with cell structure and function, these salts are compatible (Smolikova et al. 2021; Belott, Janis, and Menze 2020). Because a complete concentration of salts is often insufficient to boost the water-retaining capacity of cells, selective deprivation is most likely the major mechanism by which macromolecules in organisms protect themselves from mild water loss.

Counteracting selective deprivation is important to prevent protein denaturation and membrane attachment when the quantity of unstable molecules (including certain ions) in cells increases during water loss (Ballesteros, Pritchard, and Walters 2020). Many suitable salts, in fact, are incapable of protecting proteins and membranes against additional drying due to air or freezing drying. Only sugars, in the dry phase of water loss, can structurally and functionally retain proteins and membranes below 0.3 (g H_2O)/g dry weight by swapping water for so many other suitable salts.

16.9 Contradictory Results

Drought tolerance induction was not linked to raffinose buildup in wheat embryos, and oligosaccharides did not appear to help stabilize cytoplasmic vitreous. It was discovered that when *Impatiens walleriana* seeds were osmoprimed, the amount of oligosaccharide reduced while the amount of sucrose increased, with no change in the glass transition temperature. Despite the glass phase's durability, it has been observed that the shelf life of dried seeds is greatly shortened, and that after priming, they are stored at a low relative humidity level (Buitink et al. 2003). Therefore, it seems that oligosaccharides that accumulate on drying should play a different role than previously thought. Prevention of sucrose crystallization occurs when the water concentration decreases during the drying of orthodox seed maturation. Oligosaccharides, on the other hand, play this role in glass models. Many non-sugary components of the cytomatrix probably prevent sucrose from crystallizing (Buitink, Hoekstra, and Leprince 2002). Sucrose and even trehalose are said to place water on lipid groups, help maintain the distance of phospholipids in membranes, and lower the transition temperature of the gel phase (T_m) (Crowe et al. 1987; Khan, Ali, Zandi, et al. 2020).

16.10 Conclusions

In general, during seed growth, three developmental stages (first stage: cell division, second stage: cell elongation, and third stage: maturity or accumulation of food reserves) are known. These three stages are common during the development of orthodox and recalcitrant seeds, but orthodox seeds, in addition to these three developmental stages, enter another developmental stage called desiccation during their growth on the mother plant and at the end of the accumulation stage. It is associated with the loss of a lot of water from the seed and is a stage of seed growth that must be passed. Or more precisely, orthodox seed desiccation is part of its developmental program that allows them to enter a state of dry stagnation by means of a DT mechanism, thereby allowing them to be stored and survive in various environmental conditions.

DT, or in other words the ability of orthodox seeds to resist erosion, generally occurs during the accumulation phase, but depends on the speed of drying (the type of drying method) that affects seed survival after drying. This acquisition of DT, through several cellular processes, ultimately improves seed quality, including seed strength as well as storage capacity. The onset of DT and the achievement of maximum tolerance to desiccation can be affected by harvest time and drying method and ultimately cause changes in seed quality.

Recalcitrant seeds are those that do not experience the desiccation stage during their development on the mother plant, and their water content remains high at all stages of development until germination and cannot be stored for long periods. To be and to survive, therefore, the use and protection of recalcitrant seed varieties, which include some economically important crops, remains a major challenge.

Desiccation is an essential stage in the development of orthodox seeds that must be completed, as well as a prerequisite for completing the plant life cycle as a compatible solution to enable seeds to survive during desiccation on the mother plant, drying, storage, and environmental stress, and also to ensure better distribution of species.

ABBREVIATIONS

ABA:	abscisic acid
DT:	desiccation tolerance
GABA:	gamma aminobutyric acid
HSPs:	heat shock proteins
LEA proteins:	late embryogenesis abundant proteins
PEG:	polyethylene glycol
RFO:	raffinose family oligosaccharides
ROS:	reactive oxygen species

REFERENCES

Abbas et al. 2018. Arsenic uptake, toxicity, detoxification, and speciation in plants: physiological, biochemical, and molecular aspects. *International Journal of Environmental Research and Public Health* 15 (1):59.

Abley et al. 2020. An ABA-GA bistable switch can account for natural variation in the variability of Arabidopsis seed germination time. *BioRxiv.* 10:e59485. https://doi.org/10.1101/2020.06.05.135681

Alfei, Silvana. 2020. Nanotechnology applications to improve solubility of bioactive constituents of foods for health-promoting purposes. In: U Hebbar, S Ranjan, N Dasgupta, and R Kumar Mishra (eds.), *Nano-food Engineering: Volume One*. Cham: Springer International Publishing, pp. 189–257.

Alferez et al. 2021. Interplay between Abscisic Acid and Gibberellins, as Related to Ethylene and Sugars, in Regulating Maturation of Non-Climacteric Fruit. *International Journal of Molecular Sciences* 22 (2):669.

Alpert, Peter. 2005. The limits and frontiers of desiccation-tolerant life. *Integrative and Comparative Biology* 45 (5):685–695.

Angelovici et al. 2010. Seed desiccation: a bridge between maturation and germination. *Trends in Plant Science* 15 (4):211–218.

Badhan et al. 2018. RNA sequencing of leaf tissues from two contrasting chickpea genotypes reveals mechanisms for drought tolerance. *Plant Physiology and Biochemistry* 129:295–304.

Bai et al. 2011. Nitric oxide enhances desiccation tolerance of recalcitrant *Antiaris toxicaria* seeds via protein S-nitrosylation and carbonylation. *PloS One* 6 (6):e20714.

Bailly et al. 2000. Antioxidant systems in sunflower (*Helianthus annuus* L.) seeds as affected by priming. *Seed Science Research* 10 (1):35–42.

Ballesteros et al. 2020. Dry architecture: Towards the understanding of the variation of longevity in desiccation-tolerant germplasm. *Seed Science Research* 30 (2):142–155.

Ballesteros et al. 2021. Cryopreservation of seeds and seed embryos in orthodox-, intermediate-, and recalcitrant-seeded species. In: WF Wolkers, and H Oldenhof (eds.), *Cryopreservation and Freeze-Drying Protocols*. New York: Springer US, pp. 663–682.

Barrero et al. 2005. A mutational analysis of the ABA1 gene of *Arabidopsis thaliana* highlights the involvement of ABA in vegetative development. *Journal of Experimental Botany* 56 (418):2071–2083.

Barreto et al. 2019. Transcriptional profiling of the AFL subfamily of B3-type transcription factors during the in vitro induction of somatic embryogenesis in the model legume Medicago truncatula. *Plant Cell, Tissue and Organ Culture (PCTOC)* 139 (2):327–337.

Bartels, D. and Ramanjulu S. 2005. Drought and salt tolerance in plants. *Critical Reviews in Plant Sciences* 24 (1):23–58.

Belott et al. 2020. Liquid-liquid phase separation promotes animal desiccation tolerance. *Proceedings of the National Academy of Sciences* 117 (44):27676–27684.

Berjak, P and Pammenter NW. 2008. From *Avicennia* to *Zizania*: seed recalcitrance in perspective. *Annals of Botany* 101 (2):213–228.

Bernardy et al. 2020. Plants' genetic variation approach applied to zinc contamination: secondary metabolites and enzymes of the antioxidant system in Pfaffia glomerata accessions. *Chemosphere* 253: 126692.

Bewley et al. 2012. *Seeds: Physiology of Development, Germination and Dormancy*. Springer Science & Business Media. Springer, New York, NY.

Biharee et al. 2020. Antimicrobial flavonoids as a potential substitute for overcoming antimicrobial resistance. *Fitoterapia* 146:104720. doi: 10.1016/j.fitote.2020.104720.

Black et al. 1999. Water content, raffinose, and dehydrins in the induction of desiccation tolerance in immature wheat embryos. *Plant Physiology* 120 (2):463–472.

Boudet et al. 2006. Comparative analysis of the heat stable proteome of radicles of Medicago truncatula seeds during germination identifies late embryogenesis abundant proteins associated with desiccation tolerance. *Plant Physiology* 140 (4):1418–1436.

Bryant, G and Joe W. 1992. Interfacial forces in cryobiology and anhydrobiology. *Cryo-Letters* 13:23–36.

Bryant et al. 2001. Membrane behaviour in seeds and other systems at low water content: the various effects of solutes. *Seed Science Research* 11 (1):17–25.

Buitink et al. 2002. Biochemistry and biophysics of tolerance systems. In: IM Black and HW Pritchard (eds.), *Desiccation and Survival in Plants: Drying Without Dying*. Wallingford: CABI Publishing.

Buitink et al. 2003. The re-establishment of desiccation tolerance in germinated radicles of *Medicago truncatula* Gaertn. seeds. *Seed Science Research* 13 (4):273–286.

Campobenedetto et al. 2020. A biostimulant seed treatment improved heat stress tolerance during cucumber seed germination by acting on the antioxidant system and glyoxylate cycle. *Frontiers in Plant Science* 11: 836.

Chandra et al. 2018. Ageing-regulated changes in genetic integrity of two recalcitrant seeded species having contrasting longevity. *Trees* 32 (1):109–123.

Chatelain et al. 2012. Temporal profiling of the heat-stable proteome during late maturation of Medicago truncatula seeds identifies a restricted subset of late embryogenesis abundant proteins associated with longevity. *Plant, Cell & Environment* 35 (8):1440–1455.

Chen et al. 2020. Recent advancements in polyethyleneimine-based materials and their biomedical, biotechnology, and biomaterial applications. *Journal of Materials Chemistry B* 8 (15):2951–2973.

Crowe et al. 1987. Stabilization of dry phospholipid bilayers and proteins by sugars. *Biochemical Journal* 242 (1):1–10.

Cutler et al. 2010. Abscisic acid: emergence of a core signaling network. *Annual Review of Plant Biology* 61:651–679.

Davari et al. 2018. Effects of jasmonic acid and titanium dioxide nanoparticles on process of changes of phytochemical and antioxidant in genotypes of *Satureja hortensis* L. *Eco-Phytochemical Journal of Medicinal Plants* 5 (4):1–20.

de Abreu et al. 2008. Physiological and biochemical changes occurring in dwarf-cashew seedlings subjected to salt stress. *Brazilian Journal of Plant Physiology* 20 (2):105–118.

Dekkers et al. 2008. Interaction between sugar and abscisic acid signalling during early seedling development in Arabidopsis. *Plant Molecular Biology* 67 (1–2):151–167.

Delahaie et al. 2013. LEA polypeptide profiling of recalcitrant and orthodox legume seeds reveals ABI3-regulated LEA protein abundance linked to desiccation tolerance. *Journal of Experimental Botany* 64 (14):4559–4573.

Eklund et al. 2018. An evolutionarily conserved abscisic acid signaling pathway regulates dormancy in the liverwort Marchantia polymorpha. *Current Biology* 28 (22):3691–3699.

Ellis et al. 1987. The development of desiccation-tolerance and maximum seed quality during seed maturation in six grain legumes. *Annals of Botany* 59 (1):23–29.

Faria et al. 2005. Changes in DNA and microtubules during loss and re-establishment of desiccation tolerance in germinating Medicago truncatula seeds. *Journal of Experimental Botany* 56 (418):2119–2130.

Fazeli-Nasab et al. 2019. Plant growth-promoting rhizobacteria and salinity stress: a journey into the soil. In: RZ Sayyed, NK Arora, and MS Reddy (eds.), *Plant Growth Promoting Rhizobacteria for Sustainable Stress Management: Volume 1: Rhizobacteria in Abiotic Stress Management*. Singapore: Springer, pp. 21–34.

Fazeli-Nasab et al. 2021. Effects of culture medium and plant hormones in organogenesis in olive (CV. Kroneiki). *Journal of Plant Bioinformatics and Biotechnology* 1 (1):1–13.

Feitosa-Araujo et al. 2020. Changes in intracellular NAD status affect stomatal development in an abscisic acid-dependent manner. *The Plant Journal* 104 (5):1149–1168.

Goel, A and Sheoran IS. 2003. Lipid peroxidation and peroxide-scavenging enzymes in cotton seeds under natural ageing. *Biologia Plantarum* 46 (3):429–434.

Guimaraes et al. 2016. Changes in gene expression and soluble carbohydrate contents during the imbibition and re-induction of desiccation tolerance in Peltophorum dubium seeds. *Seed Science and Technology* 44 (1):125–137.

Han et al. 2020. ZmDREB2A regulates ZmGH3. 2 and ZmRAFS, shifting metabolism towards seed aging tolerance over seedling growth. *The Plant Journal* 104 (1):268–282.

Hasanuzzaman et al. 2020. Regulation of ROS metabolism in plants under environmental stress: A review of recent experimental evidence. *International Journal of Molecular Sciences* 21 (22):8695.

Hibshman et al. 2020. Mechanisms of desiccation tolerance: themes and variations in brine shrimp, roundworms, and tardigrades. *Frontiers in Physiology* 11.

Hijaz, F and Killiny N. 2019. Exogenous GABA is quickly metabolized to succinic acid and fed into the plant TCA cycle. *Plant Signaling & Behavior* 14 (3):e1573096.

Hincha et al. 2003. The preservation of liposomes by raffinose family oligosaccharides during drying is mediated by effects on fusion and lipid phase transitions. *Biochimica et Biophysica Acta (BBA)-Biomembranes* 1612 (2):172–177.

Horstman et al. 2017. The BABY BOOM transcription factor activates the LEC1-ABI3-FUS3-LEC2 network to induce somatic embryogenesis. *Plant Physiology* 175 (2):848–857.

Karve et al. 2012. *Arabidopsis* Hexokinase-Like1 and Hexokinase1 form a critical node in mediating plant glucose and ethylene responses. *Plant Physiology* 158 (4):1965–1975.

Katsuta et al. 2020. Arabidopsis Raf-like kinases act as positive regulators of subclass III SnRK2 in osmostress signaling. *The Plant Journal* 103 (2):634–644.

Kermode et al. 2002. *Desiccation Sensitivity in Orthodox and Recalcitrant Seeds in Relation to Development, Desiccation and Survival in Plants: Drying without Dying*. CABI Publishing, Printed and bound in the UK by Biddles Ltd, Guildford and King's Lynn.

Khan et al. 2020. PWR/HDA9/ABI4 complex epigenetically regulates ABA dependent drought stress tolerance in arabidopsis. *Frontiers in Plant Science* 11: 623.

Khan et al. 2020. Role of sugars, amino acids and organic acids in improving plant abiotic stress tolerance. *Pakistan Journal of Botany* 52 (2):355–363.

Kibinza et al. 2011. Catalase is a key enzyme in seed recovery from ageing during priming. *Plant Science* 181 (3):309–315.

Kijowska-Oberc et al. 2020. Changes in proline levels during seed development of orthodox and recalcitrant seeds of genus acer in a climate change scenario. *Forests* 11 (12):1362.

Kim et al. 2020. Promoted ABA hydroxylation by Capsicum annuum CYP707As overexpression suppresses pollen maturation in Nicotiana tabacum. *Frontiers in Plant Science* 11.

Komatsu et al. 2013. Group A PP2Cs evolved in land plants as key regulators of intrinsic desiccation tolerance. *Nature Communications* 4 (1):1–9.

Koshland, D and Tapia H. 2019. Desiccation tolerance: an unusual window into stress biology. *Molecular Biology of the Cell* 30 (6):737–741.

Koster, KL and Anderson M. 1995. Interactions between vitrified sugars and lipids-mechanism by which the fluid-to-gel phase-transition temperature is lowered. *Plant Physiology* 108 (2):110–110.

Kranner, I and Simona B. 2005. A modulating role for antioxidants in desiccation tolerance. *Integrative and Comparative Biology* 45 (5):734–740.

Kumar et al. 2017. Mechanisms of drought tolerance in rice. *Burleigh Dodds Science Publishing Limited*:131–63.

Kurek et al. 2019. Reactive oxygen species as potential drivers of the seed aging process. *Plants* 8 (6):174.

Latef et al. 2020. Impact of the static magnetic field on growth, pigments, osmolytes, nitric oxide, hydrogen sulfide phenylalanine ammonia-lyase activity, antioxidant defense system, and yield in lettuce. *Biology* 9 (7):172.

Le Signor et al. 2018. Functional genomics and seed development in *Medicago truncatula*: an overview. *Functional Genomics in Medicago Truncatula*: 175–195.

Lee et al. 2017. PFOA-induced metabolism disturbance and multi-generational reproductive toxicity in *Oryzias latipes*. *Journal of Hazardous Materials* 340:231–240.

Lehner et al. 2006. Changes in wheat seed germination ability, soluble carbohydrate and antioxidant enzyme activities in the embryo during the desiccation phase of maturation. *Journal of Cereal Science* 43 (2):175–182.

Leprince, O and Buitink J. 2010. Desiccation tolerance: from genomics to the field. *Plant Science* 179 (6):554–564.

Leprince et al. 2004. The role of sugars and hexose phosphorylation in regulating the re-establishment of desiccation tolerance in germinated radicles of Cucumis sativa and Medicago truncatula. *Physiologia Plantarum* 122 (2):200–209.

Leprince et al. 2017. Late seed maturation: drying without dying. *Journal of Experimental Botany* 68 (4):827–841.

Li et al. 2018. Desiccation does not drastically increase the accessibility of foreign DNA to the nuclear genomes: evidence from the frequency of endosymbiotic DNA transfer. *BioRxiv*:465443.

Liu et al. 2006. Comparison of phospholipids and their fatty acids in recalcitrant and orthodox seeds. *Seed Science and Technology* 34 (2):443–452.

Liu et al. 2019. Overexpression of the wheat trehalose 6-phosphate synthase 11 gene enhances cold tolerance in *Arabidopsis thaliana*. *Gene* 710:210–217.

Liu et al. 2021. The chloroplastic small heat shock protein gene KvHSP26 is induced by various abiotic stresses in kosteletzkya virginica. *International Journal of Genomics* 2021:Article ID: 6652445.

Lopez-Molina et al. 2001. A postgermination developmental arrest checkpoint is mediated by abscisic acid and requires the ABI5 transcription factor in Arabidopsis. *Proceedings of the National Academy of Sciences* 98 (8):4782–4787.

Lu et al. 2019. Revelation of microalgae's lipid production and resistance mechanism to ultra-high Cd stress by integrated transcriptome and physiochemical analyses. *Environmental Pollution* 250:186–195.

Lu et al. 2020. The Arabidopsis kinase-associated protein phosphatase KAPP, interacting with protein kinases SnRK2. 2/2.3/2.6, negatively regulates abscisic acid signaling. *Plant Molecular Biology* 102 (1):199–212.

Maia et al. 2011. The re-establishment of desiccation tolerance in germinated Arabidopsis thaliana seeds and its associated transcriptome. *PloS One* 6 (12):e29123.

Maia et al. 2014. Abscisic acid (ABA) sensitivity regulates desiccation tolerance in germinated A rabidopsis seeds. *New Phytologist* 203 (1):81–93.

Meher et al. 2010. Reactive oxygen species, antioxidants, sulfur metabolites and their agro-biotechnological potential to enhance stress resistance of crop plants. *Indian Journal of Agricultural Biochemistry* 23 (1):1–17.

Meier et al. 2017. Oxygen activation by Cu LPMOs in recalcitrant carbohydrate polysaccharide conversion to monomer sugars. *Chemical Reviews* 118 (5):2593–2635.

Menezes-Blackburn et al. 2018. Opportunities for mobilizing recalcitrant phosphorus from agricultural soils: a review. *Plant and Soil* 427 (1):5–16.

Merchante, C and Stepanova AN. 2017. The triple response assay and its use to characterize ethylene mutants in arabidopsis. In: BM Binderand G Eric Schaller (eds.), *Ethylene Signaling: Methods and Protocols*. New York: Springer, pp. 163–209.

Morales et al. 2017. Transcriptional responses of Chilean quinoa (*Chenopodium quinoa* Willd.) under water deficit conditions uncovers ABA-independent expression patterns. *Frontiers in Plant Science* 8:216.

Mtwisha et al. 2006. *Protection Mechanisms against Water Deficit Stress: Desiccation Tolerance in Seeds as a Study Case, Drought Adaptation in Cereals*. Binghamton: Food Products Press.

Mtwisha et al. 2007. ASP53, a thermostable protein from Acacia erioloba seeds that protects target proteins against thermal denaturation. *Functional Plant Biology* 34 (2):139–149.

Nakashima et al. 2009. Three Arabidopsis SnRK2 protein kinases, SRK2D/SnRK2. 2, SRK2E/SnRK2. 6/OST1 and SRK2I/SnRK2. 3, involved in ABA signaling are essential for the control of seed development and dormancy. *Plant and Cell Physiology* 50 (7):1345–1363.

Nobakht et al. 2017. Antibacterial properties of flavonoids from kino of the eucalypt tree, Corymbia torelliana. *Plants* 6 (3):39.

Ochatt, S and Abirached-Darmency M. 2019. The underlying processes governing seed size plasticity: impact of endopolidy on seed coat development and cell expansion in Medicago truncatula. *The Model Legume Medicago Truncatula*:99–116.

Oliver et al. 2005. Desiccation tolerance in bryophytes: a reflection of the primitive strategy for plant survival in dehydrating habitats? *Integrative and Comparative Biology* 45 (5):788–799.

Oliver et al. 2020. Desiccation tolerance: avoiding cellular damage during drying and rehydration. *Annual Review of Plant Biology* 71: 435–460.

Panda et al. 2021. Unraveling salt responsive metabolites and metabolic pathways using non-targeted metabolomics approach and elucidation of salt tolerance mechanisms in the xero-halophyte Haloxylon salicornicum. *Plant Physiology and Biochemistry* 158: 284–296.

Pomeisl et al. 2020. Simple syntheses of new pegylated trehalose derivatives as a chemical tool for potential evaluation of cryoprotectant effects on cell membrane. *Molecules* 25 (3):497.

Puač et al. 2018. Activity of catalase enzyme in *Paulownia tomentosa* seeds during the process of germination after treatments with low pressure plasma and plasma activated water. *Plasma Processes and Polymers* 15 (2):1700082.

Rasouli, H and Fazeli-Nasab, B. 2014. Structural validation and homology modeling of LEA 2 protein in bread wheat. *American-Eurasian Journal of Agricultural and Environmental Sciences* 14 (10):1044–1048.

Rasouli et al. 2020. Nanoparticles: a new threat to crop plants and soil rhizobia? In: S Hayat, J Pichtel, M Faizan, and Q Fariduddin (eds.), *Sustainable Agriculture Reviews 41: Nanotechnology for Plant Growth and Development*. Cham: Springer International Publishing.

Reczek, CR and Chandel NS. 2015. ROS-dependent signal transduction. *Current Opinion in Cell Biology* 33:8–13.

Rodriguez-Concepcion et al. 2019. Control of plastidial metabolism by the Clp protease complex. *Journal of Experimental Botany* 70 (7):2049–2058.

Salem, MA and Giavalisco P. 2019. Regulatory-Associated Protein of TOR 1B (RAPTOR1B) regulates hormonal switches during seed germination in *Arabidopsis thaliana*. *Plant Signaling & Behavior* 14 (7):1613130.

Schill, RO and Hengherr S. 2018. Environmental adaptations: desiccation tolerance. In: RO Schill, (eds.), *Water Bears: The Biology of Tardigrades*. Cham: Springer.

She et al. 2016. Membrane fouling in osmotically driven membrane processes: a review. *Journal of Membrane Science* 499:201–233.

Shu et al. 2020. A transcriptomic analysis reveals soybean seed pre-harvest deterioration resistance pathways under high temperature and humidity stress. *Genome* 63 (2):115–124.
Smolikova et al. 2021. Desiccation tolerance as the basis of long-term seed viability. *International Journal of Molecular Sciences* 22 (1):101.
Solís-Guzmán et al. 2017. *Arabidopsis thaliana* sucrose phosphate synthase (sps) genes are expressed differentially in organs and tissues, and their transcription is regulated by osmotic stress. *Gene Expression Patterns* 25:92–101.
Stachura et al. 2019. Does sucrose change its mechanism of stabilization of lipid bilayers during desiccation? Influences of hydration and concentration. *Langmuir* 35 (47):15389–15400.
Sun et al. 1994. The role of sugar, vitrification and membrane phase transition in seed desiccation tolerance. *Physiologia Plantarum* 90 (4):621–628.
Tagousop et al. 2018. Antimicrobial activities of flavonoid glycosides from Graptophyllum grandulosum and their mechanism of antibacterial action. *BMC Complementary and Alternative Medicine* 18 (1):1–10.
Terrasson et al. 2013. An emerging picture of the seed desiccome: confirmed regulators and newcomers identified using transcriptome comparison. *Frontiers in Plant Science* 4:497.
Tokarz et al. 2021. Stem photosynthesis—a key element of grass pea (*Lathyrus sativus* L.) Acclimatisation to Salinity. *International Journal of Molecular Sciences* 22 (2):685.
Tweddle et al. 2003. Ecological aspects of seed desiccation sensitivity. *Journal of Ecology* 91 (2):294–304.
Umezawa et al. 2010. Molecular basis of the core regulatory network in ABA responses: sensing, signaling and transport. *Plant and Cell Physiology* 51 (11):1821–1839.
van Zanten et al. 2011. Seed maturation in Arabidopsis thaliana is characterized by nuclear size reduction and increased chromatin condensation. *Proceedings of the National Academy of Sciences* 108 (50):20219–20224.
Verdier et al. 2013. A regulatory network-based approach dissects late maturation processes related to the acquisition of desiccation tolerance and longevity of *Medicago truncatula* seeds. *Plant Physiology* 163 (2):757–774.
Vertucci, CW and Farrant JM. 1995. Acquisition and loss of desiccation tolerance. *Seed Development and Germination* 237:271.
Wang et al. 2012. Proteomic analysis of embryonic axis of *Pisum sativum* seeds during germination and identification of proteins associated with loss of desiccation tolerance. *Journal of Proteomics* 77:68–86.
Wang et al. 2015. Proteomics of seed development, desiccation tolerance, germination and vigor. *Plant Physiology and Biochemistry* 86:1–15.
Waterworth et al. 2010. A plant DNA ligase is an important determinant of seed longevity. *The Plant Journal* 63 (5):848–860.
Williams, RJ and Leopold AC. 1989. The glassy state in corn embryos. *Plant Physiology* 89 (3):977–981.
Wood, AJ. 2007. The nature and distribution of vegetative desiccation-tolerance in hornworts, liverworts and mosses. *The Bryologist* 110 (2):163–177.
Yu et al. 2021. SnRK1 phosphorylation of SDH positively regulates sorbitol metabolism and promotes sugar accumulation in peach fruit. *Tree Physiology* tpaa163.
Zhang et al. 2020a. Deterioration of orthodox seeds during ageing: influencing factors, physiological alterations and the role of reactive oxygen species. *Plant Physiology and Biochemistry* 158: 475–485.
Zhang et al. 2020b. Y 2 SK 2-and SK 3-type dehydrins from Agapanthus praecox act as protectants to improve plant cell viability during cryopreservation. *Plant Cell, Tissue and Organ Culture (PCTOC)*:1–9.
Zhao et al. 2020. Plasma-activated water: Physicochemical properties, microbial inactivation mechanisms, factors influencing antimicrobial effectiveness, and applications in the food industry. *Comprehensive Reviews in Food Science and Food Safety* 19 (6):3951–3979.
Zinsmeister et al. 2020. The seed-specific heat shock factor A9 regulates the depth of dormancy in *Medicago truncatula* seeds via ABA signalling. *Plant, Cell & Environment* 43 (10):2508–2522.

17

Chemical Ecology in Belowground Plant Communication

Marcela Claudia Pagano*
Federal University of Minas Gerais

Bakhytzhan Yelikbayev
Kazakh National Agrarian University

Eduardo J. Azevedo Correa
Campo Experimental de Pitangui

Neimar F. Duarte
Instituto Federal de Minas Gerais

CONTENTS

17.1 Introduction

In recent years, the knowledge on the potentiality of microbiomes associated with plants for crop production has opened new occasions for the proposal of effective microbial combinations as the diversity of inocula and the plant growth-promoting capacity are also important, being environmentally relevant and appreciated for sustainable crop production.

Plants incessantly form networks with other organisms through release of volatile organic compounds produced by different parts, and they also release different substances via root exudation in the rhizosphere, modifying plant fitness and biotic interactions between roots and soil microbes (Delory et al., 2016). The rhizosphere represents a hot spot of microorganisms and their interactions that can affect soil characteristics and functioning (Watteau et al., 2006). Beneficial microbes associated with plants have been investigated through classical and next-generation sequencing technologies. They managed to recognize the function of the microbiome, including culturable and non-culturable microorganisms. This brings different possibilities for the use of these microbes that influence agriculture.

Agricultural practices that use low quantities of fertilizers and water, which are of interest in more sustainable systems, are increasing with increasing appreciation that the expansion of agriculture undesirably affects soil quality (Borie et al., 2005). Cultivation influences the soil physical and chemical characteristics

* Corresponding author's email: marpagano@gmail.com

DOI: 10.1201/9781003171416-17

and alter the number, diversity, and activity of the free and symbiotic populations of the soil microflora. Plant growth-promoting rhizobacteria (PGPR) are rhizospheric microorganisms that can enhance plant growth. In addition to that, new chances for the design of more efficient microbial associations to improve nutrient acquisition through the plant and its local microbiome and adjacent soil will offer new outcomes, such as the microbiome of crop management benefit for farms, without added costs. Recommendations to increase the comprehension of microbiomes in agrosystems and pastures are indicated as well as their innovations.

The diversity of a microbial inoculum is as important as its plant growth-promoting ability, to improve the reuse of nutrients and decrease the use of commercial fertilization (Çakmakçi et al., 2007). Thus, more than ten genera of bacteria are applied as biofertilizers in sustainable farms. Microbes intimately aid their host plants, increasing growth and controlling plant pathogens. Mixtures of PGPR strains contain arbuscular mycorrhizal fungi (AMF) and can enhance biological control in horticultural plants against different pathogens. Since the publication of relevant information on AMF by Sieverding (1991), Smith and Read (2008), and Siddiqui et al. (2008), the need for a clear understanding of the impact of AMF development and function on agricultural practices has been emphasized. However, the use and control of AMF in field conditions need more research.

The interest in sustainable production is rising with the need for healthy organic food, being a challenge for the scientific community and farmers in the 21st century (Agrahari et al., 2020).

Existing agricultural practices employ improper pesticides and fertilizers, creating great environmental and human well-being problems. They adversely affect soil quality, modifying the number and diversity of free and symbiotic microbial populations.

The aim of this work is to explore the last evidences on the beneficial microbes associated with plants for the progress in microbial combinations applied to the seedlings or the soil. Beneficial microbes associated with plants have been investigated through classical and next-generation approaches. Mixtures of PGPR strains include arbuscular mycorrhizae, thus opening new options for the use of these relevant biofertilizers with inherent impact on agriculture. Inferred data on plant microbiome from different sites and regions have been included. The diversity of microbes and their ability to promote growth is increasingly recognized as keys for designing more efficient microbial consortia. Crop management benefits for farmers without adding costs are emphasized to increase understanding of cropping systems.

As microbe–plant associations help plants tolerate the abiotic and biotic stresses, handling the interactions can help to obtain agro-environmental systems (Lazcano et al., 2021). To include environmental changes, soil conditions, and the degree of disturbance that affect plant and associated microorganisms (number and diversity) in the study of microbiota is crucial. Disturbance by tillage systems can modify aeration, water content, and temperature, affecting the growth and activity of soil microorganisms and consequently nutrient dynamics.

It is known that resistant plant varieties with higher abundances of recognized biocontrol microbes had developed rhizosphere-specific bacteria, such as *Pseudomonas*, under field conditions; for better disease management, this synergy needs to be understood clearly (Lazcano et al., 2021).

17.1.1 Microbiome and Arbuscular Mycorrhizal Fungi

Fertilizers and organic compost are costly due to the high amount time and labor required for commercial production. The use of symbiotic associations and the specific diversity and activity of soil microorganisms can complement the fertilization of crops having a great impact on biotic properties and free and symbiotic fungal populations (Borie et al., 2005). AMF can improve the agro-environmental and organic ecosystems based on natural nutrient cycles to obtain the nutrients required for growth (Borie et al.; 2006, Ryan and Tibbett (2008). AMF usually present a low occurrence in conventional farms (crops and pastures), attaining 10%–50% of root length colonization. However, the occurrence of AMF in organic farms can be related to the absence of phosphorus (P) fertilizers that supply soluble P, insignificant application of biocides, and different rotations (reviewed by Ryan and Tibbett, 2008) (Table 17.1). AMF do not replace fertilizers; in fact, high root colonization by AMF can indicate low soil P. Some AM associations in agroecosystems are well known (Muthukumar and Prakash 2009). In Brazil, Pagano et al. (2019) used AMF inocula in field experiments with native plants and they studied the biofertilizers from non-disturbed sites (spontaneous vegetation), showing the presence of different AM families, between field and glasshouse experiments. In most reports, Glomeraceae was commonly dominant.

TABLE 17.1

Examples of Evidence on Detected Microbiota (AMF) in Associated Plant Species

Source	Location/ Ecosystem Type	Crops/ Vegetables	SN	AMF Species Richness[a]	AMC[b]	H[c]	GC	Microbiota Species
Pagano et al. (2009)	Farms, Minas Gerais State, Brazil	Lettuce, *Allium fistulosum*	3–24	2	49.5%	ND	ND	
Wright et al. (2007)	Farming systems, Mid-Atlantic area of the USA	Corn/soybean; wheat/soybean	ND	ND	ND	ND	0.53 to 0.66	
Bedini et al. (2007)	Siena, Tuscany, Italy	Maize monoculture	464.0 to 58.5	~7	ND	ND	2	
Borie et al. (2006)	No tillage; reduced and conventional tillage, Chile	Wheat	372 to 755	NI	53%	ND	5.7 to 7.2	
Ma et al. (2005)[d]	Organic farm fields, Canada	NI	ND	0–4	ND	ND	ND	
Matias et al. (2009)	Brazil			~7			ND	*Acaulospora delicata* *Acaulospora laevis* *Acaulospora mellea* *Acaulospora spinosa* *Gigaspora margarita* *Scutellospora cerradensis* *Scutellospora verrucosa*
Pagano et al. (2010)	Brazil			10				
	Brazil							
	Brazil							
Menéndez et al. (2001)	Field site, Buenos Aires, Argentina	Wheat, barley, red clover	<20 to 50	17	ND	ND	ND	

[a]Spores.
[b]AM colonization (%) reported; SN, spore number per 100 g soil, ND=not determined in the study.
[c]Hyphae >5 µm in soil; GC=total glomalin content mg/g soil.
[d]Molecular identification.

17.1.2 Agricultural Practices and Plant Microbiome

Agricultural practices such as tillage, crop sequence, plant breeding, and fertilizer and pesticide application affect the plant microbiota such as AM fungal populations, species composition, and root colonization by AMF (Kurle and Pfleger, 1994). Conventional tillage reduces the AMF symbiosis through the disruption of their hyphal systems in the soil, which favors soil aggregation (Degens et al., 1996). The beneficial use of AMF may proportionate acceptable plant yield levels with minimal fertilizer amounts, low costs, and minimum environmental contamination threat. AMF association is affected by propagule numbers found in natural systems, when soil is minimally disturbed (McGonigle and Miller, 1996). However, it depends on the plant and fungal species, indicating a benefit of rotation on AMF populations.

Propagules (mycorrhizal roots, AM spores, and fungal mycelia) in the soil could colonize roots of the following crop of the rotation system, but active hyphae from preceding crops comprise principal sources of AMF in soil. Thus, the occurrence of active hyphae can be influenced by different crop management strategies (Borie et al., 2006).

Organic farms lack fertilizers containing readily soluble P and present higher colonization of plants by AMF and higher species diversity than conventional systems, because colonization may be limited by tillage or remaining available soil P. Organic farms can favor AMF more efficiently for plant P uptake. We don't know the need for inoculants in organic farms. On the other hand, high soil fertility and tillage as well as monocultures may stimulate the development of less advantageous communities of AMF (Ryan and Tibbett, 2008).

Plants select microorganisms by establishing a microbial community in the adjacent soil: the microbiome (Figure 17.1) which offer benefits, including stimulation of plant growth, better use of nutrients, and attenuation of plant pests. To bring innovation in the potential of plant-associated microbiome for crop production such as new methodologies for improving the nutrient fluxes among plants, the local microbiome and surrounding soils will help in the design of more efficient microbial combinations. For inoculum, the diversity of microbes is as important as its capacity to improve plant growth; thus, more holistic approaches for enhancing the productivity and restoration of soil quality can help to carry out a better plant–soil management (Ray et al. 2020; Lazcano et al., 2021). As a result, commercial agriculture has developed procedures using different volumes of chemical pesticides and fertilizers affecting the plant microbiome.

A successful colonization reflects a profitable spore formation, which is an indicator of the ecological processes and anthropic influences on the AM communities. However, spores cannot be directly linked to one vegetal species, unless in monospecific vegetation (De Souza et al., 2008). Moreover, spores are not present along the life history of the AMF. Despite the limitations of the spore count method, it was used to evaluate the effect of cultures and agriculture practices on the AM communities (Boddington and Dodd, 2000; Franke-Snyder et al., 2001; Miranda, et al., 2005; Purin et al., 2006).

To characterize AMF communities, there are traditional methods (spore morphology) and molecular techniques that objectively define intricate fungal communities in agrosystems (Ma et al., 2005).

The progress of efficient analytic techniques will facilitate the use of AMF in agro-ecological agriculture (Zimmerman et al. 2009). Inexpensive methods could also help in investigating the holistic perspective of microbiome.

In Brazil, Miranda et al. (2005) reported improved AMF in different crops under cultivation systems with Andropogon grass and soybean (at greenhouse), showing the contribution of AMF to plant growth. In the cultivated soils, the number of AM spores were increased and percent root colonization fluctuated influenced by soil moisture, crop, cropping time, and rotation systems. The number of species was influenced by harvesting and culture management being higher under annual crops in rotation. The presence of AMF in the soil effectively favored the growth of soybean and Andropogon grass.

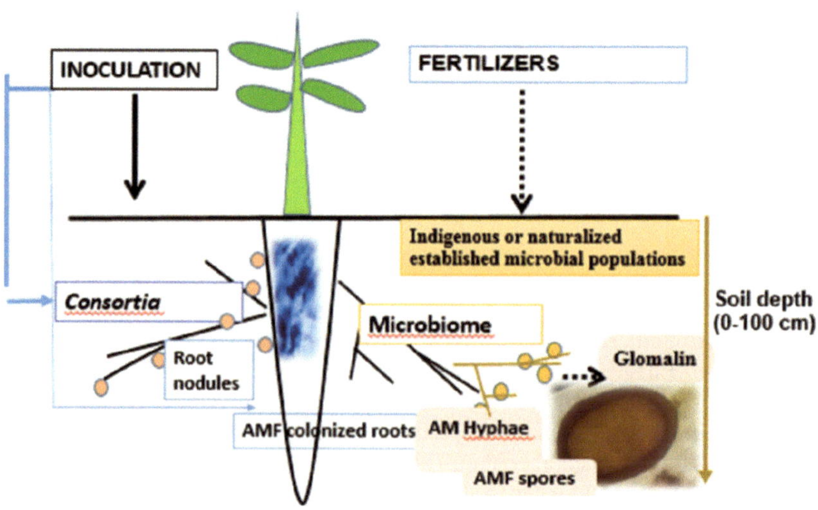

FIGURE 17.1 The plant microbiome (holistic approach).

Also in Brazil, Pagano et al. (2009) showed the mycorrhization in horticultural commercial plants in three farm sites (organic production), showing that vegetables presented AM colonization; however, a lower spore number was found in their rhizosphere, *Glomus* being dominant.

The relations between different AMF and pathogens differ with the host plant and the farming method. The protective effect of AMF inoculation may be investigated on the molecular basis promoting the use of AMF in agriculture (Siddiqui and Pichtel, 2008).

Anthropogenic modifications to improve the productivity of agrosystems (tillage, monoculture, crop rotation, irrigation, application of soil conditioners, and crop protection) disrupt the established soil microbial ecosystem. Improvements in soil quality and plant health depend on controlling such perturbations, although moderate levels of perturbation will improve ecosystem performance in the short term; higher levels of stress may decrease performance, resulting in degraded and less resilient soils (Sturz and Christie, 2003).

AMF associate with many beneficial microorganisms, such as nitrogen-fixing bacteria and phosphorus-solubilizing microbes (Barea et al.2002; de Varennes and Goss, 2007).

In the tripartite symbiosis—legume, AMF, and rhizobial bacteria—mycorrhizas improve the nodulation and the nitrogen fixed by rhizobium (Barea et al., 2002; de Varennes and Goss, 2007). AMF can also provide drought resistance, protecting bacteria, nodules, and plants.

17.1.2.1 Glomalin and Microbiome

PGPR that interact with AMF can influence glomalin production. Glomalin is a soil protein produced only by hyphae and spores of all AMF species (Wright et al., 1996) and, nowadays, is referred to as glomalin-related soil protein, whose quantity of carbon (C) is higher than that in the microbes (Rillig et al., 2004). Glomalin content is a useful indicator of soil carbon changes produced by crop systems (Rillig et al., 2004) implicated in C sequestration (Rillig et al., 1999).

The vegetative AMF occurrence can be used to predict glomalin abundance, and the availability of plant C can determine glomalin contents.

The plants can also influence soil glomalin content (Rillig et al., 2004; Wright and Upadhyaya, 1998). As glomalin presents low decomposition (Steinberg and Rillig, 2003), no tillage systems had significantly greater whole soil glomalin-related soil protein (GRSP) than more severe tillage soils for organic production, also in the soil aggregates. Similarly, GRSP concentrations and distributions in organic farming are greater than in chisel tillage (Wright et al. (2007).

The potential uses of microbial inoculants may be more tested (Pagano et al., 2010). It is well known that ectomycorrhiza (ECM) improves water balance of host plants, reduces impacts on trees from root pathogens, and mobilizes essential plant nutrients directly from soil, forming extensive mycelia connected by hyphal strands or rhizomorphs that transfer water and nutrients over extended areas (Carney, 1992).

De Souza et al. (2008) remarked that the successful use of the morphological identification method depends on the taxonomic background of the researcher as well as the previous knowledge of the diversity of the studied site, and also the use of trap cultures.

Nowadays, there is a need to further investigate the management of AMF to obtain other benefits. The use of the AMF constitute an option to high addition of fertilizers and commercial control agents in agrosystems (Gianinazzi and Schüepp, 1994). However, the knowledge on AMF inoculation is restricted due to the lack of artificial propagation in growth media.

In Brazil (Pagano and Lugo 2019), mycorrhizal inoculum consisting of spores was commonly used. The inoculum was added to the roots of the seedlings. Moreover, 10 mL rhizobial inoculum was commonly used for the native legumes. The AMF species selected were *Acaulospora scrobiculata, Gigaspora margarita, Glomus brohultii, and Scutellospora cerradensis.*

Other reports have shown that AMF species (*Gigaspora margarita*) can persist and disseminate, introduced with the coffee seedlings, and interacted with sporulation of native AMF species (Balota and Lopes, 1996).

More evidence on the benefits from plant-associated microbiota was reported in some projects, which were established through agroforest systems of mixed plant species and *Eucalyptus* in degraded sites at the Brazilian semiarid region, where it was shown that the successful growth of native plants depended on mycorrhizal fungi (Pagano et al., 2009, 2010). Increasing attention is being paid to the agroforestry potential of *E. camaldulensis* (Marcar et al., 2002), which shows AM and ECM colonization (Pagano et al., 2010).

The incidence of AMF species is influenced by host plant mixtures, and managing practices influence the survival of AM propagules (Jefwa et al., 2006).

Fertilizers and nutrient supply disparate levels of N and P decreasing the AMF occurrence and sporulation (Johnson and Gehring, 2007; (Guillemin et al., 1992; Olsson et al., 2003). However, it depends on the spore production of AMF species to different added elements (Jefwa et al., 2006). Therefore, different AMF species show different responses to fertilization. Some spores of *Glomus* (an important AMF genus in agriculture) were classified as slightly sensitive to fertilization, and other species, highly sensitive (Bhadalung et al., 2005).

Nowadays, due to the difficulty in making specific fertilizer recommendations, the requirements of native plants need to be intensively studied. The importance of agroforestry and restoration demands a complete understanding of plant life histories, including traits related to AM formation.

17.1.2.2 Soil and Microbiome

The undesirable effects of soil properties on microbiome and, consequently, on root uptake of elements and water become intense when soils become more compacted. Moreover, soil macropores decrease compared to micropores (Arvidsson, 1999).

Tillage practices negatively affect plant growth, diminishing nutrient uptake. There is scarcity in the data showing the soil compaction consequences on plants treated with AMF (Nadian et al., 1997) and in the data related to the effects of soil compaction on the growth of corn treated with AMF (Miransari et al., 2007, 2009). However, AM can diminish the stress of soil compaction enhancing nutrient uptake, as different species of AMF tolerate different ecological and abiotic conditions (Miransari et al., 2009). The effects of AMF on soil aggregation depend on the plant and AMF species (Piotrowski, 2004). Moreover, different microbial communities associate with residues depending on the location (surface or incorporated residues) (Hu et al., 2020). Furthermore, fungi act in soil aggregation than bacteria due to the higher organic carbon (OC) assimilation, extended persistence, and greater vegetative biomass (Hu et al., 2020).

As regards root traits, which maximize belowground resource acquisition in annuals and conservation and persistence in perennials, studies on roots remain limited. Correlation of root traits with plant growth strategies showed that generally, high specific root length and high root nitrogen content are found in fast-growing plant species, associated with high root respiration rate and activity (Roumet et al., 2006).

17.2 Conclusions

In this chapter, we briefly described the benefits of plant microbiome to plant hosts in agricultural systems focusing on mycorrhizal fungi. Throughout the chapter, we have shown the advantages of microbiota for attaining sustainable agricultural goals. Further research studies are required to achieve maximum benefits from microbiome as well as the relationship of plant roots, residue placement and fungal species to soil organic matter to develop surface residue for no-tillage systems at field scale. The selection of complementary rotation crops may also increase the buildup of beneficial microbiota during continuous field seasons. Additionally, management practices must be specifically adapted for each region according to its biological, social, and economic characteristics, to achieve the wise management of ecosystem services, restraining a deepening of poverty.

The choice of plant species would have great implication in the manipulation and conservation of AMF species. Despite the capacity of AMF to inhabit plants in specific environmental conditions and the damage of these fungi with disturbance, highly dependent plant hosts should be selected over mycorrhiza-independent plants.

Finally, this chapter argued that further research is necessary on agricultural systems, especially regarding belowground processes that can affect the plant microbiome, soil compaction, soil aggregation, and nutrient dynamics.

REFERENCES

Arvidsson, J, 1999. Nutrient uptake and growth of barley as affected by soil compaction. *Plant Soil.* 208, 9–19.

Agrahari et al. 2020. Plant-microbe interactions for sustainable agriculture in the post-genomic era. *Current Genomics*; 21(3), 168–178.

Balota, E.L. and Lopes, E.S. 1996. Introduction of arbuscular mycorrhizae fungus in coffee under field conditions: I. Persistence and interaction with native species. R. bras. Ci. *Solo, Campinas* 20, 217–223.

Barea et al. 2002. Mycorrhizosphere interactions to improve plant fitness and soil quality. *Antonie van Leeuwenhoek* 81, 343–351.

Bedini et al. 2007. Effects of long-term land use on arbuscular mycorrhizal fungi and glomalin-related soil protein. *Agriculture, Ecosystems and Environment* 120, 463–466.

Bergström et al. 2008. Use Efficiency and Leaching of Nutrients in Organic and Conventional Cropping Systems in Sweden. In: H. Kirchmann, L. Bergström (eds.), *Organic Crop Production – Ambitions and Limitations*, Springer, 143–159.

Bhadalung et al. 2005. Effects of long-term NP-fertilization on abundance and diversity of arbuscular mycorrhizal fungi under a maize cropping system. *Plant Soil* 270, 371–382.

Carney W.G. 1992. Translocation of solutes in ectomycorrhizal and saprophytic rhizomorfs. *Mycological Research* 96, 2, 135–141.

Çakmakçi et al. 2006. The effect of plant growth promoting rhizobacteria on barley seedling growth, nutrient uptake, some soil properties, and bacterial. *Journal of Agriculture and Forestry* 31(3).

Delory et al. 2016. Root-emitted volatile organic compounds: can they mediate belowground plant-plant interactions? *Plant Soil*, 402, 1–26.

De Souza et al. 2008. Fungos micorrízicos arbusculares: muito mais diversos do que se imaginava. In: FMS Moreira, JO Siqueira, L Brussaard, (eds.), *Biodiversidade do solo em ecossistemas brasileiros*, Lavras: Editora UFLA.

de Varennes and Goss, 2007. The tripartite symbiosis between legumes, rhizobia and indigenous mycorrhizal fungi is more efficient in undisturbed soil 2007. *Soil Biology and Biochemistry* 39(10):2603–2607.

Hu et al. 2020. Direct measurement of the in situ decomposition of microbial-derived soil organic matter. *Soil Biology and Biochemistry* 141, 2020, 107660.

Johnson, NC. and Gehring, CA. 2007. Mycorrhizas: symbiotic mediators of rhizosphere and ecosystem processes. In: ZG Cardon, JL Whitbeck, (eds.) *The Rhizosphere: An Ecological Perspective.* London: Elsevier Academic Press; 31–56.

Lazcano et al. 2021. The rhizosphere microbiome plays a role in the resistance to soil-borne pathogens and nutrient uptake of strawberry cultivars under field conditions *Scientific Reports* 11:3188.

Marcar et al. 2002. Effect of NaCl and high pH on seedling growth of 15 Eucalyptus camaldulensis Dehnh. Provenances. *New Forests* 23:193–206.

Matias et al. 2009. Effect of rhizobia, mycorrhizal fungi and phosphate-solubilizing microorganisms in the rhizosphere of native plants used to recover an iron ore area in Brazil. *European Journal of Soil Biology* 45(3): 259–266.

Nadian et al. 1997. Effects of soil compaction on plant growth phosphorus uptake and morphological characteristics of vesicular—arbuscular mycorrhizal colonization of *Trifolium subterraneum*, *New Phytologist* 135, 2, 303–311.

Olsson et al. 2003. Foraging and resource allocation strategies of Mycorrhizalfungi in a patchy environment. In: van der Heijden MGA, Sanders IR (eds), *Mycorrhizal Ecology.* Springer, Berlin.

Pagano, M. C. and Lugo, M. A. (Eds.) 2019. *Mycorrhizal Fungi in South America.* 374, 978-3-030-15228-4.

Pagano et al. 2009. Intercropping system of tropical leguminous species and Eucalyptus camaldulensis, inoculated with rhizobia and/or mycorrhizal fungi in semiarid, Brazil.

Pagano et al. 2010. Agroforestry in dry forest, Brazil: mycorrhizal fungi potential. In: LR Kellymore, ed. *Handbook on Agroforestry: Management Practices and Environmental Impact.* Nova Science Publishers, New York. Chapter 14. ISBN: 978-1-60876-359-7.

Ray et al. 2020 Microbe to microbiome: a paradigm shift in the application of microorganisms for sustainable agriculture. *Frontiers in Microbiology* 11:622926.

Rillig M. C. 2004. Arbuscular mycorrhizae, glomalin, and soil aggregation. *Canadian Journal of Soil Science* 84, 355–363.

Ryan M. H. and Tibbett M. 2008. The role of arbuscular mycorrhizas in organic farming. In: H Kirchmann, L Bergström, (eds.) *Organic Crop Production – Ambitions and Limitations*, DOI 10.1007/978-1-4020-9316-6-10, Springer, 189–229.

Siddiqui, Z. and Pichtel, J. 2008. Mycorrhiza: an overview. In: ZA Siddiqui, MS Akhtar, K Futai, (eds). *Mycorrhiza: Sustainable Agriculture and Forestry*. Springer, 359p.

Sieverding, E. 1991. *Vesicular-Arbuscular Mycorrhiza Management in Tropical Agrosystems*. Eschborn: Deutsche Gesellschaft für Technische Zusammenarbeit.(GTZ) GmnBH, Esch boran.

Smith S.E. and Read D.J. 2008. *Mycorrhizal Symbiosis*. New York: Elsevier.

Watteau et al. (2006). Root impact on the stability and constitution of the fine organo-mineral associations in a maize cultivated soil. *European Journal of Soil Science* 57, 247–257.

Zimmerman et al. 2009. Sustainable agriculture and the multigenomic model: how advances in the genetics of arbuscular mycorrhizal fungi will change soil management practices. In: K Bouarab, N Brisson, F Daayf, (eds.) *Molecular Plant-Microbe Interactions*. CABI. 269–278. Doi: 10.1079/9781845935740.0269

18

Possible Bioremediation Strategies for Arsenic Detoxification by Consortium of Beneficial Bacteria

Sathi Paul, Akansha Jain, Surbhi Shriti, and Sampa Das*
Bose Institute

CONTENTS

18.1 Introduction

Arsenic (As), a naturally occurring toxic metalloid, is introduced into the environment as well as in the food chain through natural geochemical processes and several anthropogenic activities. The multifaceted negative impact of As contamination on agriculture, environment, and eventually the human health is a matter of serious concern for academicians, researchers, and legislators. Thus, deterrence of future contamination from this element is imperative for human welfare. Recently, several remediation methods based on physical, chemical, and biological approaches have been developed to achieve complete or partial elimination of As from soil and groundwater or to reduce its bioavailability for minimizing toxicity. The most commonly used physicochemical approaches to As mitigation from soil and water include soil capping, soil washing, oxidation, co-precipitation, adsorption, ion exchange, and membrane techniques such as filtration and reverse osmosis (Shen, 1973; Cheng et al., 1994; Joshi and Chaudhuri, 1996). But the application of these conventional methods is not very economical and requires very high-tech operation and maintenance. Some of the oxidizing agents used in these physicochemical methods result in the formation of by-products (Katsoyiannis and Zouboulis, 2004). Also, the sludge generated from precipitation/coagulation treatment technologies if discarded under inappropriate Eh/pH conditions can remobilize As. Thus, these techniques require further use of various chemicals, which in turn increases the cost of treatment and leads to environmental pollution (Leist et al., 2000). Ion exchange as a treatment technology is disadvantageous as it generates toxic chemical reagents, which are released into

* Corresponding author's email: sampa@jcbose.ac.in

DOI: 10.1201/9781003171416-18

the environment (Shih, 2005). Utilization of membrane techniques is unfavorable in developing countries for its high expenditure of energy consumption to produce only a small amount of treated water (US EPA, 2000). An alternative approach to these conventional cleanup technologies is to use natural biological activity for the degradation or detoxification of substances hazardous to human and animal well-being and/or the environment. Naturally occurring microorganisms (primarily bacteria, fungi, yeast, and algae) or plants can incapacitate or translate environmental pollutants to an insipid state and/or final products through reactions that occur during their metabolism. The success of the bioremediation process of any contaminant depends upon several factors, including the existence of the type of microbes able to degrade the pollutants and the receptiveness of the pollutants to those microbes and several environmental conditions, such as the temperature, pH, type of soil, availability of oxygen or electron acceptors, and nutrients (Mahimairaja et. al., 2005). Bacteria play a key role in the biogeochemical cycle of arsenic by a network of processes of As causing it to immobilize and transform. In recent years, a number of phylogenetically diverse bacteria have been reported (Oremland and Stolz, 2003; Hong, 2006; Cai and Wang, 2009), which can deal with the toxicity of As by efflux of the heavy metal exterior the cell, inflation and complex formation of the heavy metal in the cell, and transformation by oxidation–reduction mechanism of the heavy metal ions to a less toxic state (Nies, 1999). Various soil bacteria such as *Bacillus* sp. (Bakhat et al., 2017), *Halobacterium* sp., and *Rhodococcus* sp. (Williams et al., 2013) have been reported to be capable of adsorption of different inorganic and organic species of As. In addition to that, concurrent bacterial oxidation of trivalent arsenite (As III) and Fe II leads to the production of Fe III (hydr) oxides, which in turn adsorb pentavalent arsenate (As V) produced from As III oxidation (Inskeep et al. 2004; Sun 2008). This is how As immobilization has been deliberated as a probable bioremediation plan of action of As in an anoxic condition. A wide variety of bacteria, e.g., *Alcaligenes faecalis*, *Hydrogenophaga* sp., *A. ferrooxidans*, *T. aquaticus*, and *T. thermophilus* (Gihring et al. 2001; Oremland and Stolz 2003; Stolz et al. 2006), have been characterized for their As III oxidation potential. Arsenite oxidase enzyme present in these bacteria allows them to grow chemolithotrophically by using As III as donor of electrons and oxygen as acceptor of electrons (Duquesne et al., 2008). Moreover, bacteria containing the As III-S-adenosylmethionine methyltransferase (arsM) gene are also capable of transforming inorganic As compounds by biomethylation into organic As compounds (Woolson, 1977; Cullen and Reimer, 1989; Gadd, 1993), which has actually been described as the most important detoxification process since it reduces the affinity of these As compounds for tissue in human body (Vahter and Marafante, 1988). Such metabolic processes taking place within bacteria can be utilized to design novel, efficient, and environment-friendly remediation strategies of arsenic-rich environment.

18.2 Strategies Adopted by Bacteria to Alleviate Arsenic Toxicity

The microbial detoxification of As can be accomplished by reduction, oxidation, and methylation (Figure 18.1). The key factors accountable for As resistance in bacteria include transporters, redox proteins, methyltransferases, different biosynthetic pathways for arsenosugars, lipids, and other non-harmful molecules of arsenic (Li et al., 2016; Yang and Rosen, 2016). Bioremediation prospects and suitability of the process make bacteria an upcoming basis of eco-friendly way to get rid of arsenic in the environment.

18.2.1 Arsenic Uptake and Efflux Systems

Arsenic has no nutritional or metabolic roles and is rather toxic for any life form. Therefore, none of the organism has evolved with an uptake system specific for As. Both As III and As V are moved inside most cells via different types of existing transporter molecules. The physiological action in bacteria consists of the movement of nutrients and minerals as glucose, phosphate, and glycerol. Arsenic, on the other hand, can share the uptake systems of these natural substrates for being structurally similar to them. The glycerol transporter GlpF, a part of important intrinsic protein (MIP) (Reizer et al., 1993) superfamily, is known to take up uncharged As III into *Escherichia coli* cells. *Aquaglyceroporins* are recognized as a subset of the aquaporin (AQP) (Agre et al., 2002) family of integral membrane proteins, which transport water, glycerol, and other small uncharged solutes. Arsenite in the form of arsenic trioxide (As_2O_3) dissolves to form $As(OH)_3$, which is a chemical analogue of glycerol and conducted through

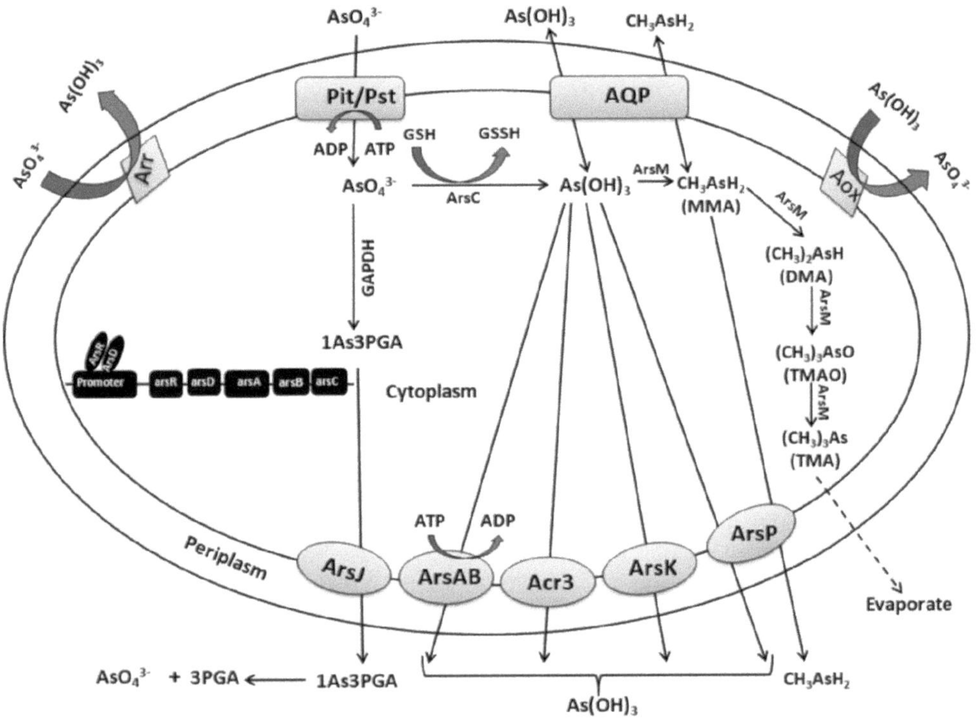

FIGURE 18.1 Schematic representation of the processes involved in arsenic metabolism within a bacterial cell. Inorganic arsenic compounds inadvertently enter the cell via nutrient transporters such as aquaglyceroporins (AqpS) or phosphate transporters (Pit/Pst). As V is either reduced to As III by ArsC and/or transformed to 1AsPGA by GAPDH prior to efflux by ArsAB and ArsJ pumps, respectively. As III directly extrudes out of the cell by ArsAB, Acr3, ArsK, and ArsP pumps and/or is altered into organic forms in a methylation cascade catalyzed by ArsM.

AQPs. Although bidirectional in nature, most AQPs allow As III movement inside the bacterial cells increasing metal toxicity, except an aquaglyceroporin (AqpS) coded by *ars* operon of *Sinorhizobium meliloti*, which is used to efflux the toxic As III (Yang et al., 2005) outside the cell. GlpF homologs are observed in other bacterial species such as *Leishmania major* or *Pseudomonas putida*, facilitating As III movement traverse the cell membrane (Gourbal et al., 2004). In water and aerobic soil, As V is the prevalent species of As. Arsenate behaves as a molecular mimic of phosphate and enters most cells by phosphate transporters. Two of the phosphate transporters (Pit and Pst) are adopted for As V uptake in *E. coli* with Pst acting as the major one (Rosen and Liu, 2009) allowing higher concentrations of As V inside the cell (Willsky and Malamy, 1980); however, Pit is primarily responsible for As V toxicity. The most prevalent system for tolerance to As toxicity in bacteria is the removal from cells or intracellular sequestration (Ben Fekih et al., 2018). Inorganic As after entering the cell either is effluxed by functional pumps such as ArsJ, ArsAB, Acr3, ArsK, and ArsP, or is transformed by enzymatic actions of arsenate reductase (ArsC), As-III-S-adenosylmethionine methyltransferase (ArsM), or glyceraldehyde 3-phosphate dehydrogenase (GAPDH) before efflux. The majority of prokaryotes use the ArsAB, while few bacteria function only with ArsB. During intracellular sequestration, the heavy metal ions form complexes with different molecules in the cell cytoplasm. Nowadays, intracellular accumulation of metals by bacteria has been exploited predominantly in effluent treatment and other cleanup practices. Ghodsi et al. (2011) suggested that the cellular accumulation is related to higher uptake followed by lower efflux and the higher resistance in these bacteria could be controlled by a regulatory protein encoded by ars operon with explicit binding site for As III. A potent As-accumulating bacterium *Marinomonas communis* with high removal efficiency has been isolated in Japan from marine and non-marine environments (Takeuchi et al., 2007). The factors known to influence As bioaccumulation in microbial cells involve soil pH, temperature, moisture, aeration, organic and inorganic matter of soil, concentration and speciation, and rhizosphere (Mahimairaja et al., 2005). Bacterial species belonging to genera *Bacillus, Enterobacter, Flavobacterium*, *Pseudomonas*, and *Micrococcus* have significant biosorption ability for their greater surface area-to-volume ratios and/or the presence of teichoic acid as crucial chemisorption sites on the cell wall (Mosa et al., 2016). Researchers have demonstrated that microbes are more functional and better fitted to survive in a consortium (Sannasi et al., 2006). Kader et al. (2007) demonstrated that the mixed cultures are metabolically better to individual bacteria for biosorption of heavy metals and more suitable for field utility.

18.2.2 Bacterial Redox Reaction

Bacteria-mediated redox reactions mainly act upon As III and As V species, resulting in changes in solubility and/or toxicity of As.

18.2.2.1 Oxidation of Arsenite to Arsenate

Oxidation of As III is an important detoxification system as As V is less soluble and strongly adsorbed onto inorganic soil components, resulting in immobilization of As. The As III oxidizing propensity to the bacteria is provided by *aox* operon, encoding AoxAB complex, which belongs to DMSO reductase family of molybdenum enzymes. The As III oxidase enzyme acts as a heterodimer (AoxA, ~14 kDa, and AoxB, 90 kDa) containing Fe and molybdenum in its catalytic site (Ellis et al., 2001). After being synthesized, AoxAB is exported to the periplasm utilizing the TAT (twin-arginine translocation) leader sequence, where it involves in oxidation of As III to As V (Silver and Phung, 2005). The expression of aox operon is controlled by a two-component signal transduction system comprised of AoxS (sensor kinase) and AoxR (response regulator) (Kashyap et al. 2006a). After a while, evidences of additional proteins such as molybdate transporter and a Na+/H+ antiporter (Kashyap et a., 2006b), as well as RpoN (alternative sigma factor, s54) and DnaJ (heat shock protein J) (Koechler, et al., 2010), have also been found to be essential for As III oxidation. Arsenite-oxidizing bacteria acquire metabolic energy from As III oxidation (Ilyaletdinov and Abdrashitova, 1981). Homologous genes encoding As III oxidase enzymes identified in the genomes of different microorganisms are designated with dissimilar names. The genes are designated as aoxA/aoxB in *Cenibacterium arsenoxidans* strain ULPAsl (Muller et al. 2003) and *Agrobacterium tumefaciens* (Kashyap et al. 2006a), whereas different nomenclatures such as asoB/asoA and aroB/aroA are designated to the genes of *Alcaligenes faecalis* (Stolz et al. 2006) and *Rhizobium* sp. strain NT-26 (Santini and vanden Hoven, 2004), respectively. Bacteria belonging to several other genera such as *Bacillus, Pseudomonas* (Frankenberger and Losi, 1995), and *Thermus* (Gihring et al., 2001) have also been reported with As III-oxidizing potential. Paul et al. (2017) identified twelve hyper-tolerant bacteria (showing a maximum tolerable concentration for As V $\geq$ 300 mM and As III $\geq$ 30 mM) under the genera of *Acinetobacter, Bacillus*, *Paenibacillus*, and *Rhodococcus* inhabiting contaminated soil and groundwater of West Bengal, India, majority of which could transform As III to As V.

18.2.2.2 Reduction of Arsenate to Arsenite

Bacterial As V reduction to As III can be achieved either by a cytosolic reductase conjoined with As III efflux pump (encoded by ars operon) that is well conserved as a detoxification system, or by employing a periplasmic respiratory As V reductase (encoded by arr operon), which functions as a terminal electron acceptor, allowing growth in anaerobic condition. Although the above-mentioned two reductases play different roles for the survival and growth of bacteria in an As-contaminated environment, the two operons encoding them often lie in close proximity, "suggestive of an arsenic metabolism island," in various bacterial species (Silver and Phung, 2005). The respiratory As V reductase is a heterodimeric protein with a native molecular mass of $\approx$123–131 kDa and has a larger subunit ArrA (mass of $\approx$ 87–95 kDa), with molybdopterin, and a smaller subunit ArsB (mass of $\approx$ 27–26 kDa), with many Fe–S clusters (Saltikov and Newman, 2003). A mutation in the arrAB gene cluster of *Shewanella* sp. strain ANA-3 has been found to inhibit the growth of this bacterium in the presence of As V (Saltikov et al., 2003). *Escherichia coli* and *Staphylococcus aureus* use the plasmid-encoded, detoxifying reductase (ArsC) present in their cytoplasm to reduce As V to As III, followed by its quick eviction from the cell (Ji et al., 1994; Diorio et al., 1995). Bacterial strains such as *Bacillus asoselenatis, B. selenitireducens, Chrysiogenes arsenatis, Pseudomonas* sp., *Sulfurospirillum barnesii, S. arsenophilum,* and *Wolinella* spp. (Ahmann et al., 1994; Newman et al., 1998) use As V as a terminal electron acceptor in an aerobic respiration, resulting in a dissimilatory reduction of As V (Stolz and Oremland, 1999).

18.2.2.3 Biovolatilization of Arsenic

Bacteria can transform the highly toxic inorganic As to relatively less toxic organic compounds through methylation (Ridley et al., 1977; Gadd, 1993). These methylated As compounds for their low boiling point and/or high vapor pressure are susceptible to volatilization and can easily be lost to the atmosphere (Braman and Foreback, 1973). Generation of arsines and methyl arsenicals is primarily mediated by As V reduction following oxidative addition of methyl groups (Dombrowski et al., 2005) from compounds such as methylcobalmin and S-adenosylmethionine (SAM) as evident in many bacterial systems (Gadd and White, 1993). The methylation pathway involving generation of a series of methylated arsenic species: monomethylarsonous acid (MMA-III), dimethylarsinic acid (DMA-V), dimethylarsinous acid (DMA-III), trimethylarsine oxide (TMAO), and trimethyl arsenite (TMA-III), is catalyzed by As-III-S-adenosylmethionine methyltransferase enzyme encoded by arsM gene. McBride and Wolfe (1971) first demonstrated the conversion of As V to a limited amount of volatile methyl arsines in the pure culture of a methanogen, *Methanobacterium bryantii.* Till date, numerous aerobic and anaerobic bacteria have been found to be capable of producing methyl arsines, including *Achromobacter* sp., *Aeromonas* sp., *Enterobacter* sp., *Nocardia* sp., *Methanobacterium formicicum, Clostridium collagenovorans, Desulfovibrio vulgaris,* and *D. gigas* (Michalke et al., 2000).

18.3 Molecular Mechanisms of Bacterial Tolerance to Arsenic Species

Getting an insight into the molecular and genetic levels of As resistance in bacteria could be of key importance for screening efficient As bioremediation approaches. The ars operon (typically arsRDABC) encoded on either the chromosome or plasmids of prokaryotes is the most well-characterized genetic system that employs resistance mechanism to As toxicity (Xu et al., 1998). The cytosolic reductase enzyme encoded by arsC gene is an important component that facilitates the reduction of As V to As III, followed by its subsequent eviction by another transmembrane protein, As III expulsion pump encoded by arsB. Arsenate binds to the Arg residues of a recognition domain on ArsC and forms disulfide bond between the cysteine residues on ArsC and the reducing equivalents. Reduction of the disulfide bond mediated via electron transfer results in the release of As III and regeneration of reduced ArsC (Silver and Phung, 2005). ArsA is an intracellular ATPase that supports ArsB for pumping As III out of the cell by deriving the necessary energy from ATP hydrolysis (Tisa and Rosen, 1990). ArsR and ArsD are upregulatory components responsible for transcriptional repression and controlled expression of the above-mentioned structural genes (Rosen, 2002). Later, several new ars genes were discovered, indicating parallel evolution and operation of complex regulatory pathways (Butcher et al., 2000).

18.4 Role of Bacteria in Arsenic Toxicity Amelioration in Plants

Nowadays, phytoremediation assisted by bacterial endophytes and rhizospheric bacteria is gaining immense interest for the remediation of As-polluted sites since these bacteria can mitigate the toxic effects of As in the plant by utilizing their metal tolerance potential and expedite plant growth in amenable metal stress (Table 18.1). Several bacterial mechanisms that enhance phytoremediation of As-contaminated soils involve advancement of plant growth by production of metabolites, such as indole-3-acetic acid, heavy metal chelation by siderophores, production of organic acids, phosphate solubilization, methylation of inorganic As, and managing As afflict by synthesis of 1-aminocyclopropane-1-carboxylic acid deaminase (ACCD) (Lebeau et al., 2008; Ma et al., 2011). In inclusion to resilience for heavy metal, such plant beneficial microbe can function as biological control agents against some phytopathogens and assure nitrogen fixation. Endophytes colonizing the internal tissues of *Pteris vittata*—an arsenic hyper-collector—were observed to revamp As V to As III as a means of As stress tolerance in the host plant and also produced IAA, siderophores, and solubilized phosphate as mechanisms of plant growth promotion. Besides, their natural host endophytes can have benign effects on obscure plant genera, which indicates that endophytes can proxy plants of economic value and plants applicable for bioremediation. Mukherjee et al. (2018) explored the possibility of As bioremediation of the polluted location at the Ganga–Brahmaputra Delta, India, using microbe-assisted phytoremediation. Arsenic-resistant endophytes from As-tolerant *Lantana camara* were isolated and formulated within *S. nigrum* as a proxy host. The As-tolerant endophytic consortium was found to improve plant growth with better phosphate assimilation, high photosynthetic efficiency, and increased glutathione values apart from elevated bioaccumulation under As stress. Single-microbe-treated plants had an inconsistent effect, some with increased growth promotion, but having negligible function in the reformation of As bioaccumulation or vice versa; however, when used as a mixture, they significantly improved As bioaccumulation in *S. nigrum*. Rhizosphere is the area where comprehensive associations take place between plants, soil, and soil microorganisms. Plant root exudates containing amino acids, various hormones, organic acids, growth promoters, and sugars (Antoun and Kloepper, 2001) are a great source of nutrition for rhizobacteria. On the other hand, they benefit the host plant by metabolically transforming the heavy metal or diluting it in their organic exudates, which makes it more applicable for plant uptake. Several As-tolerant microbes, viz. *Bacillus, Kocuria, Micrococcus, Achromobacter, Brevundimonas, Microbacterium, Ochrobactrum, Pseudomonas, Comamonas, Stenotrophomonas*, and *Ensifer*, have been observed to reduce harmful effects of As and promote plant growth by modulating arsenic assemblage and bioaccumulation (Ghosh et al., 2011; Wang et al., 2011; Yang et al., 2012; Pandey et al., 2013; Mallick et al., 2014, 2018; Mesa et al., 2017).

TABLE 18.1

Identified Microbes Involved in Amelioration of As Stress and Regulation of As Accumulation in Plants

S.No.	Arsenic Metabolising Microorganisms	Mode of Action	Habitat/Target Plant	Tolerance Limit and Respective form of Arsenic	References
	Pantoea dispersa strain As18	Improved antioxidant enzymatic activities and reduced As uptake in rice plant. Also capable of nitrogen fixation, phosphate solubilization, 1-aminocyclopropane-1-carboxylic acid deaminase (ACCD) activity, indole-3-acetic acid (IAA) production	*Oryza sativa*	3750 µg/mL As III	Ghosh et al., 2021
	Pseudomonas gessardii	Siderophore production and phosphate solubilization	*Triticum aestivum*	1500 ppm As III and 7000 ppm As V	Soto et al., 2019
	Brevundimonas intermedia	Production of indoleacetic acid (IAA)	*Triticum aestivum*	1500 ppm As III and 6000 ppm As V	Soto et al., 2019
	Kocuria flava	Reduced bioavailability of As	*Oryza sativa*	35 mM As III and 450 mM As V	Mallick et al., 2018
	Bacillus vietnamensis	Reduced bioavailability of As	*Oryza sativa*	20 mM of As III and 350 mM As V	Mallick et al., 2018
	Enterobacter sp. (LC1, LC4, and LC6), Kosakonia sp. (LC7) ,Kocuria sp. (isolated from Lantana camara)	Enhanced phosphate nutrition, photosynthetic performance, and elevated glutathione levels in plant besides increasing bioaccumulation and root-to-shoot transport under As stress.	*S. nigrum*	4000 ppm As V	Mukherjee et al., 2018

(*Continued*)

TABLE 18.1 (*Continued*)

Identified Microbes Involved in Amelioration of As Stress and Regulation of As Accumulation in Plants

S.No.	Arsenic Metabolising Microorganisms	Mode of Action	Habitat/Target Plant	Tolerance Limit and Respective form of Arsenic	References
	Acinetobacter lwoffii (RJB-2)	Siderophore and IAA production and phosphate solubilization	*Vigna radiata*	50 mM As III and 25 mM As V	Das and Sarkar., 2018
	Methylobacterium oryzae	Production of auxins, cytokinins, ACC deaminase and increasedGSH concentration andactivity of glutathione-S transferase in plant	*Acacia farnesiana*	580 μM As V	Alcántara-Martínez et al., 2018
	Ralstonia eutropha Q2-8, Rhizobium tropici Q2-13, Exiguobacteriuma-urantiacum Q3-11	IAA and siderophore producing strains	*Brassica rapa,Raphanus sativus*	20 mM As III and 300 mM As V ; 25 mM As III and 380 mM As V; 30 mM As III and 380 mM As V	Wang et al., 2017
	Ensifer, Pseudomonas	Siderophore production, IAA, ACC-deaminase activity and phosphate solubilization	*Betula celtiberian*	>5 mM As III and >100 mM As V	Mesa et al., 2017
	Brevundimonas diminuta NBRI012	Siderophore production IAA, ACC-deaminase activity and phosphate solubilization	*Orya sativa*	20 ppm As III and 150 ppm As V	Singh et al., 2016
	Bacillus flexus	Siderophore production, IAA, ACC-deaminase activity and phosphate solubilization	*Oryza sativa*	32 mM As III and 280 mM As V	Das et al., 2016
	Pseudomonas, Buttiauxella	Reduction of As(V) to As(III) was achieved via a detoxification mechanism	*Cirsium arvense, Deschampsia caespitosa*	15 mmol As III and 300 mmol As V	Cavalca et al., 2015
	Staphylococcus aureus	IAA, siderophores production and ACC deaminase activity	*Brassica juncea (L.) Czern. Var. R-46*	40 mM As III and 366 mM As V	Srivastava, et al., 2012
	Agrobacterium radiobacter D14	IAA and siderophores production	*Populus deltoides LII05-17*	14 mmol l^{-1} (1·05 g l^{-1}) for As III and 150 mmol l^{-1} (11·24 g l^{-1}) for As V	Wang et al., 2011
	Naxibacter sp. AH4, *Mesorhizobium sp.* AH5, *and Pseudomonas sp.* AH21	Siderophore production, IAA, ACC-deaminase activity and phosphate solubilization	*Pteris vittata*	400 mM As V	Huang et al.,2010

18.5 A Genomic Perspective of Arsenic Bioremediation

Microorganisms are potential candidates to create new/improved, environmentally safe bioremediation strategies. Omics can be of significant importance for understanding metabolic reactions going inside these microbes and communication between microbial communities. The advancement of high-throughput sequencing has provided a complete genome sequence of unculturable microbes. Although major issues related to sampling and annotation require further research (Teeling and Glöckner, 2012), a large number of ongoing metagenome projects are expected to generate lot more information. The advancement of genomics has allowed the comprehensive view of microbial interactions in natural remediation system and finding microbial species for devising As bioremediation plants. The molecular processes associated with As bioremediation have recently been reviewed (Andres and Bertin, 2016). The metagenomic approach applied under acidic condition in France resulted in the characterization of *aio* genes for As III oxidase in *Thiomonas* sp. and *rus* genes coding for rusticyanin in *Acidithiobacillus* spp. (Bertin et al., 2011). Functional genomics on *Herminiimonas arsenicoxydans* indicated that it activates As resistance on the induction of efflux mode before oxidation of As III (Cleiss-Arnold et al., 2010). In a recent study, genome sequencing of As III-oxidizing *Halomonas* A3H3 (Koechler et al., 2013), and *Pseudomonas xanthomarina* S11 obtained from an As-polluted gold mine (Koechler et al., 2015) has also been described.

18.6 How a Bacterial Consortium Plays a Better Role Than Individual Strains

Beneficial microbes can work in a consortium depicting natural communities working in for biocontrol, bioremediation, etc. Microbes in the consortium can act superior to single-microbe treatment as different mechanisms of action can work simultaneously for detoxification and these microbes can also promote their mutual growth. Microbial consortiums were applied for screening As elimination capacity along with iron and manganese oxidation infiltration structure (Chhetri et al., 2014). Similarly, consortiums of sulfate-reducing bacteria were also tested for As removal coupled with sulfate reduction processes from contaminated water (Teclu et al., 2008; Serrano and Leiva, 2017). Although a large number of studies reported the efficiency of As bioremediation from water by bacteria, these approaches need to be completely explored for As bioremediation, and apprehension of the diverseness and assignment of functional genes involved in As detoxification is largely discrete (Andres and Bertin, 2016).

18.7 Conclusions and Future Prospects

Sustainable agriculture is a future need for safe agri-products in As-polluted areas and bioremediation of the polluted locations. Arsenic-tolerant and detoxifying plant growth-promoting bacteria are the most anticipated candidates in such regard. Beneficial microorganisms that can mediate As toxicity alleviation in plants and surroundings are presently one of the most focused topics of research. The future relies on the development of an economically feasible PGPM-based strategy for different sites and the development of a potential consortium of these beneficial bacteria for As detoxification. Currently, emerging advanced genomic DNA sequencing techniques provide access to the genomic information of less represented and/or uncultivable microbial species involved in As bioremediation. In addition to the laboratory and field research and methodical sampling, the development of powerful computing solutions for studying the huge amounts of data associated with the microbiological processes concealed in As eradication is also absolutely required for gaining the benefit of the newer technologies.

Acknowledgments

SP acknowledges the Council of Scientific and Industrial Research for providing financial support under Research Associateship (File No: 09/015(0526)/2018-EMR-I) program. AJ is grateful to the Department of Science and Technology, Govt. of India, New Delhi, for financial assistance under DST Women Scientist-WOS-A Scheme (SR/WOS-A/LS-377/2018). SS thanks University Grants Commission, India, for providing fellowship as Senior Research Fellow. SD acknowledges Indian National Science Academy for providing Senior Scientist Fellowship.

REFERENCES

Agre et al. (2002). Aquaporin water channels–from atomic structure to clinical medicine. *The Journal of Physiology, 542*(1), 3–16.

Ahmann et al. (1994). Microbe grows by reducing arsenic. *Nature, 371*(6500), 750–750.

Alcántara-Martínez et al. (2018). An endophytic strain of Methylobacterium sp. increases arsenate tolerance in Acacia farnesiana (L.) Willd: A proteomic approach. *Science of The Total Environment, 625*, 762–774.

Andres, J. and Bertin, P. N. (2016). The microbial genomics of arsenic. *FEMS Microbiology Reviews, 40*(2), 299–322.

Antoun, H. and Kloepper, J.W. (2001). Plant growth promoting rhizobacteria. In: Brenner S, Miller JF (Eds), *Encyclopedia of Genetics*, Academic Press. 1477–1480.

Bahar et al. (2012). Arsenic bioremediation potential of a new arsenite-oxidizing bacterium Stenotrophomonas sp. MM-7 isolated from soil. *Biodegradation, 23*(6), 803–812.

Bakhat et al. (2017). Arsenic uptake, accumulation and toxicity in rice plants: possible remedies for its detoxification: a review. *Environmental Science and Pollution Research, 24*(10), 9142–9158.

Ben Fekih et al. (2018). Distribution of arsenic resistance genes in prokaryotes. *Frontiers in Microbiology, 9*, 2473.

Bertin et al. (2011). Metabolic diversity among main microorganisms inside an arsenic-rich ecosystem revealed by meta-and proteo-genomics. *The ISME Journal, 5*(11), 1735–1747.

Bin, H. O. N. G. (2006). Influence of microbes on biogeochemistry of arsenic—mechanism of arsenic mobilization in groundwater. *Advances in Earth Science, 1.*

Braman, R. S. and Foreback, C. C. (1973). Methylated forms of arsenic in the environment. *Science, 182*(4118), 1247–1249.

Butcher et al. (2000). The chromosomal arsenic resistance genes of thiobacillus ferrooxidans have an unusual arrangement and confer increased arsenic and antimony resistance to *Escherichia coli. Applied and Environmental Microbiology, 66*(5), 1826–1833.

Cai, L. and Wang, G. J. (2008). Advance on studies of arsenic-resistant microorganisms and molecular mechanisms. *Microbiology*, (08).

Cheng et al. (1994). Enhanced coagulation for arsenic removal. *Journal-American Water Works Association, 86*(9), 79–90.

Chhetri et al. (2014). Bacterial diversity in biological filtration system for the simultaneous removal of arsenic, iron and manganese from groundwater. *Journal of Water and Environment Technology, 12*(2), 135–149.

Cleiss-Arnold et al. (2010). Temporal transcriptomic response during arsenic stress in Herminiimonas arsenicoxydans. *BMC Genomics, 11*(1), 1–10.

Cullen, W.R. and Reimer, K.J. (1989). Arsenic speciation in the environment. *Chemical Reviews, 89*(4), 713–764.

Das, J. and Sarkar, P. (2018). Remediation of arsenic in mung bean (Vigna radiata) with growth enhancement by unique arsenic-resistant bacterium Acinetobacter lwoffii. *Science of the Total Environment, 624*, 1106–1118.

Das et al. (2016). Arsenite-oxidizing bacteria exhibiting plant growth promoting traits isolated from the rhi zosphere of Oryza sativa L.: implications for mitigation of arsenic contamination in paddies. *Journal of Hazardous Materials, 302*, 10–18.

Dey et al. (2016). Isolation and characterization of arsenic-resistant bacteria and possible application in bioremediation. *Biotechnology Reports, 10*, 1–7.

Diorio et al. (1995). An Escherichia coli chromosomal ars operon homolog is functional in arsenic detoxification and is conserved in gram-negative bacteria. *Journal of Bacteriology, 177*(8), 2050–2056.

Dombrowski et al. (2005). Thermodynamic analysis of arsenic methylation. *Environmental Science and Technology, 39*(7), 2169–2176.

Duquesne et al. (2008). Arsenite oxidation by a chemoautotrophic moderately acidophilic Thiomonas sp.: from the strain isolation to the gene study. *Environmental Microbiology, 10*(1), 228–237.

Ellis et al. (2001). Crystal structure of the 100 kDa arsenite oxidase from alcaligenes faecalis in two crystal forms at 1.64 Å and 2.03 Å. *Structure, 9(2)*, 125–132.

Frankenberger Jr, W. T. and Losi, M. E. (1995). Applications of bioremediation in the cleanup of heavy metals and metalloids. *Bioremediation: Science and Applications, 43*, 173–210.

Gadd, G. M. (1993). Microbial formation and transformation of organometallic and organometalloid compounds. *FEMS Microbiology Reviews, 11*(4), 297–316.

Gadd, G. M. and White, C. (1993). Microbial treatment of metal pollution—a working biotechnology? *Trends in Biotechnology, 11*(8), 353–359.

Ghodsi et al. (2011). Investigation of bioremediation of arsenic by bacteria isolated from contaminated soil. *African Journal of Microbiology Research, 5*(32), 5889–5895.

Ghosh et al. (2011). Arsenic-resistant bacteria solubilized arsenic in the growth media and increased growth of arsenic hyperaccumulator Pteris vittata L. *Bioresource technology, 102*(19), 8756–8761.

Ghosh et al. (2021). Abatement of arsenic-induced phytotoxic effects in rice seedlings by an arsenic-resistant Pantoea dispersa strain. *Environmental Science and Pollution Research*, 1–17.

Gihring et al. (2001). Rapid arsenite oxidation by Thermusaquaticus and Thermusthermophilus: field and laboratory investigations. *Environmental Science & Technology, 35*(19), 3857–3862.

Gourbal et al. (2004). Drug uptake and modulation of drug resistance in Leishmania by an aquaglyceroporin. *Journal of Biological Chemistry, 279*(30), 31010–31017.

Huang et al. (2010). Characterization of arsenic-resistant bacteria from the rhizosphere of arsenic hyperaccumulator Pteris vittata. *Canadian Journal of Microbiology, 56*(3), 236–246.

Ilyaletdinov, A. N. and Abdrashitova, S. A. (1981). Autotrophic oxidation of arsenic by *Pseudomonas arsenitoxidans. Microbiology, 50*(2), 197–205.

Inskeep et al. (2004). Biomineralization of As (V)-hydrous ferric oxyhydroxide in microbial mats of an acid-sulfate-chloride geothermal spring, Yellowstone National Park. *Geochimica et Cosmochimica Acta, 68*(15), 3141–3155.

Ji et al. (1994). Arsenate reductase of *Staphylococcus aureus* plasmid pI258. *Biochemistry, 33*(23), 7294–7299.

Joshi, A. and Chaudhuri, M. (1996). Removal of arsenic from ground water by iron oxide coated sand. *Journal of Environmental Engineering, 122*(8), 769–771.

Kader et al. (2007). Removal of Cr (VI) from aqueous solutions by growing and non-growing populations of environmental bacterial consortia. *Global Journal of Environmental Research, 1*(1), 12–17.

Kashyap et al. (2006a). Complex regulation of arsenite oxidation in *Agrobacterium tumefaciens. Journal of Bacteriology, 188*(3), 1081–1088.

Kashyap et al. (2006b). A Na+: H+ antiporter and a molybdate transporter are essential for arsenite oxidation in *Agrobacterium tumefaciens. Journal of Bacteriology, 188*(4), 1577–1584.

Katsoyiannis, I. A. and Zouboulis, A. I. (2004). Application of biological processes for the removal of arsenic from groundwaters. *Water Research, 38*(1), 17–26.

Koechler et al. (2013). Genome sequence of Halomonas sp. strain A3H3, isolated from arsenic-rich marine sediments. *Genome Announcements, 1*(5).

Koechler et al. (2015). Constitutive arsenite oxidase expression detected in arsenic-hypertolerant Pseudomonas xanthomarina S11. *Research in Microbiology, 166*(3), 205–214.

Lebeau et al. (2008). Performance of bioaugmentation-assisted phytoextraction applied to metal contaminated soils: a review. *Environmental Pollution, 153*(3), 497–522.

Leist et al. (2000). The management of arsenic wastes: problems and prospects. *Journal of Hazardous Materials, 76*(1), 125–138.

Li et al. (2016). The organoarsenical biocycle and the primordial antibiotic methylarsenite. *Metallomics, 8*(10), 1047–1055.

Ma et al. (2011). Plant growth promoting rhizobacteria and endophytes accelerate phytoremediation of metalliferous soils. *Biotechnology Advances, 29*(2), 248–258.

Mahimairaja et al. (2005). Arsenic contamination and its risk management in complex environmental settings. *Advances in Agronomy, 86*, 1–82.

Mallick et al. (2018). Effective rhizoinoculation and biofilm formation by arsenic immobilizing halophilic plant growth promoting bacteria (PGPB) isolated from mangrove rhizosphere: a step towards arsenic rhizoremediation. *Science of the Total Environment, 610*, 1239–1250.

Mallick et al. (2014). Brevibacillus sp. KUMAs2, a bacterial isolate for possible bioremediation of arsenic in rhizosphere. *Ecotoxicology and Environmental Safety, 107*, 236–244.

McBride, B.C. and Wolfe, R.S. (1971). Biosynthesis of dimethylasrine by a methanobacterium. *Biochemistry, 10 (23),* 4312–4317.

Mesa et al. (2017). Use of endophytic and rhizosphere bacteria to improve phytoremediation of arsenic-contaminated industrial soils by autochthonous Betula celtiberica. *Applied and Environmental Microbiology, 83*(8).

Michalke et al. (2000). Production of volatile derivatives of metal (loid) s by microflora involved in anaerobic digestion of sewage sludge. *Applied and Environmental Microbiology, 66*(7), 2791–2796.

Mosa et al. (2016). Potential biotechnological strategies for the cleanup of heavy metals and metalloids. *Frontiers in Plant Science, 7*, 303.

Mukherjee et al. (2018). An endophytic bacterial consortium modulates multiple strategies to improve arsenic phytoremediation efficacy in *Solanum nigrum. Scientific Reports*, *8*(1), 1–16.

Muller et al. (2003). Arsenite oxidase aox genes from a metal-resistant β-proteobacterium. *Journal of Bacteriology, 185*(1), 135–141.

Newman et al. (1998). A brief review of microbial arsenate respiration. *Geomicrobiology Journal, 15*(4), 255–268.

Nies, D. H. (1999). Microbial heavy-metal resistance. *Applied Microbiology and Biotechnology, 51*(6), 730–750.

Oremland, R.S. and Stolz, J.F. (2003). The ecology of arsenic. *Science. 300*, 939–44

Pandey et al. (2013). Role of heavy metal resistant Ochrobactrum sp. and Bacillus spp. strains in bioremediation of a rice cultivar and their PGPR like activities. *Journal of Microbiology, 51*(1), 11–17.

Paul et al. (2017). Diversity of bacterial communities inhabiting soil and groundwater of arsenic contaminated areas in West Bengal, India. *Microbiology, 86*(2), 264–275.

Reizer et al. (1993). The MIP family of integral membrane channel proteins: sequence comparisons, evolutionary relationships, reconstructed pathway of evolution, and proposed functional differentiation of the two repeated halves of the proteins. *Critical Reviews in Biochemistry and Molecular Biology, 28*(3), 235–257.

Ridley et al. (1977). Biomethylation of toxic elements in the environment. *Science, 197*(4301), 329–332.

Roman-Ponce et al. (2016). Kocuriaarsenatis sp. nov., an arsenic-resistant endophytic actinobacterium associated with Prosopis laegivata grown on high-arsenic-polluted mine tailing. *International Journal of Systematic and Evolutionary Microbiology, 66*(2), 1027–1033.

Rosen, B. P. (2002). Biochemistry of arsenic detoxification. *FEBS Letters, 529*(1), 86–92.

Rosen, B. P. and Liu, Z. (2009). Transport pathways for arsenic and selenium: a minireview. *Environment International, 35*(3), 512–515.

Saltikov, C. W. and Newman, D. K. (2003). Genetic identification of a respiratory arsenate reductase. *Proceedings of the National Academy of Sciences, 100*(19), 10983–10988.

Saltikov et al. (2003). The ars detoxification system is advantageous but not required for As(V) respiration by the genetically tractable shewanella species strain ANA-3. *Applied and Environmental Microbiology, 69*(5), 2800–2809.

Sannasi et al. (2006). Sorption of Cr (VI), Cu (II) and Pb (II) by growing and non-growing cells of a bacterial consortium. *Bioresource technology, 97*(5), 740–747.

Santini, J. M. and vanden Hoven, R. N. (2004). Molybdenum-containing arsenite oxidase of the chemolithoautotrophic arsenite oxidizer NT-26. *Journal of Bacteriology*, *186*(6), 1614–1619.

Serrano, J. and Leiva, E. (2017). Removal of arsenic using acid/metal-tolerant sulfate reducing bacteria: A new approach for bioremediation of high-arsenic acid mine waters. *Water, 9*(12), 994.

Shen, Y. S. (1973). Study of arsenic removal from drinking water. *Journal of the American Water Works Association*, *65*(8), 543–548.

Shih, M. C. (2005). An overview of arsenic removal by pressure-drivenmembrane processes. *Desalination, 172*(1), 85–97.

Silver, S. and Phung, L. T. (2005). Genes and enzymes involved in bacterial oxidation and reduction of inorganic arsenic. *Applied and Environmental Microbiology, 71(2)*, 599–608.

Singh et al. (2016). Brevundimonas diminuta mediated alleviation of arsenic toxicity and plant growth promotion in Oryza sativa L. *Ecotoxicology and Environmental Safety, 125*, 25–34.

Soto et al. (2019). Enhanced arsenic tolerance in Triticum aestivum inoculated with arsenic-resistant and plant growth promoter microorganisms from a heavy metal-polluted soil. *Microorganisms, 7*(9), 348.

Srivastava et al. (2012). Isolation and characterization of Staphylococcus sp. strain NBRIEAG-8 from arsenic contaminated site of West Bengal. *Applied Microbiology and Biotechnology, 95*(5), 1275–1291.

Stolz, J. F. and Oremland, R. S. (1999). Bacterial respiration of arsenic and selenium. *FEMS Microbiology Reviews*, *23*(5), 615–627.

Stolz et al. (2006). Arsenic and selenium in microbial metabolism. *Annual Review of Microbiology, 60*, 107–130.

Sun et al. (2009). Molecular characterization and in situ quantification of anoxic arsenite-oxidizing denitrifying enrichment cultures. *FEMS Microbiology Ecology, 68*(1), 72–85.

Takeuchi et al. (2007). Arsenic resistance and removal by marine and non-marine bacteria. *Journal of Biotechnology, 127*(3), 434–442.

Teclu et al. (2008). Bioremoval of arsenic species from contaminated waters by sulphate-reducing bacteria. *Water Research, 42*(19), 4885–4893.

Teeling, H. and Glöckner, F. O. (2012). Current opportunities and challenges in microbial metagenome analysis—a bioinformatic perspective. *Briefings in Bioinformatics, 13*(6), 728–742.

Tisa, L. S. and Rosen, B. P. (1990). Molecular characterization of an anion pump. The ArsB protein is the membrane anchor for the ArsA protein. *Journal of Biological Chemistry, 265(1)*, 190–194.

US Environment Protection Agency (US EPA) (2000). Arsenic in Drinking Water Rule Economic Analysis, EPA-815-R-00-026.

Vahter, M. and Marafante, E. (1988). In vivo methylation and detoxication of arsenic. In: Craig, P. J., Glockling, F. (Eds.) *The Biological Alkylation of Heavy Elements*. Royal Society of Chemistry, London.

Wang et al. (2011). Effect of applying an arsenic-resistant and plant growth–promoting rhizobacterium to enhance soil arsenic phytoremediation by Populus deltoides LH05-17. *Journal of Applied Microbiology, 111*(5), 1065–1074.

Wang et al. (2017). Isolation of As-tolerant bacteria and their potentials of reducing As and Cd accumulation of edible tissues of vegetables in metal (loid)-contaminated soils. *Science of the Total Environment, 579*, 179–189.

Williams et al. (2013). Biosorption and bio-kinetic properties of solar saltern Halobacterial strains for managing Zn2+, As2+ and Cd2+ Metals. *Geomicrobiology Journal, 30*(6), 497–500.

Willsky, G. R. and Malamy, M. H. (1980). Effect of arsenate on inorganic phosphate transport in Escherichia coli. *Journal of Bacteriology, 144*(1), 366–374.

Xu et al. (1998). Metalloid resistance mechanisms in prokaryotes. *Journal of Biochemistry, 123*(1), 16–23.

Yang, H. C. and Rosen, B. P. (2016). New mechanisms of bacterial arsenic resistance. *Biomedical Journal, 39*(1), 5–13.

Yang et al. (2005). Novel pathway for arsenic detoxification in the legume symbiont Sinorhizobium meliloti. *Journal of Bacteriology, 187*(20), 6991–6997.

Yang et al. (2012). Effectiveness of applying arsenate reducing bacteria to enhance arsenic removal from polluted soils by Pteris vittata L. *International Journal of Phytoremediation, 14*(1), 89–99.

Index